The 80386DX Microprocessor

The 80386DX Microprocessor

Hardware, Software, and Interfacing

WALTER A. TRIEBEL
Intel Corporation

PRENTICE HALL
Englewood Cliffs, New Jersey 07632

Library of Congress Cataloging-in-Publication Data

```
Triebel, Walter A.
    The 80386DX microprocessor : hardware, software, and interfacing /
  Walter A. Triebel.
        p.    cm.
    Includes bibliographical references and index.
    ISBN 0-13-249566-X
    1. Intel 80386DX (Microprocessor)    I. Title.
  QA76.8.I292815T75   1992
  004.165--dc20                                              91-25140
                                                                  CIP
```

Acquisitions Editor: *Holly Hodder*
Production Editor: *Gretchen K. Chenenko*
Copy Editor: *Linda L. Thompson*
Cover Designer: *Lundgren Graphics, Ltd.*
Cover Photo: *Reprinted by permission
 of Intel Corporation, © 1990.*
Prepress Buyer: *Ilene Levy*
Manufacturing Buyer: *Ed O'Dougherty*
Supplements Editor: *Judy Casillo*

© 1992 by Prentice-Hall, Inc.
A Simon & Schuster Company
Englewood Cliffs, New Jersey 07632

Portions of this text were previously published as
THE 80286 MICROPROCESSOR:
Hardware, Software, and Interfacing
  by Triebel and Singh (Prentice Hall, 1990)
*16-BIT AND 32-BIT MICROPROCESSORS:
Architecture, Software,
and Interfacing Techniques*
  by Singh and Triebel (Prentice Hall, 1991)

PAL™ is a trademark
of Monolithic Memories, Inc.
PC/AT™ is a trademark
of International Business Machines Corp.
UNIX™ is a trademark of AT&T.
Quick-Pulse Programming Algorithm™, 386™,
387™, Intelligent Programming Algorithm™,
and MCS-86™ are trademarks of Intel Corp.

All rights reserved. No part of this book may be
reproduced, in any form or by any means,
without permission in writing from the publisher.

Printed in the United States of America

10 9 8 7 6 5 4 3 2 1

ISBN   0-13-249566-X

PRENTICE-HALL INTERNATIONAL (UK) LIMITED, *London*
PRENTICE-HALL OF AUSTRALIA PTY. LIMITED, *Sydney*
PRENTICE-HALL CANADA INC., *Toronto*
PRENTICE-HALL HISPANOAMERICANA, S.A. *Mexico*
PRENTICE-HALL OF INDIA PRIVATE LIMITED, *New Delhi*
PRENTICE-HALL OF JAPAN, INC., *Tokyo*
SIMON & SCHUSTER ASIA PTE. LTD., *Singapore*
EDITORA PRENTICE-HALL DO BRASIL, LTDA., *Rio de Janeiro*

To
my daughter Lindsey

Contents

PREFACE   *xiii*

1   INTRODUCTION TO MICROPROCESSORS AND MICROCOMPUTERS   *1*

    1.1   Introduction   1

    1.2   The IBM and IBM-Compatible Personal Computers: Reprogrammable Microcomputers   2

    1.3   General Architecture of a Microcomputer System   5

    1.4   Evolution of the Intel Microprocessor Architecture   7

2   REAL-ADDRESSED-MODE SOFTWARE ARCHITECTURE OF THE 80386DX MICROPROCESSOR   *15*

    2.1   Introduction   15

    2.2   Software: The Microcomputer Program   16

    2.3   Software Model of the 80386DX Microprocessor   21

    2.4   Memory Address Space and Data Organization   23

    2.5   Data Types   27

    2.6   Segment Registers and Memory Segmentation   30

2.7 Instruction Pointer 33
2.8 General-Purpose Registers 34
2.9 Flags Register 37
2.10 Generating a Real-Mode Memory Address 38
2.11 The Stack 42
2.12 Addressing Modes of the 80386DX Microprocessor 46

3 REAL-MODE 80386DX MICROPROCESSOR PROGRAMMING 1   71

3.1 Introduction 71
3.2 The 80386DX's Instruction Set 71
3.3 Data Transfer Instructions 73
3.4 Arithmetic Instructions 86
3.5 Logic Instructions 99
3.6 Shift Instructions 100
3.7 Rotate Instructions 104
3.8 Bit Test and Bit Scan Instructions 107

4 REAL-MODE 80386DX MICROPROCESSOR PROGRAMMING 2   117

4.1 Introduction 117
4.2 Flag Control Instructions 117
4.3 Compare and Set Instructions 119
4.4 Jump Instructions 123
4.5 Subroutines and the Subroutine Handling Instructions 131
4.6 The Loop and the Loop Handling Instructions 143
4.7 Strings and the String Handling Instructions 146

5 PROTECTED-MODE SOFTWARE ARCHITECTURE OF THE 80386DX   157

5.1 Introduction 157
5.2 Protected-Mode Register Model 159

	5.3	Protected-Mode Memory Management and Address Translation 168	
	5.4	Descriptor and Page Table Entries 178	
	5.5	Protected-Mode System Control Instruction Set 185	
	5.6	Multitasking and Protection 188	
	5.7	Virtual 8086 Mode 202	
6	**THE 80386DX MICROPROCESSOR AND ITS MEMORY INTERFACE**		**206**
	6.1	Introduction 206	
	6.2	The 80386DX Microprocessor 207	
	6.3	Signal Interfaces of the 80386DX 209	
	6.4	System Clock 217	
	6.5	Bus States and Pipelined and Nonpipelined Bus Cycles 217	
	6.6	Read and Write Bus Cycle Timing 221	
	6.7	Hardware Organization of the Memory Address Space 226	
	6.8	Memory Interface Circuits 231	
	6.9	Programmable Logic Arrays: Bus Control Logic 243	
	6.10	Program Storage Memory: ROM, PROM, and EPROM 250	
	6.11	Data Storage Memory: SRAM and DRAM 258	
	6.12	80386DX Microcomputer System Memory Interface Circuitry 266	
	6.13	Cache Memory 270	
	6.14	The 82385DX Cache Controller and the Cache Memory Subsystem 275	
7	**INPUT/OUTPUT INTERFACE OF THE 80386DX MICROPROCESSOR**		**295**
	7.1	Introduction 295	
	7.2	Types of Input/Output 296	
	7.3	The Isolated Input/Output Interface 298	

	7.4	Input and Output Bus Cycle Timing 304
	7.5	Input/Output Instructions 306
	7.6	82C55A Programmable Peripheral Interface 312
	7.7	Implementing Isolated I/O Parallel Input/Output Ports Using the 82C55A 325
	7.8	Implementing Memory-Mapped I/O Parallel Input/Output Ports Using the 82C55A 329
	7.9	Input/Output Polling and Handshaking 333
	7.10	The 82C54 Programmable Interval Timer 338
	7.11	The 82C37A Programmable Direct Memory Access Controller 351
	7.12	80386DX Microcomputer System I/O Circuitry 366

8 INTERRUPT AND EXCEPTION PROCESSING OF THE 80386DX MICROPROCESSOR 374

	8.1	Introduction 374
	8.2	Types of Interrupts and Exceptions 375
	8.3	Interrupt Vector and Interrupt Descriptor Tables 377
	8.4	Interrupt Instructions 381
	8.5	Enabling/Disabling of Interrupts 384
	8.6	External Hardware Interrupt Interface 385
	8.7	External Hardware Interrupt Sequence 387
	8.8	The 82C59A Programmable Interrupt Controller 394
	8.9	Interrupt Interface Circuits Using the 82C59A 408
	8.10	Internal Interrupts and Exception Functions 411

9 80386DX PC/AT MICROCOMPUTER SYSTEM HARDWARE 421

	9.1	Introduction 421
	9.2	Architecture of the System Processor Board in the Original IBM PC/AT 421
	9.3	High-Integration PC/AT-Compatible Peripheral ICs 427
	9.4	Core 80386DX Microcomputer 429

9.5   82345 Data Buffer   440

9.6   82346 System Controller   447

9.7   82344 ISA Controller   458

9.8   82341 High-Integration Peripheral Combo   473

9.9   82077 Floppy Disk Controller   481

BIBLIOGRAPHY *493*

ANSWERS TO SELECTED ASSIGNMENTS *495*

INDEX *509*

Preface

In the last several years, the 80386 family have become the most widely used microprocessors in modern *personal computer advanced technology* (PC/AT™)-compatible computers. For instance, it is the microprocessor employed in Compaq's popular PC/AT-compatible family of 32-bit personal computers. Intel Corporation has two compatible 32-bit architectures available in their popular 8086 family, the 80386 architecture and 80486 architecture. Today, the two members of the 80386 family, the 80386DX and 80386SX, are the most widely used 32-bit microprocessors in modern microcomputer systems. The 80386DX is a 32-bit data bus version of the 80386 and the 80386SX is a 16-bit bus version. The 80386 architecture is used in a large number of manufacturer's personal computers as well as a wide variety of other electronic equipment. This book represents a thorough study of the 80386DX microprocessor and its microcomputer system.

The 80386DX Microprocessor: Hardware, Software, and Interfacing is written for use as a textbook in courses on 32-bit microprocessors at colleges and universities. It is intended for use in a one-semester course in microprocessor technology that emphasizes both 80386DX assembly language software and microcomputer circuit design. Reviewers who contributed their insight to the text include R. T. George, Duke University, Durham, NC; Howard Atwell, Fullerton College, Fullerton, CA; Michael Chen, Dutchess Community College, Poughkeepsie, NY.

Individuals involved in the design of microprocessor-based electronic equipment need a systems-level understanding of the 80386DX microcomputer, that is, a thorough understanding of both its software and hardware. The first half of *The 80386DX Microprocessor: Hardware, Software, and Interfacing* explores the *real-*

addressed-mode and *protected-addressed mode* software architectures of the 80386DX microprocessor and teaches the reader how to write assembly language programs. The things that are needed to learn to be successful at writing assembly language programs for the 80386DX are the following:

1. *Software architecture:* the internal registers, flags, memory organization, and stack, and how they are used from a software point of view.

2. *Instruction set:* the function of each of the instructions in the instruction set, the allowed operand variations, and how to write statements using the instructions.

3. *Programming techniques:* basic techniques of programming, such as flowcharting, jumps, loops, strings, subroutines, parameter passing, and so on.

4. *Applications:* the reader is lead step by step through the process of writing a program for a practical application.

All of this material is developed in detail in Chapters 2 through 4.

Chapter 5 is a detailed study of the 80386DX's protected-address mode architecture. Here we begin by introducing the protected-mode register model. This is followed by a detailed study of the function of the 80386DX's memory management unit, segmentation and paging, virtual addressing, and the translation of virtual addresses to physical addresses. The various types of descriptor table and page table entries supported by the 80386DX are covered and how they relate to memory management and the protection model. The instructions of the protected-mode system control instruction set are described. Finally, the concepts of protection, the task, task switching, the multitasking system environment, and virtual 8086 mode are explored.

The second half of the book examines the hardware architecture of microcomputers built with the 80386DX microprocessor. To understand the hardware design of an 80386DX-based microcomputer system, the reader must begin by first understanding the function and operation of each of its hardware interfaces: memory, input/output, and interrupt. After this, the role of each of these subsystems can be explored relative to overall microcomputer system operation. It is this material that is presented in Chapters 6 through 9.

We begin in Chapter 6 by examining the architecture of the 80386DX microprocessor from a hardware point of view. This includes information such as pin layout, primary signal interfaces, signal functions, and clock requirements. The latter part of Chapter 6 covers the memory interface of the 80386DX. This material includes extensive coverage of memory bus cycles, pipelining, memory interface circuits (address latches and buffers, data bus transceivers, and address decoders), use of programmable logic devices in implementing the bus control logic, program storage memory devices (ROM, PROM, and EPROM), data storage memory devices (SRAM and DRAM), and program and data storage memory circuitry.

This hardware introduction is followed by separate studies of the architectural characteristics, operation, and circuit designs for the input/output and interrupt interfaces of the 80386DX-based microcomputer in Chapters 7 and 8, respectively. This material includes information such as isolated and memory-mapped I/O, the function of the signals at each of the interfaces, input/output and interrupt acknowledge bus cycle activity, real- and protected-mode operation, input polling, handshaking I/O interfaces, and examples of typical input, output, and interrupt interface circuit designs. Included in these chapters is detailed coverage of LSI peripheral ICs, such as the 82C55A, 82C54, 82C37A, and 82C59A.

The hardware design section closes in Chapter 9 with a thorough study of the 80386DX microcomputer design used for a main processor board of a PC/AT-compatible personal computer. The microcomputer design examined employs Intel Corporation's 82340 high-integration PC/AT-compatible chip set. Each of the ICs in the 82340 chip set (82345 data buffer, 82346 system controller, 82344 ISA controller, and 82341 peripheral combo) are described in detail from a hardware point of view. This material includes their block diagram, signal interfaces, and interconnection to implement the PC/AT-compatible microcomputer. Moreover, the circuitry used in the design of the cache memory subsystem, DRAM array and BIOS EPROMs, floppy disk interface, IDE hard disk interface, serial communication interfaces, and parallel printer interface is described. In this chapter we demonstrate a practical implementation of the material presented in the prior chapters on microcomputer interfacing techniques.

Walter A. Triebel

The 80386DX Microprocessor

1

Introduction to Microprocessors and Microcomputers

1.1 INTRODUCTION

In the last decade, most of the important advances in computer system technology have been closely related to the development of high-performance 16- and 32-bit microprocessor architectures and the microcomputer systems built with them. During this period, there has been a major change in the direction of businesses away from minicomputers to smaller, lower-cost microcomputers. The *IBM personal computer* (the PC as it has become known), which was introduced in mid-1981, was one of the earliest microcomputers that used a 16-bit microprocessor, the 8088, as its central processing unit. Several years later it was followed by another IBM personal computer, the PC/AT (personal computer advanced technology). This new system was implemented using the more powerful 80286 microprocessor. The PC and PC/AT quickly became cornerstones of the evolutionary process from minicomputer to microcomputer. In 1985 an even more powerful microprocessor, the 80386DX, was introduced. The 80386DX was Intel Corporation's first 32-bit member of the 8086 family of microprocessors. Availability of the 80386DX quickly lead to a new generation of very high performance PC/ATs. Today the PC and PC/AT architectures represent industry standards in the personal computer marketplace.

Since the introduction of the IBM PC, the microprocessor market has matured significantly. Today, several complete families of 16- and 32-bit microprocessors are available. They all include support products such as very large scale integrated (VLSI) peripheral devices, emulators, and high-level languages. Over the same period of time, these higher-performance microprocessors have become more widely used in the design of new electronic equipment and computers. This book presents a detailed study of the 32-bit microprocessor used in high-performance PC/AT compatibles, the 80386DX from Intel Corporation.

In this chapter we begin our study with an introduction to microprocessors and microcomputers. The following topics are discussed:

1. The IBM and IBM-compatible personal computers: reprogrammable microcomputers
2. General architecture of a microcomputer system
3. Evolution of the Intel microprocessor architecture

1.2 THE IBM AND IBM-COMPATIBLE PERSONAL COMPUTERS: REPROGRAMMABLE MICROCOMPUTERS

The IBM personal computer (the PC), which is shown in Fig. 1.1, was IBM's first entry into the microcomputer market. After its introduction in mid-1981, market acceptance of the PC grew by leaps and bounds so that it quickly became the leading personal computer architecture. One of the important keys to its success is that an enormous amount of application software became available for the machine. Today, there are more than 20,000 off-the-shelf software packages available for use with the PC. They include business applications, software languages, educational programs, games, and even alternative operating systems.

The success of the PC caused IBM to spawn additional family members. IBM's *PCXT* is shown in Fig. 1.2 and an 80286 based *PC/AT* is illustrated in Fig. 1.3. The PCXT employed the same system architecture as that of the original PC. It was also designed with the 8088 microprocessor, but had one of the floppy disk drives replaced by a 10M-byte hard disk drive. The original PC/AT was designed

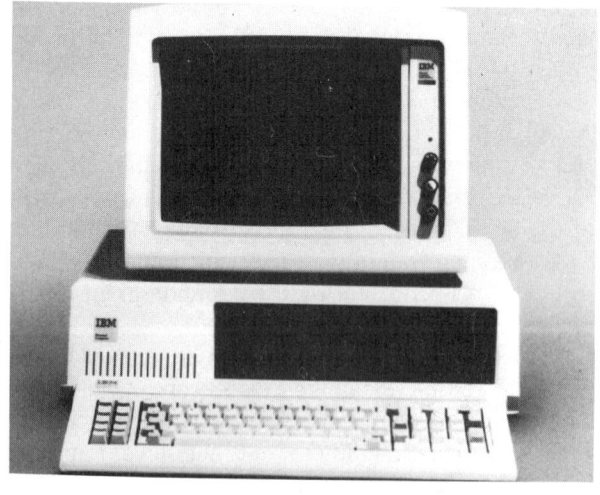

Figure 1.1 Original IBM personal computer (PC). (Courtesy of International Business Machines Corporation)

Figure 1.2 IBM's PCXT. (Courtesy of International Business Machines Corporation)

with a 6-MHz 80286 microprocessor and defined a new architecture that today is called the *industry standard architecture* (ISA). This architecture provides a new, higher-performance I/O expansion bus. Today, 80286- and 80386DX-based ISA PCs, not PCXTs, are the mainstays of the personal computer marketplace. Figure 1.4 shows a popular 80386DX-based PC/AT-compatible personal computer.

Figure 1.3 Original IBM PC/AT. (Courtesy of International Business Machines Corporation)

Figure 1.4 80386DX microprocessor-based PC/AT-compatible personal computer. (Reprinted by permission of Compaq Computer Corporation. All rights reserved.)

In 1987, IBM began introducing a new family of personal computers called the *Personal System/2*, which originally included five models: Model 30, Model 50, Model 60, Model 70, and Model 80. All but the Model 30 employ a newer, higher-performance architecture known as the *Micro Channel Architecture*. These machines offer a wide variety of computing capabilities, range of performance, and software base for use in business and at home.

The 80386DX-based PC/AT compatible is an example of a *reprogrammable microcomputer*. By "reprogrammable microcomputer" we mean that it is intended to run programs for a wide variety of applications. For example, one user could use the PC/AT with a standard application package for accounting or inventory control. In this type of application, the primary task of the microcomputer is to analyze and process a large amount of data, known as the *data base*. Another user could be running a word-processing software package. This is an example of a data input/output-intensive task. The user enters text information, which is reorganized by the microcomputer and then output to a diskette or printer. In a third example, a programmer uses a language, such as FORTRAN, to write programs for a scientific application. Here the primary function of the microcomputer is to solve complex mathematical problems. The PC/AT used for each of these applications is the same; the difference is in the software application that the microcomputer is running. That is, the microcomputer is simply reprogrammed to run the new application. We have already mentioned that the PC/AT is a *microcomputer*. Let us now look at what a microcomputer is and how it differs from the other classes of computers.

Evolution of the computer marketplace over the last 25 years has taken us from very large *mainframe computers* to smaller *minicomputers*, and now to even smaller *microcomputers*. These three classes of computers did not replace each other. They all coexist today in the marketplace. Computer users have the opportunity to select the computer that best meets their needs. For instance, a large university or institution would still select a mainframe computer for its data processing center. On the other hand, a department at a university or in business might select a minicomputer for a multiuser-dedicated need such as application software development. Moreover, managers may select a microcomputer, such as the PC/AT, for their personal needs, such as word processing and data base management.

Along the evolutionary path from mainframes to microcomputers, the basic concepts of computer architecture have not changed. Just like the mainframe and minicomputer, the microcomputer is a general-purpose electronic data processing system intended for use in a wide variety of applications. The key difference is that microcomputers, such as the PC/AT, employ the newest *very large scale integrated* (VLSI) circuit technology to implement a smaller, reduced-capability computer system, but at a much lower cost than a minicomputer. Microcomputers such as an 80386DX-based PC/AT, which are designed for the high-performance end of the microcomputer market, are currently capturing much of the market that was traditionally supported with low-to-medium-performance minicomputers.

1.3 GENERAL ARCHITECTURE OF A MICROCOMPUTER SYSTEM

The *hardware* of a microcomputer system can be divided into four functional sections: the *input unit, microprocessing unit, memory unit,* and *output unit*, shown in the block diagram of Fig. 1.5. Each of these units has a special function in terms of overall system operation. Next we look at each of these sections in more detail.

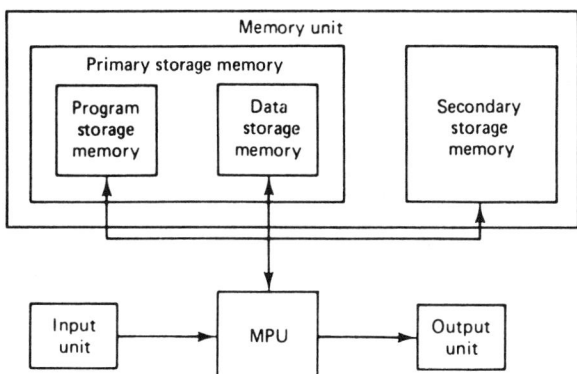

Figure 1.5 General architecture of a microcomputer system.

The heart of a microcomputer is its microprocessing unit (MPU). The MPU of a microcomputer is implemented with a VLSI device known as a *microprocessor*. A microprocessor is a general-purpose processing unit built into a single integrated circuit (IC). The microprocessor used in most high-performance PC/ATs is Intel Corporation's 80386DX, which is shown in Fig. 1.6.

The 80386DX MPU is the part of the microcomputer that executes instructions of the program and processes data. It is responsible for performing all arith-

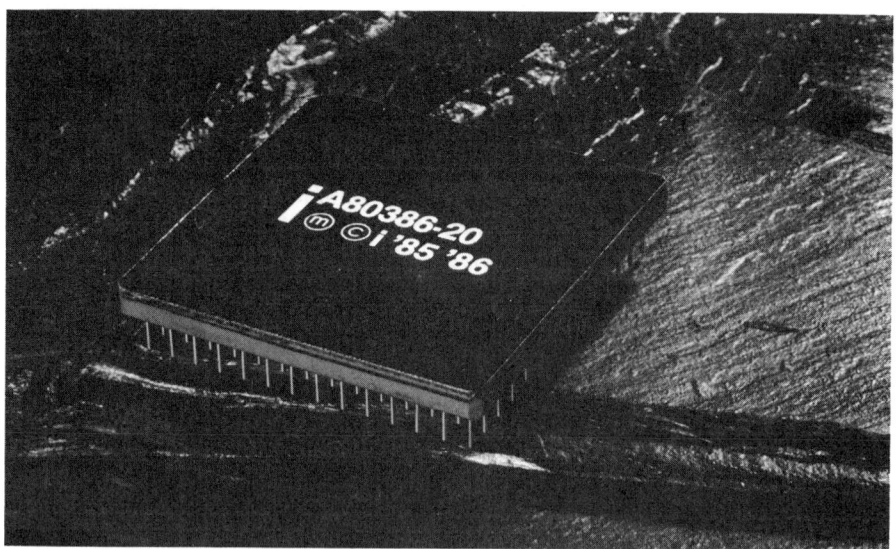

Figure 1.6 80386DX microprocessor. (Reprinted by permission of Intel Corp. copyright Intel Corp. 1990.)

metic operations and making the logical decisions initiated by the computer's program. In addition to arithmetic and logic functions, the MPU controls overall system operation.

The input and output units are the means by which the MPU communicates with the outside world. Input units, such as the *keyboard* on the PC/AT, allow the user to input information or commands to the MPU. For instance, a programmer could key in the lines of a BASIC program from the keyboard. Many other input devices are available for the PC/AT, two examples being a *mouse*, for implementing a more user-friendly input interface, and a *joystick*, for use when playing video games.

The most widely used output devices of a PC/AT are the *display* and *printer*. The output unit in a microcomputer is used to give feedback to the user. For example, key entries from the keyboard are echoed back to the display. This lets the user confirm that the correct entry was made. Moreover, the results produced by the MPU's processing can be either displayed or printed. For our earlier example of a BASIC program, once it is entered and corrected a listing of the instructions could be printed. Alternative output devices are also available for the microcomputer. For instance, it can be equipped with a color video display instead of the standard monochrome video display.

The memory unit in a microcomputer is used to *store* information, such as number or character data. By "store" we mean that memory has the ability to hold this information for processing or for outputting at a later time. Programs that define how the computer is to operate and process data also reside in memory.

In the microcomputer system, memory can be divided into two different sections, called *primary storage memory* and *secondary storage memory*. Secondary storage memory is used for long-term storage of information that is not currently being used. For example, it can hold programs, files of data, and files of information. In the original IBM PC/AT, the *floppy disk drive* is one of the secondary storage memory subsystems. It is a 5¼-inch drive that uses double-sided, quad-density floppy diskette media that can store up to 1.2M (1,200,000) bytes of data. The IBM PC/AT also employs a second type of secondary storage device, called a *hard disk drive*. Typical hard disk sizes are 20M bytes (20 million bytes), 40M bytes, and 80M bytes. The original PC/AT was equipped with a 20M-byte drive.

Primary storage memory is normally smaller in size and used for temporary storage of active information, such as the operating system of the microcomputer, the program that is currently being executed, and the data that it is processing. In Fig. 1.5 we see that primary storage memory is further subdivided into *program storage memory* and *data storage memory*. The program section of memory is used to store instructions of the operating system and application programs. The data section normally contains data that are to be processed by the programs as they are executed: for example, text files for a word processor program or a data base for a data base management program. However, programs can also be loaded into data memory for execution.

Typically, primary storage memory is implemented with both *read-only memory* (ROM) and *random-access read/write memory* (RAM) integrated circuits. The original IBM PC/AT has 64K bytes of ROM and can be configured with either 256K bytes or 512K bytes of RAM without adding a memory expansion board.

Data, whether they represent numbers, characters, or instructions, can be stored in either ROM or RAM. In the IBM PC/AT a small part of the operating system and BASIC language are made resident to the computer by supplying them in ROM. By using ROM, this information is made *nonvolatile*—that is, the information is not lost if power is turned off. This type of memory can only be read from; it cannot be written into.

On the other hand, data that are to be processed and information that frequently changes must be stored in a type of primary storage memory from which they can be read by the microprocessor, modified through processing, and written back for storage. For this reason, they are stored in RAM instead of ROM. For instance, the *DOS 4.1 operating system* for the PC/AT is provided on diskettes; to be used it must be loaded into the RAM of the microcomputer either from a diskette or from the hard disk. RAM is an example of a *volatile* memory. That is, when power is turned off, the data that it holds are lost. This is why the DOS operating system must be reloaded into the PC/AT each time power is turned on.

1.4 EVOLUTION OF THE INTEL MICROPROCESSOR ARCHITECTURE

The principal way in which microprocessors and microcomputers are categorized is in terms of the maximum number of binary bits in the data they process, that is, their word length. Over time, four standard data widths have evolved for microprocessors and microcomputers: *4-bit*, *8-bit*, *16-bit*, and *32-bit*.

Figure 1.7 illustrates the evolution of Intel's microprocessors since their introduction in 1972. The first microprocessor, the *4004*, was designed to process data arranged as 4-bit words. This organization is also referred to as a *nibble* of data.

The 4004 implemented a very low performance microcomputer by today's standards. This low performance and limited system capability restricted its use to simpler, special-purpose applications. A common use was in electronic calculators.

In 1974 a second generation of microprocessors began to be introduced. These devices, the *8008, 8080,* and *8085*, are 8-bit microprocessors. That is, they were all designed to process 8-bit (one-byte-wide) data instead of 4-bit data. The 8080, identified in Fig. 1.6, was introduced in 1975.

These newer 8-bit microprocessors were characterized by higher-performance operation, larger system capabilities, and greater ease of programming. They were able to provide the system requirements for many applications that could not be satisfied with the earlier 4-bit microprocessors. These extended capabilities let to widespread acceptance of multichip 8-bit microcomputers for special-purpose system designs. Examples of these dedicated applications are electronic instruments, cash registers, and printers.

Plans for the development of third-generation 16-bit microprocessors were announced by many of the leading semiconductor manufacturers in the mid-1970s. Looking at Fig. 1.7, we see that Intel's first 16-bit microprocessor, the *8086*, first became available in 1979 and was followed the next year by its 8-bit bus version,

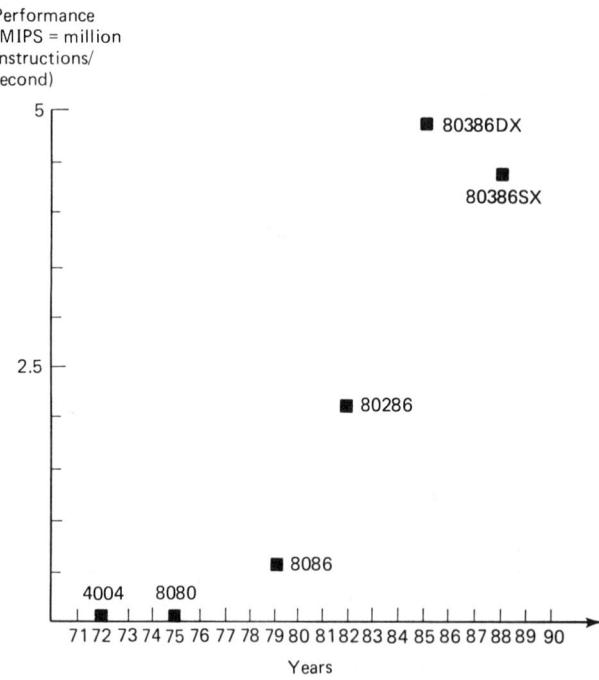

Figure 1.7 Evolution of the Intel microprocessor architecture.

the *8088*. This was the birth of Intel's 8086 family architecture. Other family members, such as the *80286, 80186,* and *80188,* were introduced in the years that followed.

These 16-bit microprocessors provide high performance and have the ability to satisfy a broad scope of special-purpose and general-purpose microcomputer applications. They all have the ability to handle 8-bit, 16-bit, and special-purpose data types. Moreover, their powerful instruction set is more in line with those provided by a minicomputer.

In 1985, Intel Corporation introduced its first 32-bit microprocessor, the *80386DX*. The 80386DX microprocessor brings true minicomputer-level performance to the microcomputer system. This device was followed by a 16-bit external bus version, the 80386SX in 1988.

This evolution of microprocessors is made possible by advances in semiconductor process technology. Semiconductor device geometries decreased from about 5 microns in the early 70s to submicron levels today. The smaller geometry permits integration of an order-of-magnitude more transistors into the same-size chip and at the same time has lead to higher operating speeds. In Fig. 1.8 we see that the 4004 contained about *10,000* transistors. Transistor density was increased to about *30,000* with the development of the 8086 in 1979; with the introduction

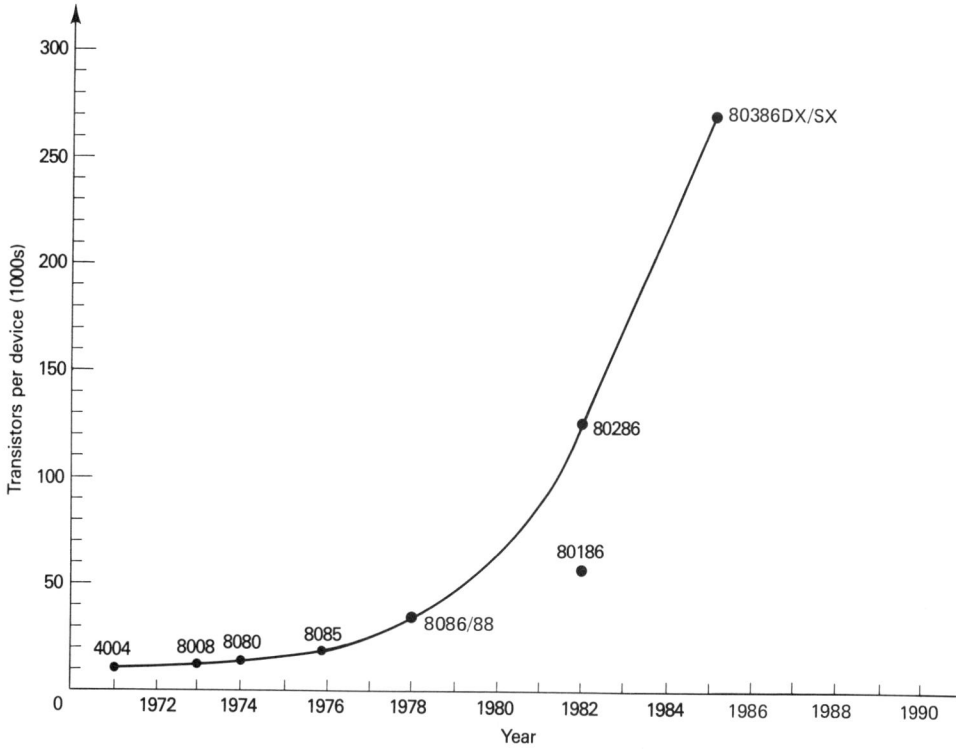

Figure 1.8 Device complexity.

Sec. 1.4   Evolution of the Intel Microprocessor Architecture of the 80286, the transistor count was increased to approximately *140,000*; and finally, the 80386DX contains more than *275,000* transistors.

Microprocessors can be classified based on the type of application for which they have been designed. In Fig. 1.9 we have classified Intel microprocessors into

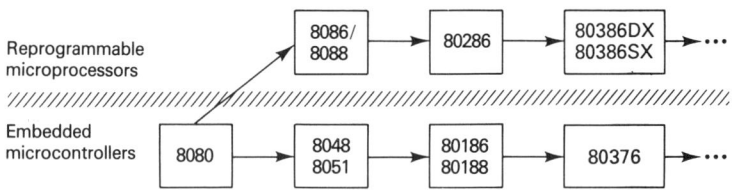

Figure 1.9 Reprogrammable and embedded control applications.

two application-oriented categories called *reprogrammable microprocessors* and *embedded microcontrollers*. Devices such as the 8080 were initially most widely used as *special-purpose microcomputers*. By "special-purpose microcomputer" we mean a system that has been tailored to meet the needs of a specific application. These special-purpose microcomputers were used in *embedded control applications*, that is, an application in which the microcomputer performs a dedicated control function.

Embedded control applications are further divided into those that involve primarily *event control* and those that require *data control*. An example of an embedded control application that is primarily event control is a microcomputer used for industrial process control. Here the program of the microprocessor is used to initiate a timed sequence of events. On the other hand, an application that focuses more on data control than event control is a hard disk controller interface. In this case, a block of data that is to be processed, for example a data base, must be quickly transferred from secondary storage memory to primary storage memory.

The spectrum of embedded control applications requires a wide variety of system features and performance levels. Devices developed specifically for the needs of this marketplace have stressed low cost and high integration. In Fig. 1.9 we see that the earlier multichip 8080 solutions were initially replaced by highly integrated 8-bit single-chip microcomputer devices such as the 8048 and 8051. These devices were tailored to work best as event controllers. For instance, the 8051 offers one-order-of-magnitude-higher performance than the 8080, a more powerful instruction set, and special on-chip functions such as ROM, RAM, interval/event timer, universal asynchronous receiver/transmitter (UART), and programmable parallel I/O ports. Today these type of devices are called *microcontrollers*.

Later devices, such as the 80186 and 80188, were designed to better meet the needs of data control applications. They are also highly integrated, but with added features, such as string instructions and DMA channels, that better handle the movement of data.

The category of reprogrammable microprocessors represents the class of applications in which a microprocessor is used to implement a *general-purpose microcomputer.* Unlike a special-purpose microcomputer, a general-purpose microcomputer is intended to run a wide variety of applications. That is, while in use it can easily be reprogrammed to run a different application. Two examples of reprogrammable microcomputers are the personal computer and the minicomputer. In Fig. 1.9 we see that the 8086, 8088, 80286, and 80386DX are the Intel microprocessors intended for use in this type of application.

Architectural compatibility is a critical need of microprocessors developed for use in reprogrammable applications. As shown in Fig. 1.10, the 80386DX

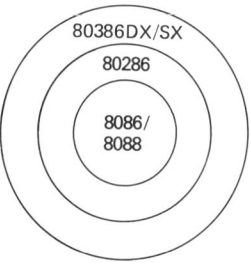

Figure 1.10 Code- and system-level compatibility.

provides a superset of the 80286 architecture, and the 80286 is a superset of the original 8086/8088 architecture. The 80386DX can operate in either of two modes: the *real-addressed mode* or *protected-addressed mode*. When in the real mode, it operates like a superperformance 8086. It can be used to execute the base instruction set, which is object-code compatible with the 8086/8088. For this reason, operating systems and programs written for the 8086 and 8088 are run on the 80386DX without modification. Moreover, a number of new instructions have been added in the instruction sets of the 80286 and 80386DX to enhance their performance and functionality. We say that object code is *upward compatible* within the 8086 architecture. This means that 8086/8088 code will run on the 80286 and 80386DX, but the reverse is not true if any of the new instructions are in use.

A microprocessor such as the 80386DX, which is designed for implementing general-purpose microcomputers, must offer more advanced system features than those of a microcontroller. For example, it needs to support and manage a large memory subsystem. The 80386DX is capable of managing a *64 terabyte* (64T-byte) *address space*. Moreover, a reprogrammable microcomputer, such as a personal computer, normally runs an operating system. The architecture of the 80386DX has been enhanced with on-chip support for operating system functions such as *memory management, protection,* and *multitasking.* These new features become active only when the 80386DX is operated in the protected mode. The 80386DX also has a special mode of operation known as the *virtual 8086 mode* that permits 8086/8088 code to be run in the protected mode.

Reprogrammable microcomputers require a wide variety of input/output resources. Figure 1.11 shows the kinds of interfaces that are frequently implemented in a personal computer or minicomputer system. A large family of VLSI peripheral ICs are needed to support a reprogrammable microprocessor such as the 80386DX. Examples are floppy disk controllers, hard disk controllers, local area network controllers, and communication controllers. For this reason, the 8086,

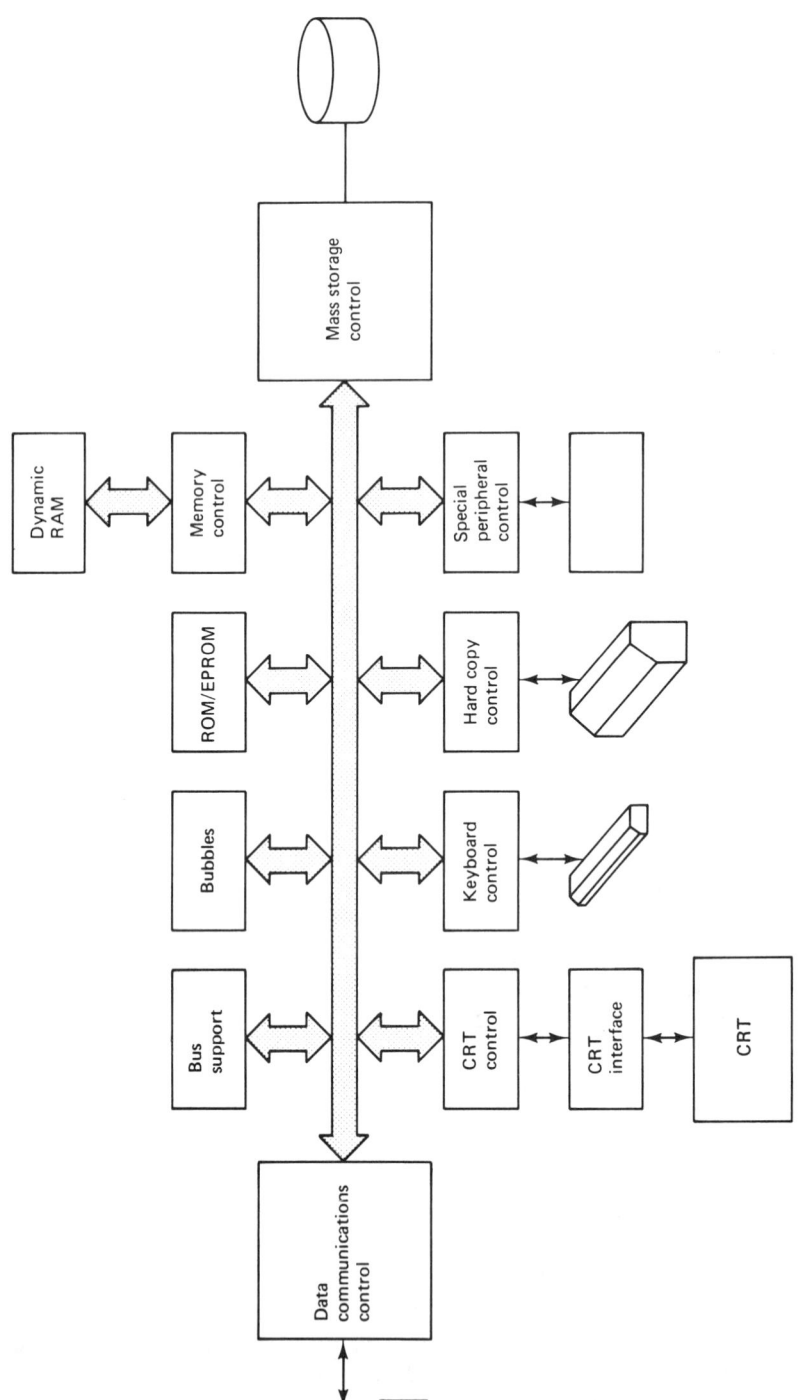

Figure 1.11 Peripheral support for the MPU.

8088, 80286, and 80386DX are designed to implement a multichip microcomputer system. In this way a system can easily be configured with the appropriate set of I/O interfaces.

ASSIGNMENT

Section 1.2

1. What does *PC/AT* stand for?
2. What architecture is used in the Model 80 Personal System/2?
3. What is a reprogrammable microcomputer?
4. Name the three classes of computers.
5. What are the main similarities and differences between the minicomputer and the microcomputer?
6. What does VLSI stand for?

Section 1.3

7. What are the four building blocks of a microcomputer system?
8. What is the heart of the microcomputer system called?
9. What is an 80386DX?
10. What is the primary input unit of the PC/AT? Give two other examples of input units available for the PC/AT.
11. What are the primary output devices of the PC/AT?
12. Into what two sections is the memory of a PC/AT partitioned?
13. What is the storage capacity of the standard 5¼-inch floppy diskette for the original PC/AT? What is the storage capacity of its standard hard disk?
14. What do *ROM* and *RAM* stand for?
15. How much ROM is provided in the original PC/AT's processor board? What is the maximum amount of RAM that can be implemented on this processor board?
16. Why must the operating system be reloaded from the DOS diskette each time power is turned on?

Section 1.4

17. What are the standard data word lengths for which microprocessors have been developed?
18. What was the first 4-bit microprocessor introduced by Intel Corporation? Eight-bit microprocessor? Sixteen-bit microprocessor? Thirty-two bit microprocessor?
19. Name six members of the 8086 family architecture.
20. Approximately how many transistors are used to implement the 8088 microprocessor? The 80286 microprocessor? The 80386DX microprocessor?
21. What is an embedded microcontroller?
22. Name the two types of embedded microcontrollers.

23. What is the difference between a multichip microcomputer and a single-chip microcomputer?
24. Name four 8086 family microprocessors intended for use in reprogrammable microcomputer applications.
25. Give the names for the 80386DX's two modes of operation.
26. What is meant by *upward software compatibility* relative to 8086 architecture microprocessors?
27. List three advanced architectural features provided by the 80386DX microprocessor.
28. Give three types of VLSI peripheral support needed in a reprogrammable microcomputer system.

2

Real-Addressed-Mode Software Architecture of the 80386DX Microprocessor

2.1 INTRODUCTION

In this chapter we begin our study of the 80386DX microprocessor and its assembly language programming. To program the 80386DX with assembly language, we must understand how the microprocessor and its memory subsystem operate from a software point of view. Remember that the 80386DX can operate in either of two modes, called the *real-addressed mode* (real mode) and the *protected-address mode* (protected mode). Moreover, we pointed out in Chapter 1 that when in the real mode, the 80386DX operates like a very high performance 8086 microprocessor. In fact, a 16-MHz 80386DX provides more than 10 times higher performance than that of the standard 5-MHz 8086. Here we will examine just the real-mode software architecture. The 80386DX's protected mode is presented in Chapter 5. The following topics are covered in the chapter.

1. Software: the microcomputer program
2. Real-mode software model of the 80386DX microprocessor
3. Real-mode memory address space and data organization
4. Data types
5. Segment registers and memory segmentation
6. Instruction pointer
7. General-purpose registers

8. Flags register
9. Generating a real-mode memory address
10. The stack
11. Addressing modes of the 80386DX microprocessor

2.2 SOFTWARE: THE MICROCOMPUTER PROGRAM

In this section we begin our study of the 80386DX's software architecture with the topics of software and the microcomputer program. A microcomputer does not know how to process data. It must be told exactly what to do, where to get data, what to do with the data, and where to put the results when it is finished. This is the job of *software* in a microcomputer system.

The sequence of commands that are used to tell a microcomputer what to do is called a *program*. Each command in a program is an *instruction*. A program may be simple and include just a few instructions, or very complex and contain more than 100,000 instructions. When the microcomputer is operating, it fetches and executes one instruction of the program after the other. In this way, the instructions of the program guide it step by step through the task that is to be performed.

Software is a general name used to refer to a wide variety of programs that can be run by a microcomputer. Examples are *languages*, *operating systems*, *application programs*, and *diagnostics*.

The native language of an 80386DX-based PC/AT is *machine language*. Programs must always be coded in this machine language before they can be executed by the MPU. The 80386DX microprocessor understands and performs operations for more than 150 basic instructions. When expressed in machine code, an instruction is encoded using 0s and 1s. A single machine language instruction can take up one or more bytes of code. Even though the 80386DX understands only machine code, it is almost impossible to write programs directly in machine language. For this reason, programs are normally written in other languages, such as *80386DX assembly language* or a high-level language such as *C*.

In 80386DX assembly language each of the operations that can be performed by the 80386DX microprocessor is described with alphanumeric symbols instead of with zeros and ones. Each instruction in a program is represented by a single *assembly language statement*. This statement must specify which operation is to be performed and what data are to be processed. For this reason, an instruction is divided into two parts: its *operation code (opcode)* and its *operands*. The opcode is the part of the instruction that identifies the operation that is to be performed. For example, typical operations are add, subtract, and move. Each opcode is assigned a unique letter combination called a *mnemonic*. The mnemonics for the earlier mentioned operations are ADD, SUB, and MOV. Operands describe the data that are to be processed as the microprocessor carries out the operation specified by the opcode. They identify whether the source and destination of the data are registers within the MPU or storage locations in data memory.

An example of an instruction written in 80386DX assembly language is

```
ADD   EAX, EBX
```

This instruction says, "Add the contents of registers EBX and EAX together and put the sum in register EAX." EAX is called the *destination operand*, because it is the place where the result ends up, and EBX is called the *source operand*.

An example of a complete assembly language statement is

```
START:   MOV   EAX, EBX        ;COPY EBX INTO EAX
```

This statement begins with the word START:. START is an address identifier for the instruction MOV   EAX, EBX. This type of identifier is known as a *label*. The instruction is followed by ;COPY EBX INTO EAX. This part of the statement is called a *comment*. Thus a general format for an assembly language statement is

```
LABEL:   INSTRUCTION     ;COMMENT
```

Programs written in assembly language are referred to as *source code*. An example of a short 80386DX assembly language program is shown in Fig. 2.1(a). The assembly language instructions are located toward the left. Notice that labels are not used in most statements. On the other hand, a comment describing the statement is usually included on the right. This type of documentation makes it easier for a program to be read and debugged.

Assembly language programs cannot be run directly on the 80386DX. They must still be converted to an equivalent machine language program for execution by the 80386DX. This conversion is done automatically by running the source code through a program known as an *assembler*. The machine language output produced by the assembler is called *object code*.

Figure 2.1(b) is the *listing* produced by assembling the assembly language source code in Fig. 2.1(a) with IBM's macroassembler for the PC/AT. Reading from left to right, this listing contains line numbers, addresses of memory locations, the machine language instructions, the original assembly language statements, and comments. For example, line 53, which is

```
0013  8A 24       NXTPT:   MOV  AH, [SI]      ;MOVE A BYTE
```

shows that the assembly language instruction MOV   AH, [SI] is encoded as 8A24 in machine language and that this two-byte instruction is loaded into memory starting at address 0013_{16} and ending at address 0014_{16}. Note that for simplicity the machine language instructions are expressed in hexadecimal notation, not in binary form. Use of assembly language makes it much easier to write a program. But notice that there is still a one-to-one relationship between assembly and machine language instructions.

```
TITLE    BLOCK-MOVE PROGRAM

         PAGE      ,132

COMMENT *This program moves a block of specified number of bytes
         from one place to another place*


;Define constants used in this program

         N           =       16         ;Bytes to be moved
         BLK1ADDR=           100H       ;Source block offset address
         BLK2ADDR=           120H       ;Destination block offset addr
         DATASEGADDR=        1020H      ;Data segment start address


STACK_SEG     SEGMENT      STACK 'STACK'
              DB           64 DUP(?)
STACK_SEG     ENDS


CODE_SEG      SEGMENT      'CODE'
BLOCK         PROC         FAR
         ASSUME   CS:CODE_SEG,SS:STACK_SEG

;To return to DEBUG program put return address on the stack

         PUSH     DS
         MOV      AX, 0
         PUSH     AX

;Set up the data segment address

         MOV      AX, DATASEGADDR
         MOV      DS, AX

;Set up the source and destination offset addresses

         MOV      SI, BLK1ADDR
         MOV      DI, BLK2ADDR

;Set up the count of bytes to be moved

         MOV      CX, N

;Copy source block to destination block

NXTPT:   MOV      AH, [SI]       ;Move a byte
         MOV      [DI], AH
         INC      SI             ;Update pointers
         INC      DI
         DEC      CX             ;Update byte counter
         JNZ      NXTPT          ;Repeat for next byte
         RET                     ;Return to DEBUG program
BLOCK         ENDP
CODE_SEG      ENDS
         END      BLOCK          ;End of program

                    (a)
```

Figure 2.1 (a) Example of an 80386DX assembly language program; (b) assembled version of the program.

The IBM Personal Computer MACRO Assembler 02-22-88   PAGE   1-1
BLOCK-MOVE PROGRAM

```
  1
  2
  3                                     TITLE    BLOCK-MOVE PROGRAM
  4
  5                                              PAGE     ,132
  6
  7                                     COMMENT *This program moves a block of specified number of bytes
  8                                              from one place to another place*
  9
 10
 11                                     ;Define constants used in this program
 12
 13      = 0010                                 N        =       16              ;Bytes to be moved
 14      = 0100                                 BLK1ADDR=        100H            ;Source block offset address
 15      = 0120                                 BLK2ADDR=        120H            ;Destination block offset addr
 16      = 1020                                 DATASEGADDR=     1020H           ;Data segment start address
 17
 18
 19      0000                            STACK_SEG       SEGMENT         STACK 'STACK'
 20      0000    40 [                                    DB              64 DUP(?)
 21                      ??
 22                   ]
 23
 24      0040                            STACK_SEG       ENDS
 25
 26
 27      0000                            CODE_SEG        SEGMENT         'CODE'
 28      0000                            BLOCK           PROC            FAR
 29                                              ASSUME  CS:CODE_SEG,SS:STACK_SEG
 30
 31                                     ;To return to DEBUG program put return address on the stack
 32
 33      0000   1E                              PUSH    DS
 34      0001   B8 0000                         MOV     AX, 0
 35      0004   50                              PUSH    AX
 36
 37                                     ;Set up the data segment address
 38
 39      0005   B8 1020                         MOV     AX, DATASEGADDR
 40      0008   8E D8                           MOV     DS, AX
 41
 42                                     ;Set up the source and destination offset adresses
 43
 44      000A   BE 0100                         MOV     SI, BLK1ADDR
 45      000D   BF 0120                         MOV     DI, BLK2ADDR
 46
 47                                     ;Set up the count of bytes to be moved
 48
 49      0010   B9 0010                         MOV     CX, N
 50
 51                                     ;Copy source block to destination block
 52
 53      0013   8A 24                   NXTPT:  MOV     AH, [SI]                ;Move a byte
```

(b)

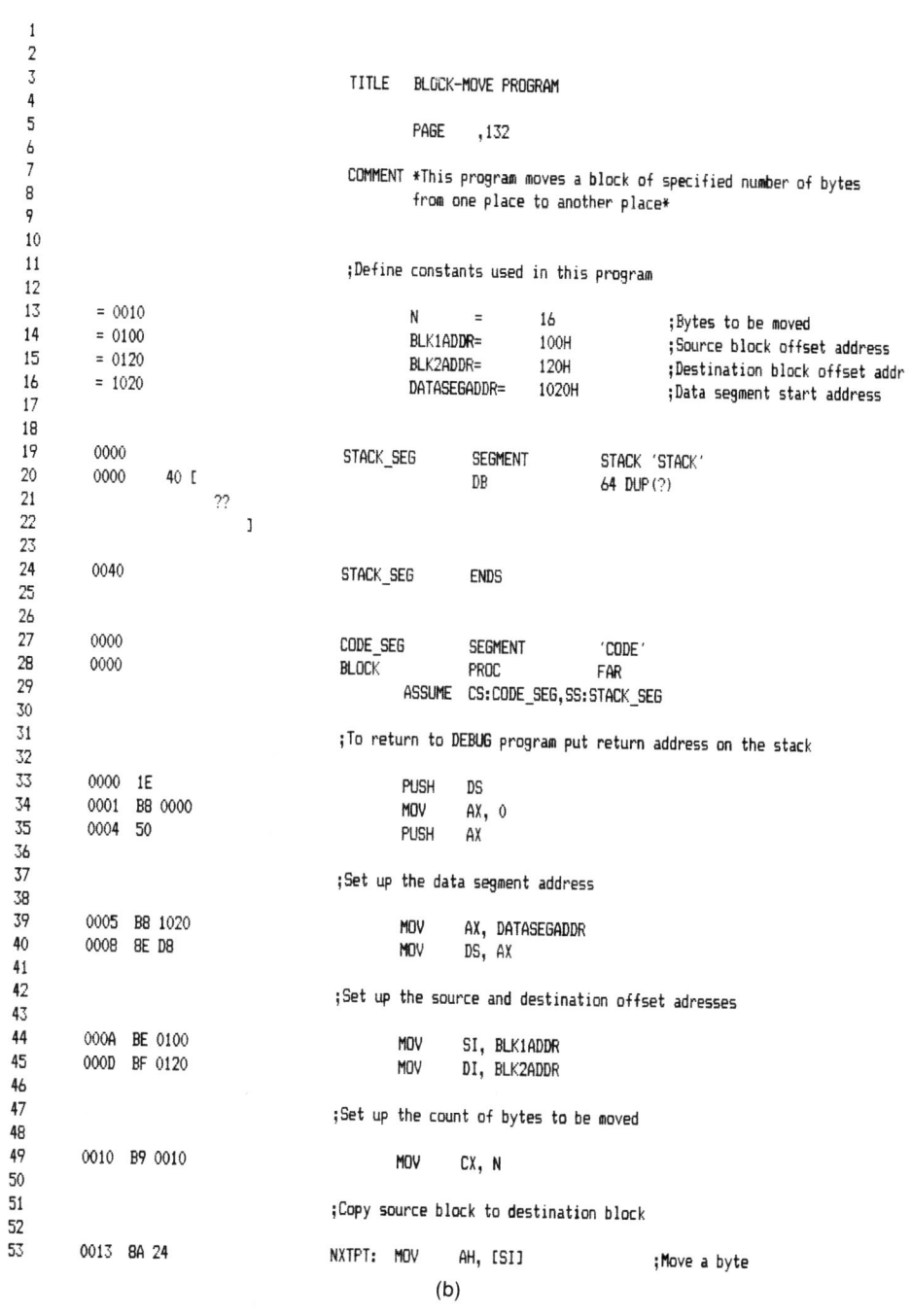

Figure 2.1 (*Continued*)

```
The IBM Personal Computer MACRO Assembler 02-22-88      PAGE    1-2
BLOCK-MOVE PROGRAM

54      0015  88 25                   MOV     [DI], AH
55      0017  46                      INC     SI              ;Update pointers
56      0018  47                      INC     DI
57      0019  49                      DEC     CX              ;Update byte counter
58      001A  75 F7                   JNZ     NXTPT           ;Repeat for next byte
59      001C  CB                      RET                     ;Return to DEBUG program
60      001D                  BLOCK           ENDP
61      001D                  CODE_SEG        ENDS
62                                    END     BLOCK           ;End of program
```

(b)

Figure 2.1 (*Continued*)

High-level languages make writing programs even easier. In a language such as BASIC, high-level commands such as FOR, NEXT, and GO are provided. These commands no longer correspond to a single machine language statement. Instead, they implement functions that may require many assembly language statements. Again, the statements must be converted to machine code before they can be run on the 80386DX. The program that converts high-level-language statements to machine code instructions is called a *compiler*.

Some languages—BASIC, for instance—are not always compiled. Instead, *interpretive* versions of the language are available. When a program written in an interpretive form of BASIC is executed, each line of the program is interpreted just before it is executed and at that moment replaced with a corresponding machine language routine. It is this machine code routine that is executed by the 80386DX.

The question you may be asking yourself right now is: If it is so much easier to write programs with a high-level language, why is it important to know how to program the 80386DX in its assembly language? This question will now be answered.

We just pointed out that if a program is written in a high-level language such as C, it must be compiled into machine code before it can be run on the 80386DX. The general nature with which compilers must be designed usually results in inefficient machine code. That is, the quality of the machine code that is produced for the program depends on the quality of the compiler program in use. What is found is that a compiled machine code implementation of a program that was written in a high-level language results in many more machine language instructions than an assembled version of an equivalent handwritten assembly language program. This leads us to the two key benefits derived from writing programs in assembly language: first, the machine code program that is produced will take up less memory space than the compiled version of the program; and second, it will execute much faster.

Now we know the benefits attained by writing programs in assembly language, but we still do not know when these benefits are important. To be important, they must outweigh the additional effort that must be put into writing the program in assembly language instead of a high-level language. One of the major uses of assembly language programming is in *real-time applications*. By "real time" we mean that the task required by the application must be completed before any other input to the program can occur that will alter its operation.

For example, the *device service routine* that controls the operation of the floppy disk drive of the PC/AT is a good example of the kind of program that is usually written in assembly language. This is because it is a segment of program that must closely control the microcomputer hardware in real time. In this case, a program that is written in a high-level language probably could not respond quick enough to control the hardware, and even if it could, operations performed with the disk subsystem would be much slower. Some other examples of hardware-related operations typically performed by routines written in assembler are communication routines such as those that drive the display and printer in a personal computer and the I/O routine that scans the keyboard.

Assembly language is important not only for controlling hardware devices of the microcomputer system, but also when performing pure software operations. For example, applications frequently require the microcomputer to search through a large table of data in memory looking for a special string of characters: say, a person's name. This type of operation can easily be performed by writing a program in a high-level language; however, for very large tables of data the search will take very long. By implementing the search routine through assembly language, the performance of the search operation is greatly improved. Other examples of software operations that may require implementation with high-performance routines derived with assembly language are *code translations*, such as from ASCII to EBCDIC, *table sort routines*, such as a bubble sort, and *mathematical routines*, such as those for floating-point arithmetic.

Not all parts of an application require real-time performance. For this reason it is a common practice to mix in the same program, routines developed through a high-level language and routines developed with assembly language. That is, assembler is used to code those parts of the application that must perform real-time operations, and high-level language is used to write those parts that are not time critical. The machine code obtained by assembling or compiling the two types of program segments are linked together to form the final application program.

2.3 SOFTWARE MODEL OF THE 80386DX MICROPROCESSOR

The purpose of a *software model* is to aid the programmer in understanding the operation of the microcomputer system from a software point of view. To be able to program a microprocessor, one does not necessarily need to know the function of the signals at its various pins, their electrical connections, or their electrical switching characteristics. The function, interconnection, and operation of the internal circuits of the microprocessor also need not normally be considered. What is important to the programmer is to know the various registers within the device and to understand their purpose, functions, operating capabilities, and limitations. Furthermore, it is essential to know how external memory is organized and how it is addressed to obtain instructions and data.

The software architecture of the 80386DX microprocessor is illustrated with the software model shown in Fig. 2.2. Looking at this diagram, we find that it includes six 16-bit registers and 24 32-bit registers. The registers (or parts of registers) that are important to real-mode application programming are highlighted in the diagram. Nine of them—the data registers (EAX, EBX, ECX, and EDX), the pointer registers (EBP and ESP), the index registers (ESI and EDI), and the flag register (FLAGS)—are identical to the corresponding registers in the 8086 and 80286 software models except that they are now all 32 bits in length. On the other hand, the instruction pointer (IP) and segment registers (CS, DS, SS, and ES) are still 16 bits in length. From a software point of view, all of these registers serve functions similar to those they performed in the 8086 and 80286 architectures. For example, CS:IP points to the next instruction that is to be fetched.

Several new registers are found in the 80386DX's software model. For example, it has two more data segment registers, denoted FS and GS. These registers are not implemented in either the 8086 or 80286 microprocessors. Another new register is called *control register zero* (CR_0). The five least significant bits of this register are called the *machine status word* (MSW) and are identical to the MSW of the 80286 microprocessor. However, only one bit, bit 0, in the MSW is active in real mode. This is the *protection enable* (PE) bit. PE is used to switch the 80386DX from real to protected mode. At reset, PE is set to 0 and selects the real-addressed mode of operation.

Looking at the software model in Fig. 2.2, we see that the 80386DX architecture implements independent memory and I/O address spaces. Notice that the memory address space is 1,048,576 bytes (1M byte) in length and the I/O address space is 65,536 bytes (64K bytes) in length. We are concerned here with what can be done with this architecture and how to do it through software. For this purpose, we now begin a detailed study of the elements of the model and their relationship to software.

2.4 MEMORY ADDRESS SPACE AND DATA ORGANIZATION

Now that we have introduced the idea of a software model, let us look at how information such as numbers, character, and instructions are stored in memory. As shown in Fig. 2.3, the 80386DX microcomputer supports 1M byte (1,048,576 bytes) of external memory. This memory space is organized from a software point of view as individual bytes of data stored at consecutive addresses over the address range 00000_{16} (0H) through $FFFFF_{16}$ (FFFFFH). However, the 80386DX can access any two consecutive bytes of data as a word of data or any four consecutive bytes as a double word of data.

The real-mode memory address space is partitioned into general use and dedicated use areas in the same way as for the 8086 microprocessor. For instance, in Fig. 2.3 we find that the first 1024 bytes of memory address space, addresses 0_{16} through $3FF_{16}$, are dedicated. They are again used for storage of the microcomputer's interrupt vector table. This table contains pointers that define the starting point of interrupt-service routines. Each pointer in the table requires four

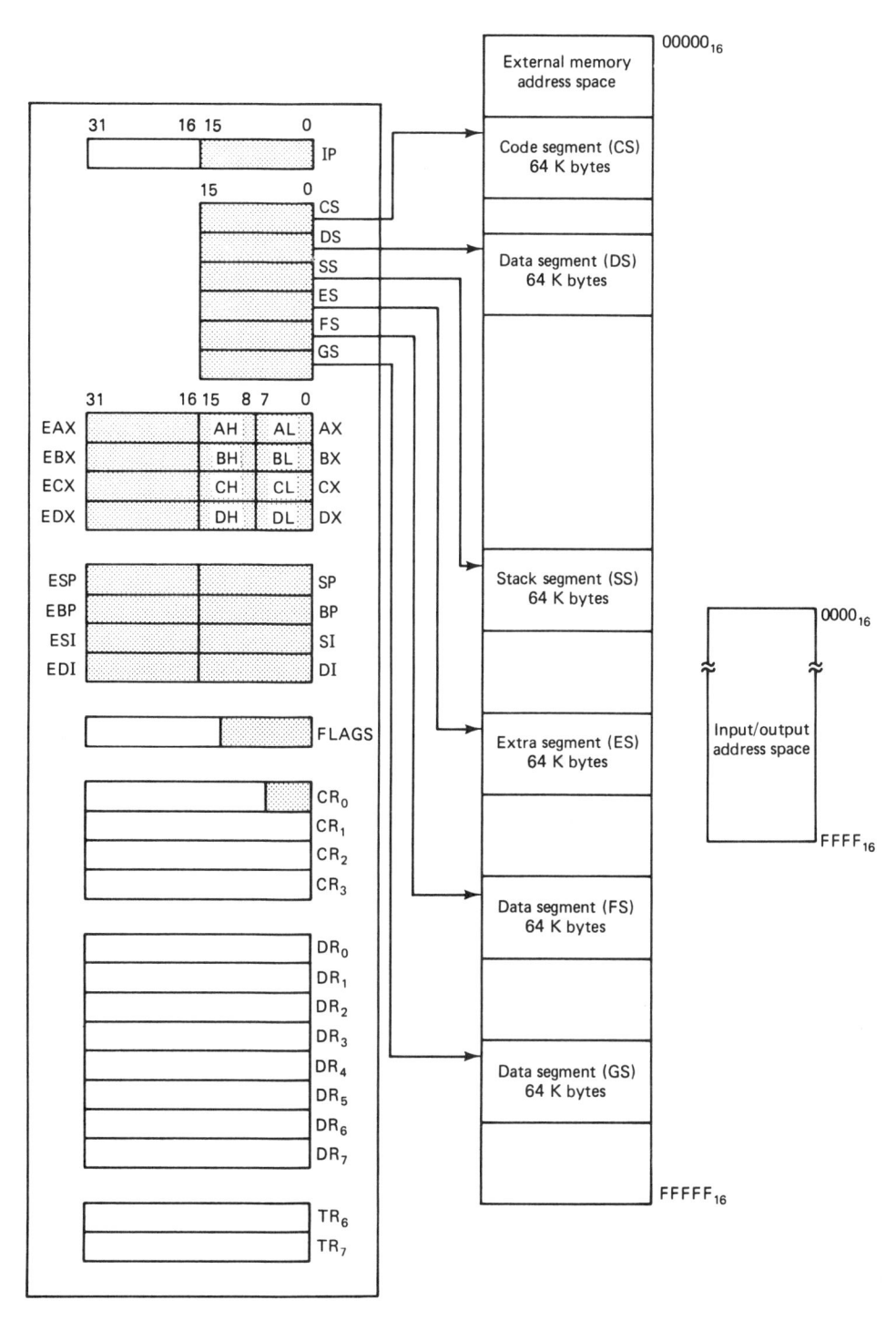

Figure 2.2 Real-mode software model of the 80386DX microprocessor.

Sec. 2.4   Memory Address Space and Data Organization

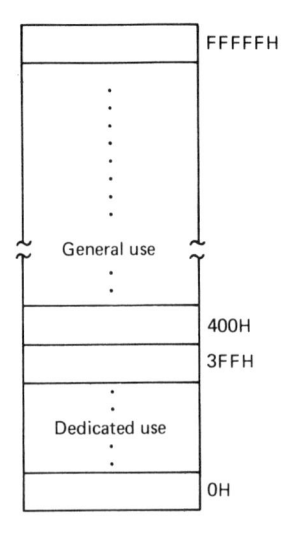

Figure 2.3 Real-mode memory address space.

bytes of memory. Therefore, it can contain up to 256 interrupt pointers. A pointer is a two-word address element. The word of this pointer at the higher address is called the *segment base address*, and the word at the lower address is the *offset*. *General-use memory* is where data or instructions of the program are stored. In Fig. 2.3, we see that the general-use area of memory is the range from address 400_{16} through $FFFFF_{16}$.

To permit efficient use of memory, words and double words of data can be stored at what are called either aligned double-word boundaries or unaligned double-word boundaries. Aligned double-word boundaries correspond to addresses that are a multiple of 4: for instance, 00000_{16}, 00004_{16}, 00008_{16}, and so on. Figure 2.4 shows a number of words and double words that are stored at

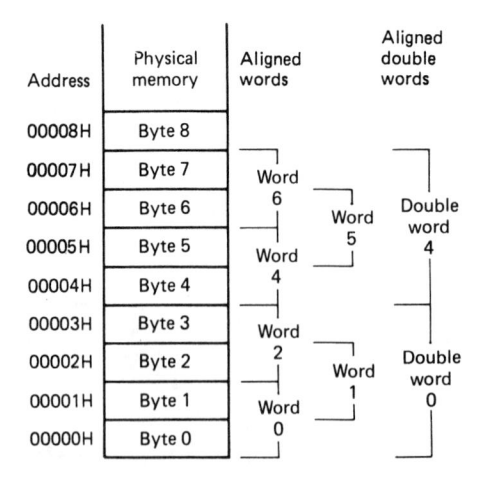

Figure 2.4 Examples of aligned data words and double words.

24   Real-Addressed-Mode Software Architecture   Chap. 2

aligned double-word address boundaries. For each of these pieces of data all of the bytes of data exist within the same double word.

For example, the value $5AF0_{16}$ is stored in memory at double-word aligned address 02000_{16}, as shown in Fig. 2.5(a). Notice that the lower addressed byte

Address	Memory (binary)	Memory (hexadecimal)	Address	Memory (binary)
02001_{16}	0101 1010	5A	$0200E_{16}$	0010 1100
02000_{16}	1111 0000	F0	$0200D_{16}$	1001 0110

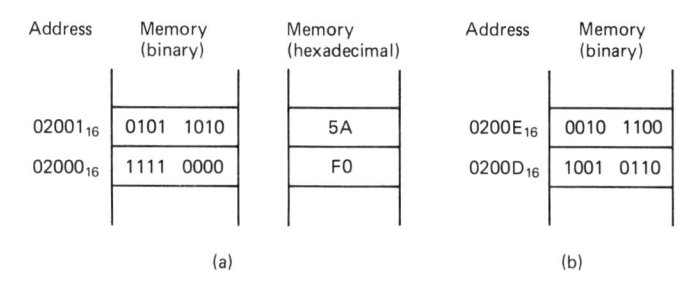

(a)   (b)

Figure 2.5 (a) Storing an aligned word of data in memory; (b) an example.

storage location, 02000_{16}, contains the value $11110000_2 = F0_{16}$. Moreover, the contents of the next-higher addressed byte, which is in storage location 02001_{16}, are $01011010_2 = 5A_{16}$. These two bytes represent the word $0101101011110000_2 = 5AF0_{16}$.

EXAMPLE 2.1

What is the data word shown in Fig. 2.5(b)? Express the result in hexadecimal form. Is it stored at an aligned double-word boundary?

SOLUTION The most significant byte of the word is stored at address $0200E_{16}$ and equals

$$00101100_2 = 2C_{16}$$

Its least significant byte is stored at address $0200D_{16}$ and is

$$10010110_2 = 96_{16}$$

Together these bytes give the word

$$0010110010010110_2 = 2C96_{16}$$

This word is aligned within double-word address boundary $0200C_{16}$.

It is not always possible to have all words or double words of data aligned at double-word boundaries. Figure 2.6 shows some examples of misaligned words and double words of data that can be accessed by the 80386DX. Notice that word 3 consists of byte 3 from aligned double word 0 and byte 4 from aligned double word 4. An example that shows a double word of data that is stored in memory at an unaligned double-word boundary is given in Fig. 2.7(a). Here we see that the higher-addressed word, which equals 0123_{16}, is stored in memory starting at address 02104_{16} in aligned double-word address 02104_{16}, and the lower-addressed

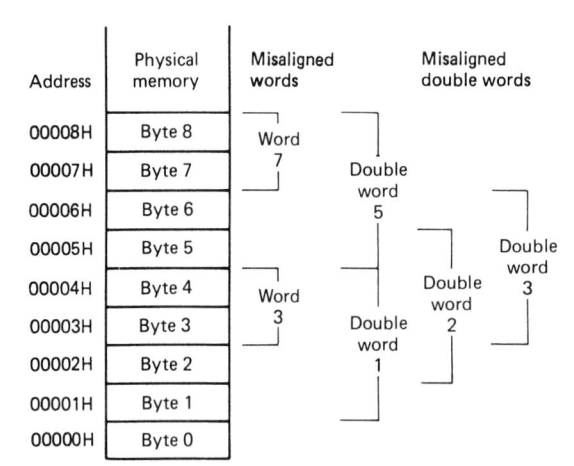

Figure 2.6 Examples of misaligned data words and double words.

word, whose value is $ABCD_{16}$, begins at address 02102_{16} in the aligned double-word address 02100_{16}. The complete double word equals $0123ABCD_{16}$.

EXAMPLE 2.2

How should the double word $A00055FF_{16}$ be stored in memory starting at address $0210C_{16}$? Is the double word aligned or misaligned?

SOLUTION Storage of the double word requires four consecutive byte locations in memory, starting at address $0210C_{16}$. The least significant byte is stored at address $0210C_{16}$. This value is FF_{16} in Fig. 2.7(b). The second byte, which equals 55_{16}, is stored at address $0210D_{16}$. These two bytes are followed by the values 00_{16} and $A0_{16}$ at addresses $0210E_{16}$ and $0210F_{16}$, respectively. This double word is aligned at double-word address $0210C_{16}$.

Figure 2.7 (a) Storing a misaligned double word in memory; (b) an example of an aligned double word.

26 Real-Addressed-Mode Software Architecture   Chap. 2

2.5 DATA TYPES

In the preceding section we identified the fundamental data formats of the 80386DX as the byte, word, and double word and showed how each of these elements is stored in memory. Here we will continue by examining the types of data that can be coded into these formats for processing by the 80386DX microprocessor.

Let us begin with the *integer* data type. The 80386DX can process data as either *unsigned* or *signed integer* numbers. Moreover, both types of integer can be byte wide, word wide, or double-word wide. Figure 2.8(a) represents an *un-*

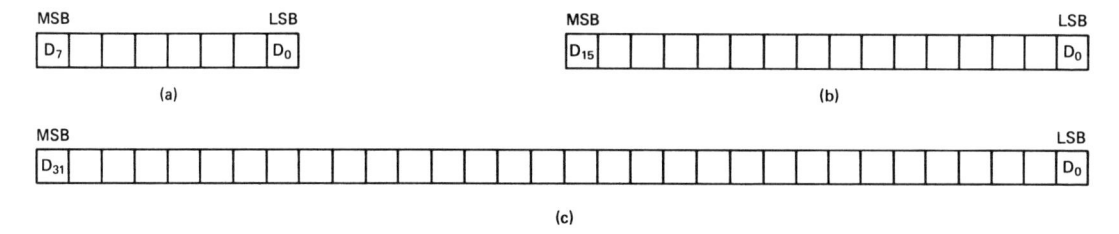

Figure 2.8 (a) Unsigned byte integer; (b) unsigned word integer; (c) unsigned double-word integer.

signed byte integer. This data type can be used to represent decimal numbers in the range 0 through 255. The *unsigned word integer* is shown in Fig. 2.8(b). It can be used to represent decimal numbers in the range 0 through 65,535. Finally, Fig. 2.8(c) illustrates an *unsigned double-word integer*.

EXAMPLE 2.3

What value does the unsigned double-word integer 00010000_{16} represent?

SOLUTION   First, the hexadecimal integer is converted to binary form:

$$00010000_{16} = 00000000000000010000000000000000_2$$

Next, we find the decimal value for the binary number:

$$00000000000000010000000000000000_2 = 2^{16} = 65,536$$

The *signed integer* byte, word, and double word of Fig. 2.9(a), (b), and (c) are similar to the unsigned integer data types we just introduced; however, here the most significant bit is a sign bit. A zero in this bit position identifies a positive number. For this reason, the signed integer byte can represent decimal numbers in the range +127 through −128, and the signed integer word permits numbers in the range +32767 through −32768. For example, the number +3 expressed as a signed integer byte is 00000011 (03_{16}). On the other hand, the 80386DX always expresses negative numbers in 2's-complement notation. Therefore, −3 is coded as 11111101 (FD_{16}).

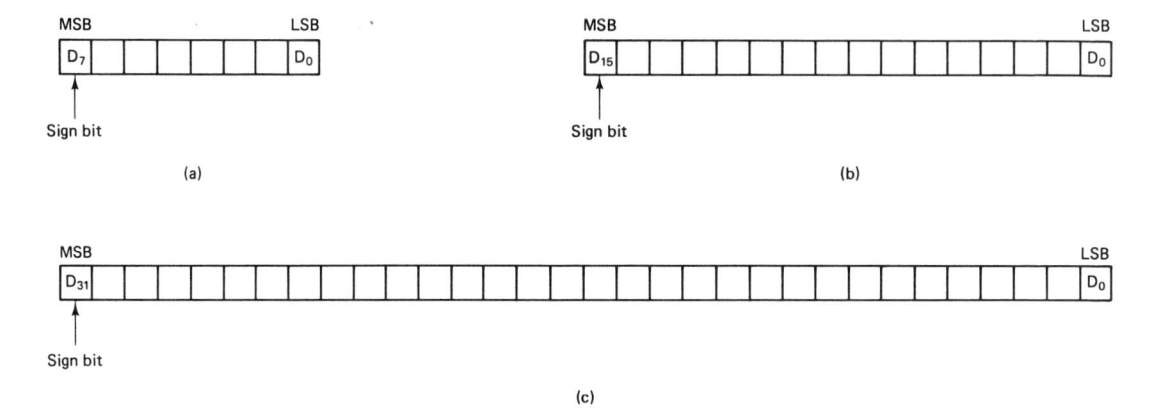

Figure 2.9 (a) Signed byte integer; (b) signed word integer; (c) signed double-word integer.

EXAMPLE 2.4

A double-word signed integer equals $FFFEFFFF_{16}$. What decimal number does it represent?

SOLUTION Expressing the hexadecimal number in binary form, we get

$$FFFEFFFF_{16} = 11111111111111101111111111111111_2$$

Since the most significant bit is 1, the number is negative and is in 2's-complement form. Converting to its binary equivalent by subtracting 1 from the least significant bit and then complementing all bits gives

$$FFFEFFFF_{16} = -00000000000000010000000000000001_2$$
$$= -65537$$

The 80386DX can also process data coded as *BCD numbers*. Figure 2.10(a) lists the BCD values for decimal numbers 0 through 9. BCD data can be stored in either unpacked or packed form. For instance, the *unpacked BCD byte* in Fig. 2.10(b) shows that a single BCD digit is stored in the four least significant bits and the upper four bits are set to zero. Figure 2.10(c) shows a byte with packed BCD digits. Here two BCD numbers are stored in a byte. The upper four bits represent the most significant digit of a two-digit BCD number.

EXAMPLE 2.5

The packed BCD data stored at byte address 01000_{16} equals 10010001_2. What is the two-digit decimal number?

SOLUTION Writing the value 10010001_2 as separate BCD digits gives

$$10010001_2 = 1001_{BCD}\ 0001_{BCD} = 91_{10}$$

Information expressed in *ASCII* (*American Standard Code for Information Interchange*) can also be directly processed by the 80386DX microprocessor. The

Decimal	BCD
0	0000
1	0001
2	0010
3	0011
4	0100
5	0101
6	0110
7	0111
8	1000
9	1001

(a)

MSB .... LSB
[ ][ ][ ][D_3][ ][ ][ ][D_0]
         BCD Digit

(b)

MSB .... LSB
[D_7][ ][ ][D_4][D_3][ ][ ][D_0]
   BCD Digit 1   BCD Digit 0

(c)

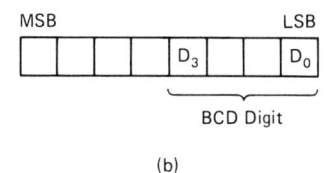

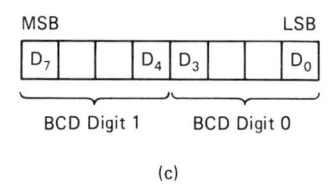

Figure 2.10 (a) BCD numbers; (b) unpacked BCD byte; (c) packed BCD digits.

chart in Fig. 2.11(a) shows how numbers, letters, and control characters are coded in ASCII. For instance, the number 5 is coded as

$$H_1 H_0 = 0110101 = 35H$$

where H denotes that the ASCII-coded number is in hexadecimal form. As shown in Fig. 2.11(b), ASCII data are stored one character per byte.

EXAMPLE 2.6

Byte addresses 01100_{16} through 01104_{16} contain the ASCII data 01000001, 01010011, 01000011, 01001001, and 01001001, respectively. What do the data stand for?

SOLUTION Using the chart in Fig. 2.11(a), the data are converted to ASCII as follows:

$$(01100H) = 01000001_{ASCII} = A$$

$$(01101H) = 01010011_{ASCII} = S$$

$$(01102H) = 01000011_{ASCII} = C$$

$$(01103H) = 01001001_{ASCII} = I$$

$$(01104H) = 01001001_{ASCII} = I$$

	b₇	0	0	0	0	1	1	1	1
	b₆	0	0	1	1	0	0	1	1
	b₅	0	1	0	1	0	1	0	1
b₄ b₃ b₂ b₁ H₁/H₀		0	1	2	3	4	5	6	7
0 0 0 0	0	NUL	DLE	SP	0	@	P	`	p
0 0 0 1	1	SOH	DC1	!	1	A	Q	a	q
0 0 1 0	2	STX	DC2	"	2	B	R	b	r
0 0 1 1	3	ETX	DC3	#	3	C	S	c	s
0 1 0 0	4	EOT	DC4	$	4	D	T	d	t
0 1 0 1	5	ENQ	NAK	%	5	E	U	e	u
0 1 1 0	6	ACK	SYN	&	6	F	V	f	v
0 1 1 1	7	BEL	ETB	'	7	G	W	g	w
1 0 0 0	8	BS	CAN	(	8	H	X	h	x
1 0 0 1	9	HT	EM	)	9	I	Y	i	y
1 0 1 0	A	LF	SUB	*	:	J	Z	j	z
1 0 1 1	B	V	ESC	+	;	K	[	k	}
1 1 0 0	C	FF	FS	,	<	L	\	l	|
1 1 0 1	D	CR	GS	-	=	M	]	m	}
1 1 1 0	E	SO	RS	.	>	N	∧	n	~
1 1 1 1	F	SI	US	/	?	O	-	o	DEL

(a)

```
MSB                    LSB
┌──┬──┬──┬──┬──┬──┬──┬──┐
│D₇│  │  │  │  │  │  │D₀│
└──┴──┴──┴──┴──┴──┴──┴──┘
      ASCII Digit
```

(b)

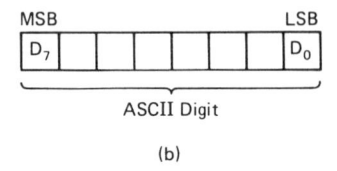

Figure 2.11 (a) ASCII table; (b) ASCII byte.

2.6 SEGMENT REGISTERS AND MEMORY SEGMENTATION

Even though the 80386DX has a 1M-byte address space in real mode, not all of this memory can be active at one time. Actually, the 1M bytes of memory can be partitioned into 64K (65,536)-byte *segments*. A segment represents an independently addressable unit of memory consisting of 64K consecutive byte-wide storage locations. Each segment is assigned a *segment base address* that identifies its starting point, that is, its lowest-addressed byte storage location.

Only six of these 64K-byte segments can be active at a time: the *code segment*, *stack segment*, and *data segments D, E, F*, and *G*. The location of the segments of memory that are active, as shown in Fig. 2.12, are identified by the value of address held in the 80386DX's six internal segment registers: *CS* (*code segment*), *SS* (*stack segment*), *DS* (*data segment*), *ES* (*extra segment*), *FS* (*data segment F*), and *GS* (*data segment G*). Each of these registers contains a 16-bit base address that points to the lowest-addressed byte of the segment in memory. Six segments give a maximum of 384K bytes of active memory. Of this, 64K bytes

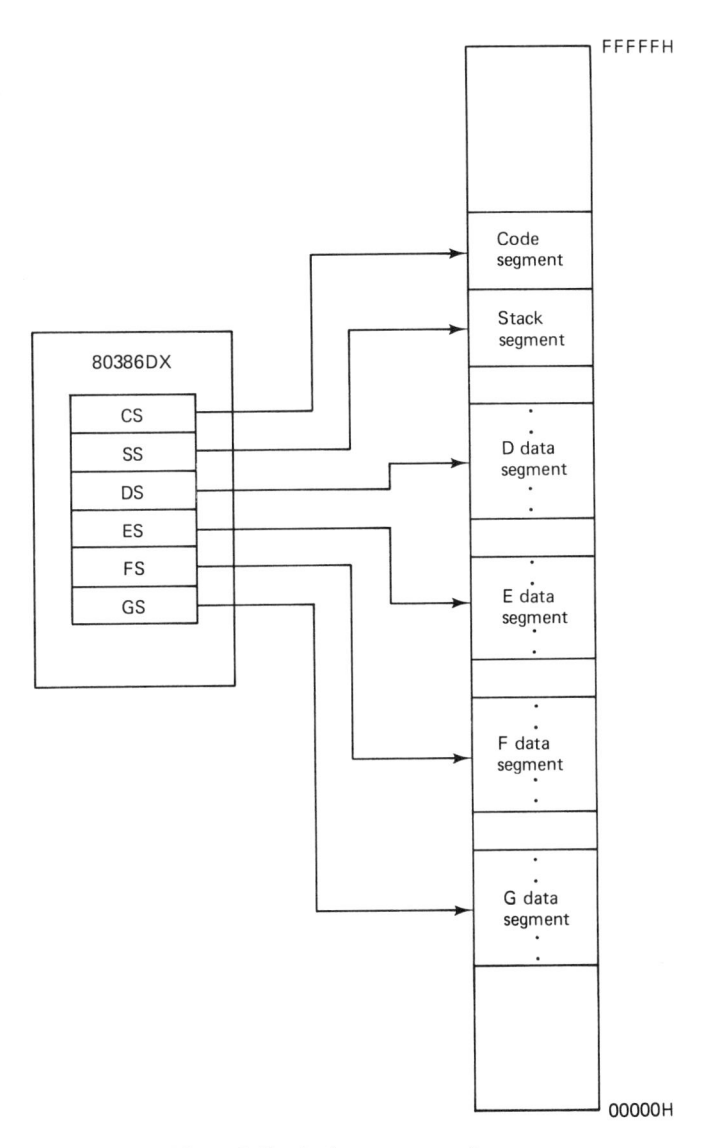

Figure 2.12 Active segments of memory.

Sec. 2.6   Segment Registers and Memory Segmentation

are for code (*program storage*), 64K bytes for a *stack*, and 256K bytes are for *data storage*.

The values held in these registers are usually referred to as the *current segment register values*. For example, the word in CS points to the first double-word storage location in the current code segment. Code is always fetched as double words, not as words or bytes.

Figure 2.13 illustrates the segmentation of memory. In this diagram we have

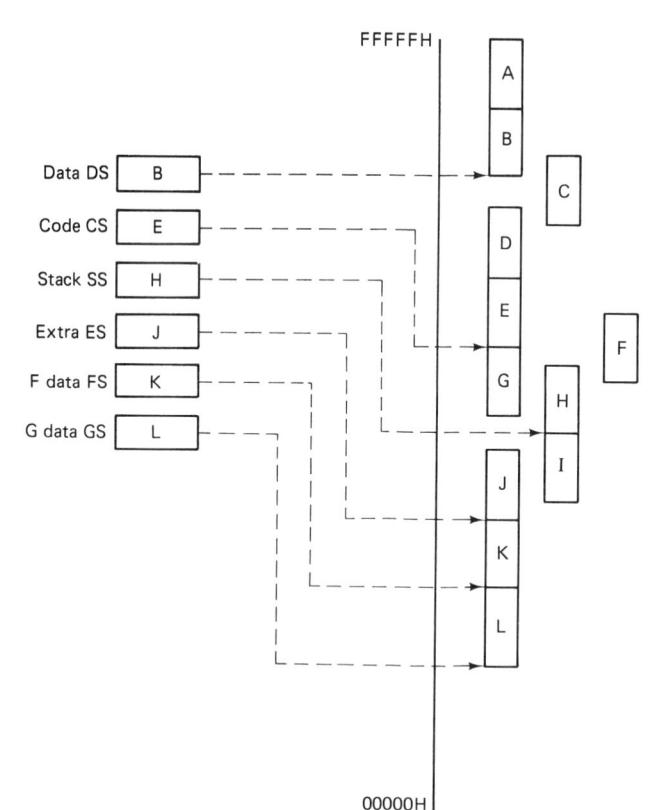

Figure 2.13 Contiguous, adjacent, disjointed, and overlapping segments. (Reprinted by permission of Intel Corp. copyright Intel Corp. 1979.)

identified 64K-byte segments with letters such as A, B, and C. The data segment (DS) register contains the value B. Therefore, the second 64K-byte segment of memory from the top, which is labeled B, acts as the current data storage segment. This is one of the segments in which data that are to be processed by the microcomputer are stored. For this reason this part of the microcomputer's memory address space must contain read/write storage locations that can be accessed by instructions as source and destination operands. Segment E is selected by CS as the code segment. It is this segment of memory from which instructions of the program are currently being fetched for execution. The stack segment (SS) register contains H, thereby selecting the 64K-byte segment labeled as H for use as a stack. Finally, the extra segment register, F data segment register, and G data segment register are loaded with the values J, K, and L such that segments J, K, and L of memory can function as additional 64K-byte data storage segments.

The segment registers are said to be *user accessible*. This means that the programmer can change the value they hold through software. Therefore, for a program to gain access to another part of memory, it just has to change the value of the appropriate segment register or registers. For instance, a new data space, with up to 256K bytes, can be brought in simply by changing the values in DS, ES, FS, and GS.

There is one restriction on the value that can be assigned to a segment as a base address: It must reside on a 16-byte address boundary. Valid examples are 00000_{16}, 000010_{16}, 00020_{16}, and so on. Except for this restriction, segments can be set up to be contiguous, adjacent, disjoint, or even overlapping. For example, in Fig. 2.13 segments A and B are contiguous, whereas segments B and C are overlapping.

2.7 INSTRUCTION POINTER

The register that we consider next in the 80386DX's real-mode software model of Fig. 2.2 is the *instruction pointer* (IP). IP is also 16 bits in length and identifies the location of the next double word of instruction code to be fetched from the code segment of memory. It is similar to a program counter; however, IP contains the offset of the next double word of code in the current code segment instead of its actual address. This is because in the real-mode 80386DX, IP and CS are both 16 bits in length, but a 20-bit address is needed to access memory. Internal to the 80386DX, the offset in IP is combined with the current value in CS to generate the address of the instruction code. Therefore, the value of the address for the next code access is often denoted as CS:IP.

During normal operation, the 80386DX fetches instructions from memory, stores them in its instruction queue, and executes them one after the other. Every time a double word of instruction code is fetched from memory, the 80386DX updates the value in IP such that it points to the next sequential double word of code. That is, IP is incremented by four. Actually, the 80386DX prefetches up to 16 bytes of instruction code into its internal code queue and holds them there waiting for execution. After an instruction is read from the output of the code queue, it is decoded and, if necessary, operands are read from either memory or internal registers. Next, the operation specified in the instruction is performed on the operands and the result is written back to either a storage location in memory or an internal register. The 80386DX is now ready to execute the next instruction in the code queue.

The active code segment can be changed simply by executing an instruction that loads a new value into the CS register. For this reason, we can use any of the 64K-byte segments in memory for storage of instruction code.

2.8 GENERAL-PURPOSE REGISTERS

As shown in Fig. 2.2 eight general-purpose registers are located within the 80386DX. During program execution, they are used for temporary storage of frequently used intermediate results. Their contents can be read, loaded, or modified through software. Any of the general-purpose registers can be used as the source or destination of an operand during an arithmetic operation such as ADD or a logic operation such as AND. For instance, the values of two pieces of data, called A and B, could be moved from memory into separate data registers and then operations such as addition, subtraction, and multiplication performed on them. The advantage of storing these data in internal registers instead of memory during processing is that they can be accessed much faster.

The first four registers, known as the *data registers*, are shown in more detail in Fig. 2.14(a). Notice that they are referred to as the *accumulator register* (A), *base register* (B), *count register* (C), and *data register* (D). These names identify special functions that they perform in the 8086, 80286, and 80386DX microprocessors. Figure 2.14(b) summarizes these operations. Notice that the C register is used as a count register during string, loop, rotate, and shift operations. For example, the value in the C register is the number of bits by which the contents of an operand must be shifted or rotated during execution of the multibit shift or rotate instructions.

Each of the data registers can be accessed as a whole, 32 bits, for double-word data operations, the lower 16 bits as a word of data, two 8-bit registers for byte-wide data operations, or as 32 individual bits. References to a register as a word are identified by an X after the register letter. For instance, the 16-bit accumulator register is referenced as AX. Similarly, the other three word registers are referred to as BX, CX, and DX.

On the other hand, when referencing all 32 bits of a register, the word register name is prefixed by the letter E. For example, the extended accumulator is denoted as EAX and the extended count register as ECX. Finally, the high byte and low byte of a word register are identified by following the register name with the letter H or L, respectively. For the A register, the more significant byte is referred to as AH and the less significant byte as AL.

Actually, these registers can also be used to store address information such as a base address or index address. Even though these registers are 32 bits in length, address operands are always 16 bits in length. For example, AX could hold a 16-bit index address.

The other four general-purpose registers are the two *index registers*, ESI and EDI, and the two *pointer registers*, EBP and ESP. In the 8086 and 80286 architectures, these registers serve special functions. They are used to store offset addresses of memory locations relative to the segment registers. The index registers are used to reference data relative to the data segment or extra segment register, and the pointer registers are used to store offset addresses of memory locations relative to the stack segment register. For the 80386DX, the functions of these registers are also generalized. They can be used to hold data operands, base addresses, or index addresses. One exception is that the stack pointer cannot

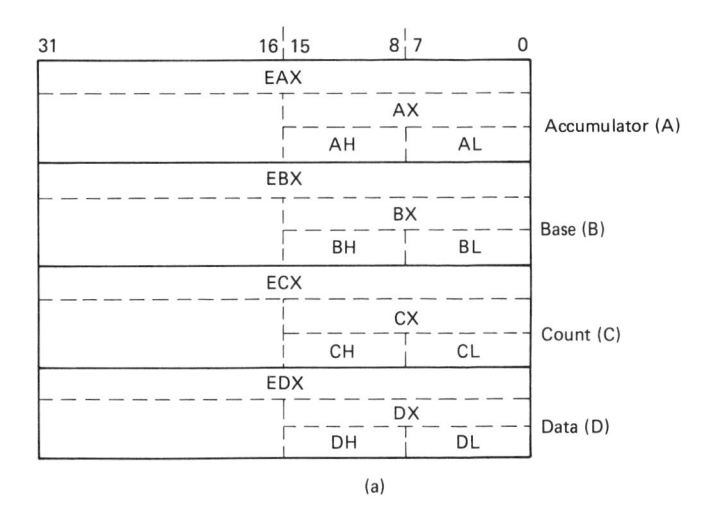

Register	Operations
EAX, AX, AH, AL	ASCII adjust for addition/subtraction Convert byte to word/word to double word/ double word to quad word Decimal adjust for addition/subtraction Unsigned multiply/divide Signed divide Input/output operations Load/store flags Load/compare/store string operations Table-lookup translations
EBX, BX, BH, BL	Table-lookup translations
ECX, CX, CH, CL	Loop operations Repeat string operations Variable shift/rotate operations
EDX, DX, DH, DL	Indirect input/output operations Input/output string operations Unsigned word/double word multiply Signed word/double word divide Unsigned word/double word divide

(b)

Figure 2.14 (a) General-purpose data registers; (b) special register functions.

be used as an index. It can only be used as a base. When used as an operand, these registers can hold either word or double-word data, but not bytes. Moreover, in real mode, the base or index address is always 16 bits in length.

Figure 2.15 shows that the two pointer registers are the *extended stack pointer* (ESP) and *extended base pointer* (EBP). When used to hold real-mode address information, the length of the base is always 16 bits. Therefore, the registers are identified as SP and BP. The values in SP and BP are normally used as offsets from the current value of SS. In fact, during the execution of an instruction that involves SP or BP as a base address, the value in the base register

Sec. 2.8 General-Purpose Registers 35

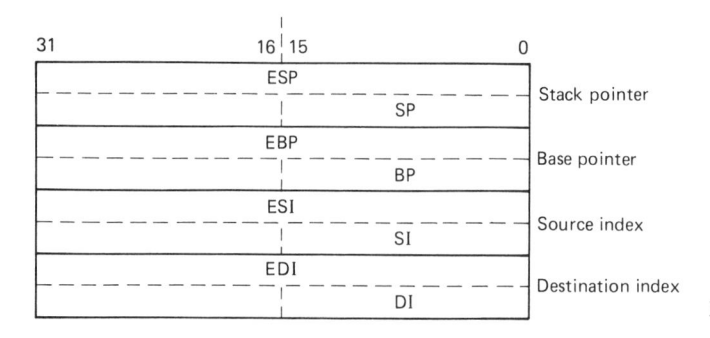

Figure 2.15 Pointer and index registers.

is combined with the contents of the stack segment register to produce the physical memory address. The value in SP always represents the offset of the next stack location that is to be accessed. That is, in real mode, combining SP with the value in SS (SS:SP) results in a 20-bit address that points to the *top of the stack* (TOS).

BP is used to access storage locations within the stack segment of memory. To do this it is employed as the offset in an addressing mode called the based addressing mode. If BP is used as an index instead of a base, it can serve as an offset with respect to any of the segment registers. One common use of BP is to reference parameters that were passed to a subroutine by way of the stack. In this case instructions are included in the subroutine that use based addressing to read the values of the parameters from the stack.

The index registers are used to hold offset addresses for instructions that access data stored in the data segment part of memory. For this reason, they are automatically combined with the value in the DS register. In instructions that use indexed addressing, the *extended source index register* (ESI) is used to hold an offset address for a source operand, and the *extended destination index register* (EDI) holds an offset that identifies the location of a destination operand. For example, a string instruction that requires offsets to the location of its source and destination operands uses these registers. Even though these registers are 32 bits in length, in real mode just the lower 16 bits SI and DI are used in indexed addressing. On the other hand, when ESI and EDI are used to hold operands, they can be accessed either as a word or a double word of data.

Earlier we pointed out that any of the general-purpose registers can be used as the source or destination of an operand during an arithmetic operation such as an ADD or a logic operation such as AND. However, for some operations, such as those performed by string instructions, a specific register must be used. For instance, in string instructions SI and DI are used as the pointers to the source and destination locations in memory, respectively. Moreover, in the case of a repeated string operation, the C register is used to store a count representing the number of times the string operation is to be repeated.

2.9 FLAGS REGISTER

In Fig. 2.2 we find that the flags register (FLAGS) is another 32-bit register within the 80386DX. This register is shown in more detail in Fig. 2.16. Notice that just

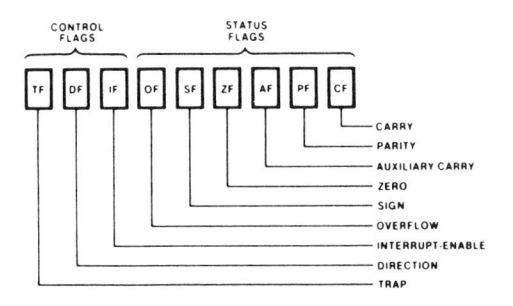

Figure 2.16 Status and control flags. (Reprinted by permission of Intel Corp. copyright Intel Corp. 1979.)

nine of its bits are active in the real mode. These bits are the same as those implemented in the FLAGS registers of the 8086 and real-mode 80286 microprocessors. Six of these bits represent *status flags*: the *carry flag* (CF), *parity flag* (PF), *auxiliary carry flag* (AF), *zero flag* (ZF), *sign flag* (SF), and *overflow flag* (OF). The logic states of these status flags indicate conditions that are produced as the result of executing an instruction. That is, after executing an instruction such as ADD, specific flag bits are reset (logic 0) or set (logic 1) based on the result that is produced.

Let us first summarize the operation of these flags:

1. The carry flag (CF): CF is set if there is a carry-out or a borrow-in for the most significant bit of the result during the execution of an instruction. Otherwise, CF is reset.

2. The parity flag (PF): PF is set if the result produced by the instruction has even parity, that is, if it contains an even number of bits at the 1 logic level. If parity is odd, PF is reset.

3. The auxiliary carry flag (AF): AF is set if there is a carry-out from the low nibble into the high nibble or a borrow-in from the high nibble into the low nibble of the lower byte in an 8-, 16-, or 32-bit word. Otherwise, AF is reset.

4. The zero flag (ZF): ZF is set if the result of execution of an instruction is zero. Otherwise, ZF is reset.

5. The sign flag (SF): The MSB of the result is copied into SF. Thus SF is set if the result is a negative number or reset if it is positive.

6. The overflow flag (OF): When OF is set, it indicates that the signed result is out of range. If the result is not out of range, OF remains reset.

For example, at the completion of execution of a byte-addition instruction, the carry flag (CF) could be set to indicate that the sum of the operands caused a carry-out condition. The auxiliary carry flag (AF) could also set due to the execution of the instruction. This depends on whether or not a carry-out occurred from the least significant nibble to the most significant nibble when the byte operands are added. The sign flag (SF) is also affected and it will reflect the logic level of the MSB of the result. The overflow flag (OF) is set if there is a carry-out of the sign bit, but no carry into the sign bit.

The 80386DX provides instructions in its instruction set that are able to use these flags to alter the sequence in which the program is executed. For instance, the condition ZF equal to logic 1 could be tested for to initiate a jump to another part of the program. This operation is called *jump on zero.*

The other three implemented flag bits—*direction flag* (DF), *interrupt enable flag* (IF), and *trap flag* (TF)—are control flags. These three flags are provided to control functions of the 80386DX as follows:

1. The trap flag (TF): If TF is set, the 80386DX goes into the single-step mode. When in single-step mode, it executes one instruction at a time. This type of operation is very useful for debugging programs.

2. The interrupt flag (IF): For the 80386DX to recognize *maskable interrupt requests* at the INT input, the IF flag must be set. When IF is reset, requests at INT are ignored and the maskable interrupt interface is disabled.

3. The direction flag (DF): The logic level of DF determines the direction in which string operations occur. When set, the string instruction automatically decrements the address. Therefore, the string data transfers proceed from high address to low address. On the other hand, resetting DF causes the string address to be incremented. In this way, data transfers proceed from low address to high address.

The instruction set of the 80386DX includes instructions for saving, loading, and manipulating the flags. For instance, special instructions are provided to permit user software to set or reset CF, DF, and IF at any point in the program. For example, just prior to the beginning of a string operation, DF could be set so that the string address automatically decrements.

2.10 GENERATING A REAL-MODE MEMORY ADDRESS

A *logical address* is described by a segment address and an offset address. As shown in Fig. 2.17, both the segment and offset are 16-bit quantities. This is because the segment registers, instruction pointer, and base or index registers used in address calculations are always 16 bits long in the 80386DX's real-mode software architecture. However, the *physical address* that is used to access memory is 20 bits in length. Notice that this physical address is generated by combining the values of the 16-bit offset address and a 16-bit base address that is located in one of the segment registers.

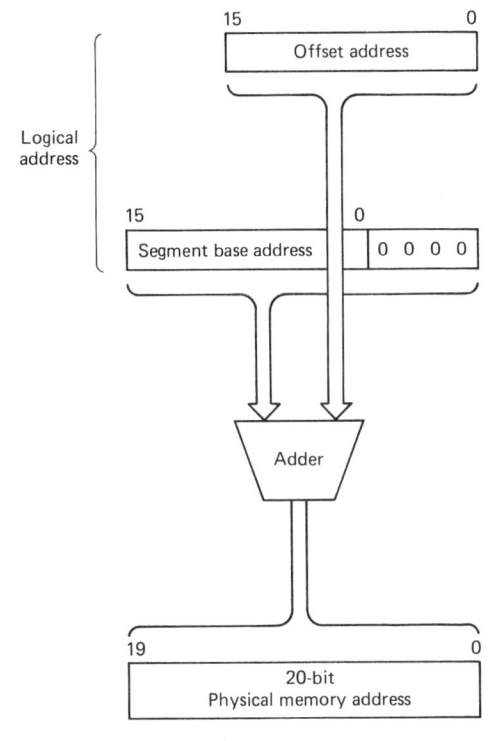

Figure 2.17 Real-address mode physical address generation. (Reprinted by permission of Intel Corp. copyright Intel Corp. 1981.)

The source of the offset address depends on which type of memory reference is taking place. For instance, it could be the base pointer (BP) register, the base (BX) register, the source index (SI) register, the destination index (DI) register, or the instruction pointer (IP). An offset can even be formed from the contents of several of these registers. On the other hand, the segment base address always resides in one of the 80386DX's segment registers: CS, DS, SS, ES, FS, or GS.

During code accesses, data accesses, and stack accesses of memory, the 80386DX selects the appropriate segment registers based on the rules shown in Fig. 2.18. For instance, when an instruction acquisition takes place, the source of the segment base address is always CS and the source of the offset address is always IP. This physical address can be denoted as CS:IP. On the other hand, if the value of a variable is being written to memory during execution of an instruction (local data), the segment base address defaults to the DS register and the offset will be in a register, such as DI or BX. An example is the physical address DS:DI. Segment override prefixes can be used to change the segment from which the variable is accessed. For instance, a prefix could be used to make a data access occur in which the segment base address is in the FS register.

Another example is the stack address that is needed when pushing parameters onto the stack. This address is formed from the value of the segment base in the SS register and offset in the SP register and is described as SS:SP.

Type of Reference	Segment Used Register Used	Default Selection Rule
Instructions	Code Segment CS register	Automatic with instruction fetch.
Stack	Stack Segment LSS register	All stack pushes and pops. Any memory reference which uses ESP or EBP as a base register.
Local Data	Data Segment DS register	All data references except when relative to stack or string destination.
Destination Strings	E-Space Segment ES register	Destination of string instructions.

Figure 2.18 Segment register references for memory accesses. (Reprinted by permission of Intel Corp. copyright Intel Corp. 1989.)

Remember that the segment base address represents the starting location of the 64K-byte segment in memory, that is, the lowest addressed byte in the segment. Figure 2.19 shows that the offset identifies the distance in bytes that the

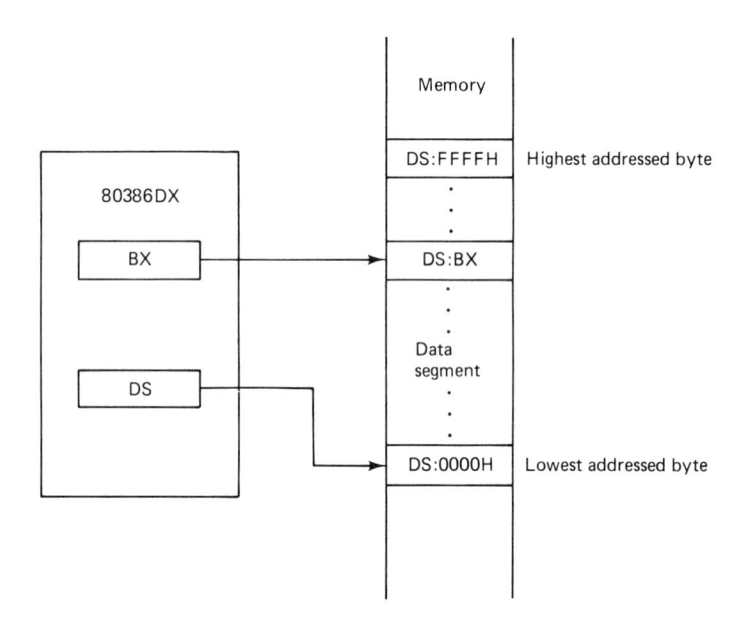

Figure 2.19 Boundaries of a segment.

storage location of interest resides from this starting address. Therefore, the lowest-addressed byte in a segment has an offset of 0000_{16} and the highest-addressed byte has an offset of $FFFF_{16}$.

Figure 2.20 illustrates how a segment base address and offset value are combined to give a physical address. What happens is that the value in the segment

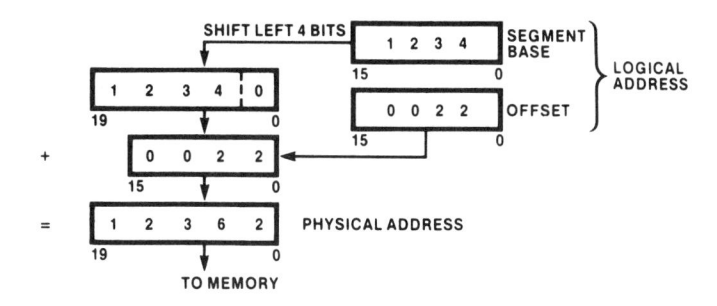

Figure 2.20 Generating a real-mode physical address. (Reprinted by permission of Intel Corp. copyright Intel Corp. 1979.)

register is shifted left by four bit positions, with its LSBs being filled with zeros. Then the offset value is added to the 16 LSBs of the shifted segment base address. The result of this addition is the 20-bit physical address. This is identical to how the 8086 and real-mode 80286 microprocessors perform the physical address calculation.

The example in Fig. 2.20 represents a segment base address of 1234_{16} and an offset address of 0022_{16}. First let us express the base address in binary form. This gives

$$1234_{16} = 0001001000110100_2$$

Shifting left four times and filling with zeros results in

$$00010010001101000000_2 = 12340_{16}$$

The offset in binary form is

$$0022_{16} = 0000000000100010_2$$

Adding the shifted segment base address and offset, we get

$$00010010001101000000_2 + 0000000000100010_2 = 00010010001101100010_2$$
$$= 12362_{16}$$

This type of address calculation is done automatically within the 80386DX each time a memory access is initiated.

EXAMPLE 2.7

What would be the offset required to map to physical address location $002C3_{16}$ if the segment base address is $002A_{16}$?

SOLUTION The offset value can be obtained by shifting the segment base address left four bit positions and then subtracting it from the physical address. Shifting left gives

$$002A0_{16}$$

Now subtracting, we get the value of the offset:

$$002C3_{16} - 002A0_{16} = 0023_{16}$$

Sec. 2.10 Generating a Real-Mode Memory Address

Actually, many different logical addresses can map to the same physical address in memory. This is done simply by changing the values of the base address in the segment register and corresponding offset. The diagram in Fig. 2.21 demonstrates this idea. Notice that base $002B_{16}$ with offset 0013_{16} maps to physical address $002C3_{16}$ in memory. However, if the segment base address is changed to $002C_{16}$ with a new offset of 0003_{16}, the physical address is still $002C3_{16}$.

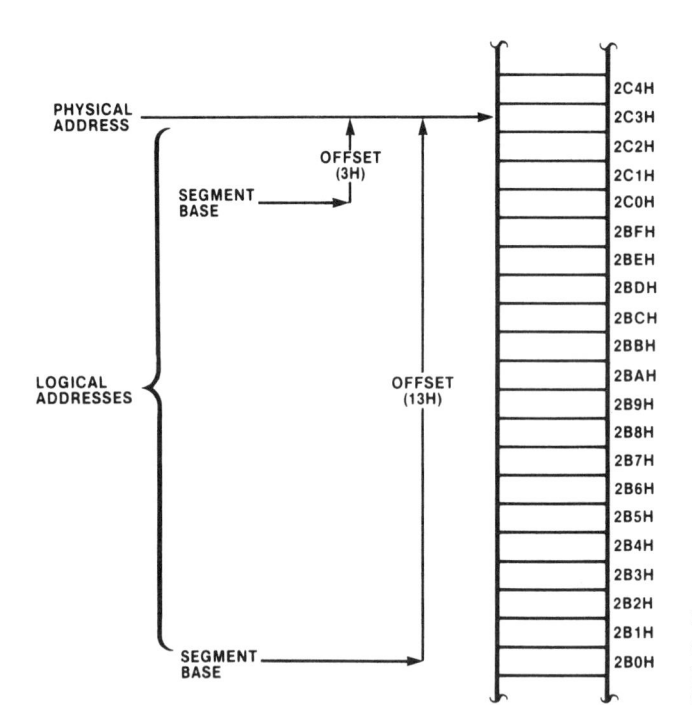

Figure 2.21 Relationship between logical and physical addresses. (Reprinted by permission of Intel Corp. copyright Intel Corp. 1979.)

2.11 THE STACK

As indicated earlier, the *stack* is implemented in the memory of the 80386DX microcomputer system. Stack is used for temporary storage of information such as data or addresses. For instance, when a *call* instruction is executed, the 80386DX automatically pushes the current values in CS and IP onto the stack. As part of the subroutine, the contents of other registers can also be saved on the stack by executing *push* instructions. An example is the instruction PUSH ESI. When executed it causes the contents of the extended source index register to be pushed onto the stack. At the end of the subroutine, *pop* instructions can be included to pop values from the stack back into their corresponding internal registers. For example, POP ESI causes the value at the top of the stack to be popped back into the extended source index register. At the completion of the subroutine, a return instruction causes the values of CS and IP to be popped off the stack and put back into the internal register where they originally resided.

From a real-mode software point of view, the stack is 64K bytes long and is organized as 32K words. Figure 2.22 shows that the lowest-addressed word in the current stack is pointed to by the segment base address in the SS register. The contents of the SP and BP registers are used as offsets into the stack segment of memory.

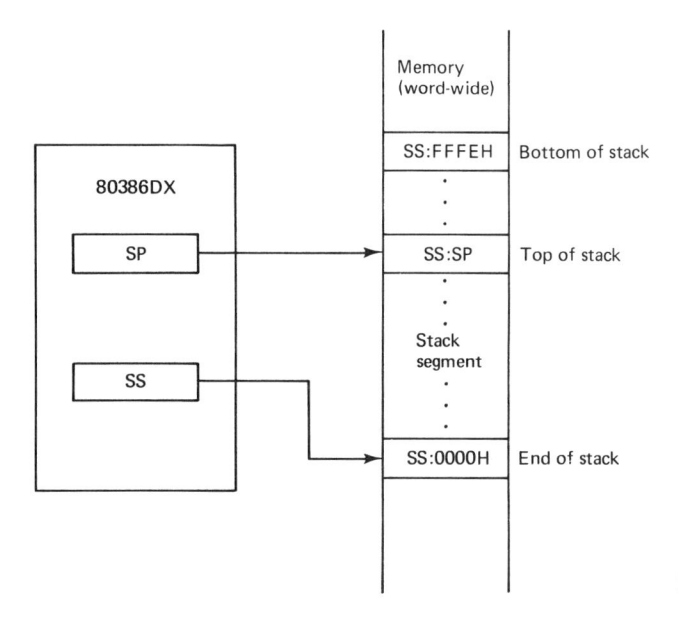

```
                              Memory
                              (word-wide)

                              SS:FFFEH    Bottom of stack
     ┌─────────────┐               ·
     │  80386DX    │               ·
     │             │               ·
     │  ┌──────┐   │          SS:SP        Top of stack
     │  │  SP  │───┼─────►         ·
     │  └──────┘   │               ·
     │             │              Stack
     │             │             segment
     │  ┌──────┐   │               ·
     │  │  SS  │───┼──┐            ·
     │  └──────┘   │  │            ·
     │             │  └──►     SS:0000H    End of stack
     └─────────────┘
```

Figure 2.22 Stack segment of memory.

The address obtained from the contents of SS and SP (SS:SP) is the physical address of the last storage location in the stack to which data were pushed. In Fig. 2.22 we find that this location is known as the *top of the stack*. After power-up of the microcomputer, the value in SP can be initialized to $FFFE_{16}$. Combining this value with the current value in SS gives the highest-address word location in the stack (SS:FFFEH), that is, the *bottom of the stack*.

The 80386DX can push data and address information onto the stack from its internal registers, a storage location in memory, or the immediate operand of an instruction. These data must be either a word or a double word, not a byte. Each time a word or double word is to be pushed onto the top of the stack, the value in SP is first automatically decremented by two or four, respectively, and then the contents of the register are written into the stack part of memory. In this way we see that the stack grows down in memory from the bottom of the stack, which corresponds to the physical address SS:FFFEH, toward the *end of the stack*, which corresponds to the physical address obtained from SS and offset 0000_{16}.

When a value is popped from the top of the stack, the reverse of this sequence occurs. The physical address defined by SS and SP points to the location of the last value pushed onto the stack. Its contents are first popped off the stack and put into the specified register within the 80386DX; then SP is automatically incremented by two or four, depending on whether a word or double word has been popped. The top of the stack now corresponds to the previous value pushed onto the stack.

An example that shows how the contents of a word-wide register are pushed onto the stack is shown in Fig. 2.23(a). Here we find the state of the stack prior

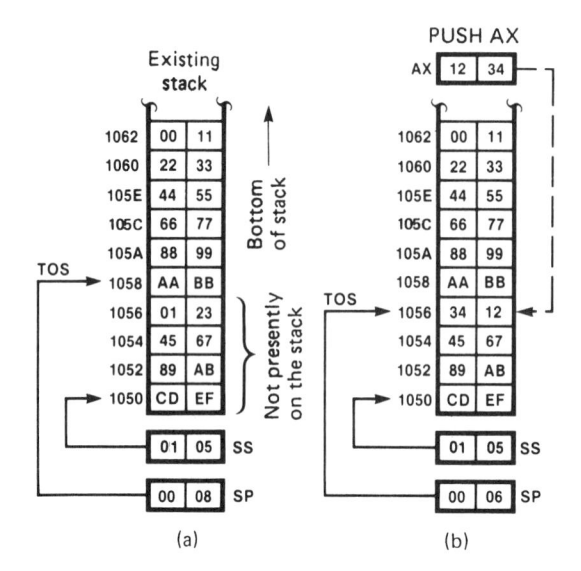

Figure 2.23 (a) Stack just prior to push operation. (Reprinted by permission of Intel Corp. copyright Intel Corp. 1979.) (b) Stack after execution of the PUSH AX instruction. (Reprinted by permission of Intel Corp. copyright Intel Corp. 1979.)

to execution of the PUSH AX instruction. Notice that the stack segment register contains 0105_{16}. As indicated, the bottom of the stack resides at the physical address derived from SS and offset $FFFE_{16}$. This gives the bottom of stack address A_{BOS} as

$$A_{BOS} = 1050_{16} + FFFE_{16}$$
$$= 1104E_{16}$$

Furthermore, the stack pointer, which represents the offset from the beginning of the stack specified by the contents of SS to the top of the stack, equals 0008_{16}. Therefore, the current top of the stack is at physical address A_{TOS}, which equals

$$A_{TOS} = 1050_{16} + 0008_{16}$$
$$= 1058_{16}$$

Addresses with higher values than that of the top of the stack, 1058_{16}, contain valid stack data. Those with lower addresses do not yet contain valid stack data. Notice that the last value pushed to the stack in Fig. 2.23(a) was $BBAA_{16}$.

Figure 2.23(b) demonstrates what happens when the PUSH AX instruction is executed. Here we see that AX contains the value 1234_{16}. Notice that execution of the push instruction causes the stack pointer to be decremented by two but does not affect the contents of the stack segment register. Therefore, the next location to be accessed in the stack corresponds to address 1056_{16}. It is to this location that the value in AX is pushed. Notice that the most significant byte of AX, which equals 12_{16}, now resides in the least significant byte of the word in stack, and the least significant byte of AX, which is 34_{16}, is held in the most significant byte.

Now let us look at an example in which data are popped from the stack back into the register from which they were pushed. Figure 2.24 illustrates this oper-

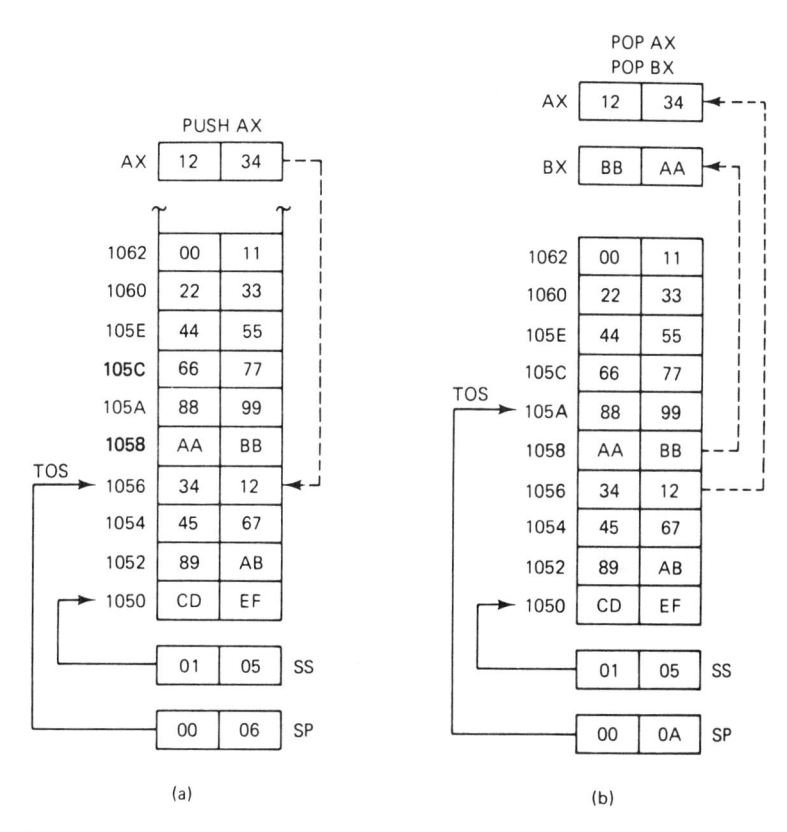

Figure 2.24 (a) Stack just prior to pop operation. (Reprinted by permission of Intel Corp. copyright Intel Corp. 1979.) (b) Stack after execution of the POP AX and POP BX instructions. (Reprinted by permission of Intel Corp. copyright Intel Corp. 1979.)

ation. In Fig. 2.24(a) the stack is shown to be in the state that resulted due to our prior PUSH AX example. That is, SP equals 0006_{16}, SS equals 0105_{16}, the address of the top of the stack equals 1056_{16}, and the word at the top of the stack equals 1234_{16}.

Looking at Fig. 2.24(b), we see what happens when the instructions POP AX and POP BX are executed in that order. Here we see that execution of the

Sec. 2.11   The Stack   45

first instruction causes the 80386DX to read the value from the top of the stack and put it into the AX register as 1234_{16}. Next, SP is incremented to give 0008_{16} and another read operation is initiated from the stack. This second read corresponds to the POP BX instruction, and it causes the value $BBAA_{16}$ to be loaded into the BX register. SP is incremented once more and now equals $000A_{16}$.

In Fig. 2.24(b) we see that the values read out of addresses 1056_{16} and 1058_{16} remain at these addresses. But now they reside at locations that are considered to be above the top of the stack. Therefore, they no longer represent valid data. If new data are pushed to the stack, these values are written over.

> **EXAMPLE 2.8**
>
> Assume that the stack is in the state shown in Fig. 2.24(a) and that the instruction POP ECX is executed instead of the instructions POP AX and POP BX. What value would be popped from the stack? Into which register would it be popped? What is the new address of the top of the stack?
>
> **SOLUTION** Since a 32-bit register is specified in the POP instruction, the value popped from the stack is $1234BBAA_{16}$. This value is popped into the ECX register. After the value is popped, the contents of SP are incremented by four. Therefore, it equals $000A_{16}$ and the new top of stack is at address $105A_{16}$.

Any number of stacks may exist in an 80386DX microcomputer. A new stack can be brought in simply by changing the value in the SS register. For instance, executing the instruction MOV SS, DX loads a new value from DX into SS. Even though many stacks can exist, only one can be active at a time.

2.12 ADDRESSING MODES OF THE 80386DX MICROPROCESSOR

When the 80386DX executes an instruction, it performs the specified function on data. These data, called its operands, may be part of the instruction, may reside in one of the internal registers of the microprocessor, may be stored at an address in memory, or may be held at an I/O port. To access these different types of operands, the 80386DX is provided with various *addressing modes*. Addressing modes are categorized into three types: *register operand addressing*, *immediate operand addressing*, and *memory operand addressing*. Let us now consider in detail the addressing modes in each of these categories.

Register Operand Addressing Mode

With the *register addressing mode*, the operand is specified as residing in an internal register of the 80386DX. Figure 2.25 lists the internal registers that can be used as a source or destination operand. Notice that only the data registers can be accessed in byte, word, or double-word sizes.

An example of an instruction that uses this addressing mode is

MOV AX, BX

Register	Operand size		
	Byte (Reg8)	Word (Reg16)	Double word (Reg32)
Accumulator	AL, AH	AX	EAX
Base	BL, BH	BX	EBX
Count	CL, CH	CX	ECX
Data	DL, DH	DX	EDX
Stack pointer	–	SP	ESP
Base pointer	–	BP	EBP
Source index	–	SI	ESI
Destination index	–	DI	EDI
Code segment	–	CS	–
Data segment	–	DS	–
Stack segment	–	SS	–
E data segment	–	ES	–
F data segment	–	FS	–
G data segment	–	GS	–

Figure 2.25 Direct addressing registers and their sizes.

This stands for "move the word-wide contents of EBX, which is specified as the source operand BX, to the word location in EAX, which is identified by the destination operand BX." In this way, both the source and destination operands have been specified as the contents of internal registers of the 80386DX.

Let us now look at the effect of executing the register addressing mode MOV instruction. In Fig. 2.26(a) we see the state of the 80386DX just prior to fetching the instruction. Notice that the physical address formed from CS and IP (CS:IP) points to the MOV AX, BX instruction at address 01000_{16}. This instruction is fetched into the 80386DX's internal instruction queue, where it is held waiting to be executed. Prior to execution of this instruction, the contents of BX are $ABCD_{16}$ and the contents of AX represent a don't-care state. The instruction is read from the output side of the queue, decoded, and executed. As shown in Fig. 2.26(b), the result produced by executing this instruction is that the value $ABCD_{16}$ is copied into AX.

Immediate Operand Addressing Mode

If an operand is part of the instruction, it represents what is called an *immediate operand* and is accessed using the *immediate addressing mode*. Figure 2.27 shows that the operand, which can be 8 bits (Imm8), 16 bits (Imm16), or 32 bits (Imm32) in length, is encoded into the instruction following the opcode. Since the data are coded directly into the instruction, immediate operands normally represent constant data. This addressing mode can only be used to specify a source operand.

In the instruction

```
MOV AL, 15H
```

the source operand 15_{16} is an example of a byte-wide immediate source operand. The destination operand, which is the contents of AL, uses register addressing. Thus this instruction employs both the immediate and register addressing modes.

	Address	Memory content	Instruction
	01000	8B	MOV AX, BX
	01001	C3	
	01002	XX	Next instruction

80386DX MPU

IP 0000
CS 0100
DS
SS
ES
FS
GS

EAX — XXXX — AX
EBX — ABCD — BX
ECX — — CX
EDX — — DX

ESP — — SP
EBP — — BP
ESI — — SI
EDI — — DI

(a)

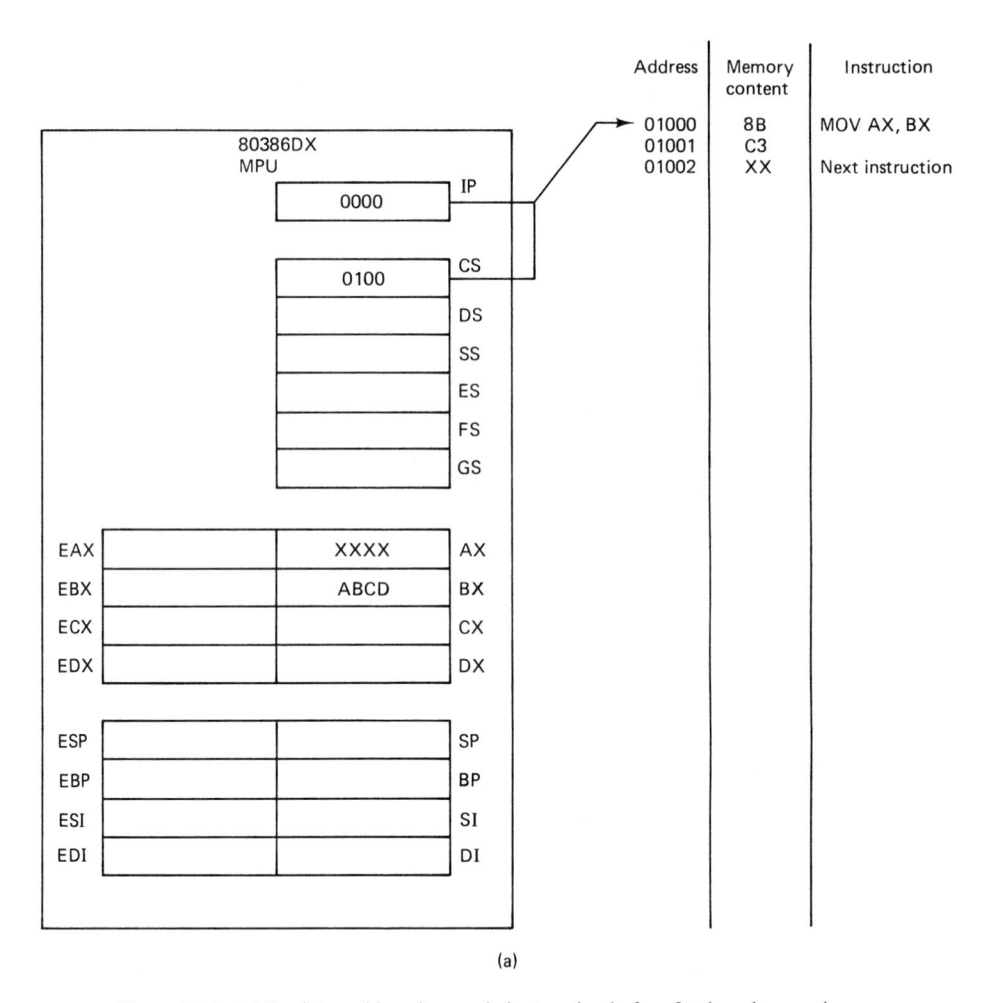

Figure 2.26 (a) Register addressing mode instruction before fetch and execution; (b) after execution.

Figure 2.28(a) and (b) illustrate the fetch and execution of this instruction. Here we find that the immediate operand 15_{16} is stored in the code segment of memory in the byte location following the opcode of the instruction. This value is fetched, along with the opcode for MOV, into the instruction queue within the 80386DX. When it performs the move operation, the source operand is fetched from the queue, not from memory, and no external memory operations are performed. Notice that the result produced by executing this instruction is that the immediate operand, which equals 15_{16}, is loaded into the lower byte part of the accumulator (AL).

Memory Operand Addressing Modes

To reference an operand in memory, the 80386DX must calculate the physical address (PA) of the operand and then initiate a read or write operation to this

	Address	Memory content	Instruction
	01000	8B	MOV AX, BX
	01001	C3	
	01002	XX	Next instruction

80386DX MPU

IP 0002
CS 0100
DS
SS
ES
FS
GS

EAX		ABCD	AX
EBX		ABCD	BX
ECX			CX
EDX			DX

ESP			SP
EBP			BP
ESI			SI
EDI			DI

(b)

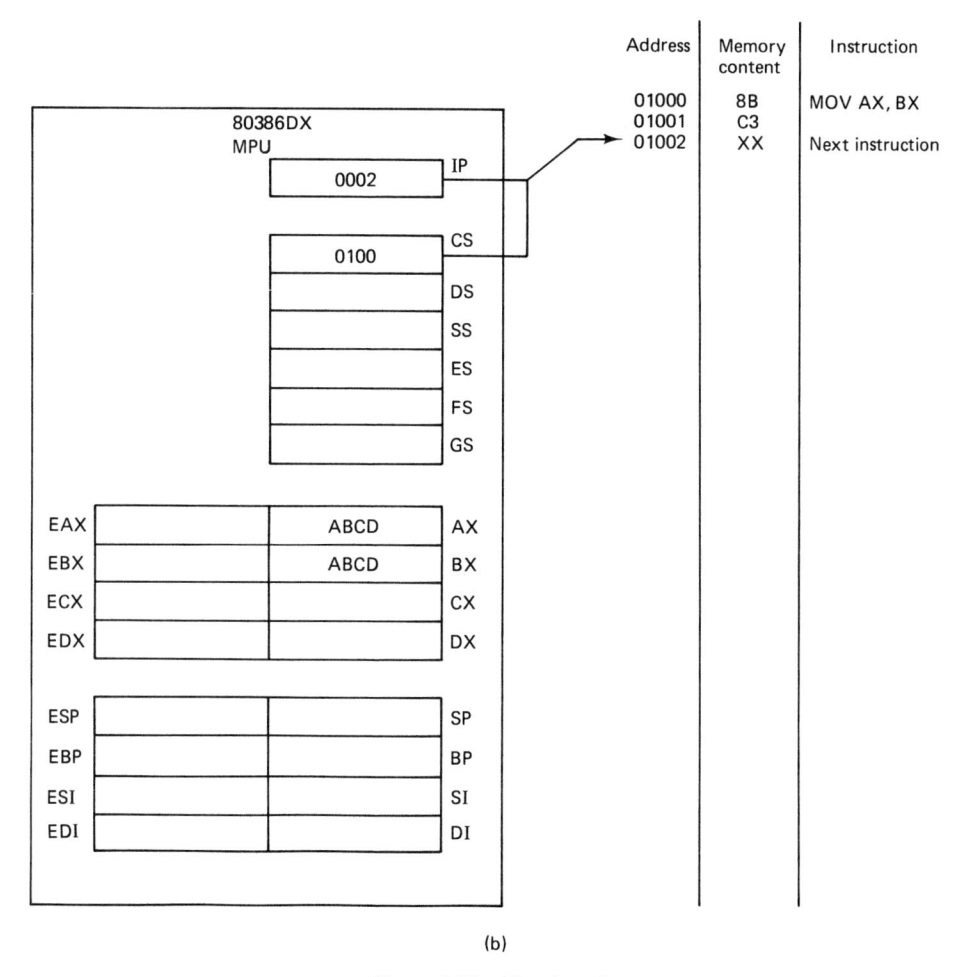

Figure 2.26 (*Continued*)

storage location. Looking at Fig. 2.29 we see that the physical address is formed from a *segment base address* (SBA) and an *effective address* (EA). SBA identifies the starting location of the segment in memory and EA represents the offset of the operand from the beginning of this segment of memory. Earlier we showed how SBA and EA are combined within the 80386DX to form the real-mode physical address SBA:EA.

The value of the EA can be specified in a variety of ways. One way is to encode the effective address of the operand directly in the instruction. This rep-

Opcode	Immediate operand

Figure 2.27 Instruction encoded with an immediate operand.

Sec. 2.12　Addressing Modes of the 80386DX Microprocessor

[Figure 2.28 diagram showing 80386DX MPU registers: IP=0000, CS=0100, DS, SS, ES, FS, GS segment registers; EAX (AX=XX), EBX, ECX, EDX; ESP, EBP, ESI, EDI. Memory content at addresses 01000=B0, 01001=15, 01002=XX, 01003=XX with instructions MOV AL, 15H and Next instruction.]

(a)

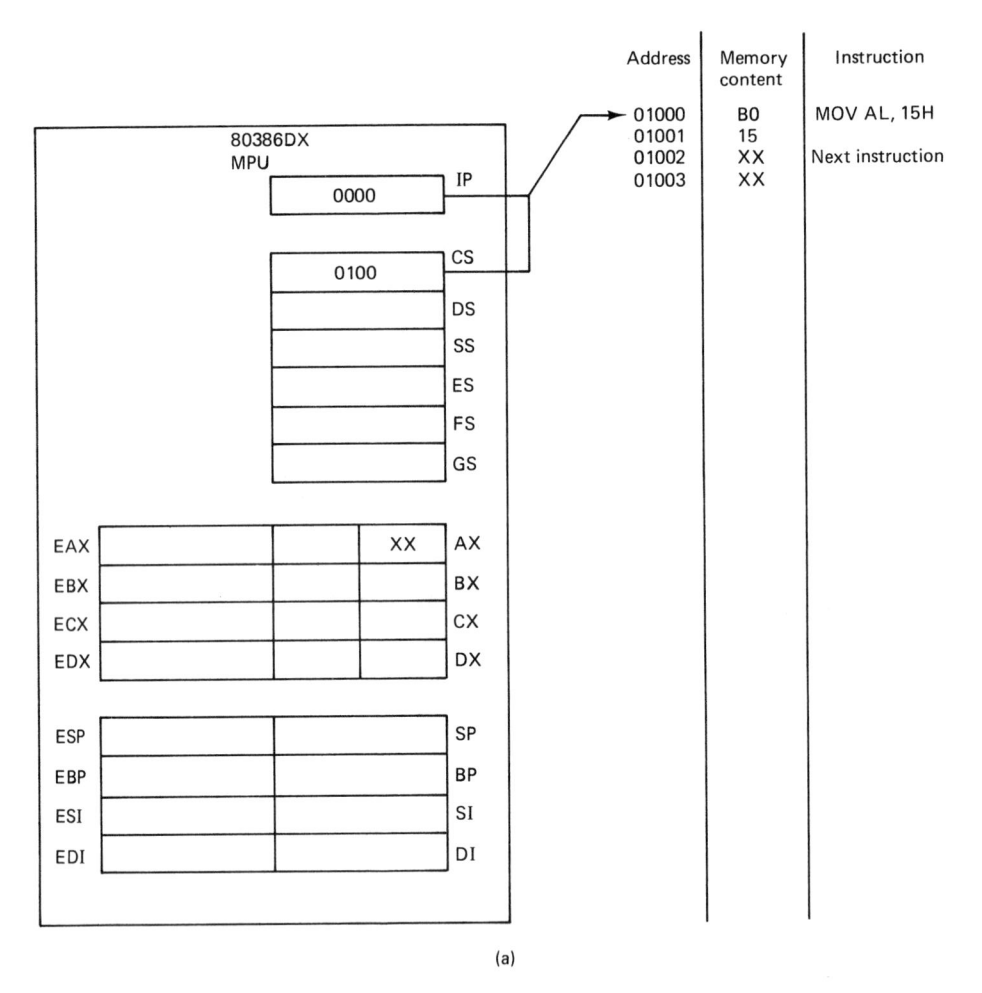

Figure 2.28 (a) Immediate addressing mode instruction before fetch and execution; (b) after execution.

resents the simplest type of memory addressing, known as the *direct addressing mode*. Figure 2.29 shows that an effective address can also be made up from as many as four elements: the *base*, *index*, *scale factor*, and *displacement*. Using these elements, the effective address calculation is made by the general formula

$$EA = base + (index \times scale\ factor) + displacement$$

Figure 2.29 also identifies the register that can be used to hold the values of segment base, base, and index. For example, it tells us that any of the six segment registers can be the source of the segment base for the physical address calculation and that the value of base for the effective address can be in any of the general-purpose registers. Also identified in Fig. 2.29 are the values that are permitted for scale and the sizes of displacement. Notice that scale factors equal to 1, 2, 4, and 8 can be used to multiply the value of index.

Address	Memory content	Instruction
01000	B0	MOV AL, 15H
01001	15	
01002	XX	Next instruction
01003	XX	

(b)

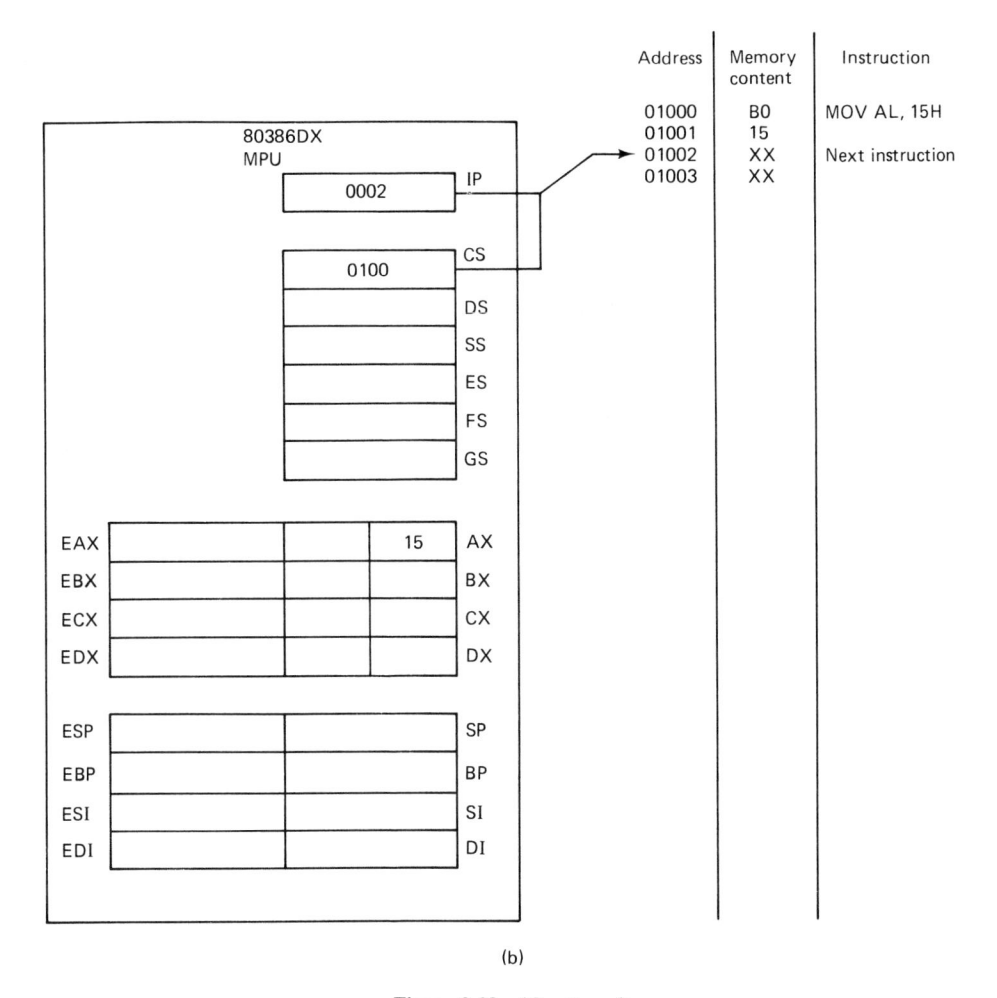

Figure 2.28 (*Continued*)

PA = SBA:EA

PA = Segment base: { Base + (Index × Scale factor) + Displacement }

$$PA = \left\{\begin{matrix}CS\\SS\\DS\\ES\\FS\\GS\end{matrix}\right\} : \left\{\begin{matrix}AX\\BX\\CX\\DX\\SP\\BP\\SI\\DI\end{matrix}\right\} + \left\{\begin{matrix}AX\\BX\\CX\\DX\\BP\\SI\\DI\end{matrix}\right\} \times \left\{\begin{matrix}1\\2\\4\\8\end{matrix}\right\} + \left\{\begin{matrix}\text{8-bit displacement}\\\text{16-bit displacement}\end{matrix}\right\}$$

Figure 2.29 Real-mode physical and effective address computation for memory operands.

Not all of these elements are always used in the effective address calculation. In fact, a number of memory addressing mode types are defined by using various combinations of these elements. They are called *register indirect addressing*, *based addressing*, *indexed addressing*, and *based indexed addressing*. For instance, using based addressing mode, the effective address calculation includes just a base. These addressing modes provide the programmer with different ways of computing the effective address of an operand in memory. Next we examine each of the memory operand addressing modes in detail.

Direct Addressing Mode

Direct addressing is similar to immediate addressing in that information is encoded directly into the instruction. However, in this case, the instruction opcode is followed by an effective address instead of the data. As shown in Fig. 2.30, this

PA = Segment base:Direct address

$$PA = \left\{ \begin{array}{l} CS \\ DS \\ SS \\ ES \\ FS \\ GS \end{array} \right\} : \left\{ \text{Direct address} \right\}$$

Figure 2.30 Computation of a direct memory address.

effective address is used directly as the 16-bit offset of the storage location of the operand from the location specified by the current value in the segment register selected. The default segment register is always DS. Therefore, the 20-bit physical address of the operand in memory is normally obtained as DS:EA. But by using a segment override prefix (SEG) in the instruction, any of the six segment registers can be referenced.

An example of an instruction that uses direct addressing for its source operand is

```
MOV  CX, [BETA]
```

This stands for "move the contents of the memory location, which is labeled as BETA in the current data segment, into internal register CX." The assembler computes the offset of BETA from the beginning of the data segment and encodes it as part of the instruction's machine code.

In Fig. 2.31(a) we find that the value of the offset is stored in the two byte locations that follow the instruction's opcode. Notice that the value assigned to BETA is 1234_{16}. As the instruction is executed, the 80386DX combines 1234_{16} with 0200_{16} to get the physical address of the source operand. This gives

$$PA = 02000_{16} + 1234_{16}$$

$$= 03234_{16}$$

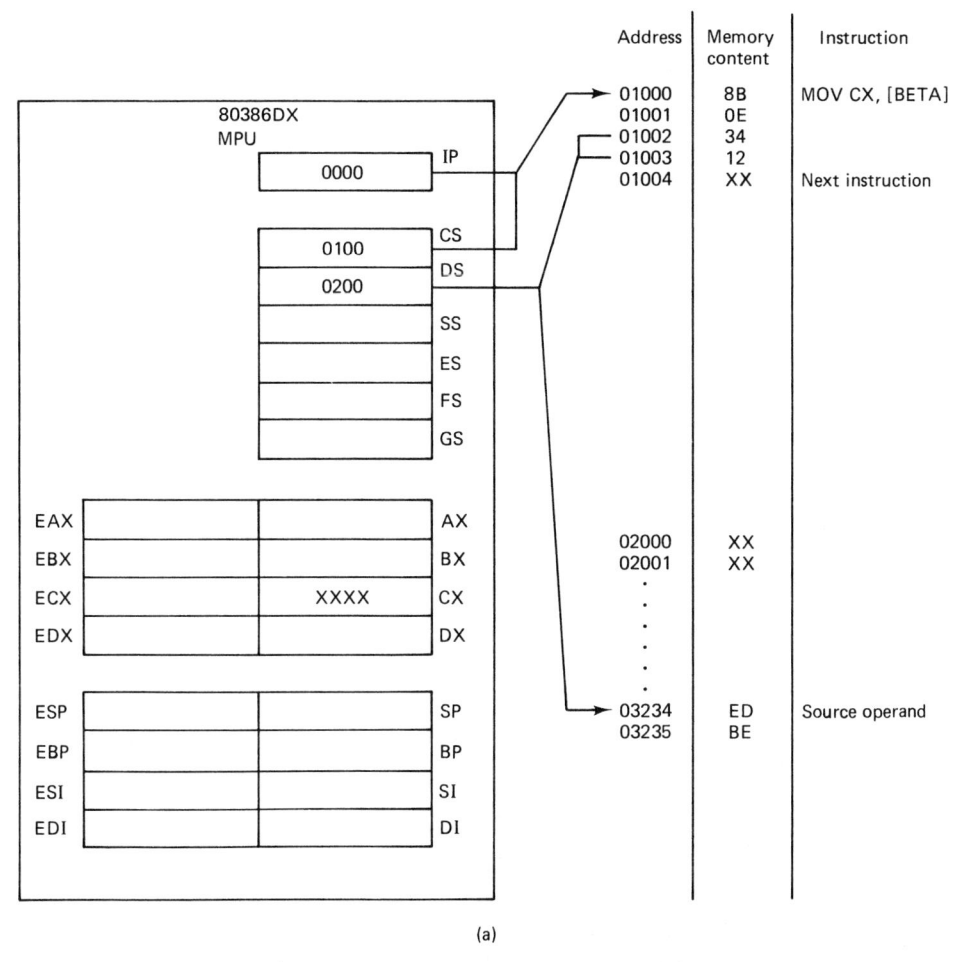

Figure 2.31 (a) Direct addressing mode instruction before fetch and execution; (b) after execution.

Then it reads the word of data starting at this address, which is $BEED_{16}$, and loads it into the CX register. This result is illustrated in Fig. 2.31(b).

Register Indirect Addressing Mode

Register indirect addressing is similar to the direct address we just described, in that an effective address is combined with the contents of a segment register to obtain a physical address. However, it differs in the way the offset is specified. Figure 2.32 shows that this time the 16-bit EA resides in one of the 80386DX's general-purpose registers. Again, DS is the default segment register. Another segment register can be referenced by using a segment override prefix.

An example of an instruction that uses register indirect addressing for its source operand is

```
MOV  AX, [SI]
```

Sec. 2.12   Addressing Modes of the 80386DX Microprocessor   53

	Address	Memory content	Instruction
	01000	8B	MOV CX, [BETA]
	01001	0E	
	01002	34	
	01003	12	
	01004	XX	Next instruction

80386DX MPU

IP: 0004
CS: 0100
DS: 0200
SS, ES, FS, GS

EAX / AX
EBX / BX
ECX / CX — BEED
EDX / DX
ESP / SP
EBP / BP
ESI / SI
EDI / DI

Address	Memory content
02000	XX
02001	XX
⋮	
03234	ED
03235	BE

(b)

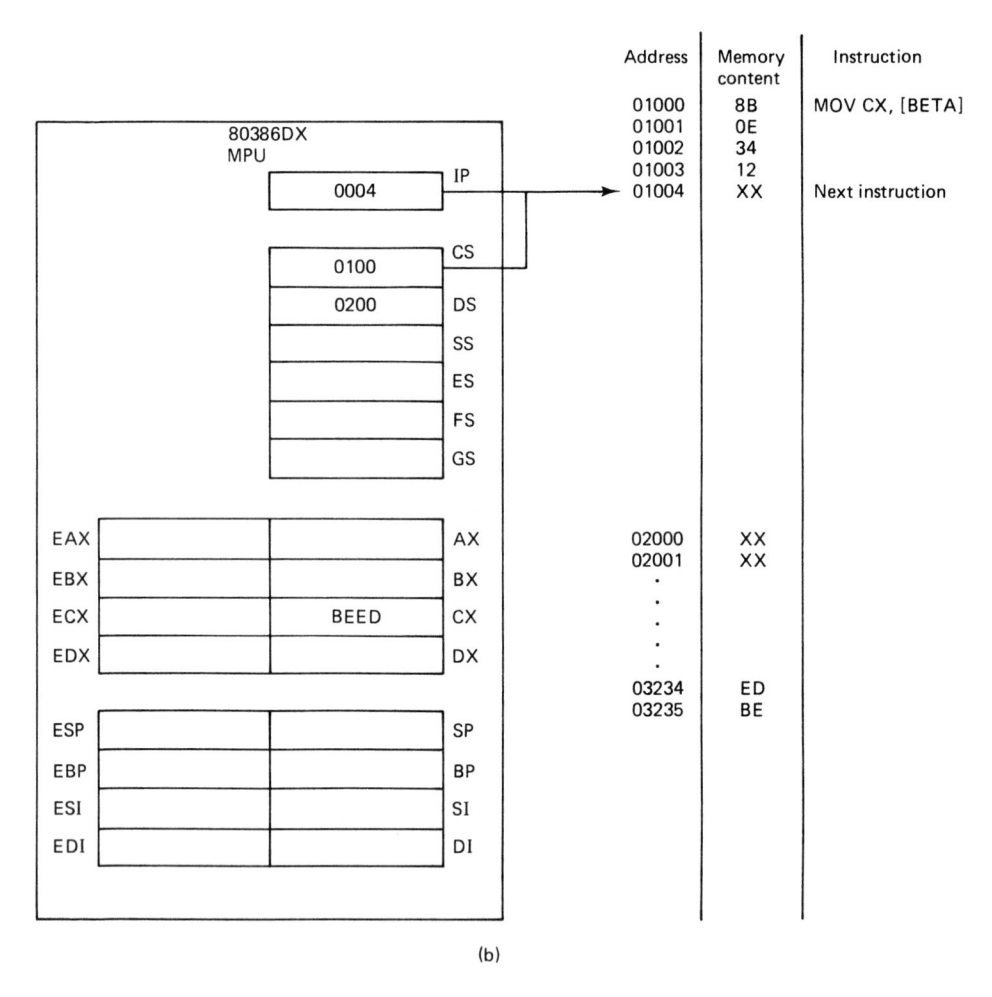

Figure 2.31 (*Continued*)

PA = Segment base : Indirect address

$$PA = \left\{ \begin{array}{c} CS \\ DS \\ SS \\ ES \\ FS \\ GS \end{array} \right\} : \left\{ \begin{array}{c} AX \\ BX \\ CX \\ DX \\ SP \\ BP \\ SI \\ DI \end{array} \right\}$$

Figure 2.32 Computation of an indirect memory address.

Execution of this instruction moves the contents of the memory location that is offset from the beginning of the current data segment by the value of EA in register SI into the AX register.

For instance, as shown in Fig. 2.33(a) and (b), if SI contains 1234_{16} and DS contains 0200_{16}, the result produced by executing the instruction is that the contents of memory location

$$PA = 02000_{16} + 1234_{16}$$
$$= 03234_{16}$$

are moved into the AX register. Notice in Fig. 2.33(b) that this value is $BEED_{16}$. In this example the value 1234_{16} that was found in the SI register had been loaded with another instruction prior to executing the move instruction.

(a)

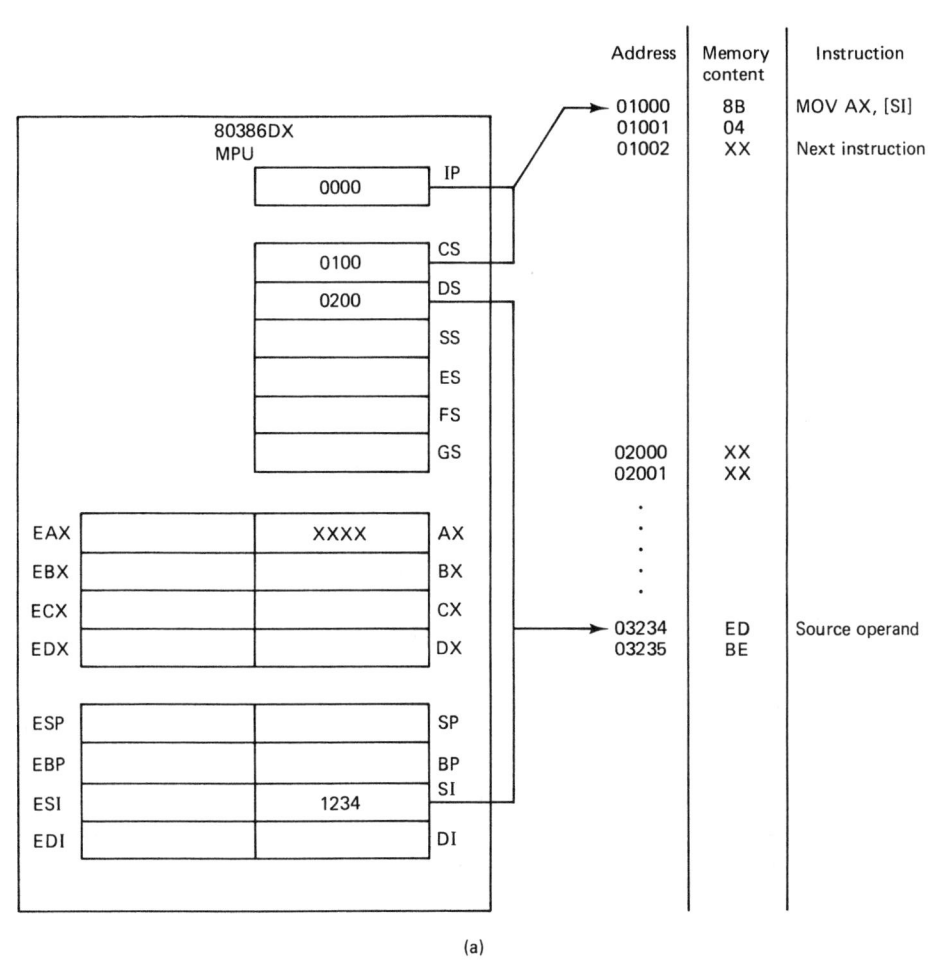

Figure 2.33 (a) Instruction using register indirect addressing before fetch and execution; (b) after execution.

Address	Memory content	Instruction
01000	8B	MOV AX, [SI]
01001	04	
01002	XX	Next instruction
02000	XX	
02001	XX	
⋮		
03234	ED	
03235	BE	

80386DX MPU registers:
- IP: 0002
- CS: 0100
- DS: 0200
- SS, ES, FS, GS
- AX: BEED
- SI: 1234

(b)

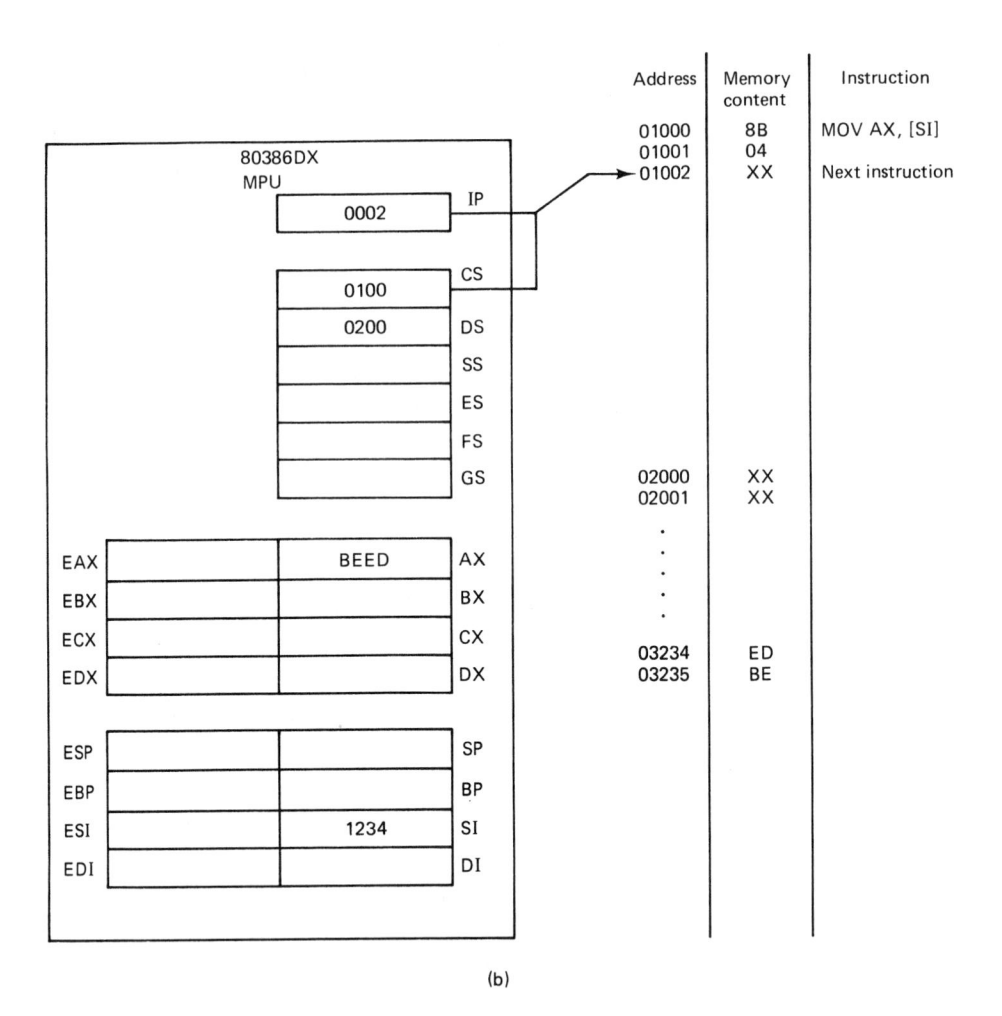

Figure 2.33 (*Continued*)

Notice that the result produced by executing this instruction and the example for the direct addressing mode are the same. However, they differ in the way in which the physical address was generated. The direct addressing method lends itself to applications where the value of EA is a constant. On the other hand, register indirect addressing could be used when the value of EA is calculated and stored in a register by previous instructions. That is, EA is a variable. For example, the instructions executed just before our example instruction could have incremented the address in SI by 2.

Based Addressing Mode

In the based addressing mode, the physical address of the operand is obtained by adding a direct or indirect displacement to the contents of a base register.

Looking at Fig. 2.34(a), we see that the value in the base register defines a data structure, such as a record, in memory and the displacement selects the element of data within this structure. To access a different element in the record, the programmer simply changes the value of the displacement. On the other hand, to access the same element in another record, the programmer can change the value in the base register so that it points to the new record. As shown in Fig. 2.34(b), the base register can be any of the general-purpose registers of the 80386DX. Moreover, in the real mode, the displacement can be either 8 or 16 bits in length.

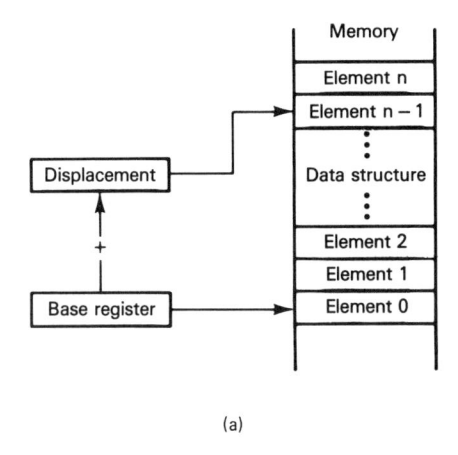

(a)

PA = Segment base:Base + Displacement

$$PA = \begin{Bmatrix} CS \\ DS \\ SS \\ ES \\ FS \\ GS \end{Bmatrix} : \begin{Bmatrix} AX \\ BX \\ CX \\ DX \\ SP \\ BP \\ SI \\ DI \end{Bmatrix} + \begin{Bmatrix} \text{8-bit displacement} \\ \text{16-bit displacement} \end{Bmatrix}$$

(b)

Figure 2.34 (a) Based addressing of a structure of data; (b) computation of a based address.

A move instruction that uses based addressing to specify the location of its destination operand is as follows:

                MOV   [BX] + BETA, AL

This instruction uses base register BX and direct displacement BETA to derive the EA of the destination operand. The based addressing mode is implemented by specifying the base register in brackets followed by a + sign and the direct displacement.

Figure 2.35(a) and (b) show that the fetch and execution of this instruction

Sec. 2.12   Addressing Modes of the 80386DX Microprocessor   57

```
                                               Address  Memory   Instruction
                                                        content
                                                01000    88     MOV [BX] + BETA, AL
                      80386DX                   01001    07
                       MPU                      01002    34
                                      IP        01003    12
                         0000                   01004    XX     Next instruction

                                 CS
                     0100
                                 DS
                     0200
                                 SS
                                 ES
                                 FS
                                 GS             02000    XX
                                                02001    XX

  EAX                 BE    ED    AX
  EBX                    1000     BX
  ECX                              CX
  EDX                              DX

  ESP                              SP
  EBP                              BP
  ESI                              SI           04234    XX     Destination operand
  EDI                              DI           04235    XX
```

(a)

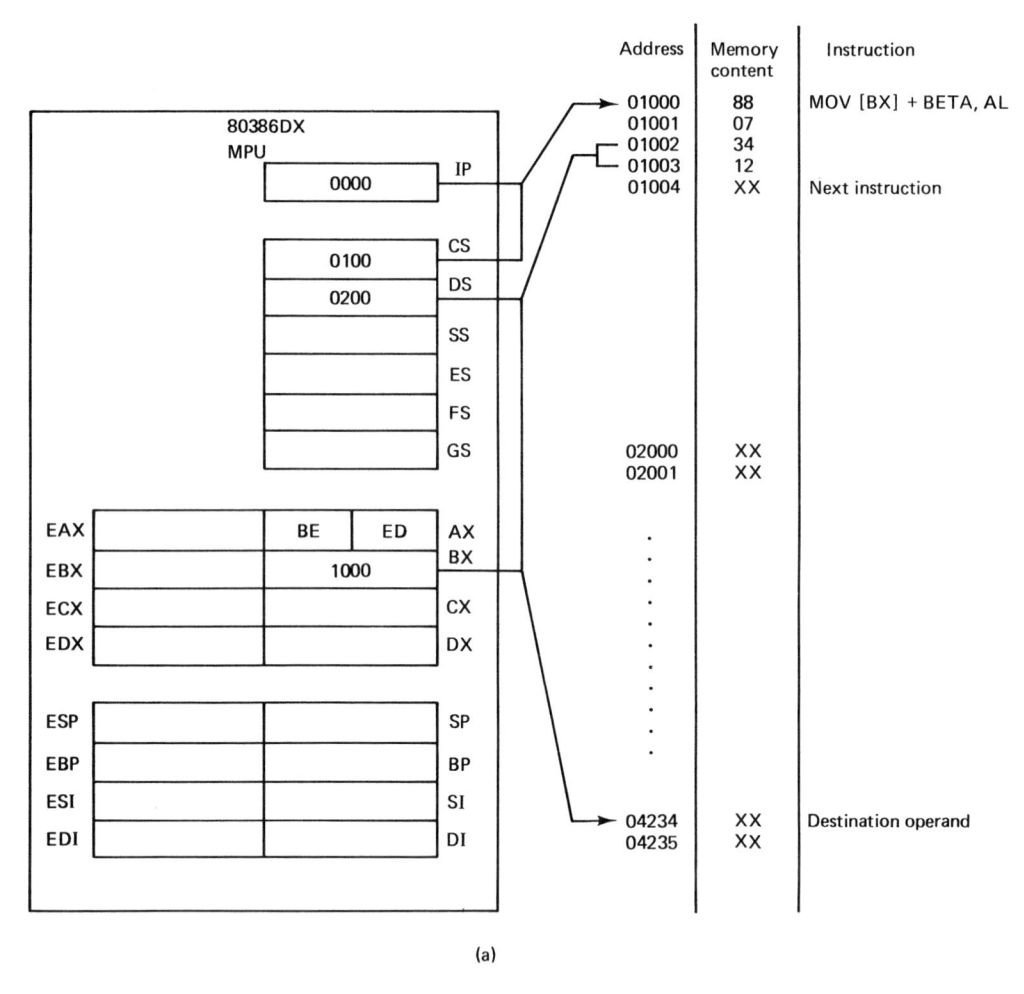

Figure 2.35 (a) Instruction using direct base pointer addressing before fetch and execution; (b) after execution.

causes the 80386DX to calculate the physical address of the destination operand from the contents of DS, BX, and the direct displacement. The result is

$$PA = 02000_{16} + 1000_{16} + 1234_{16}$$
$$= 04234_{16}$$

Then it writes the contents of source operand AL into the storage location at 04234_{16}. The result is that ED_{16} is written into the destination memory location.

Notice that DS is again the default segment register for the physical address calculation. The choice of segment register can be changed with a segment override prefix. Moreover, if either SP or BP is used as the base register, calculation

(b)

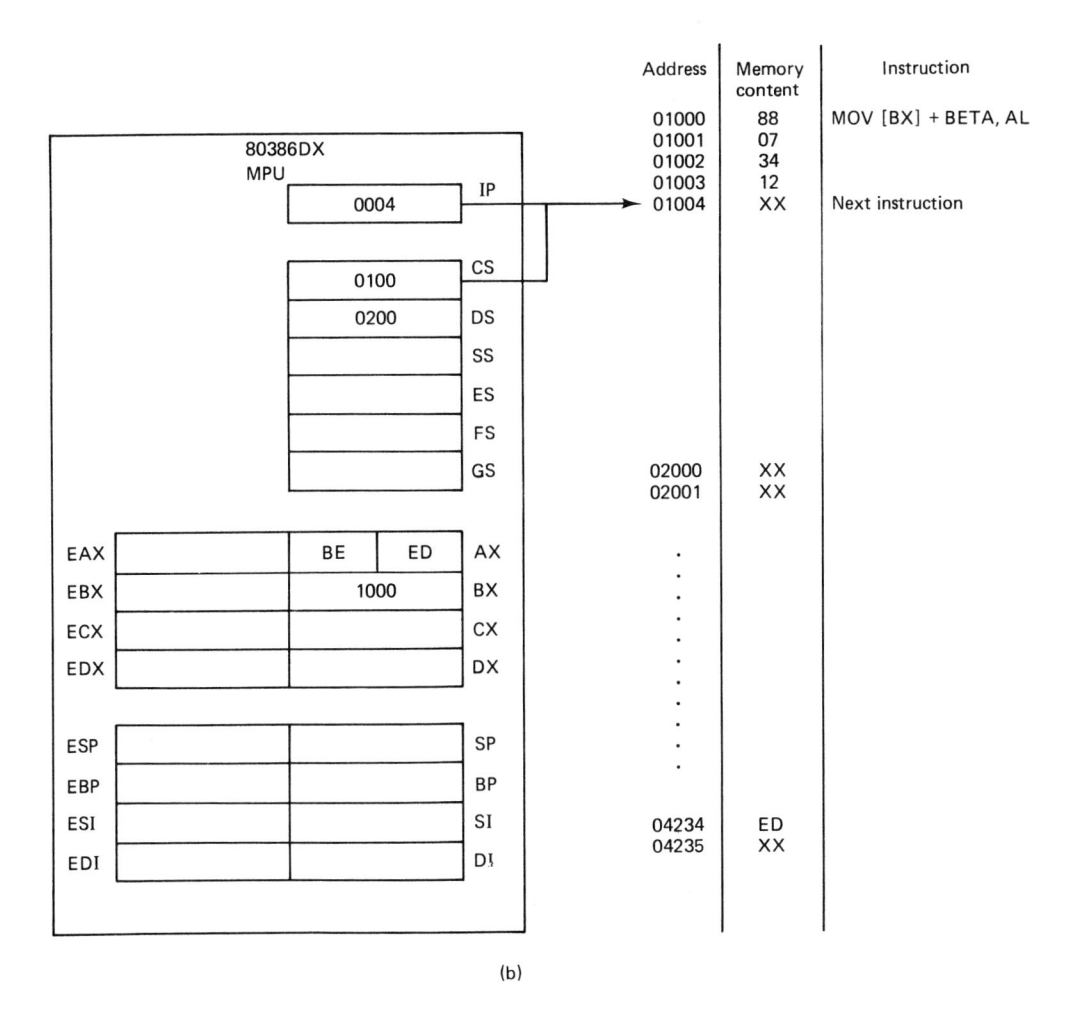

Figure 2.35 (*Continued*)

of the physical address is performed automatically using the contents of the stack segment (SS) register instead of DS. This permits access to data in the stack segment of memory.

Indexed Addressing Mode

Indexed addressing mode works in a manner similar to that of the based addressing mode we just described. However, as shown in Fig. 2.36(a), indexed addressing mode uses the value of the displacement as a pointer to the starting point of an array of data in memory and the contents of the specified register as an index that selects the specific element in the array that is to be accessed. For instance,

Sec. 2.12 Addressing Modes of the 80386DX Microprocessor

Memory
Element n + 1
Element n
⋮
Array of data
⋮
Element 2
Element 1
Element 0

Index register

Displacement

(a)

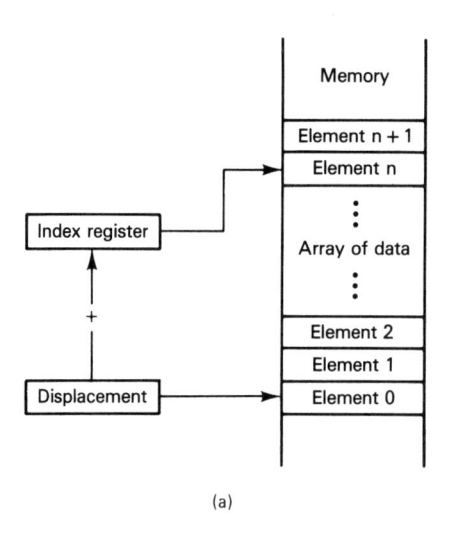

PA = Segment base : Index + Displacement

$$PA = \begin{Bmatrix} CS \\ DS \\ SS \\ ES \\ FS \\ GS \end{Bmatrix} : \begin{Bmatrix} AX \\ BX \\ CX \\ DX \\ BP \\ SI \\ DI \end{Bmatrix} + \begin{Bmatrix} \text{8-bit displacement} \\ \text{16-bit displacement} \end{Bmatrix}$$

(b)

PA = Segment base : (Index × Scale factor) + Displacement

$$PA = \begin{Bmatrix} CS \\ DS \\ SS \\ ES \\ FS \\ GS \end{Bmatrix} : \begin{Bmatrix} AX & & 1 \\ BX & & \\ CX & & 2 \\ DX & \times & \\ BP & & 4 \\ SI & & \\ DI & & 8 \end{Bmatrix} + \begin{Bmatrix} \text{8-bit displacement} \\ \text{16-bit displacement} \end{Bmatrix}$$

(c)

Figure 2.36 (a) Indexed addressing of an array of data elements; (b) computation of an indexed address; (c) scaled index addressing.

in Fig. 2.36(a), the index register holds the value n. In this way it selects data element n in the array. Figure 2.36(b) shows the registers that can be used as an index. Notice that the stack pointer (SP) register is not included in this list.

The instruction

MOV   AL, ARRAY + [SI]

employs direct indexed addressing for the source operand. Notice that the notation this time is such that the index selected, which is in register SI, is added to

ARRAY, which is a *direct displacement*. Just as for the base register in based addressing, the index register is enclosed in brackets. In this example the physical address is calculated as

$$PA = DS:\{ARRAY + (SI)\}$$

The example in Fig. 2.37(a) and (b) shows the result of executing the move instruction. First the physical address of the source operand is calculated from the values in DS, SI, and the direct displacement:

$$PA = 02000_{16} + 1234_{16} + 2000_{16}$$

$$= 05234_{16}$$

(a)

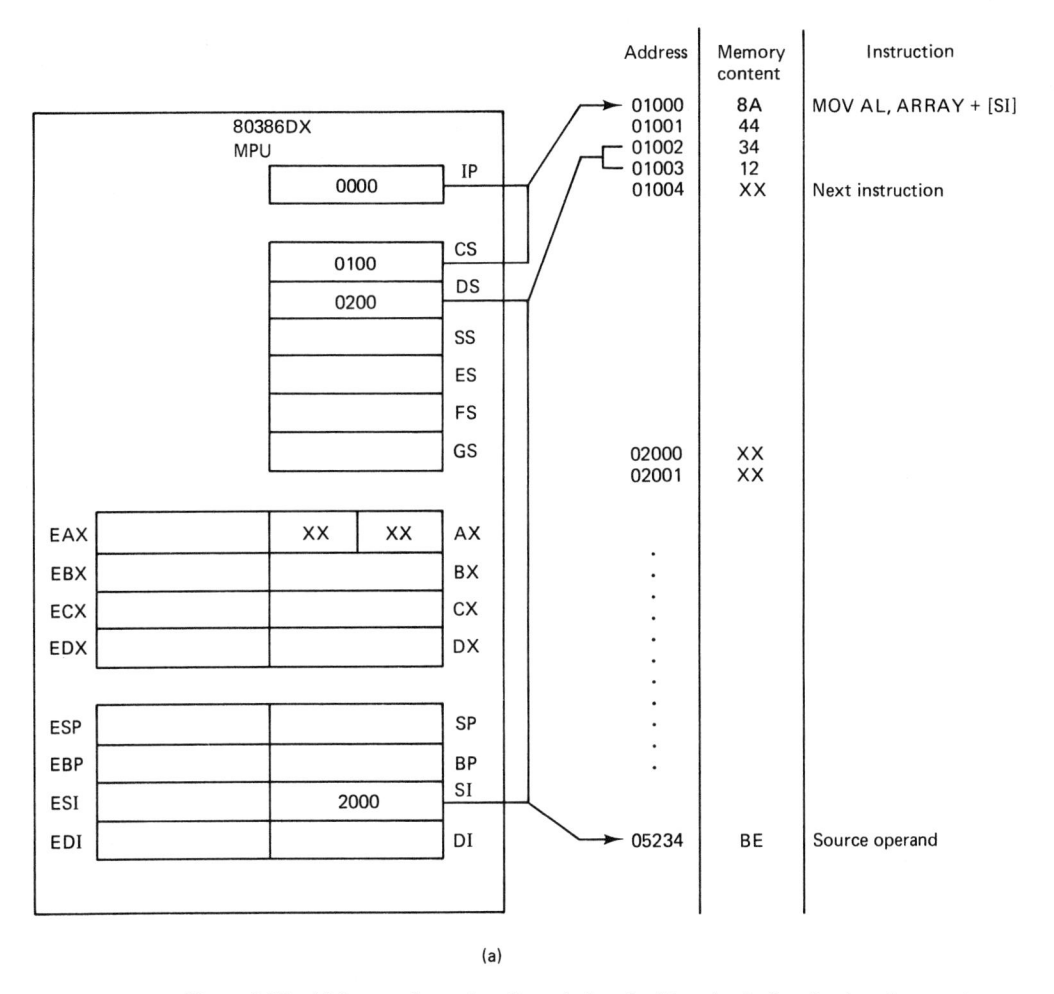

Figure 2.37 (a) Instruction using direct indexed addressing before fetch and execution; (b) after execution.

Sec. 2.12 Addressing Modes of the 80386DX Microprocessor

	Address	Memory content	Instruction
	01000	8A	MOV AL, ARRAY + [SI]
	01001	44	
	01002	34	
	01003	12	
	01004	XX	Next instruction
	02000	XX	
	02001	XX	
	05234	BE	

80386DX MPU

IP: 0004
CS: 0100
DS: 0200
SS, ES, FS, GS

EAX: XX BE (AX)
EBX (BX)
ECX (CX)
EDX (DX)
ESP (SP)
EBP (BP)
ESI: 2000 (SI)
EDI (DI)

(b)

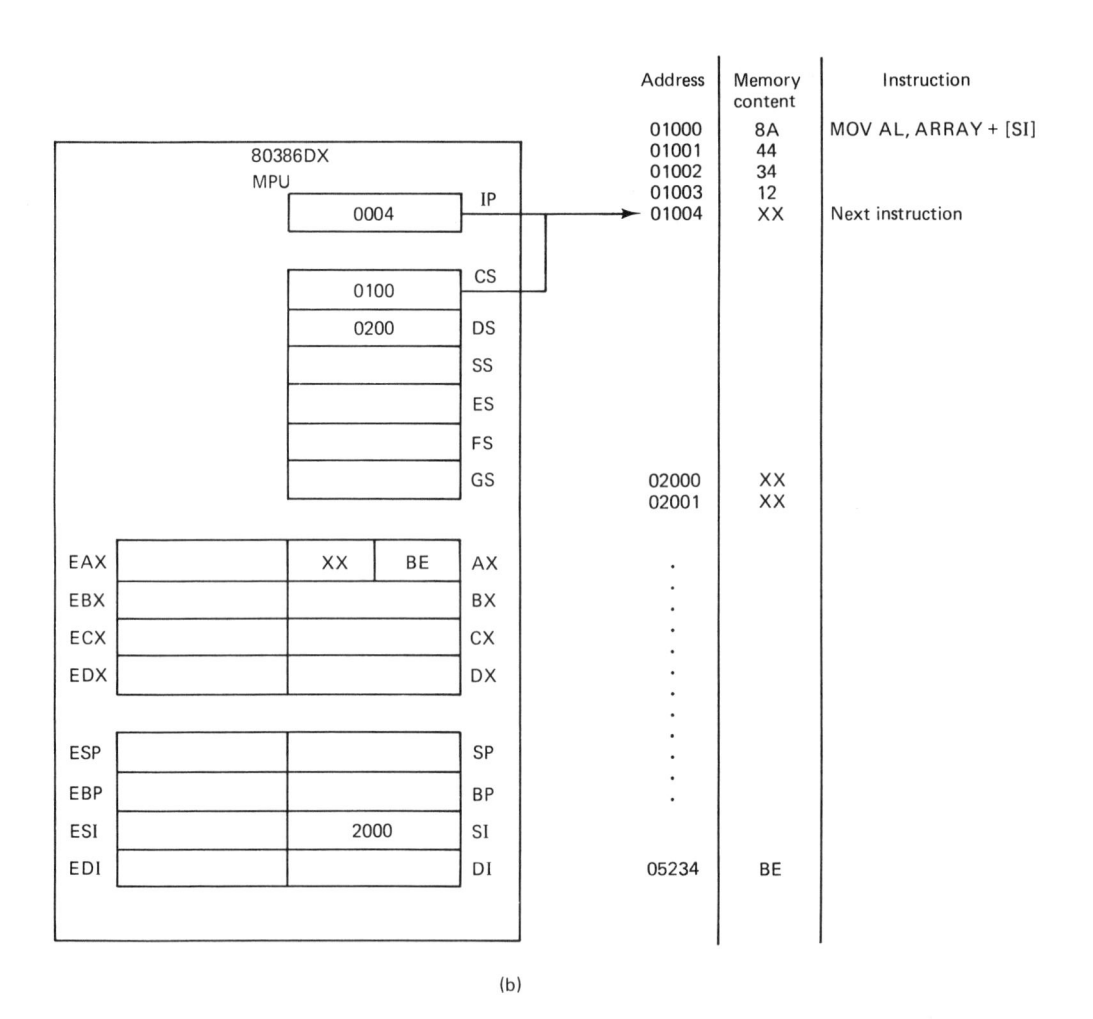

Figure 2.37 (*Continued*)

Then the byte of data stored at this location, which is BE_{16}, is read into the lower byte (AL) of the accumulator register.

The 80386DX also supports what is called *scaling* of indexes. Figure 2.36(c) shows that a scale factor can be included with the index of an indexed address. Notice that the scale factor can take on the values 1, 2, 4, or 8. When the effective address computation is made, the value in the index register is first multiplied by the scale factor and then added to the displacement. For example, if the instruction in Fig. 2.37(a) was changed to

```
MOV   AL, ARRAY + [SI × 2]
```

the physical address becomes

$$PA = 02000_{16} + 1234_{16} + 2000_{16} \times 2$$
$$= 07234_{16}$$

Based–Indexed Addressing Mode

Combining the based addressing mode and the indexed addressing mode together results in a new, more powerful mode known as based indexed addressing. This addressing mode can be used to access complex data structures such as two-dimensional arrays. Figure 2.38(a) shows how it can be used to access elements in an $m \times n$ array of data. Notice that the displacement, which is a fixed value, locates the array in memory. The base register specifies the m coordinate of the array, and the index register identifies the n coordinate. Any element in the array can be accessed simply by changing the values in the base and index registers. The registers permitted in the based–indexed address computation are shown in Fig. 2.38(b).

Let us consider an example of a move instruction using this type of addressing:

```
MOV  AH, [BX][SI] + BETA
```

Notice that the source operand is accessed using the based–indexed addressing mode. Here BX is the base register, SI the index register, and BETA the displacement. Therefore, the effective address of the source operand is obtained as

$$EA = (BX) + (SI) + BETA$$

and the physical address of the operand is formed from the current DS and the calculated EA as

$$PA = DS:\{(BX) + (SI) + BETA\}$$

An example of executing this instruction is illustrated in Fig. 2.39(a) and (b). The address of the source operand is calculated as

$$PA = 02000_{16} + 1000_{16} + 2000_{16} + 1234_{16}$$
$$= 6234_{16}$$

Execution of the instruction causes the value stored at this location to be read into AH.

Just as for indexed addressing, the index in a based–indexed address can be scaled. Figure 2.38(c) shows how the computation for a scaled address is performed.

(a)

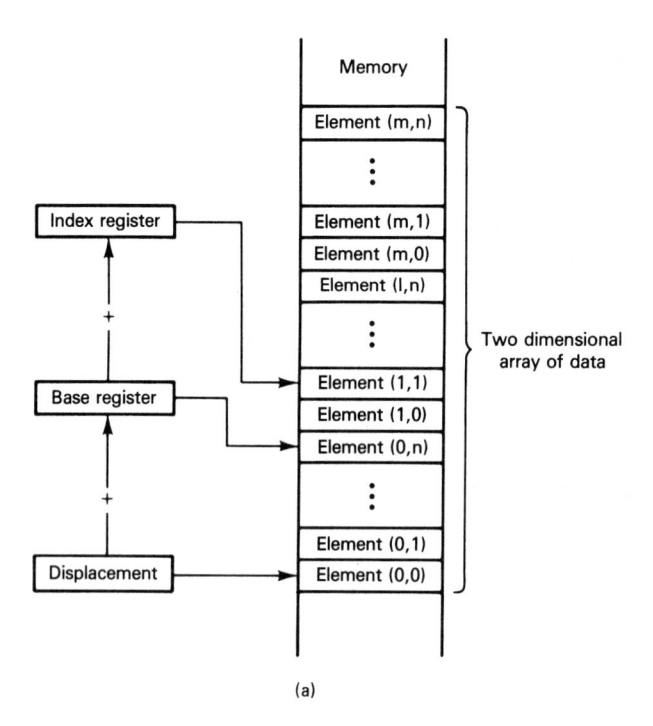

PA = Segment base : Base + Index + Displacement

$$PA = \begin{Bmatrix} CS \\ DS \\ SS \\ ES \\ FS \\ GS \end{Bmatrix} : \begin{Bmatrix} AX \\ BX \\ CX \\ DX \\ SP \\ BP \\ SI \\ DI \end{Bmatrix} + \begin{Bmatrix} AX \\ BX \\ CX \\ DX \\ BP \\ SI \\ DI \end{Bmatrix} + \begin{Bmatrix} \text{8-bit displacement} \\ \text{16-bit displacement} \end{Bmatrix}$$

(b)

PA = Segment base : Base + { Index × Scale factor } + Displacement

$$PA = \begin{Bmatrix} CS \\ DS \\ SS \\ ES \\ FS \\ GS \end{Bmatrix} : \begin{Bmatrix} AX \\ BX \\ CX \\ DX \\ SP \\ BP \\ SI \\ DI \end{Bmatrix} + \begin{Bmatrix} AX & 1 \\ BX & \\ CX & 2 \\ DX & \times \\ BP & 4 \\ SI & \\ DI & 8 \end{Bmatrix} + \begin{Bmatrix} \text{8-bit displacement} \\ \text{16-bit displacement} \end{Bmatrix}$$

(c)

Figure 2.38 (a) Based-indexed addressing of a two-dimensional array of data; (b) computation of a based-indexed address; (c) scaled based-indexed addressing.

Address	Memory content	Instruction
01000	8A	MOV AH, [BX] [SI] + BETA
01001	20	
01002	34	
01003	12	
01004	XX	Next instruction
02000	XX	
02001	XX	
06234	BE	Source operand

(a)

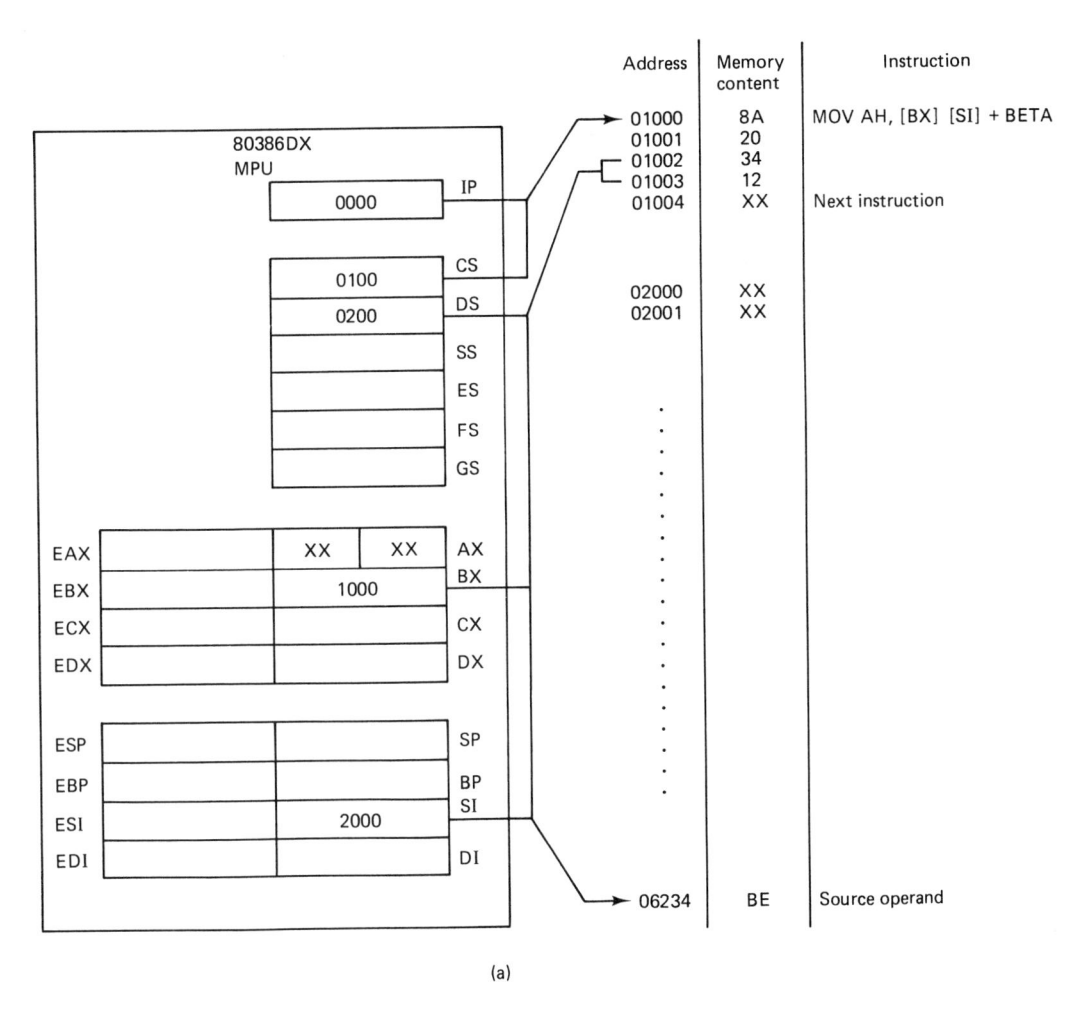

Figure 2.39 (a) Instruction using based indexed addressing before fetch and execution; (b) after execution.

ASSIGNMENT

Section 2.2

1. What tells a microcomputer what to do, where to get data, how to process the data, and where to put the results when done?
2. What is the name given to a sequence of instructions used to guide a computer through a task?

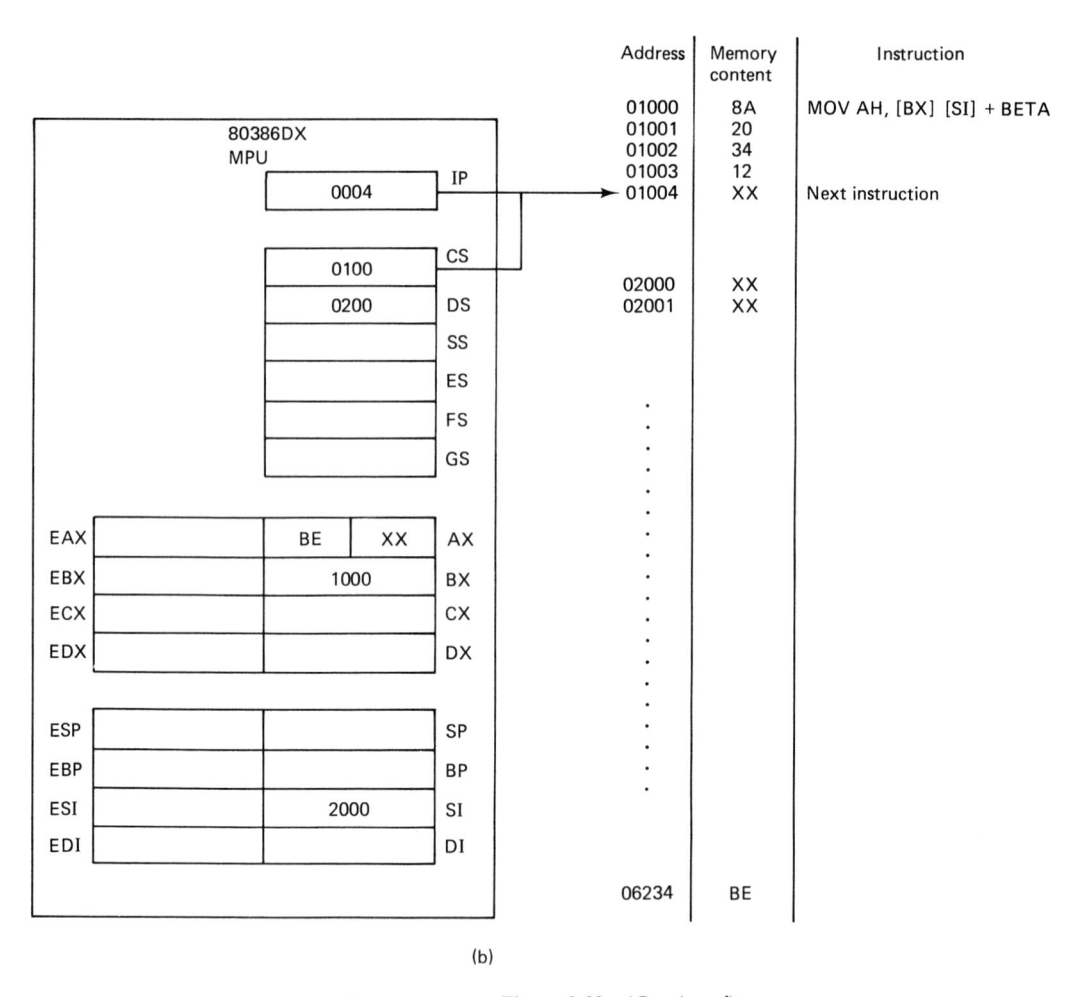

Figure 2.39 (*Continued*)

3. What is the native language of the 80386DX?
4. How does machine language differ from assembly language?
5. What is an opcode? Give two examples.
6. What is an operand? Give two types.
7. In the assembly language statement

   ```
   START:   ADD   EAX, EBX        ; ADD EBX TO EAX
   ```

   what is the label? What is the comment?
8. What is the function of an assembler? A compiler?
9. What is source code? Object code?
10. Give two benefits derived from writing programs in assembly language instead of a high-level language.

11. What is meant by the phrase *real-time application*?
12. List two hardware-related applications where performance can be increased by using assembly language programming. Name two software-related applications.

Section 2.3

13. What is the purpose of a software model of the 80386DX microprocessor?
14. What must an assembly language programmer know about the registers within the 80386DX microprocessor?
15. How many of the 80386DX's internal registers are important to the real-mode application programmer?
16. What are the five least significant bits of CR_0 called?
17. What does PE stand for? What is the purpose of this bit of CR_0?
18. How large is the 80386DX's real-mode memory address space?

Section 2.4

19. What is the highest address in the 80386DX's real-mode address space? The lowest address?
20. Is memory in the 80386DX microcomputer organized as bytes, words, or double words?
21. How much of the 80386DX's real-mode address space is dedicated to storage of the interrupt vector table?
22. The contents of memory location $B0000_{16}$ is FF_{16}, and that at $B0001_{16}$ is 00_{16}. What is the data word stored starting at address $B0000_{16}$? Is the word aligned or misaligned?
23. What is the value of the double word stored in memory starting at address $B0003_{16}$ if the contents of memory locations $B0003_{16}$, $B0004_{16}$, $B0005_{16}$, and $B0006_{16}$ are 11_{16}, 22_{16}, 33_{16}, and 44_{16}, respectively? Is this an example of an aligned double word or a misaligned double word?
24. Show how the word $ABCD_{16}$ is stored in memory starting at address $0A002_{16}$. Is the word aligned or misaligned?
25. Show how the double word 12345678_{16} will be stored in memory starting at address $0A001_{16}$. Is the double word aligned or misaligned?

Section 2.5

26. List five data types processed directly by the 80386DX.
27. Express each of the following signed decimal integers as either a byte or a word hexadecimal number (use 2's-complement notation for negative numbers).
    (a) +127   (b) −10   (c) −128   (d) +500
28. How would the integer in Problem 27(d) be stored in memory starting at address $0A000_{16}$?
29. How would the decimal number −1000 be expressed for processing by the 80386DX?
30. Express the following decimal numbers as unpacked and packed BCD bytes.
    (a) 29   (b) 88

31. How would the number in Problem 27(a) be stored in memory starting at address $0B000_{16}$? Assume that it is coded in packed BCD form and that the least significant digits are stored at the lower address.
32. What is the statement coded in ASCII by the following binary strings?

$$1001110$$
$$1000101$$
$$1011000$$
$$1010100$$
$$0100000$$
$$1001001$$

33. How would the decimal number 1234 be coded in ASCII and stored in memory starting at address $0C000_{16}$? Assume that the least significant digit is stored at the lower addressed memory location.

Section 2.6

34. How large is a real-mode memory segment?
35. Which of the 80386DX's internal registers are used for memory segmentation?
36. What register defines the beginning of the current code segment in memory?
37. How much memory can be active at a time in the 80386DX microcomputer?
38. How much of the 80386DX's active memory is available as general-purpose data storage memory?
39. Which range of the 80386DX's memory address space can be used to store program instructions?

Section 2.7

40. What is the function of the instruction pointer register?
41. Provide an overview of the fetch and execution of an instruction by the 80386DX.
42. What happens to the value in IP each time the 80386DX completes an instruction fetch?
43. How large is the 80386DX's instruction prefetch queue?

Section 2.8

44. Make a list of the data registers of the 80386DX.
45. How is the word value of the base register labeled? The double-word value?
46. How are the upper and lower bytes of the data register denoted?
47. Name two dedicated operations that are assigned to the CX registers.
48. Name the two pointer registers.
49. For which segment register is the contents of the pointer registers used as an offset?
50. What do *SI* and *DI* stand for?

51. Which sizes of data can be stored in the base and index registers? What size real-mode offset addresses?

Section 2.9

52. Categorize each flag bit of the 80386DX as either a control flag or a flag that monitors the status due to execution of an instruction.
53. Describe the function of each status flag.
54. How are the status flags used by software?
55. What does *TF* stand for?
56. Which flag determines whether the address for a string operation is incremented or decremented?
57. Can the state of the flags be modified through software?

Section 2.10

58. What are the word lengths of the 80386DX's real-mode logical and physical addresses?
59. What two address elements are combined to form a physical address?
60. What is the default segment register for the segment base address in a string instruction destination operand access?
61. Calculate the value of each of the physical addresses that follow.
    (a) 1000H:1234H
    (b) 0100H:ABCDH
    (c) A200H:12CFH
    (d) B2C0H:FA12H
62. Find the unknown value for each of the following physical addresses.
    (a) A000H:? = A0123H
    (b) ?:14DAH = 235DAH
    (c) D765H:? = DABC0H
    (d) ?:CD21H = 32D21H
63. If the current values in the code segment register and the instruction pointer are 0200_{16} and $01AC_{16}$, respectively, what physical address is used in the next instruction fetch?
64. A data segment is to be located from address $A0000_{16}$ to $AFFFF_{16}$, what value must be loaded into DS?
65. If the data segment register contains the value found in Problem 64, what value must be loaded into DI if it is to point to a destination operand stored in memory at address $A1234_{16}$?

Section 2.11

66. What is the function of the stack?
67. If the current values in the stack segment register and stack pointer are $C000_{16}$ and $FF00_{16}$, respectively, what is the address of the top of the stack?
68. For the segment base and offset addresses in Problem 67, how many words of data are currently held in the stack?

69. Show how the value EE11$_{16}$ from register AX would be pushed onto the top of the stack as it exists in Problem 67.

Section 2.12

70. List the names of the three types of addressing modes available on the 80386DX.

71. What four elements can be used to form the effective address of an operand in memory?

72. Name the five memory operand addressing modes.

73. Identify the addressing modes used for the source and the destination operands in the following instructions.
  (a) MOV   AL, BL
  (b) MOV   AX, FFH
  (c) MOV   [DI], AX
  (d) MOV   DI, [SI]
  (e) MOV   [BX] + XYZ, CX
  (f) MOV   XYZ + [DI], AH
  (g) MOV   [BX][DI] + XYZ, AL

74. Compute the physical address for the specified operand in each of the following instructions. The register and memory contents are as follows: CS = 0A00$_{16}$, DS = 0B00$_{16}$, ESI = 00000100$_{16}$, EDI = 00000200$_{16}$, EBX = 00000300$_{16}$, and XYZ = 0400$_{16}$.
  (a) Destination operand of the instruction in Problem 73(c)
  (b) Source operand of the instruction in Problem 73(d)
  (c) Destination operand of the instruction in Problem 73(e)
  (d) Destination operand of the instruction in Problem 73(f)
  (e) Destination operand of the instruction in Problem 73(g)

3

Real-Mode 80386DX Microprocessor Programming 1

3.1 INTRODUCTION

Up to this point we have studied the real-mode software architecture of the 80386DX microprocessor. In this chapter we begin a detailed study of the 80386DX's real-mode instruction set. A large part of the instruction set is covered in this chapter. These instructions provide the ability to write straight-line programs. The rest of the real-mode instruction set and some more sophisticated programming concepts are covered in Chapter 4. The following topics are presented in this chapter.

1. The 80386DX's instruction set
2. Data transfer instructions
3. Arithmetic instructions
4. Logic instructions
5. Shift instructions
6. Rotate instructions
7. Bit test and bit scan instructions

3.2 THE 80386DX'S INSTRUCTION SET

The instruction set of a microprocessor defines the basic operations that a programmer can make the device perform. Earlier we pointed out that the 80386DX microprocessor provides a powerful instruction set that contains more than 150 basic instructions. The wide range of operands and addressing modes permitted for use with these instructions further expands the instruction set into many more executable instructions at the machine code level. For instance, the basic MOV instruction expands into more than 30 different machine-level instructions.

Figure 3.1 shows the evolution of the instruction set for the 8086 architectures. The instruction set of the 8086 and 8088 microprocessors, called the *basic instruction set*, was enhanced in the 80286 microprocessor to implement what is known as the *extended instruction set*. This extended instruction set includes several new instructions and implements additional addressing modes for a few of the instructions already available in the basic instruction set. For example, two instructions added as extensions to the basic instruction set are (*push all*) PUSHA and (*pop all*) POPA. Moreover, the PUSH and IMUL instructions have been enhanced to permit the use of immediate operand addressing. In this way we see that the extended instruction set is a super set of the basic instruction set. These instructions are all executable in the real mode.

Finally, Fig. 3.1 shows that in the 80386DX the real-mode instruction set is further enhanced by a group of instructions called the *80386DX specific instruction set*. For example, it includes instructions to load a pointer directly into the FS, GS, and SS registers and bit test and bit scan instructions. Moreover, it contains additional forms of existing instructions that have been added to perform the identical operation in a more general way, on special registers, or on a double word of data. In this way we see that the 80386DX's real-mode instruction set is a super set of the 8086's basic instruction set.

In Fig. 3.1 we see that the instruction set of the 80286 is further enhanced with system control instructions to give the *system control instruction set*. These instructions are provided to control the multitasking, memory management, and

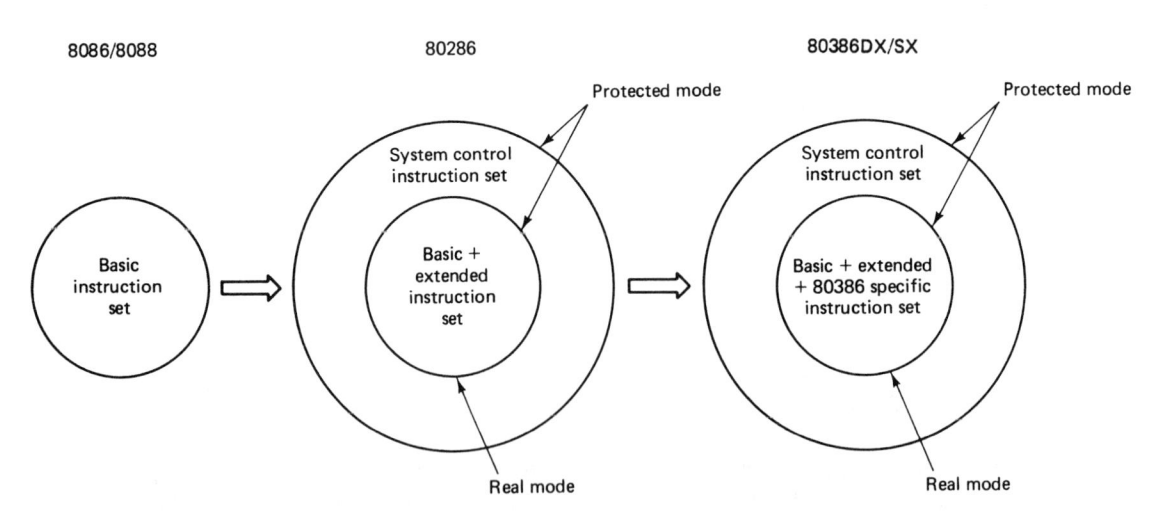

Figure 3.1 Evolution of the 8086-family instruction set.

protection mechanisms of the protected-mode 80286. For instance, the instruction (*clear task switched flag*) CLTS can be used to clear the task switch flag of the machine status word. The 80386DX's system control instruction set is identical to that of the 80286.

For the purpose of discussion, the 80386DX's real-mode instruction set will be divided into a number of groups of functionally related instructions. In this chapter we consider the data transfer instructions, arithmetic instructions, logic instructions, shift instructions, rotate instructions, and bit test and bit scan instructions. Advanced instructions such as those for program and processor control are described in Chapter 4. The system control instructions are covered in Chapter 5.

3.3 DATA TRANSFER INSTRUCTIONS

The 80386DX microprocessor has a group of *data transfer instructions* that are provided to move data either between its internal registers or between an internal register and a storage location in memory. This group includes the *move* (MOV) *instruction, sign-extend and move* (MOVSX) *instruction, zero-extend and move* (MOVZX) *instruction, exchange* (XCHG) *instruction, translate* (XLAT) *instruction, load effective address* (LEA) *instruction, load data segment* (LDS) *instruction, load extra segment* (LES) *instruction, load register and SS* (LSS) *instruction, load register and FS* (LFS) *instruction,* and *load register and GS* (LGS) *instruction*. These instructions are all discussed in this section.

The MOV Instruction

The MOV instruction shown in Fig. 3.2(a) is used to transfer a byte, a word, or a double word of data from a source operand to a destination operand. These operands can be internal registers of the 80386DX and storage locations in memory. Figure 3.2(b) shows the valid source and destination operand variations. This large choice of operands results in many different MOV instructions. Looking at this list of operands, we see that data can be moved between general-purpose registers, between a general-purpose register and a segment register, between a general-purpose register or segment register and memory, between a memory location and the accumulator, or between a general-purpose register and a special register.

Mnemonic	Meaning	Format	Operation	Flags affected
MOV	Move	MOV D, S	(S) → (D)	None

(a)

Figure 3.2 (a) Move data transfer instruction; (b) allowed operands; (c) MOV DX,CS instruction before fetch and execution; (d) after execution; (e) special registers.

Destination	Source
Memory	Accumulator
Accumulator	Memory
Register	Register
Register	Memory
Memory	Register
Register	Immediate
Memory	Immediate
Seg-reg	Reg16
Seg-reg	Mem16
Reg16	Seg-reg
Mem16	Seg-reg
Register	Spec-reg
Spec-reg	Register

(b)

Address	Memory content	Instruction
01100	8C	MOV DX, CS
01101	CA	
01102	XX	Next instruction
02000	XX	
02001	XX	

80386DX MPU

IP: 0100
CS: 0100
DS: 0200
SS:
ES:
FS:
GS:

EAX / AX
EBX / BX
ECX / CX
EDX / XXXX / DX

ESP / SP
EBP / BP
ESI / SI
EDI / DI

(c)

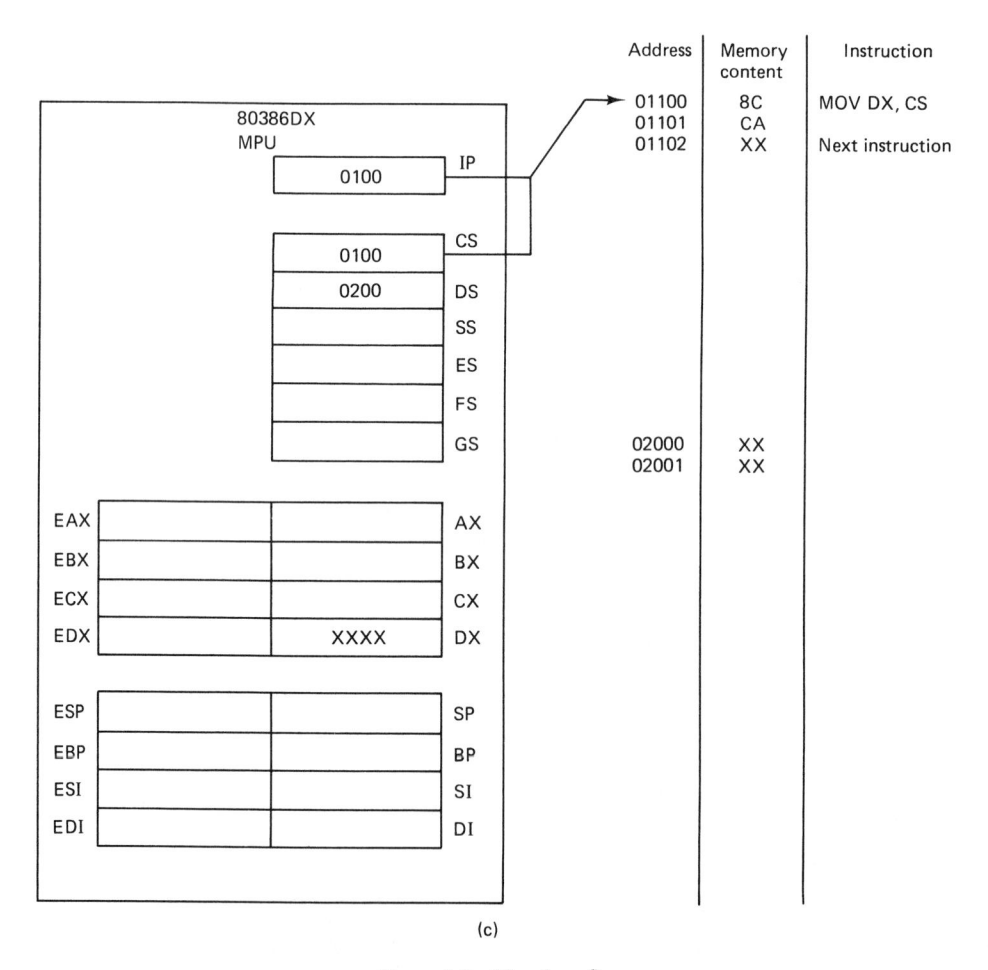

Figure 3.2 (*Continued*)

74   Real-Mode 80386DX Microprocessor Programming 1   Chap. 3

	Address	Memory content	Instruction
	01100	8C	MOV DX, CS
	01101	CA	
	01102	XX	Next instruction
	02000	XX	
	02001	XX	

80386DX MPU

IP: 0102
CS: 0100
DS: 0200
SS:
ES:
FS:
GS:

EAX / AX
EBX / BX
ECX / CX
EDX: 0100 / DX

ESP / SP
EBP / BP
ESI / SI
EDI / DI

(d)

Mnemonic	Name
CR_0	Control register 0
CR_2	Control register 2
CR_3	Control register 3
DR_0	Debug register 0
DR_1	Debug register 1
DR_2	Debug register 2
DR_3	Debug register 3
DR_6	Debug register 6
DR_7	Debug register 7
TR_6	Test register 6
TR_7	Test register 7

(e)

Figure 3.2 (*Continued*)

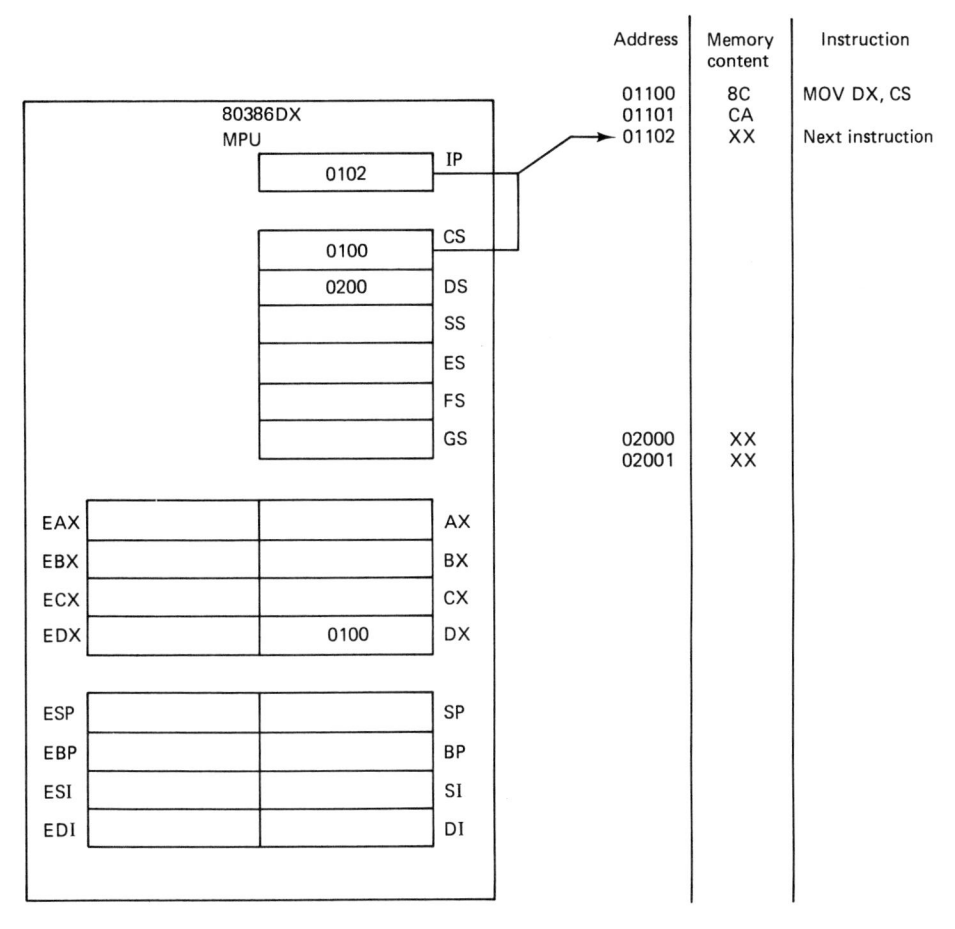

Notice that the MOV instruction cannot transfer data directly between a source and a destination that both reside in external memory. Instead, the data must first be moved from memory into an internal register, such as to the accumulator (EAX), with one move instruction and then moved to the new location in memory with a second move instruction.

All transfers between general-purpose registers and memory can involve either a byte, word, or double word of data. The fact that the instruction corresponds to byte, word, or double word data is designated by the way in which its operands are specified. For instance, AL or AH would be used to specify a byte operand, AX, a word operand, and EAX, a double-word operand. On the other hand, data moved between one of the general-purpose registers and a segment register or between a segment register and a memory location must always be word wide. In Fig. 3.2(a), we also find additional important information. For instance, flag bits within the 80386DX are not modified by execution of a MOV instruction.

The example of a segment register to general-purpose register MOV instruction shown in Fig. 3.2(c) is

$$MOV \quad DX, CS$$

In this instruction the code segment register is the source operand and the data register is the destination. It stands for "move the contents of CS into DX." That is,

$$(CS) \rightarrow (DX)$$

For example, if the contents of CS are 0100_{16}, execution of the instruction MOV DX,CS as shown in Fig. 3.2(d) makes

$$(DX) = (CS) = 0100_{16}$$

In all memory reference MOV instructions, the machine code for the instruction includes an offset address relative to the current contents of the data segment register. An example of this type of instruction is

$$MOV \quad [SUM], AX$$

In this instruction, the memory location identified by SUM is specified using direct addressing. That is, the value of the offset is included in the two byte locations that follow its opcode in program memory.

Let us assume that the content of DS equals 0200_{16} and that SUM corresponds to a displacement of 1212_{16}. Then this instruction means "move the contents of accumulator AX to the memory location offset by 1212_{16} from the starting location of the current data segment." The physical address of this location is obtained as

$$PA = 02000_{16} + 1212_{16} = 03212_{16}$$

Thus the effect of the instruction is

$$(AL) \to (\text{Memory Location } 03212_{16})$$

and

$$(AH) \to (\text{Memory Location } 03213_{16})$$

EXAMPLE 3.1

What is the effect of executing the instruction

```
MOV   CX, [SOURCE_MEM]
```

where SOURCE_MEM is the memory location offset by 20_{16} relative to the data segment starting at address $1A000_{16}$?

SOLUTION Execution of this instruction results in the following:

$$((DS)0 + 20_{16}) \to (CL)$$

$$((DS)0 + 20_{16} + 1_{16}) \to (CH)$$

In other words, CL is loaded with the contents held at memory address

$$1A000_{16} + 20_{16} = 1A020_{16}$$

and CH is loaded with the contents of memory address

$$1A000_{16} + 20_{16} + 1_{16} = 1A021_{16}$$

The move instruction can also be used to load or save the contents of one of the 80386DX's special registers. The names and mnemonics of these registers are listed in Fig. 3.2(e). Notice that they include four control registers (CR_0 through CR_3) and eight debug registers (DR_0 through DR_7). These registers provide special functions for the 80386DX microcomputer. For instance, control register CR_0 contains a number of system control flags. Two examples are the emulation (EM) control bit and the protection enable (PE) bit. An example is the instruction

```
MOV   CR0, EAX
```

When executed, this instruction causes the 32-bit value in EAX to be loaded into control register 0. On the other hand, the instruction

```
MOV   EAX, CR0
```

saves the contents of CR_0 in the accumulator.

Sign-Extend and Zero-Extend Move Instructions: MOVSX and MOVZX

In Fig. 3.3(a) we find that a number of special-purpose move instructions have been included in the instruction set of the 80386DX. The instructions *move with sign-extend* (MOVSX) and *move with zero-extend* (MOVZX) are used to sign extend or zero extend, respectively, a source operand as it is moved to the destination operand location. Figure 3.3(b) shows that the source operand is either

Mnemonic	Meaning	Format	Operation	Flags affected
MOVSX	Move with sign-extend	MOVSX D, S	(S) → (D) MSBs of D are filled with sign bit of S	None
MOVZX	Move with zero-extend	MOVZX D, S	(S) → (D) MSBs of D are filled with 0	None

(a)

Destination	Source
Reg16	Reg8
Reg32	Reg8
Reg32	Reg16
Reg16	Mem8
Reg32	Mem8
Reg32	Mem16

(b)

Figure 3.3 (a) Sign-extend and zero-extend move instructions; (b) allowed operands.

a byte or a word of data in a register or a storage location in memory, while the destination operand is either a 16- or a 32-bit register. For example, the instruction

```
MOVSX   EBX, AX
```

is used to copy the 16-bit value in AX into EBX. As the copy is performed, the value in the sign bit, which is bit 15 of AX, is extended into the 16 higher-order bits of EBX. If AX contains $FFFF_{16}$, the sign bit is logic 1. Therefore, after execution of the MOVSX instruction, the value that results in EBX is $FFFFFFFF_{16}$. The MOVZX instruction performs a similar function to the MOVSX instruction except that it extends the value moved to the destination operand location with zeros.

EXAMPLE 3.2

Explain the operation performed by the instruction

        MOVZX   CX, BYTE PTR   [DATA_BYTE]

if the value of data at memory address DATA_BYTE is FF_{16}.

SOLUTION When the MOVZX instruction is executed, the value FF_{16} is copied into the lower byte of CX and the upper eight bits are filled with zeros. This gives

$$CX = 00FF_{16}$$

The XCHG Instruction

In our study of the move instruction, we found that it could be used to copy the contents of a register or memory location into a register or contents of a register into a storage location in memory. In all cases the original contents of the source location are preserved and the original contents of the destination are destroyed. In some applications it is required to change the contents of two registers. For instance, we might want to exchange the data in the AX and BX registers.

This could be done using multiple move instructions and storage of the data in a temporary register such as DX. However, to perform the exchange function more efficiently, a special instruction has been provided in the instruction set of the 80386DX. This is the exchange (XCHG) instruction. The form of the XCHG instruction and its allowed operands are shown in Fig. 3.4(a) and (b). Here we see that it can be used to swap data between two general-purpose registers or between a general-purpose register and a storage location in memory. In particular, it allows for the exchange of a word or double word of data between one of the general-purpose registers and the accumulator (AX or EAX), exchange of a byte, word, or double word of data between one of the general-purpose registers and a location in memory, or between two of the general-purpose registers.

Let us consider an example of an exchange between two internal registers. Here is a typical instruction:

        XCHG   AX, DX

Its execution by the 80386DX swaps the contents of AX with that of DX. That is,

$$(AX \text{ original}) \rightarrow (DX)$$
$$(DX \text{ original}) \rightarrow (AX)$$

or

$$(AX) \leftrightarrow (DX)$$

Mnemonic	Meaning	Format	Operation	Flags affected
XCHG	Exchange	XCHG D, S	(D) ↔ (S)	None

(a)

Destination	Source
Accumulator	Reg16
Accumulator	Reg32
Memory	Register
Register	Register

(b)

Address	Memory content	Instruction
11101	87	XCHG [SUM], BX
11102	1E	
11103	34	
11104	12	
11105	XX	Next instruction
...		
12000	XX	
12001	XX	
...		
13234	FF	Variable "SUM"
13235	00	

80386DX MPU
IP: 0101
CS: 1100
DS: 1200
SS, ES, FS, GS
EBX: 11 AA (BX)

(c)

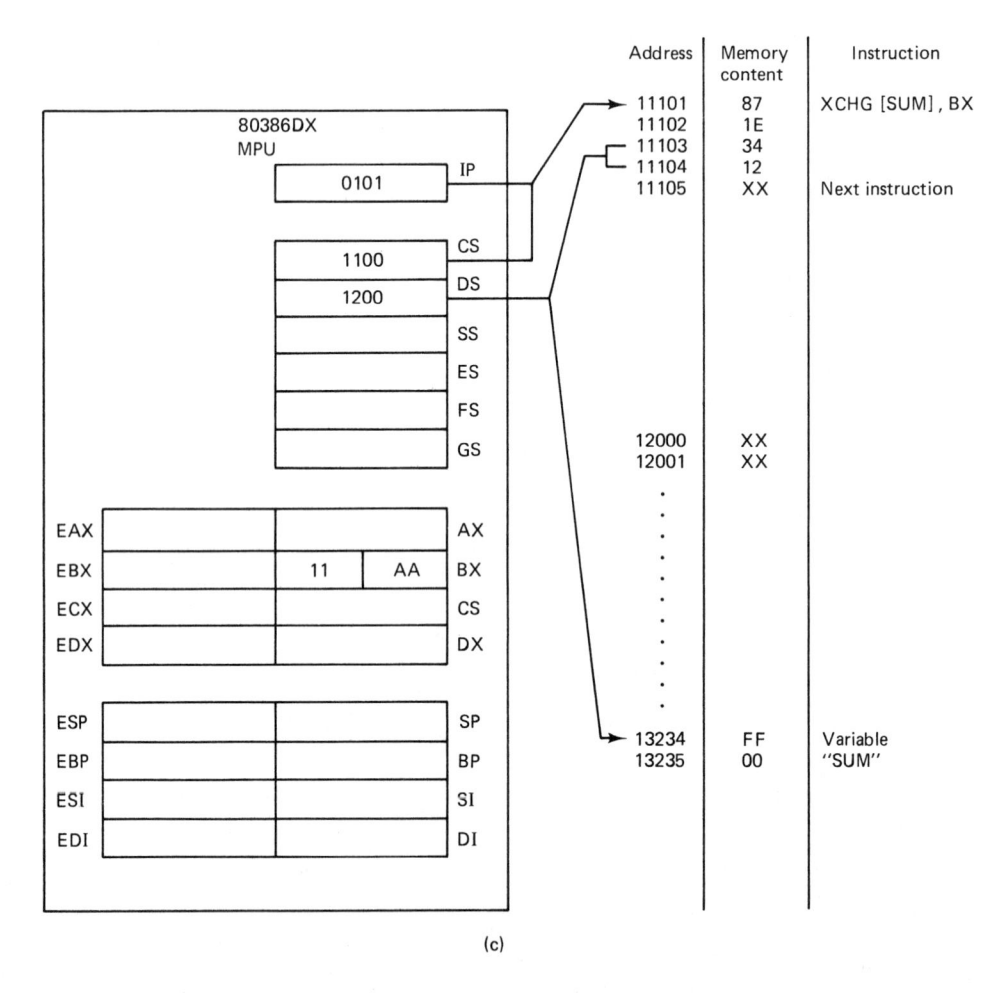

Figure 3.4 (a) Exchange data transfer instruction; (b) allowed operands; (c) XCHG [SUM], BX instruction before fetch and execution; (d) after execution.

(d)

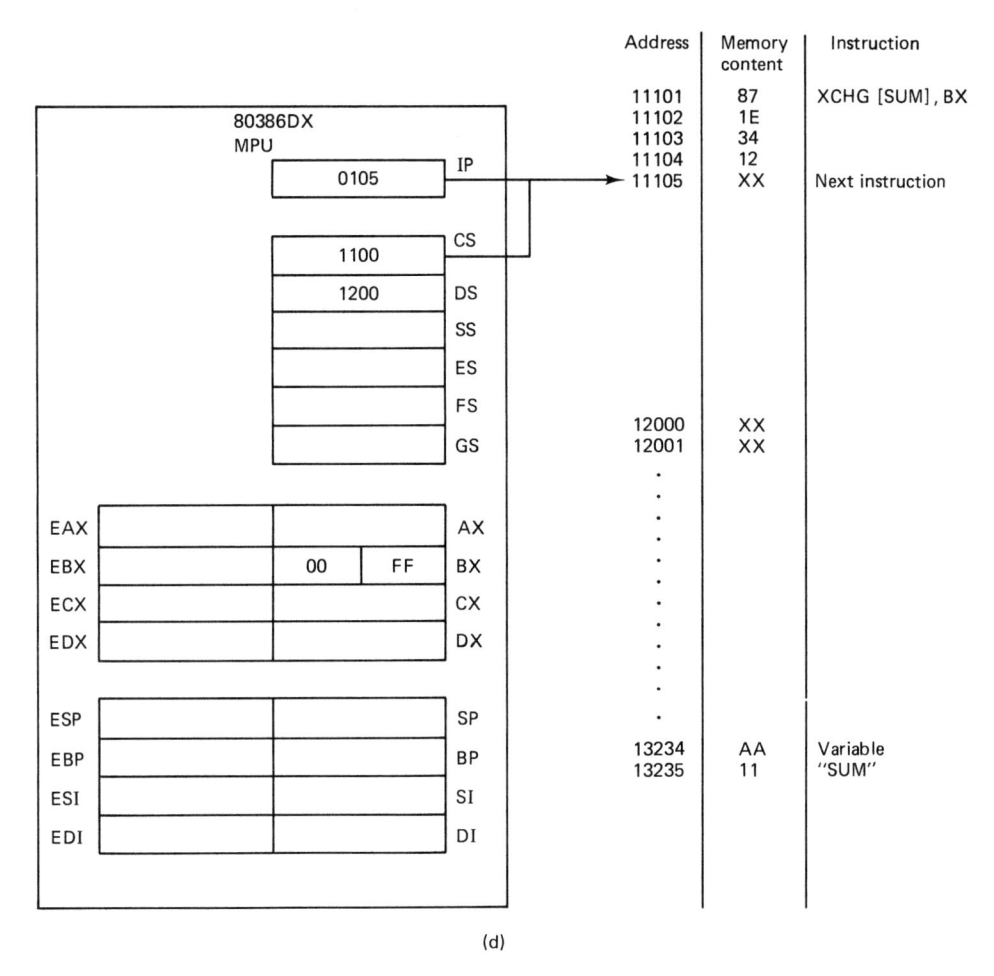

Figure 3.4 (*Continued*)

EXAMPLE 3.3

For the data shown in Fig. 3.4(c), what is the result of executing the instruction

$$\text{XCHG} \quad \text{[SUM], BX}$$

SOLUTION   Execution of this instruction performs the function

$$((DS) 0 + SUM) \leftrightarrow (BX)$$

In Fig. 3.4(c) we see that $(DS) = 1200_{16}$ and the direct address $SUM = 1234_{16}$. Therefore, the physical address is

$$PA = 12000_{16} + 1234_{16} = 13234_{16}$$

Sec. 3.3   Data Transfer Instructions

Notice that this location contains FF_{16} and the address that follows contains 00_{16}. Moreover, note that BL contains AA_{16} and BH contains 11_{16}.

Execution of the instruction performs the following 16-bit swap:

$$(13234_{16}) \leftrightarrow (BL)$$

$$(13235_{16}) \leftrightarrow (BH)$$

As shown in Fig. 3.4(d), we get

$$(BX) = 00FF_{16}$$

$$(SUM) = 11AA_{16}$$

The XLAT and XLATB Instructions

The translate instructions have been provided in the instruction set of the 80386DX to simplify implementation of lookup-table operation. These instructions are described in Fig. 3.5. When using XLAT, the contents of register BX represent the offset of the starting address of the lookup table from the beginning of the current data segment. Also, the contents of AL represent the offset of the element to be accessed from the beginning of the lookup table. This 8-bit element address permits a table with up to 256 elements. The values in both of these registers must be initialized prior to execution of the XLAT instruction.

Mnemonic	Meaning	Format	Operation	Flags affected
XLAT	Translate	XLAT Source-table	((AL) + (BX) + (DS)0) → (AL)	None
XLATB	Translate	XLATB	((AL) + (BX) + (DS)0) → (AL)	None

Figure 3.5 Translate data transfer instruction.

Execution of XLAT replaces the contents of AL by the contents of the accessed lookup table location. The physical address of this element in the table is derived as

$$PA = (DS)0 + (BX) + (AL)$$

An example of the use of this instruction would be for software code conversions, for instance, an ASCII-to-EBCDIC conversion. This requires an EBCDIC table in memory. The individual EBCDIC codes are located in the table at element displacements (AL) equal to their equivalent ASCII character values. That is, the EBCDIC code $C1_{16}$, which represents letter A, would be positioned at displacement 41_{16}, which equals ASCII A, from the start of the table. The start of this ASCII-to-EBCDIC table in the current data segment is identified by the contents of BX.

As an illustration of XLAT, let us assume that the $(DS) = 0300_{16}$, $(BX) = 0100_{16}$, and $(AL) = 0D_{16}$. $0D_{16}$ represents the ASCII character CR (carriage return). Execution of XLAT replaces the contents of AL by the contents of the memory location given by

$$PA = (DS)0 + (BX) + (AL)$$
$$= 03000_{16} + 0100_{16} + 0D_{16} = 0310D_{16}$$

Thus the execution can be described by

$$(0310D_{16}) \rightarrow (AL)$$

Assuming that this memory location contains 52_{16} (EBCDIC carriage return), this value is placed in AL.

$$(AL) = 52_{16}$$

When using the XLAT instruction, the use of the value in DS in the address calculation can be overridden with a segment prefix. This leads to the difference between XLAT and XLATB. The only difference between the two forms of the instruction is that XLATB does not permit the use of the SEG prefix. It must always use the contents of DS in the address calculation.

The Load Effective Address and Load Full Pointer Instructions

Another type of data transfer operation that is important is to load a segment or general-purpose register with an effective address or pointer directly from memory. Special instructions are provided in the instruction set of the 80386DX to give a programmer this capability. These instructions are described in Fig. 3.6(a). They are load register with effective address (LEA), load register and the data segment register (LDS), load register and the stack segment register (LSS), load register and the extra segment register (LES), load register and the F segment register (LFS), and load register and the G segment register (LGS).

Looking at Fig. 3.6(a), we see that these instructions provide the ability to manipulate memory addresses by loading either a 16-bit or 32-bit offset address into a general-purpose register, or a 16- or 32-bit offset address into a general-purpose register together with a 16-bit segment address into a specific segment register.

The LEA instruction is used to load a specified register with a 16- or 32-bit offset address. An example of this instruction is

```
LEA   SI, INPUT
```

When executed, it loads the SI register with an offset address value. The value of this offset is represented by the value of INPUT. INPUT is encoded following the instruction opcode in the code segment of memory.

The other five instructions, LDS, LSS, LES, LFS, and LGS, are similar to LEA except that they load the specified register as well as DS, SS, ES, FS, or GS, respectively. That is, they load a full pointer. Notice that executing the *load*

Mnemonic	Meaning	Format	Operation	Flags affected
LEA	Load effective address	LEA Reg16, EA	(EA) → (Reg16)	None
		LEA Reg32, EA	(EA) → (Reg32)	None
LDS	Load register and DS	LDS Reg16, EA	(EA) → (Reg16)	None
			(EA + 2) → (DS)	
		LDS Reg32, EA	(EA) → (Reg32)	None
			(EA + 4) → (DS)	
LSS	Load register and SS	LSS Reg16, EA	(EA) → (Reg16)	None
			(EA + 2) → (SS)	
		LSS Reg32, EA	(EA) → (Reg32)	None
			(EA + 4) → (SS)	
LES	Load register and ES	LES Reg16, EA	(EA) → (Reg16)	None
			(EA + 2) → (ES)	
		LES Reg32, EA	(EA) → (Reg32)	None
			(EA + 4) → (DS)	
LFS	Load register and FS	LFS Reg16, EA	(EA) → (Reg16)	None
			(EA + 2) → (FS)	
		LFS Reg32, EA	(EA) → (Reg32)	None
			(EA + 4) → (FS)	
LGS	Load register and GS	LGS Reg16, EA	(EA) → (Reg16)	None
			(EA + 2) → (GS)	
		LGS Reg32, EA	(EA) → (Reg32)	None
			(EA + 4) → (GS)	

(a)

Figure 3.6 (a) Load effective address and full pointer data transfer instructions; (b) LDS SI,[200H] instruction before fetch and execution; (c) after execution.

register and SS (LSS) instruction causes both the register specified in the instruction and the stack segment register to be loaded from the source operand. For example, the instruction

```
LSS  ESP, [STACK_POINTER]
```

causes the first 32 bits starting at memory address STACK_POINTER to be loaded into the 32-bit register ESP and the next 16 bits into the SS register.

EXAMPLE 3.4

Assuming that the 80386DX is set up as shown in Fig. 3.6(b), what is the result of executing the following instruction?

```
LDS  SI, [200H]
```

SOLUTION Execution of the instruction loads the SI register from the word location in memory whose offset address with respect to the current data segment is 200_{16}. Figure 3.6(b) shows that the contents of DS are 1200_{16}. This gives a physical address of

$$PA = 12000_{16} + 0200_{16} = 12200_{16}$$

	Address	Memory content	Instruction
	11100	C5	LDS SI, [200H]
	11101	36	
	11102	00	
	11103	20	
	11104	XX	Next instruction
	12000	XX	
	12001	XX	
	12200	20	
	12201	00	
	12202	00	
	12203	13	

80386DX MPU
IP: 0100
CS: 1100
DS: 1200
SI: XXXX

(b)

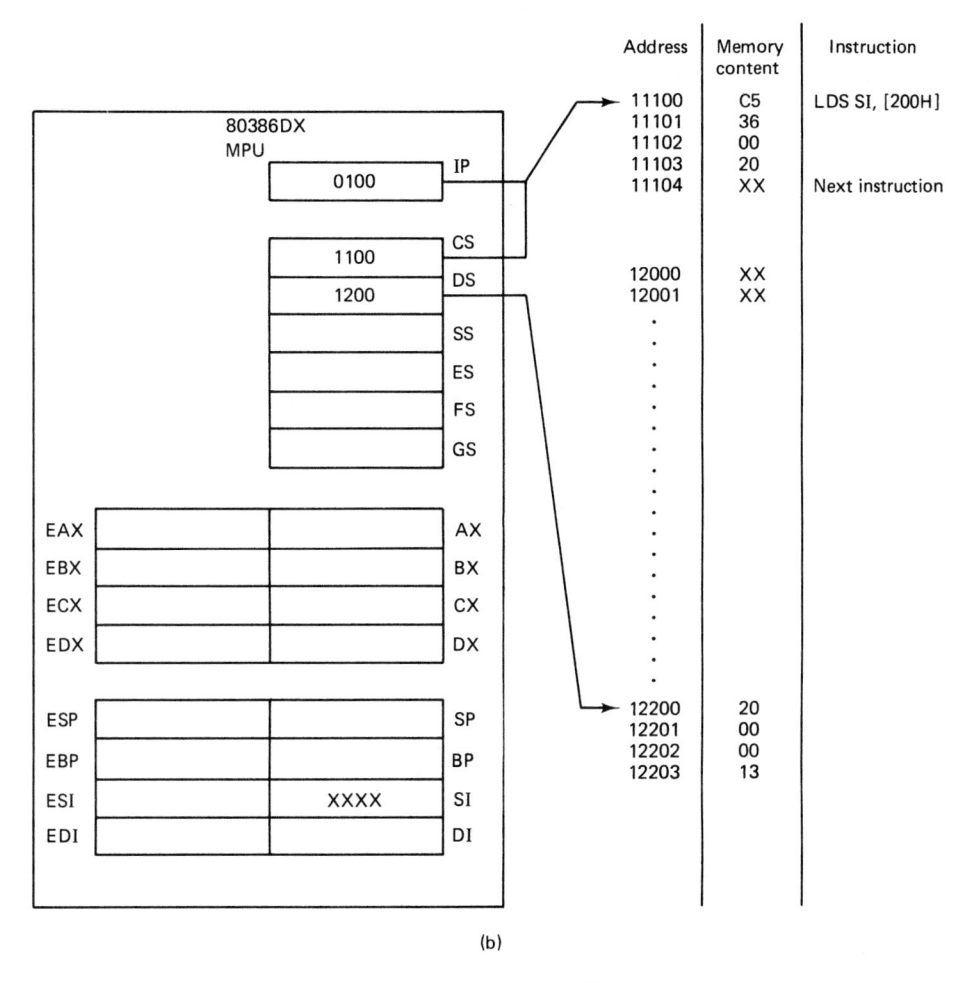

Figure 3.6 (*Continued*)

It is the contents of this location and the one that follows that are loaded into SI. Therefore, in Fig. 3.6(c) we find that SI contains 0020_{16}. The next two bytes, that is, the contents of addresses 12202_{16} and 12203_{16}, are loaded into the DS register. As shown, this defines a new data segment address of 13000_{16}.

EXAMPLE 3.5

Write an instruction that will load the 48-bit pointer starting at memory address DATA_G_ADDRESS into the ESI and GS registers.

SOLUTION This operation is performed with the instruction

```
LGS   ESI, [DATA_G_ADDRESS]
```

Address	Memory content	Instruction
11100	C5	LDS SI, [200H]
11101	36	
11102	00	
11103	20	
11104	XX	Next instruction
.	.	
12000	XX	
12001	XX	
.	.	
12200	20	
12201	00	
12202	00	
12203	13	
.	.	
13000	XX	New data segment
13001	XX	

80386DX MPU registers:
- IP: 0104
- CS: 1100
- DS: 1300
- SS, ES, FS, GS
- EAX/AX, EBX/BX, ECX/CX, EDX/DX
- ESP/SP, EBP/BP
- ESI/SI: 0020
- EDI/DI

(c)

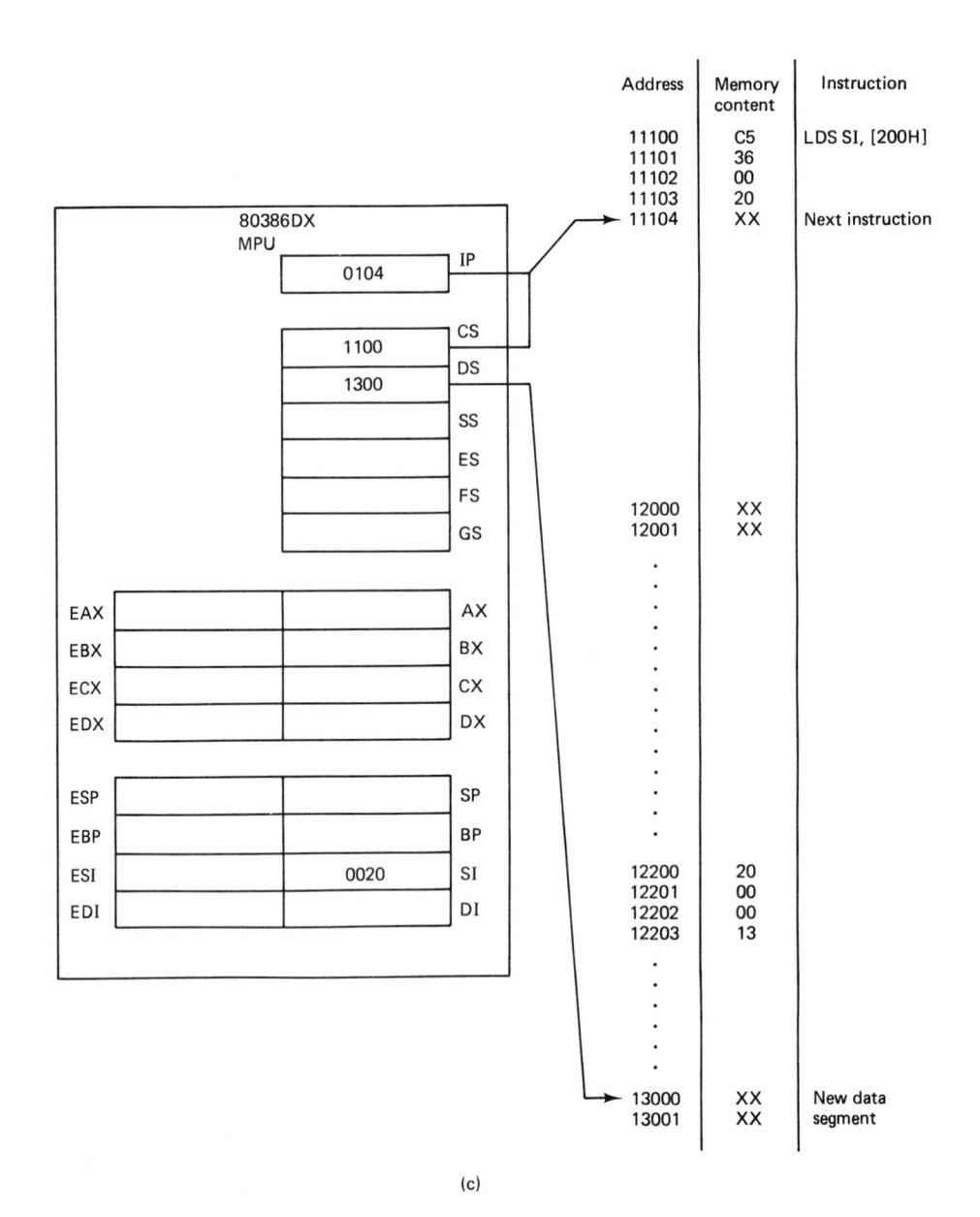

Figure 3.6 (*Continued*)

3.4 ARITHMETIC INSTRUCTIONS

The instruction set of the 80386DX microprocessor contains a variety of *arithmetic instructions*. They include instructions for the *addition, subtraction, multiplication,* and *division* operations. These operations can be performed on numbers expressed in a variety of numeric data formats. They include *unsigned or signed binary bytes, words, or double words, unpacked or packed decimal bytes,* or *ASCII numbers.* Remember that by *packed decimal* we mean that two BCD digits are packed into a byte register or memory location. *Unpacked decimal numbers* are stored one BCD digit per byte. The decimal numbers are always unsigned. Moreover, *ASCII numbers* are expressed in ASCII code and stored one number per byte.

The status that results from the execution of an arithmetic instruction is recorded in the flags of the 80386DX. The flags that are affected by the arithmetic instructions are carry flag (CF), auxiliary carry flag (AF), sign flag (SF), zero flag (ZF), parity flag (PF), and overflow flag (OF). Each of these flags was described in Chapter 2. For the purpose of discussion, we divide the arithmetic instructions into the subgroups shown in Fig. 3.7.

Addition Instructions: ADD, ADC, INC, AAA, and DAA

The form of each of the instructions in the *addition group* is shown in Fig. 3.8(a); the allowed operand variations, for the ADD and ADC instructions, are shown in Fig. 3.8(b), and the allowed operands for the INC instruction are shown in Fig. 3.8(c). Let us begin by looking more closely at the operation of the *add* (ADD) instruction. Notice in Fig. 3.8(b) that it can be used to add an immediate operand to the contents of the accumulator, another register, or a storage location in memory. It also allows us to add the contents of two registers together or the contents of a register and a memory location.

Addition	
ADD	Add byte or word
ADC	Add byte or word with carry
INC	Increment byte or word by 1
AAA	ASCII adjust for addition
DAA	Decimal adjust for addition
Subtraction	
SUB	Subtract byte or word
SBB	Subtract byte or word with borrow
DEC	Decrement byte or word by 1
NEG	Negate byte or word
AAS	ASCII adjust for subtraction
DAS	Decimal adjust for subtraction
Multiplication/Division	
MUL	Multiply byte or word unsigned
IMUL	Integer multiply byte or word
AAM	ASCII adjust for multiply
DIV	Divide byte or word unsigned
IDIV	Integer divide byte or word
AAD	ASCII adjust for division
CBW	Convert byte to word
CWDE	Convert word to double word in EAX
CWD	Convert word to double word in DX and AX
CDQ	Convert double word to quad word

Figure 3.7 Arithmetic instructions.

Mnemonic	Meaning	Format	Operation	Flags affected
ADD	Addition	ADD D, S	(S) + (D) → (D) carry → (CF)	OF, SF, ZF, AF, PF, CF
ADC	Add with carry	ADC D, S	(S) + (D) + (CF) → (D) carry → (CF)	OF, SF, ZF, AF, PF, CF
INC	Increment by 1	INC D	(D) + 1 → (D)	OF, SF, ZF, AF, PF
DAA	Decimal adjust for addition	DAA		SF, ZF, AF, PF, CF OF undefined
AAA	ASCII adjust for addition	AAA		AF, CF OF, SF, ZF, PF undefined

(a)

Destination	Source
Register	Register
Register	Memory
Memory	Register
Register	Immediate
Memory	Immediate
Accumulator	Immediate

(b)

Destination
Reg16
Reg8
Memory

(c)

Figure 3.8 (a) Addition arithmetic instructions; (b) allowed operands for ADD and ADC instructions; (c) allowed operands for INC instruction.

In general, the result of executing the instruction is expressed as

$$(S) + (D) \rightarrow (D)$$

That is, the contents of the source operand are added to those of the destination operand and the sum that results is put into the location of the destination operand.

EXAMPLE 3.6

Assume that the AX and BX registers contain 1100_{16} and $0ABC_{16}$, respectively. What are the results of executing the instruction ADD AX,BX?

SOLUTION Execution of the ADD instruction causes the contents of source operand BX to be added to the contents of destination register AX. This gives

$$(BX) + (AX) = 0ABC_{16} + 1100_{16} = 1BBC_{16}$$

This sum ends up in destination register AX.

$$(AX) = 1BBC_{16}$$

Execution of this instruction is illustrated in Fig. 3.9(a) and (b).

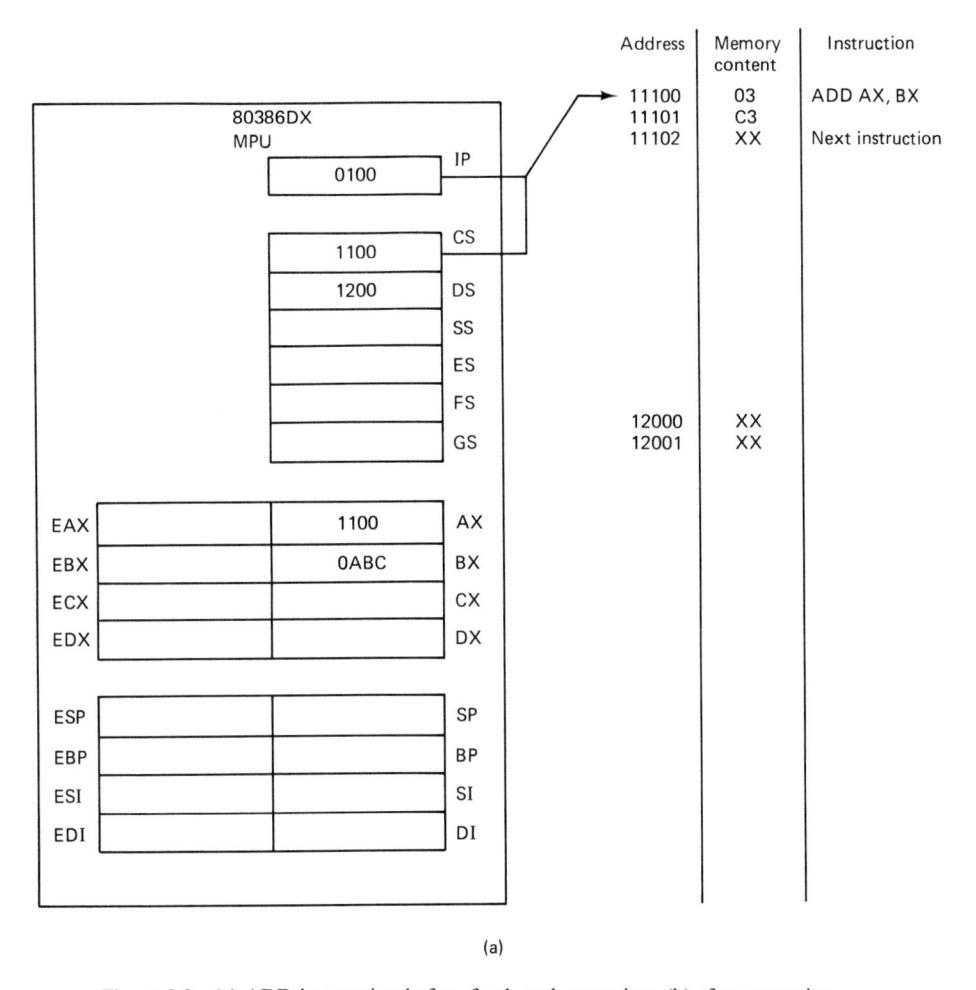

Figure 3.9 (a) ADD instruction before fetch and execution; (b) after execution.

The instruction *add with carry* (ADC) works similarly to ADD. But in this case, the content of the carry flag is also added; that is,

$$(S) + (D) + (CF) \rightarrow (D)$$

The valid operand combinations are the same as those for the ADD instruction.

Another instruction that can be considered as part of the addition subgroup of arithmetic instructions is the *increment* (INC) instruction. As shown in Fig. 3.8(c), its operands can be the contents of an 8-, 16-, or 32-bit internal register or a byte, word, or double word storage location in memory. Execution of the

Sec. 3.4 Arithmetic Instructions

```
                                                    Address | Memory  | Instruction
                                                            | content |
                                                    11100   | 03      | ADD AX, BX
                        80386DX                     11101   | C3      |
                          MPU                       11102   | XX      | Next instruction
                                          IP
                              0102

                                          CS
                              1100
                              1200        DS
                                          SS
                                          ES
                                          FS
                                                    12000   | XX
                                          GS        12001   | XX

    EAX                     1BBC          AX
    EBX                     0ABC          BX
    ECX                                   CX
    EDX                                   DX

    ESP                                   SP
    EBP                                   BP
    ESI                                   SI
    EDI                                   DI
```

(b)

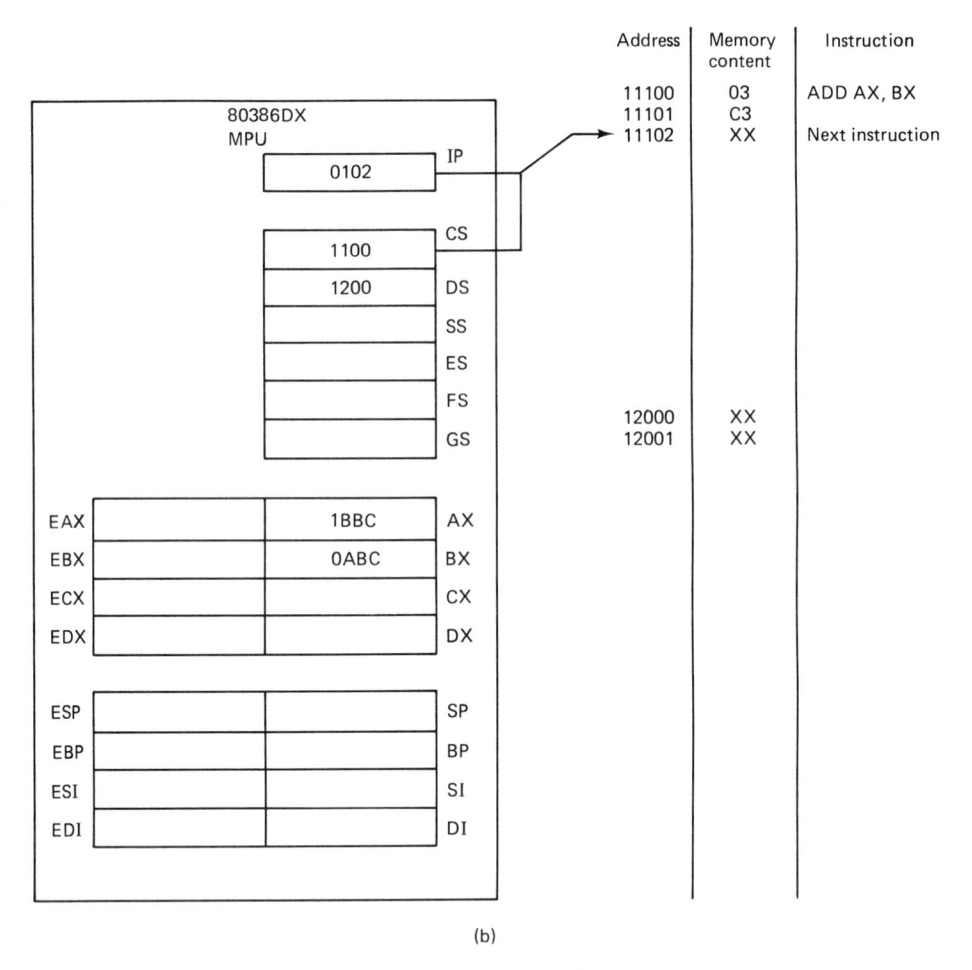

Figure 3.9 (*Continued*)

INC instruction adds one to the specified operand. An example of an instruction that increments the high byte of AX is

INC   AH

Figure 3.8(a) shows which flags are affected by execution of these three instructions.

EXAMPLE 3.7

The original contents of AX, BL, memory location SUM, and carry flag (CF) are 1234_{16}, AB_{16}, $00CD_{16}$, and 0_2, respectively. Describe the results of executing the following sequence of instructions.

```
                    ADD   AX, [SUM]
                    ADC   BL, 05H
                    INC   [SUM]
```

SOLUTION By executing the first instruction, we add the word in the accumulator and the word in the memory location identified as SUM. The result is placed in the accumulator. That is,

$$(AX) \leftarrow (AX) + (SUM) = 1234_{16} + 00CD_{16} = 1301_{16}$$

The carry flag remains reset.

The second instruction adds to the lower byte of the base register (BL), the immediate operand 5_{16}, and the carry flag, which is 0_2. This gives

$$(BL) \leftarrow (BL) + IOP + (CF) = AB_{16} + 5_{16} + 0_2 = B0_{16}$$

Since no carry is generated, CF stays reset.

The last instruction increments the contents of memory location SUM by one. That is,

$$(SUM) \leftarrow (SUM) + 1_2 = 00CD_{16} + 1_2 = 00CE_{16}$$

These results are summarized in Fig. 3.10.

Instruction	(AX)	(BL)	(SUM)	(CF)
Initial state	1234	AB	00CD	0
ADD AX, [SUM]	1301	AB	00CD	0
ADC BL, 05H	1301	B0	00CD	0
INC [SUM]	1301	B0	00CE	0

Figure 3.10 Results due to execution of arithmetic instructions in Example 3.7.

The addition instructions we just covered can also be used to directly add numbers expressed in ASCII code. This eliminates the need for doing a code conversion on ASCII data prior to processing it with addition operations. Whenever the 80386DX does an addition on ASCII format data, an adjustment must be performed on the result to convert it to a decimal number. It is specifically for this purpose that the *ASCII adjust for addition* (AAA) instruction is provided in the instruction set of the 80386DX. The AAA instruction should be executed immediately after the instruction that adds ASCII data.

Assuming that AL contains the result produced by adding two ASCII coded numbers, execution of the AAA instruction causes the contents of AL to be replaced by its equivalent decimal value. If the sum is greater than nine, AL contains the LSD and AH is incremented by one. Otherwise, AL contains the sum and AH is unchanged. Figure 3.8(a) shows that the AF and CF flags can be affected. Since AAA can only adjust data that are in AL, the destination register for ADD instructions that process ASCII numbers should be AL.

Sec. 3.4   Arithmetic Instructions

EXAMPLE 3.8

What is the result of executing the following instruction sequence?

```
ADD   AL, BL
AAA
```

Assume that AL contains 32_{16}, which is the ASCII code for number 2, BL contains 34_{16}, which is the ASCII code for number 4, and AH has been cleared.

SOLUTION Executing the ADD instruction gives

$$(AL) \leftarrow (AL) + (BL) = 32_{16} + 34_{16} = 66_{16}$$

Next, the result is adjusted to give its equivalent decimal number. This is done by execution of the AAA instruction. The equivalent of adding 2 and 4 is decimal 6 with no carry. Therefore, the result after the AAA instruction is

$$(AL) = 06_{16}$$

$$(AH) = 00_{16}$$

and both AF and CF remain cleared.

The instruction set of the 80386DX includes another instruction, called *decimal adjust for addition* (DAA). This instruction is used to perform an adjust operation similar to that performed by AAA but for the addition of packed BCD numbers instead of ASCII numbers. Information about this instruction is also provided in Fig. 3.8(a). Similar to AAA, DAA performs an adjustment on the value in AL. A typical instruction sequence is

```
ADD   AL, BL
DAA
```

Remember that the contents of AL and BL must be packed BCD numbers, that is, two BCD digits packed into a byte. The adjusted result in AL is again a packed BCD byte.

Subtraction Instructions: SUB, SBB, DEC, NEG, AAS, and DAS

The instruction set of the 80386DX includes an extensive set of instructions provided for implementing subtraction. As shown in Fig. 3.11(a), the subtraction subgroup is similar to the addition subgroup. It includes instructions for subtracting a source and destination operand, decrementing an operand, and adjusting the result of subtractions of ASCII and BCD data. An additional instruction in this subgroup is negate.

The *subtract* (SUB) instruction is used to subtract the value of a source operand from a destination operand. The result of this operation in general is given as

$$(D) \leftarrow (D) - (S)$$

Mnemonic	Meaning	Format	Operation	Flags affected
SUB	Subtract	SUB D,S	(D) − (S) → (D) Borrow → (CF)	OF, SF, ZF, AF, PF, CF
SBB	Subtract with borrow	SBB D,S	(D) − (S) − (CF) → (D)	OF, SF, ZF, AF, PF, CF
DEC	Decrement by 1	DEC D	(D) − 1 → (D)	OF, SF, ZF, AF, PF
NEG	Negate	NEG D	0 − (D) → (D) 1 → (CF)	OF, SF, ZF, AF, PF, CF
DAS	Decimal adjust for subtraction	DAS		SF, ZF, AF, PF, CF OF undefined
AAS	ASCII adjust for subtraction	AAS		AF, CF OF, SF, ZF, PF undefined

(a)

Destination	Source
Register	Register
Register	Memory
Memory	Register
Accumulator	Immediate
Register	Immediate
Memory	Immediate

(b)

Destination
Register
Memory

(c)

Figure 3.11 (a) Subtraction arithmetic instructions; (b) allowed operands for SUB and SBB instructions; (c) allowed operands for DEC and NEG instructions.

As shown in Fig. 3.11(b), it can employ the identical operand combinations as the ADD instruction.

The *subtract with borrow* (SBB) instruction is similar to SUB; however, it also subtracts the contents of the carry flag from the destination. That is,

$$(D) \leftarrow (D) - (S) - (CF)$$

EXAMPLE 3.9

Assuming that the contents of registers BX and CX are 1234_{16} and 0123_{16}, respectively, and the carry flag is zero, what will be the result of executing the following instruction?

```
        SBB   BX,CX
```

SOLUTION Since the instruction implements the operation

$$(BX) - (CX) - (CF) \rightarrow (BX)$$

we get

$$(BX) = 1234_{16} - 0123_{16} - 0_2$$
$$= 1111_{16}$$

Sec. 3.4 Arithmetic Instructions

Just as the INC instruction could be used to add one to an operand, the *decrement* (DEC) instruction can be used to subtract one from its operand. The allowed operands are shown in Fig 3.11(c).

In Fig. 3.11(c), we see that the *negate* (NEG) instruction can operate on operands in a general-purpose register or a storage location in memory. Execution of this instruction causes the value of its operand to be replaced by its negative. The way this is actually done is through subtraction. That is, the contents of the specified operand are subtracted from zero using 2's-complement arithmetic, and the result is returned to the operand location.

EXAMPLE 3.10

Assuming that register BX contains $003A_{16}$, what is the result of executing the following instruction?

$$\text{NEG} \quad \text{BX}$$

SOLUTION Executing the NEG instruction causes the 2's-complement subtraction that follows:

$$0000_{16} - (BX) = 0000_{16} + \text{2's complement of } 003A_{16}$$

$$= 0000_{16} + FFC6_{16}$$

$$= FFC6_{16}$$

The value is placed in BX.

$$(BX) = FFC6_{16}$$

In our study of the addition instruction subgroup, we found that the 80386DX is capable of directly adding ASCII and BCD numbers. The SUB and SBB instructions can also subtract numbers represented in these formats. Just as for addition, the results that are obtained must be adjusted to produce their corresponding ASCII or decimal number. In the case of ASCII subtraction, we use the *ASCII adjust for subtraction* (AAS) instruction, and for packed BCD subtraction we use the *decimal adjust for subtract* (DAS) instruction.

An example of an instruction sequence for direct ASCII subtraction is

```
SUB   AL, BL
AAS
```

ASCII numbers must be loaded into AL and BL before execution of the subtract instruction. Notice that the destination of the subtraction should be AL. After execution of AAS, AL contains the difference of the two numbers, and AH is unchanged if no borrow takes place or is decremented by one if a borrow occurs.

Multiplication and Division Instructions: MUL, DIV, IMUL, IDIV, AAM, AAD, CBW, CWDE, CWD, and CDQ

The 80386DX has instructions to support multiplication and division of binary and BCD numbers. Two basic types of multiplication and division instructions, those for the processing of unsigned numbers and signed numbers, are available. To do these operations on unsigned numbers, the instructions are MUL and DIV. On the other hand, to multiply or divide signed numbers, the instructions are IMUL and IDIV.

Figure 3.12(a) describes these instructions. The allowed operands for the MUL, DIV, and IDIV instructions are shown in Fig. 3.12(b). Notice that a byte-wide, word-wide, or double-word-wide register or memory operand is specified in an unsigned multiplication instruction. As shown in Fig. 3.12(b), the other operand, which is the destination, must already be in AL for 8-bit multiplications, in AX for 16-bit multiplications, or in EAX for 32-bit multiplications. The allowed operands for IMUL are shown in Fig. 3.12(c).

The result of executing a MUL or IMUL instruction on byte data can be represented as

$$(AX) \leftarrow (AL) \times (8\text{-bit operand})$$

That is, the resulting 16-bit product is produced in the AX register. On the other hand, for multiplications of data words, the 32-bit result is given by

$$(DX,AX) \leftarrow (AX) \times (16\text{-bit operand})$$

where AX contains the 16 LSBs and DX the 16 MSBs. Finally, multiplying double-word data produces the result

$$(EDX,EAX) \leftarrow (EAX) \times (32\text{-bit operand})$$

Here EAX contains the 32 LSBs of the product and EDX the 32 MSBs.

For the division operation, again just the source operand is specified. The other operand is either the contents of AX for 16-bit dividends, the contents of both DX and AX for 32-bit dividends, or EAX and EDX for 64-bit dividends. The result of a DIV or IDIV instruction for an 8-bit divisor is represented by

$$(AH),(AL) \leftarrow (AX)/(8\text{-bit operand})$$

where (AH) is the remainder and (AL) the quotient. For 16-bit divisions, we get

$$(DX),(AX) \leftarrow (DX,AX)/(16\text{-bit operand})$$

Here AX contains the quotient and DX contains the remainder. Finally, for a 32-bit division, the result is given by

$$(EDX),(EAX) \leftarrow (EDX,EAX)/(32\text{-bit operand})$$

The quotient is in EAX and the remainder is in EDX.

Mnemonic	Meaning	Format	Operation	Flags affected
MUL	Multiply (unsigned)	MUL D, S	(AL) · (S8) → (AX) (AX) · (S16) → (DX), (AX) (EAX) · (S32) → (EDX), (EAX)	OF, CF SF, ZF, AF, PF undefined
DIV	Division (unsigned)	DIV D, S	(1) Q((AX)/(S8)) → (AL) R((AX)/(S8)) → (AH) (2) Q((DX, AX)/(S16)) → (AX) R((DX, AX)/(S16)) → (DX) (3) Q((EDX, EAX)/(S32)) → (EAX) R((EDX, EAX)/(S32)) → (EDX) If Q is FF_{16} in case (1), $FFFF_{16}$ in case (2) or $FFFFFFFF_{16}$ in case (3), then type 0 interrupt occurs	All flags undefined
IMUL	Integer multiply (signed)	IMUL S	(AL) · (S8) → (AX) (AX) · (S16) → (DX), (AX) (EAX) · (S32) → (EDX), (EAX)	OF, CF SF, ZF, AF, PF undefined
		IMUL R, I	(R16) · (Imm8) → (R16) (R32) · (Imm8) → (R32) (R16) · (Imm16) → (R16) (R32) · (Imm32) → (R32)	OF, CF SF, ZF, AF, PF undefined
		IMUL R, S, I	(S16) · (Imm8) → (R16) (S32) · (Imm8) → (R32) (S16) · (Imm16) → (R16) (S32) · (Imm32) → (R32)	OF, CF SF, ZF, AF, PF undefined
		IMUL R, S	(R16) · (S16) → (R16) (R32) · (S32) → (R32)	OF, CF SF, ZF, AF, PF undefined
IDIV	Integer divide (signed)	IDIV S	(1) Q((AX)/(S8)) → (AL) R((AX)/(S8)) → (AH) (2) Q((DX, AX)/(S16)) → (AX) R((DX, AX)/(S16)) → (DX) (3) Q((EDX, EAX)/(S32)) → (EAX) R((EDX, EAX)/(S32)) → (EDX) If Q is $7F_{16}$ in case (1), $7FFF_{16}$ in case (2) or $7FFFFFFF_{16}$ in case (3), then type 0 interrupt occurs	All flags undefined
AAM	Adjust AL after multiplication	AAM	Q((AL)/10) → AH R((AL)/10) → AL	SF, ZF, PF OF, AF, CF undefined
AAD	Adjust AX before division	AAD	(AH) · 10 + AL → AL 00 → AH	SF, ZF, PF OF, AF, CF undefined
CBW	Convert byte to word	CBW	(MSB of AL) → (All bits of AH)	None
CWDE	Convert word to double word	CWDE	(MSB of AX) → (16 MSBs of EAX)	None
CWD	Convert word to double word	CWD	(MSB of AX) → (All bits of DX)	None
CDQ	Convert double word to quad word	CDQ	(MSB of EAX) → (All bits of EDX)	None

(a)

Figure 3.12 (a) Multiplication and division arithmetic instructions; (b) allowed operands for MUL, DIV, and IDIV instructions; (c) allowed operands for IMUL instruction.

Destination	Source
Reg16	Reg8
Reg32	Reg16
	Reg32
Reg16	Imm8
Reg32	Imm8
Reg16	Imm16
Reg32	Imm32
Reg16	Reg16, Imm8
Reg16	Mem16, Imm8
Reg32	Reg32, Imm8
Reg32	Mem32, Imm8
Reg16	Reg16, Imm16
Reg16	Mem16, Imm16
Reg32	Reg16, Imm32
Reg32	Mem16 Imm32
Reg16	Reg16
Reg16	Mem16
Reg32	Reg32
Reg32	Mem32

(c)

Destination	Source
AL	Reg8
AX	Reg16
EAX	Reg32
AL	Mem8
AX	Mem16
EAX	Mem32

(b)

Figure 3.12 (*Continued*)

EXAMPLE 3.11

If the contents of AL equals -1 and the contents of CL are -2, what will be the result produced in AX by executing the following instructions?

$$\text{MUL} \quad \text{CL}$$

and

$$\text{IMUL} \quad \text{CL}$$

SOLUTION The first instruction multiplies the contents of AL and CL as unsigned numbers.

$$-1 = 11111111_2 = FF_{16}$$
$$-2 = 11111110_2 = FE_{16}$$

Thus, executing the MUL instruction, we get

$$(AX) = 11111111_2 \times 11111110_2 = 1111110100000010_2$$
$$= FD02_{16}$$

The second instruction multiplies the same two numbers as signed numbers and gives

$$(AX) = -1_{16} \times -2_{16}$$
$$= 2_{16}$$

Sec. 3.4 Arithmetic Instructions

As shown in Fig. 3.12(a), adjust instructions for BCD multiplication and division are also provided. They are *adjust AX after multiply* (AAM) and *adjust AX before divide* (AAD). The multiplication performed just before execution of the AAM instruction is assumed to have been performed on two unpacked BCD numbers with the product produced in AL. The AAD instruction assumes that AH and AL contain unpacked BCD numbers.

The division instructions can also be used to divide an 8-bit dividend in AL by an 8-bit divisor. However, to do this, the sign of the dividend must first be extended to fill the AX register. That is, AH is filled with zeros if the number in AL is positive or with ones if it is negative. This conversion is automatically done by executing the *convert byte to word* (CBW) instruction.

In a similar way, the 32-bit by 16-bit division instructions can be used to divide a 16-bit dividend in AX by a 16-bit divisor. In this case the sign bit of AX must be extended by 16 bits into the DX register. This can be done by another instruction, which is known as *convert word to double word* (CWD). The sign extension instructions are also shown in Fig. 3.12(a).

Notice that the CBW, CWDE, CWD, and CDQ instructions are provided to handle operations where the result or intermediate results of an operation cannot be held in the correct word length for use in other arithmetic operations. Using these instructions, we can extend a byte, word, or double word of data to its equivalent word, double word, or quad word.

EXAMPLE 3.12

What is the result of executing the following sequence of instructions?

```
MOV   AL, A1H
CBW
CWD
```

SOLUTION The first instruction loads AL with $A1_{16}$. This gives

$$(AL) = A1_{16} = 10100001_2$$

Executing the second instruction extends the most significant bit of AL, which is 1, into all bits of AH. The result is

$$(AH) = 11111111_2 = FF_{16}$$

$$(AX) = 1111111110100001_2 = FFA1_{16}$$

This completes conversion of the byte in AL to a word in AX.

The last instruction loads each bit of DX with the most significant bit of AX. This bit is also 1. Therefore, we get

$$(DX) = 1111111111111111_2 = FFFF_{16}$$

Now the word in AX has been extended to the double word

$$(AX) = FFA1_{16}$$

$$(DX) = FFFF_{16}$$

3.5 LOGIC INSTRUCTIONS

The 80386DX has instructions for performing the logic operations *AND, OR, exclusive-OR*, and *NOT*. As shown in Fig. 3.13(a), the AND, OR, and XOR instructions perform their respective logic operations bit by bit on the specified source and destination operands, the result being represented by the final contents of the destination operand. Figure 3.13(b) shows the allowed operand combinations for the AND, OR, and XOR instructions.

For example, the instruction

$$\text{AND} \quad \text{AX, BX}$$

causes the contents of BX to be ANDed with the contents of AX. The result is reflected by the new contents of AX. If AX contains 1234_{16} and BX contains $000F_{16}$, the result produced by the instruction is

$$1234_{16} \cdot 000F_{16} = 0001001000110100_2 \cdot 0000000000001111_2$$
$$= 0000000000000100_2$$
$$= 0004_{16}$$

This result is stored in the destination operand.

$$(AX) = 0004_{16}$$

In this way we see that the AND instruction was used to mask off the 12 most significant bits of the destination operand.

Mnemonic	Meaning	Format	Operation	Flags Affected
AND	Logical AND	AND D,S	$(S) \cdot (D) \rightarrow (D)$	OF, SF, ZF, PF, CF AF undefined
OR	Logical Inclusive-OR	OR D,S	$(S) + (D) \rightarrow (D)$	OF, SF, ZF, PF, CF AF undefined
XOR	Logical Exclusive-OR	XOR D,S	$(S) \oplus (D) \rightarrow (D)$	OF, SF, ZF, PF, CF AF undefined
NOT	Logical NOT	NOT D	$(\overline{D}) \rightarrow (D)$	None

(a)

Destination	Source
Register	Register
Register	Memory
Memory	Register
Register	Immediate
Memory	Immediate
Accumulator	Immediate

(b)

Destination
Register
Memory

(c)

Figure 3.13 (a) Logic instructions; (b) allowed operands for AND, OR, and XOR; (c) allowed operands for NOT instruction.

The NOT logic instruction differs from those for AND, OR, and exclusive-OR in that it operates on a single operand. Looking at Fig. 3.13(c), which shows the allowed operands of the NOT instruction, we see that this operand can be the contents of an internal register or a location in memory.

EXAMPLE 3.13

Describe the result of executing the following sequence of instructions.

```
MOV   AL, 01010101B
AND   AL, 00011111B
OR    AL, 11000000B
XOR   AL, 00001111B
NOT   AL
```

SOLUTION The first instruction moves the immediate operand 01010101_2 into the AL register. This loads the data that are to be manipulated with the logic instructions. The next instruction performs a bit-by-bit AND operation of the contents of AL with immediate operand 00011111_2. This gives

$$01010101_2 \cdot 00011111_2 = 00010101_2$$

$$(AL) = 00010101_2 = 15_{16}$$

This result is produced in destination register AL. Note that this operation has masked off the three most significant bits of AL. The next instruction performs a bit-by-bit logical OR of the present contents of AL with immediate operand $C0_{16}$. This gives

$$00010101_2 + 11000000_2 = 11010101_2$$

$$(AL) = 11010101_2 = D5_{16}$$

This operation is equivalent to setting the two most significant bits of AL.
The fourth instruction is an exclusive-OR operation of the contents of AL with immediate operand 00001111_2. We get

$$11010101_2 \oplus 00001111_2 = 11011010_2$$

$$(AL) = 11011010_2 = DA_{16}$$

Note that this operation complements the logic state of those bits in AL that are ones in the immediate operand.
The last instruction, NOT AL, inverts each bit of AL. Therefore, the final contents of AL become

$$(AL) = \overline{11011010}_2 = 00100101_2 = 25_{16}$$

These results are summarized in Fig. 3.14.

3.6 SHIFT INSTRUCTIONS

The six *shift instructions* of the 80386DX can perform two basic types of shift operations. They are the *logical shift* and the *arithmetic shift*. Moreover, each of

Instruction	(AL)
MOV AL,01010101B	01010101
AND AL,00011111B	00010101
OR AL,11000000B	11010101
XOR AL,00001111B	11011010
NOT AL	00100101

Figure 3.14 Results of example program using logic instructions.

these operations can be performed to the right or to the left. The shift instructions are *shift logical left* (SHL), *shift arithmetic left* (SAL), *shift logical right* (SHR), *shift arithmetic right* (SAR), *double precision shift left* (SHLD), and *double precision shift right* (SHRD).

The logical shift instructions, SHL and SHR, are described in Fig. 3.15(a). Notice in Fig. 3.15(b) that the destination operand, the data whose bits are to be shifted, can be either the contents of an internal register or a storage location in memory. Moreover, the source operand can be specified in three ways. If it is assigned the value of 1, a one-bit shift will take place. For instance, as illustrated in Fig. 3.16(a), executing

    SHL   AX, 1

causes the 16-bit contents of the AX register to be shifted one bit position to the left. Here we see that the vacated LSB location is filled with zero and the bit shifted out of the MSB is saved in CF.

On the other hand, if the source operand is specified as CL instead of one, the count in this register represents the number of bit positions the contents of the operand are to be shifted. This permits the count to be defined under software control and allows a range of shifts from 1 to 255 bits.

An example of an instruction specified in this way is

    SHR   AX, CL

Assuming that CL contains the value 02_{16}, the logical shift right that occurs is as shown in Fig. 3.16(b). Notice that the two MSBs have been filled with zeros and the last bit shifted out at the LSB, which is zero, is placed in the carry flag.

The third way of specifying the count is with an 8-bit immediate operand. Again this permits a shift range of 1 to 255 bits to be specified.

In an arithmetic shift to the left, SAL operation, the vacated bits at the right of the operand are filled with zeros, whereas in an arithmetic shift to the right, SAR operation, the vacated bits at the left are filled with the value of the original MSB of the operand. Thus, in an arithmetic shift to the right, the original sign of the number is extended. This operation is equivalent to division by powers of 2 as long as the bits shifted out of the LSB are zeros.

Mnemonic	Meaning	Format	Operation	Flags affected
SAL/SHL	Shift arithmetic left/shift logical left	SAL/SHL D, Count	Shift the (D) left by the number of bit positions equal to Count and fill the vacated bit positions on the right with zeros. The last bit shifted out at MSB end is in CF.	SF, ZF, PF, CF AF undefined OF undefined if Count ≠ 1
SHR	Shift logical right	SHR D, Count	Shift the (D) right by the number of bit positions equal to Count and fill the vacated bit positions on the left with zeros. The last bit shifted out at the LSB end is in CF.	SF, ZF, PF, CF AF undefined OF undefined if Count ≠ 1
SAR	Shift arithmetic right	SAR D, Count	Shift the (D) right by the number of bit positions equal to Count and fill the vacated bit positions on the left with the original most significant bit. The last bit shifted out of the LSB end is in CF.	SF, ZF, PF, CF AF undefined OF undefined if Count ≠ 1
SHLD	Double precision shift left	SHLD D1, D2, Count	Shift (D1) left by the number of bit positions equal to Count. (D2) supplies the bits to be loaded into D1. Bits are shifted from the MSB of D2 to the LSB of D1. The value in D2 is unchanged. The last bit shifted out of D1 is in CF.	SF, ZF, PF, CF OF, AF undefined
SHRD	Double precision shift right	SHLD D1, D2, Count	Shift (D1) right by the number of bit positions equal to Count. (D2) supplies the bits to be loaded into D1. Bits are shifted from the LSB of D2 into the MSB of D1. The value in D2 is unchanged. The last bit shifted out of D1 is in CF.	SF, ZF, PF, CF OF, AF undefined

(a)

Destination	Count
Register	1
Register	CL
Register	Imm8
Memory	1
Memory	CL
Memory	Imm8

(b)

Destination 1	Destination 2	Count
Reg16	Reg16	Imm8
Reg16	Reg16	CL
Reg32	Reg32	Imm8
Reg32	Reg32	CL
Mem16	Reg16	Imm8
Mem16	Reg16	CL
Mem32	Reg32	Imm8
Mem32	Reg32	CL

(c)

Figure 3.15 (a) Shift instructions; (b) allowed operands for basic shift instructions; (c) allowed operands for double precision shift instructions.

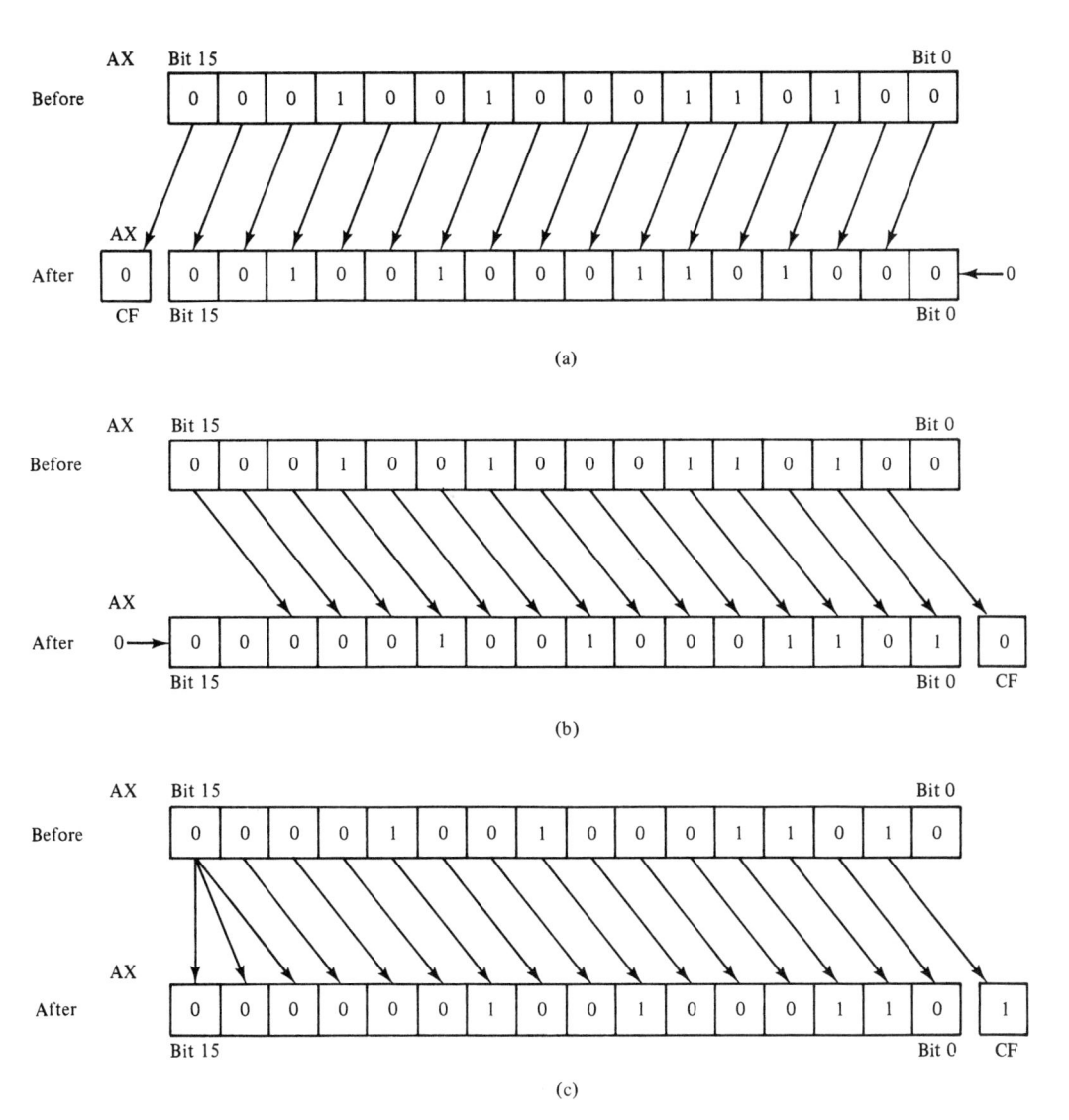

Figure 3.16 (a) Results of executing SHL AX,1; (b) results of executing SHR AX,CL; (c) results of executing SAR AX, CL.

EXAMPLE 3.14

Assume that CL contains 02_{16} and AX contains $091A_{16}$. Determine the new contents of AX and the carry flag after the instruction

```
        SAR   AX, CL
```

is executed.

Sec. 3.6   Shift Instructions   103

SOLUTION Figure 3.16(c) shows the effect of executing the instruction. Here we see that since CL contains 02_{16} a shift right by two bit locations takes place and the original sign bit, which is logic 0, is extended to the two vacated bit positions. Moreover, the last bit shifted out from the LSB location is placed in CF. This makes CF equal to 1. Therefore, the results produced by execution of the instruction are

$$(AX) = 0246_{16}$$

and

$$(CF) = 1_2$$

Now that we have described the basic shift instructions of the 80386DX, let us continue with the double precision shift instructions. Looking at Fig. 3.15(a), we find two double precision instructions: *double precision shift left* (SHLD) and *double precision shift right* (SHRD). These instructions employ two destination operands. In Fig. 3.15(c), we see that D_1 is either a register or storage location in memory. This is the operand that has its contents shifted. D_2 identifies a register that contains the bits that are to be shifted into D_1 during the shift operation. Finally, the count can be specified by either an 8-bit immediate operand or a count in CL. Even though the immediate operand or CL register is 8 bits in length, the shift performed with these instructions is limited to from 0 to 31 bits.

For instance, when the instruction

```
SHLD  EAX, EBX, 3
```

is executed, the contents of EAX are shifted left three bit positions, the three vacated bits in EAX (bits 0, 1, and 2) are filled with values from the three MSBs of EBX (bits 29, 30, and 31), and the last value shifted out of the MSB end of EAX (bit 29) is saved in CF. The value held in EBX remains unchanged during the shift operation.

3.7 ROTATE INSTRUCTIONS

Another group of instructions, known as the *rotate instructions*, are similar to the shift instructions we just introduced. This group, as shown in Fig. 3.17(a), includes the *rotate left* (ROL), *rotate right* (ROR), *rotate left through carry* (RCL), and *rotate right through carry* (RCR) instructions.

As shown in Fig. 3.17(b), the rotate instructions are similar to the shift instructions in several ways. They have the ability to rotate the contents of either an internal register or storage location in memory. Also, the rotation that takes place can be from 1 to 255 bit positions to the left or to the right. Moreover, in the case of a multibit rotate, the number of bit positions to be rotated is again specified by either the contents of CL or an 8-bit immediate operand. Their dif-

Mnemonic	Meaning	Format	Operation	Flags affected
ROL	Rotate left	ROL D, Count	Rotate the (D) left by the number of bit positions equal to Count. Each bit shifted out from the leftmost bit goes back into the rightmost bit position	CF OF undefined if Count ≠1
ROR	Rotate right	ROR D, Count	Rotate the (D) right by the number of bit positions equal to Count. Each bit shifted out from the rightmost bit goes into the leftmost bit position	CF OF undefined if Count ≠1
RCL	Rotate left through carry	RCL D, Count	Same as ROL except carry is attached to (D) for rotation	CF OF undefined if Count ≠1
RCR	Rotate right through carry	RCR D, Count	Same as ROR except carry is attached to (D) for rotation	CF OF undefined if Count ≠1

(a)

Destination	Count
Register	1
Register	CL
Register	Imm8
Memory	1
Memory	CL
Memory	Imm8

(b)

Figure 3.17 (a) Rotate instructions; (b) allowed operands.

ference from the shift instructions lies in the fact that the bits moved out at either the MSB or LSB end are not lost; instead, they are reloaded at the other end.

As an example, let us look at the operation of the ROL instruction. Execution of ROL causes the contents of the selected operand to be rotated left the specified number of bit positions. Each bit shifted out at the MSB end is reloaded at the LSB end. Moreover, the content of CF reflects the state of the last bit that was shifted out. For instance, the instruction

ROL AX, 1

causes a 1-bit rotate to the left. Figure 3.18(a) shows the result produced by executing this instruction. Notice that the original value of bit 15 is zero. This value has been rotated into CF and bit 0 of AX. All other bits have been rotated one bit position to the left.

The ROR instruction operates the same way as ROL except that it causes data to be rotated to the right instead of to the left. For example, execution of

ROR AX, CL

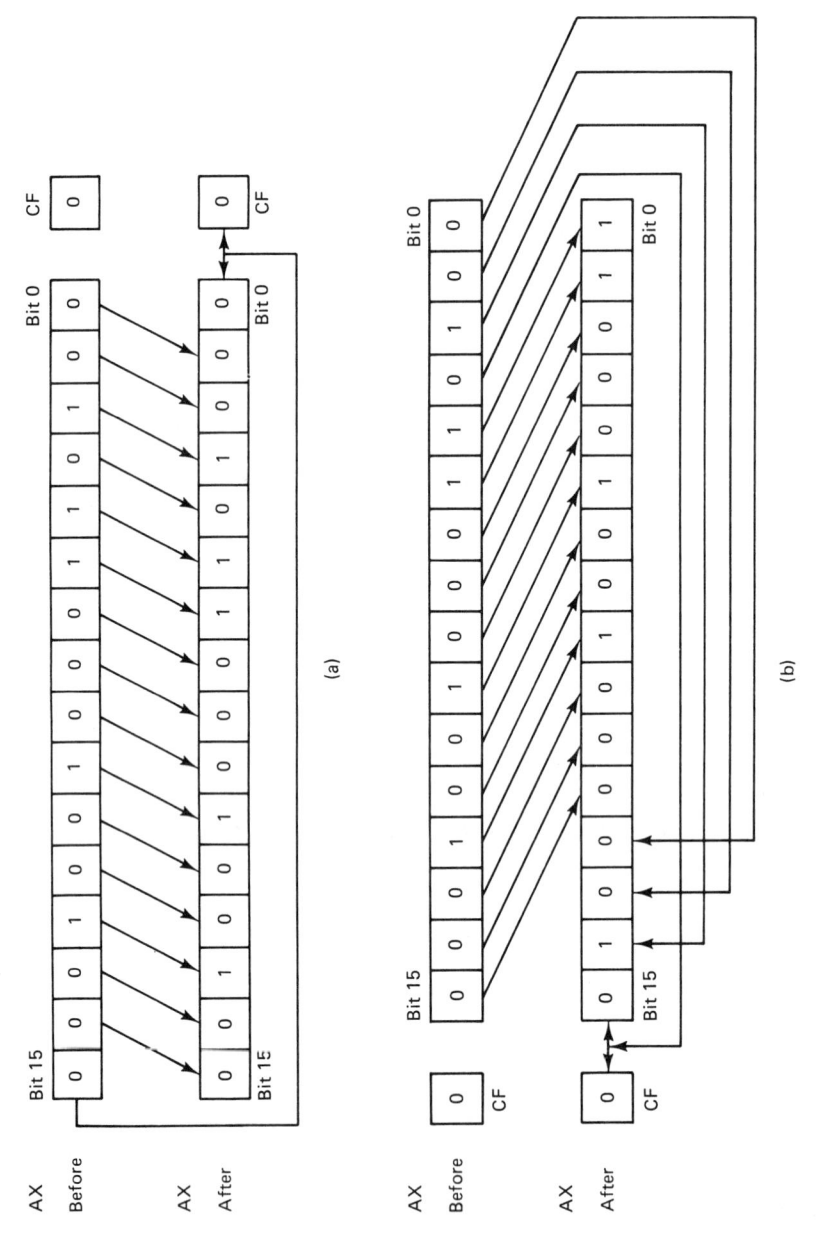

Figure 3.18 (a) Result of executing ROL AX,1; (b) results of executing ROR AX,CL.

causes the contents of AX to be rotated right by the number of bit positions specified in CL. The result for CL equal to 4 is illustrated in Fig. 3.18(b).

The other two rotate instructions, RCL and RCR, differ from ROL and ROR in that the bits are rotated through the carry flag. Figure 3.19 illustrates the rotation that takes place due to execution of the RCL instruction. Notice that the value returned to bit 0 is the prior contents of CF and not bit 31. The value rotated out of bit 31 goes into the carry flag. Thus the bits rotate through carry.

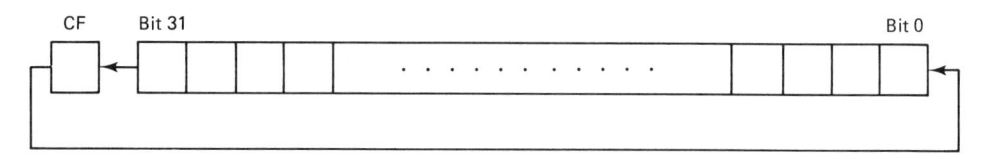

Figure 3.19 Rotation caused by execution of an RCL instruction.

EXAMPLE 3.15

What is the result in BX and CF after execution of the following instruction?

$$\text{RCR} \quad \text{BX, CL}$$

Assume that prior to execution of the instruction, $(CL) = 04_{16}$, $(BX) = 1234_{16}$, and $(CF) = 0_2$.

SOLUTION The original contents of BX are

$$(BX) = 0001001000110100_2 = 1234_{16}$$

Execution of the RCR instruction causes a 4-bit rotate right through carry to take place on the data in BX. Therefore, the original content of bit 3, which is zero, resides in carry; $CF = 0_2$ and 1000_2 has been reloaded from bit 15. The resulting contents of BX are

$$(BX) = 1000000100100011_2 = 8123_{16}$$

$$(CF) = 0_2$$

3.8 BIT TEST AND BIT SCAN INSTRUCTIONS

The *bit test and bit scan instructions* of the 80386DX enable a programmer to test the logic level of a bit in either a register or storage location in memory. Let us begin by examining the bit test instructions. They are used to test the state of a single bit in a register or memory location. When the instruction is executed, the value of the tested bit is saved in the carry flag. Moreover, instructions are provided that can also reset, set, or complement the contents of the bit tested during the execution of the instruction.

In Fig. 3.20(a) we see that the *bit test* (BT) instruction has two operands. Figure 3.20(b) shows that the destination operand identifies the register or memory location that contains the bit that is to be tested. The source operand contains an index that selects the bit that is to be tested. Notice in Fig. 3.20(b) that the index may be either an 8-bit immediate operand or the value in a 16- or 32-bit register. When this instruction is executed, the state of the tested bit is simply copied into the carry flag.

Mnemonic	Meaning	Format	Operation	Flags affected
BT	Bit test	BT D, S	Saves the value of the bit in D specified by the value in S in CF.	CF OF, SF, ZF, AF, PF undefined
BTR	Bit test and reset	BTR D, S	Saves the value of the bit in D specified by the value in S in CF and then resets the bit in D.	CF OF, SF, ZF, AF, PF undefined
BTS	Bit test and set	BTS D, S	Saves the value of the bit in D specified by the value in S in CF and then sets the bit in D.	CF OF, SF, ZF, AF, PF undefined
BTC	Bit test and complement	BTC D, S	Saves the value of the bit in D specified by the value in S in CF and then complements the bit in D.	CF OF, SF, ZF, AF, PF undefined
BSF	Bit scan forward	BSF D, S	Scan the source operand starting from bit 0. ZF = 0 if all bits are 0, else ZF = 1 and the destination operand is loaded with the bit index of the first set bit.	ZF OF, SF, AF, PF, CF undefined
BSR	Bit scan reverse	BSR D, S	Scan the source operand starting from the MSB. ZF = 0 if all bits are 0, else ZF = 1 and the destination operand is loaded with the bit index of the first set bit.	ZF OF, SF, AF, PF, CF undefined

(a)

Destination	Source
Reg16	Reg16
Reg16	Imm8
Reg32	Reg32
Reg32	Imm8
Mem16	Reg16
Mem16	Imm8
Mem32	Reg32
Mem32	Imm8

(b)

Destination	Source
Reg16	Reg16
Reg16	Mem16
Reg32	Reg32
Reg32	Mem32

(c)

Figure 3.20 (a) Bit test and bit scan instructions; (b) allowed operands for bit test instructions; (c) allowed operands for bit scan instructions.

Once the state of the bit is saved in CF, further processing can be performed through software. For instance, a conditional jump instruction could be used to test the value in CF, and if CF equals 1, program control could be passed to a service routine. On the other hand, if CF equals 0, the value of the index could be incremented, a jump performed back to the BT instruction, and the next bit in the operand tested. Notice that this routine scans the bits of the operand looking for the first bit that is logic 1.

Another example is the instruction

$$\text{BTR} \quad \text{EAX, EDI}$$

Execution of this instruction causes the bit in 32-bit register EAX that is selected by the index in EDI to be tested. The value of the tested bit is first saved in the carry flag and then this bit in the EAX register is reset.

EXAMPLE 3.16

Describe the operation that is performed by the instruction

$$\text{BTC} \quad \text{BX, 7}$$

Assume that register BX contains the value $03F0_{16}$.

SOLUTION Let us first express the value in BX in binary form. This gives

$$(BX) = 0000001111110000_2$$

Execution of the *bit test and complement* instruction causes the value of the eighth bit to be first tested and then complemented. Since this bit is logic 1, CF is set to 1. This gives

$$(CF) = 1_2$$
$$(BX) = 0000001101110000_2 = 0370_{16}$$

The *bit scan forward* (BSF) and *bit scan reverse* (BSR) instructions are used to scan through the bits of the source operand register or storage location in memory to determine whether or not they are all 0. For example, by executing the instruction

$$\text{BSF} \quad \text{ESI, EDX}$$

the bits of 32-bit register EDX are tested one after the other starting from bit 0. If all bits are found to be 0, the ZF is cleared. On the other hand, if the contents of EDX are not zero, ZF is set to 1 and the index value of the first bit that tested as 1 is copied into the destination register ESI. This index equals the value of the bit position plus 1.

Sec. 3.8   Bit Test and Bit Scan Instructions

ASSIGNMENT

Section 3.2

1. What is the name given to the part of the 80386DX's real-mode instruction set that is common to that of the 8086, 8088, and 80286 microprocessors? Common to the 80286 real-mode instruction set, but not the 8086 or 8088?
2. What is the part of the 80386DX's real-mode instruction set that is unique to the 80386DX called?
3. What is the name of the protected-mode instruction set of the 80386DX?

Section 3.3

4. Explain what operation is performed by each of the following instructions.
   - (a) MOV   AX, 0110H
   - (b) MOV   DI, AX
   - (c) MOV   BL, AL
   - (d) MOV   [0100H], AX
   - (e) MOV   [BX + DI], AX
   - (f) MOV   [DI] + [0004H], AX
   - (g) MOV   [BX][DI] + [0004H], AX
5. Assume that registers EAX, EBX, and EDI are all initialized to 00000000_{16} and that all data storage memory has been cleared. Determine the location and value of the destination operand as instructions (a) through (g) of Problem 4 are executed as a sequence.
6. Write an instruction sequence that will initialize the ES register with the immediate value 1010_{16}.
7. Write an instruction that will save the contents of the ES register in memory at address DS:1000H.
8. Why will the instruction MOV   CL, AX result in an error when it is assembled?
9. Describe the operation performed by the instruction

    MOVSX   EAX, BL

10. Write an instruction that will zero-extend the word of data at address DATA_WORD and copy it into register EAX.
11. Describe the operation performed by each of the following instructions.
    - (a) XCHG   AX, BX
    - (b) XCHG   BX, DI
    - (c) XCHG   [DATA], AX
    - (d) XCHG   [BX + DI], AX
12. If register EBX contains the value 00000100_{16}, register EDI contains 00000010_{16}, and register DS contains 1075_{16}, what physical memory location is swapped when the instruction in Problem 13(d) is executed?
13. Assume that EAX = 00000010_{16}, EBX = 00000100_{16}, and DS = 1000_{16}; what happens if the XLAT instruction is executed?

14. Write a single instruction that will load AX from address 0200_{16} and DS from address 0202_{16}.
15. What operation is performed when the instruction

    ```
    LFS   EDI, [DATA_F_ADDRESS]
    ```

    is executed?
16. Two code-conversion tables starting with offsets TABL1 and TABL2 in the current data segment are to be accessed. Write a routine that initializes the needed registers and then replaces the contents of memory locations MEM1 and MEM2 (offsets in the data segment) by the equivalent converted codes from the code-conversion tables.

Section 3.4

17. What operation is performed by each of the following instructions?
    - (a) ADD   AX, 00FFH
    - (b) ADC   SI, AX
    - (c) INC   BYTE PTR [0100H]
    - (d) SUB   DL, BL
    - (e) SBB   DL, [0200H]
    - (f) DEC   BYTE PTR [DI + BX]
    - (g) NEG   BYTE PTR [DI] + [0010H]
    - (h) MUL   DX
    - (i) IMUL   BYTE PTR [BX + SI]
    - (j) DIV   BYTE PTR [SI] + [0030H]
    - (k) IDIV   BYTE PTR [BX][SI] + [0030H]
18. Assume that the state of the 80386DX's registers and memory just prior to execution of each instruction in Problem 17 is as follows:

    $$EAX = 00000010H$$
    $$EBX = 00000020H$$
    $$ECX = 00000030H$$
    $$EDX = 00000040H$$
    $$ESI = 00000100H$$
    $$EDI = 00000200H$$
    $$CF = 1$$
    $$DS:100H = 10H$$
    $$DS:101H = 00H$$
    $$DS:120H = FFH$$
    $$DS:121H = FFH$$
    $$DS:130H = 08H$$

$$\text{DS:131H} = \text{00H}$$
$$\text{DS:150H} = \text{02H}$$
$$\text{DS:151H} = \text{00H}$$
$$\text{DS:200H} = \text{30H}$$
$$\text{DS:201H} = \text{00H}$$
$$\text{DS:210H} = \text{40H}$$
$$\text{DS:211H} = \text{00H}$$
$$\text{DS:220H} = \text{30H}$$
$$\text{DS:221H} = \text{00H}$$

What is the result produced in the destination operand by executing instructions (a) through (k) of Problem 17?

19. Write an instruction that will add the immediate value $111F_{16}$ and the carry flag to the contents of the extended data register.
20. Write an instruction that will subtract the word contents of the storage location pointed to by the base register and the carry flag from the word contents of the accumulator.
21. Two word-wide unsigned integers are stored in memory addresses $A00_{16}$ and $A02_{16}$, respectively. Write an instruction sequence that computes and stores their sum, difference, product, and quotient. Store these results at consecutive memory locations starting at address $A10_{16}$ in memory. To obtain the difference, subtract the integer at $A02_{16}$ from the integer at $A00_{16}$. For the division, divide the integer at $A00_{16}$ by the integer at $A02_{16}$. Use register indirect addressing mode to store the various results.
22. Assuming that $(AX) = 0123_{16}$ and $(BL) = 10_{16}$, what will be the new contents of AX after executing the instruction DIV BL?
23. What instruction is used to adjust the result of an addition that processed packed BCD numbers?
24. Which instruction is provided in the instruction set of the 80386DX to adjust the result of a subtraction that involved ASCII coded numbers?
25. If AL contains $A0_{16}$, what happens when the instruction CBW is executed?
26. If the value in AX is $7FFF_{16}$, what happens when the instruction CWD is executed?
27. Two byte-sized BCD integers are stored at the symbolic addresses NUM1 and NUM2, respectively. Write an instruction sequence to generate their difference and store it at NUM3. The difference is to be formed by subtracting the value at NUM1 from that at NUM2.

Section 3.5

28. Describe the operation performed by each of the following instructions.
    (a) AND   BYTE PTR [0300H], 0FH
    (b) AND   DX, [SI]
    (c) OR   [BX + DI], AX
    (d) OR   BYTE PTR [BX][DI] + [10H], F0H
    (e) XOR   AX, [SI + BX]

(f) NOT   BYTE PTR [0300H]
(g) NOT   WORD PTR [BX + DI]

29. Assume that the state of the 80386DX's registers and memory just prior to execution of each instruction in Problem 28 is as follows:

$$EAX = 00005555H$$
$$EBX = 00000010H$$
$$ECX = 00000010H$$
$$EDX = 0000AAAAH$$
$$ESI = 00000100H$$
$$EDI = 00000200H$$
$$DS:100H = 0FH$$
$$DS:101H = F0H$$
$$DS:110H = 00H$$
$$DS:111H = FFH$$
$$DS:200H = 30H$$
$$DS:201H = 00H$$
$$DS:210H = AAH$$
$$DS:211H = AAH$$
$$DS:220H = 55H$$
$$DS:221H = 55H$$
$$DS:300H = AAH$$
$$DS:301H = 55H$$

What is the result produced in the destination operand by executing instructions (a) through (g) of Problem 28?

30. Write an instruction that when executed will mask off all but bit 7 of the contents of the extended data register.

31. Write an instruction that will mask off all but bit 7 of the word of data stored at address DS:0100H.

32. Specify the relation between the old and new contents of AX after executing the following instructions:

```
NOT   AX
ADD   AX, 1
```

33. Write an instruction sequence that generates a byte-sized integer in the memory location identified by the label RESULT. The value of the byte integer is to be calculated as follows:

$$RESULT = AL \cdot NUM1 + \overline{(NUM2 \cdot AL + BL)}$$

Assume that all parameters are byte-sized.

Chap. 3 Assignment

Section 3.6

34. Explain what operation is performed by each of the following instructions.
  (a) SHL DX, CL
  (b) SHL EDX, 7
  (c) SHL BYTE PTR [0400H], CL
  (d) SHR BYTE PTR [DI], 1
  (e) SHR DWORD PTR [DI], 3
  (f) SHR BYTE PTR [DI + BX], CL
  (g) SAR WORD PTR [BX + DI], 1
  (h) SAR WORD PTR [BX][DI] + [0010H], CL

35. Assume that the state of the 80386DX's registers and memory just before execution of each instruction in Problem 34 is as follows:

$$EAX = 00000000H$$
$$EBX = 00000010H$$
$$ECX = 00000105H$$
$$EDX = 00001111H$$
$$ESI = 00000100H$$
$$EDI = 00000200H$$
$$CF = 0$$
$$DS:100H = 0FH$$
$$DS:200H = 22H$$
$$DS:201H = 44H$$
$$DS:202H = 00H$$
$$DS:203H = 00H$$
$$DS:210H = 55H$$
$$DS:211H = AAH$$
$$DS:220H = AAH$$
$$DS:221H = 55H$$
$$DS:400H = AAH$$
$$DS:401H = 55H$$

What is the result produced in the destination operand by executing instructions (a) through (h) of Problem 34?

36. Write an instruction that shifts the contents of the extended count register left by one bit position.

37. Write an instruction sequence that when executed shifts the contents of the word-wide memory location pointed to by the address in the destination index register left by eight bit positions.

38. Identify the condition under which the contents of AX would remain unchanged after executing any of the instructions that follow.

```
MOV  CL, 4H
SHL  AX, CL
SHR  AX, CL
```

39. Implement the following operation using shift and arithmetic instructions.

$$7(AX) - 5(BX) - \tfrac{1}{8}(BX) \to (AX)$$

Assume that all parameters are word-sized.

40. What instruction does the mnemonic SHRD stand for?

41. What happens when the instruction

```
SHRD  DWORD PTR [DI], EAX, CL
```

is executed?

Section 3.7

42. Describe what happens as each of the following instructions is executed by the 80386DX.
- (a) ROL DX, CL
- (b) RCL BYTE PTR [0400H], CL
- (c) ROR BYTE PTR [DI], 1
- (d) ROR BYTE PTR [DI + BX], CL
- (e) RCR WORD PTR [BX + DI], 1
- (f) RCR WORD PTR [BX][DI] + [0010H], CL

43. Assume that the state of the 80386DX's registers and memory are as follows:

$$EAX = 00000000H$$
$$EBX = 00000010H$$
$$ECX = 00000105H$$
$$EDX = 00001111H$$
$$ESI = 00000100H$$
$$EDI = 00000200H$$
$$CF = 1$$
$$DS:100H = 0FH$$
$$DS:200H = 22H$$
$$DS:201H = 44H$$
$$DS:210H = 55H$$
$$DS:211H = AAH$$

Chap. 3 Assignment   115

$$DS:220H = AAH$$
$$DS:221H = 55H$$
$$DS:400H = AAH$$
$$DS:401H = 55H$$

just prior to execution of each of the instructions in Problem 42. What is the result produced in the destination operand by executing instructions (a) through (f)?

44. Write an instruction sequence that when executed rotates the contents of the word-wide memory location pointed to by the address in the base register left through carry by one bit position.

45. Write a program that saves the contents of bit 5 in AL as a word in BX.

Section 3.8

46. What does BTS stand for?

47. If the values in AX and CX are $F0F0_{16}$ and 0004_{16}, what is the result in AX and CF after execution of each of the following instructions?
  (a) BT   AX, CX
  (b) BTR   AX, CX
  (c) BTC   AX, CX

48. Describe the operation performed by executing the instruction BTR WORD PTR [0100H], 3.

49. If the word contents of DS:100H is $00FF_{16}$ and CF = 0 just before the instruction in Problem 48 is executed, what is the new value of the word in memory and the carry flag?

50. Write an instruction that will test bit 7 of the word storage location DS:DI in memory and save this value in the carry flag.

51. Write an instruction that will scan the double-word contents of the memory location pointed to by SI and save the index of the MSB that is logic 1 in register EAX.

4

Real-Mode 80386DX Microprocessor Programming 2

4.1 INTRODUCTION

In Chapter 3 we discussed many of the instructions that can be executed by the real-mode 80386DX microprocessor. Furthermore, we used these instructions in simple programs. In this chapter, we introduce the rest of the real-mode instruction set and at the same time cover some more complicated programming techniques. The following topics are discussed in this chapter:

1. Flag control instructions
2. Compare instruction
3. Jump instructions
4. Subroutines and subroutine handling instructions
5. The loop and loop handling instructions
6. Strings and string handling instructions

4.2 FLAG CONTROL INSTRUCTIONS

The 80386DX microprocessor has a set of flags that either monitors the status of executing instructions or controls options available in its operation. These flags were described in detail in Chapter 2. The instruction set includes a group of instructions that when executed directly affects the state of the flags. These instructions, shown in Fig. 4.1(a), are *load AH from flags* (LAHF), *store AH into*

Mnemonic	Meaning	Operation	Flags affected
LAHF	Load AH from flags	(AH) ← (Flags)	None
SAHF	Store AH into flags	(Flags) ← (AH)	SF,ZF,AF,PF,CF
CLC	Clear carry flag	(CF) ← 0	CF
STC	Set carry flag	(CF) ← 1	CF
CMC	Complement carry flag	(CF) ← $\overline{\text{CF}}$	CF
CLI	Clear interrupt flag	(IF) ← 0	IF
STI	Set interrupt flag	(IF) ← 1	IF

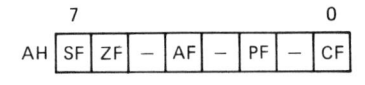

(a)

```
     7                            0
AH | SF | ZF | - | AF | - | PF | - | CF |
```

SF = Sign flag
ZF = Zero flag
AF = Auxiliary
PF = Parity flag
CF = Carry flag
— = Undefined (do not use)

(b)

Figure 4.1 (a) Flag control instructions; (b) format of the AH register for the LAHF and SAHF instructions.

flags (SAHF), *clear carry* (CLC), *set carry* (STC), *complement carry* (CMC), *clear interrupt* (CLI), and *set interrupt* (STI). A few more instructions exist that can directly affect the flags; however, we will not cover them until later in the chapter when we introduce the subroutine and string instructions.

Looking at Fig. 4.1(a), we see that the first two instructions, LAHF and SAHF, can be used either to read the flags or to change them, respectively. Notice that the data transfer that takes place is always between the AH register and the flag register. Figure 4.1(b) shows the format of the flag information in AH. Notice that bits 1, 3, and 5 are undefined. For instance, we may want to start an operation with certain flags set or reset. Assume that we want to preset all flags to logic 1. To do this, we can first load AH with $D5_{16}$ and then execute the SAHF instruction.

EXAMPLE 4.1

Write an instruction sequence to save the current contents of the 80386DX flags in memory location MEM1 and then reload the flags with the contents of memory location MEM2.

SOLUTION To save the current flags, we must first load them into the AH register and then move them to the location MEM1. The instructions that do this are

```
        LAHF
        MOV    MEM1, AH
```

Similarly, to load the flags with the contents of MEM2, we must first copy the contents of MEM2 into AH and then store the contents of AH into the flags. The instructions for this are

```
        MOV    AH, MEM2
        SAHF
```

The entire instruction sequence is shown in Fig. 4.2.

```
LAHF
MOV    MEM1,AH
MOV    AH,MEM2
SAHF
```

Figure 4.2 Instruction sequence for saving the contents of the flag register and reloading it from memory.

The next three instructions, CLC, STC, and CMC, as shown in Fig. 4.1(a), are used to manipulate the carry flag. They permit CF to be cleared, set, or complemented to its inverse logic level, respectively. For example, if CF is 1 when a CMC instruction is executed, it becomes 0.

The last two instructions are used to manipulate the interrupt flag. Executing the clear interrupt (CLI) instruction sets IF to logic 0 and disables the interrupt interface. On the other hand, executing the STI instruction sets IF to 1, and the microprocessor starts accepting interrupts from that point on.

EXAMPLE 4.2

Of the three carry flag instructions CLC, STC, and CMC, only one is really an independent instruction. That is, the operation that it provides cannot be performed by a series of the other two instructions. Determine which one of the carry instructions is the independent instruction.

SOLUTION Let us begin with the CLC instruction. The clear carry operation can be performed by an STC instruction followed by a CMC instruction. Therefore, CLC is not an independent instruction. Moreover, the operation of the set carry (STC) instruction is equivalent to the operation performed by a CLC instruction, followed by a CMC instruction. Thus STC is also not an independent instruction. On the other hand, the operation performed by the last instruction, complement carry (CMC), cannot be expressed in terms of the CLC and STC instructions. Therefore, it is the independent instruction.

4.3 COMPARE AND SET INSTRUCTIONS

An instruction is included in the instruction set of the 80386DX that can be used to compare two 8-, 16-, or 32-bit numbers. It is the *compare* (CMP) instruction of Fig. 4.3(a). Figure 4.3(b) shows that the operands can reside in a storage location

Mnemonic	Meaning	Format	Operation	Flags affected
CMP	Compare	CMP D,S	(D) − (S) is used in setting or resetting the flags	CF,AF,OF,PF,SF,ZF

(a)

Destination	Source
Register	Register
Register	Memory
Memory	Register
Register	Immediate
Memory	Immediate
Accumulator	Immediate

(b)

Figure 4.3 (a) Compare instruction; (b) allowed operands.

in memory, a register within the MPU, or as part of the instruction. For instance, a byte-wide number in a register such as BL can be compared to a second byte-wide number that is supplied as immediate data.

The result of the comparison is reflected by changes in six of the status flags of the 80386DX. Notice in Fig. 4.3(a) that it affects the overflow flag, sign flag, zero flag, auxiliary carry flag, parity flag, and carry flag. The new logic state of these flags can be used by instructions in order to make a decision whether or not to alter the sequence in which the program executes.

The process of comparison performed by the CMP instruction is basically a subtraction operation. The source operand is subtracted from the destination operand. However, the result of this subtraction is not saved. Instead, based on the result, the appropriate flags are set or reset.

The subtraction is done using 2's-complement arithmetic. For example, let us assume that the destination operand equals $10011001_2 = -103_{10}$ and the source operand equals $00011011_2 = +27_{10}$. Subtracting the source from the destination, we get

$$10011001_2 = -103_{10}$$
$$-00011011_2 = -(+27_{10})$$
$$101111110_2 = +126_{10}$$

In the process of obtaining this result, we get the following status:

1. No borrow is needed from bit 4 to bit 3; therefore, the auxiliary carry flag AF is at logic 1.
2. There is no borrow to bit 7. Thus, carry flag CF is reset.

3. Even though a borrow to bit 7 is not needed, there is a borrow from bit 7 to bit 6. This is an overflow condition and the OF flag is set.
4. There is an even number of 1s; therefore, this makes parity flag PF equal to 1.
5. Bit 7 is zero and therefore sign flag SF is at logic 0.
6. The result that is produced is nonzero, which makes zero flag ZF logic 0.

Notice that the result produced by the subtraction of the two 8-bit numbers is not correct. This condition was indicated by setting the overflow flag.

EXAMPLE 4.3

Describe what happens to the status flags as the following sequence of instructions is executed.

```
MOV   AX, 1234H
MOV   BX, ABCDH
CMP   AX, BX
```

Assume that flags ZF, SF, CF, AF, OF, and PF are all initially reset.

SOLUTION The first instruction loads AX with 1234_{16}. No status flags are affected by the execution of a MOV instruction. The second instruction puts $ABCD_{16}$ into the BX register. Again, status is not affected. Thus after execution of these two move instructions, the contents of AX and BX are

$$(AX) = 1234_{16} = 0001001000110100_2$$

and

$$(BX) = ABCD_{16} = 1010101111001101_2$$

The third instruction is a 16-bit comparison with AX representing the destination and BX the source. Therefore, the contents of BX are subtracted from that of AX.

$$(AX) - (BX) = 0001001000110100_2 - 1010101111001101_2$$

$$= 0110011001100111_2$$

The flags are either set or reset based on the result of this subtraction. Notice that the result is nonzero and positive. This makes ZF and SF equal to zero. Moreover, the carry, auxiliary carry, and no overflow conditions occur. Therefore, while CF and AF are logic 1, OF is at logic 0. Finally, the result has odd parity; therefore, PF is 0. These results are summarized in Fig. 4.4.

Byte Set On Condition: SETcc

Earlier we pointed out that the flag bits set or reset by the compare instruction are examined through software to decide whether or not branching should take place in the program. One way of using these bits is to test them directly with

Instruction	ZF	SF	CF	AF	OF	PF
Initial state	0	0	0	0	0	0
MOV AX,1234H	0	0	0	0	0	0
MOV BX,ABCDH	0	0	0	0	0	0
CMP AX,BX	0	0	1	1	0	0

Figure 4.4 Effect on flags of executing instructions.

the jump instruction. Another approach is to test them for a specific condition and then save a flag value representing whether the tested condition is true or false. An instruction that performs this operation is *byte set on condition* (SETcc). This flag value can then be used later for program branching decisions.

The *byte set on condition* (SETcc) instruction can be used to test for various states of the flags. In Fig. 4.5(a) we see that the general form of the instruction is denoted as

$$SETcc \quad D$$

Here the cc part of the mnemonic stands for a general flag relationship and must be replaced with a specific relationship when writing the instruction. Figure 4.5(b) is a list of the mnemonics that can be used to replace cc and their corresponding flag relationships. For instance, replacing cc by A gives the mnemonic SETA. This stands for *set byte if above* and tests the flags to determine if

$$(CF) = 0 \cdot (ZF) = 0$$

If these conditions are satisfied, a byte of ones is written to the register or memory location specified as the destination operand. Notice in Fig. 4.5(c) that the allowed operands are a byte-wide internal register (Reg8) or a byte storage location in memory (Mem8). On the other hand, if the conditions are not valid, a byte of zeros is written to the destination operand.

An example is the instruction

$$SETE \quad AL$$

Looking at Fig. 4.5(b), we find that execution of this instruction causes the ZF to be tested. If ZF equals 1, 11111111_2 is written into AL; otherwise, it is loaded with 00000000_2.

As another example, let us write an instruction that will load memory location EVEN_PARITY with the value FF_{16} if the result produced by the last instruction had even parity. In Fig. 4.5(b), the instruction that tests for (PF) = 1 is SETPE and making its destination operand the memory location EVEN_PARITY gives

$$SETPE \quad [EVEN\_PARITY]$$

4.4 JUMP INSTRUCTIONS

The purpose of a *jump* instruction is to alter the execution path of instructions in the program. In the 80386DX microprocessor, the code segment register and instruction pointer keep track of the next instruction to be fetched for execution. Thus a jump instruction involves altering the contents of these registers. In this

Mnemonic	Meaning	Format	Operation	Affected flags
SETcc	Byte set on condition	SETcc D	11111111 → D if cc is true 00000000 → D if cc is false	None

(a)

Instruction	Meaning	Conditions code relationship
SETA r/m8	Set byte if above	CF = 0 · ZF = 0
SETAE r/m8	Set byte if above or equal	CF = 0
SETB r/m8	Set byte if below	CF = 1
SETBE r/m8	Set byte if below or equal	CF = 1 + ZF = 1
SETC r/m8	Set if carry	CF = 1
SETE r/m8	Set byte if equal	ZF = 1
SETG r/m8	Set byte if greater	ZF = 0 + SF = OF
SETGE r/m8	Set byte if greater	SF = OF
SETL r/m8	Set byte if less	SF <> OF
SETLE r/m8	Set byte if less or equal	ZF = 1 · SF <> OF
SETNA r/m8	Set byte if not above	CF = 1
SETNAE r/m8	Set byte if not above	CF = 1
SETNB r/m8	Set byte if not below	CF = 0
SETNBE r/m8	Set byte if not below	CF = 0 · ZF = 0
SETNC r/m8	Set byte if not carry	CF = 0
SETNE r/m8	Set byte if not equal	ZF = 0
SETNG r/m8	Set byte if not greater	ZF = 1 + SF <> OF
SETNGE r/m8	Set if not greater or equal	SF <> OF
SETNL r/m8	Set byte if not less	SF = OF
SETNLE r/m8	Set byte if not less or equal	ZF = 1 · SF <> OF
SETNO r/m8	Set byte if not overflow	OF = 0
SETNP r/m8	Set byte if not parity	PF = 0
SETNS r/m8	Set byte if not sign	SF = 0
SETNZ r/m8	Set byte if not zero	ZF = 0
SETO r/m8	Set byte if overflow	OF = 1
SETP r/m8	Set byte if parity	PF = 1
SETPE r/m8	Set byte if parity even	PF = 1
SETPO r/m8	Set byte if parity odd	PF = 0
SETS r/m8	Set byte if sign	SF = 1
SETZ r/m8	Set byte if zero	ZF = 1

(b)

Source
Reg8
Mem8

(c)

Figure 4.5 (a) Byte set on condition instruction; (b) flag relationships; (c) allowed operands.

way execution continues at an address other than that of the next sequential instruction. That is, a jump occurs to another part of the program. Typically, program execution is not intended to return to the next sequential instruction after the jump instruction. Therefore, no return linkage is saved when the jump takes place.

The Unconditional and Conditional Jump

The 80386DX microprocessor allows two different types of jump instructions. They are the *unconditional jump* and the *conditional jump*. In an unconditional jump, no status requirements are imposed for the jump to occur. That is, as the instruction is executed, the jump always takes place to change the execution sequence.

This concept is illustrated in Fig. 4.6(a). Notice that when the instruction JMP AA in part I is executed, program control is passed to a point in part III identified by the label AA. Execution resumes with the instruction corresponding to AA. In this way the instructions in part II of the program have been bypassed; that is, they have been jumped over.

On the other hand, for a conditional jump instruction, status conditions that exist at the moment the jump instruction is executed decide whether or not the jump will occur. If this condition or conditions are met, the jump takes place; otherwise, execution continues with the next sequential instruction of the program. The conditions that can be referenced by a conditional jump instruction are status flags such as carry (CF), parity (PF), and overflow (OF).

Looking at Fig. 4.6(b), we see that execution of the conditional jump instruction in part I causes a test to be initiated. If the conditions of the test are not met, the NO path is taken and execution continues with the next sequential instruction. This corresponds to the first instruction in part II. However, if the result of the conditional test is YES, a jump is initiated to the segment of program identified as part III and the instructions in part II are bypassed.

Unconditional Jump Instruction

The unconditional jump instruction of the 80386DX is shown in Fig. 4.7(a) together with its valid operand combinations in Fig. 4.7(b). There are two basic kinds of unconditional jumps. The first, called an *intrasegment jump*, is limited to addresses within the current code segment. This type of jump is achieved by just modifying the value in IP. The other kind of jump, the *intersegment jump*, permits jumps from one code segment to another. Implementation of this type of jump requires modification of the contents of both CS and IP.

Jump instructions specified with a *Short-label, Near-label, Memptr16*, or *Regptr16 operand* represent intrasegment jumps. The Short-label and Near-label operands specify the jump relative to the address of the jump instruction itself. For example, in a Short-label jump instruction an 8-bit number is coded as an immediate operand to specify the *signed displacement* of the next instruction to be executed from the location of the jump instruction. When the jump instruction is executed, IP is reloaded with a new value equal to the updated value in IP, which is (IP) +2, plus the signed displacement. The new value of IP and current value in CS give the address of the next instruction to be fetched from memory. With an 8-bit displacement, the Short-label operand can only be used to initiate a jump in the range from −126 to +129 bytes from the location of the jump instruction.

On the other hand, Near-label operands specify the displacement with a 16-bit immediate operand. This corresponds to a range equal to 32K bytes forward or backward from the jump instruction. The displacement is automatically calculated by the 80386DX's assembler. Thus a programmer can use symbolic labels as operands.

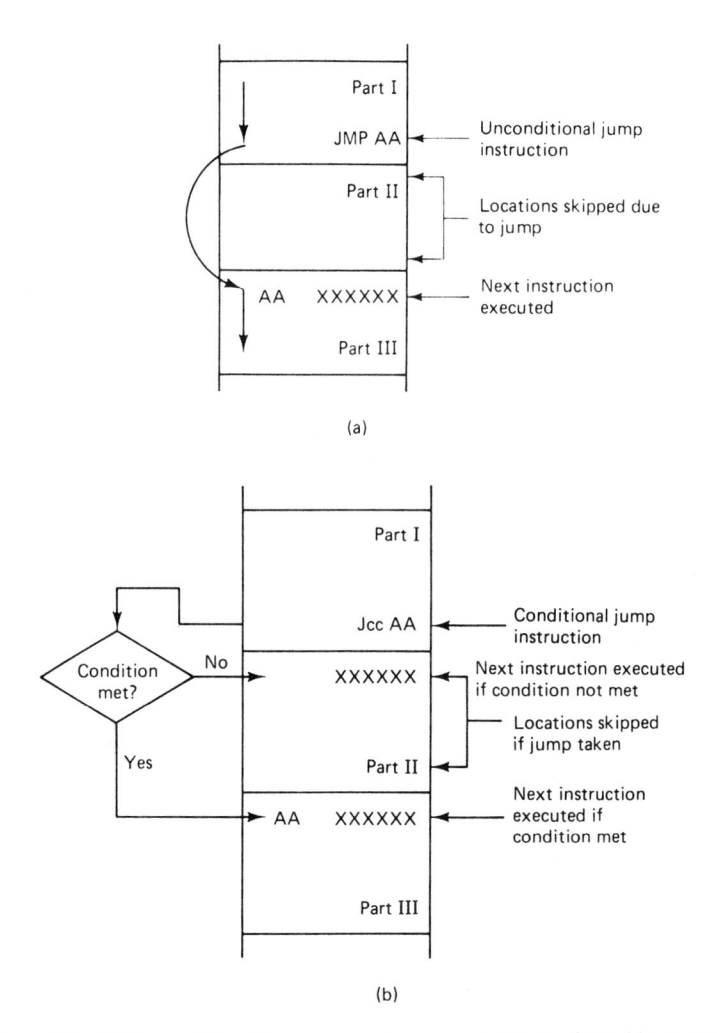

Figure 4.6 (a) Unconditional jump program sequence; (b) conditional jump program sequence.

Sec. 4.4   Jump Instructions

Mnemonic	Meaning	Format	Operation	Affected flags
JMP	Unconditional jump	JMP Operand	Jump is initiated to the address specified by the operand	None

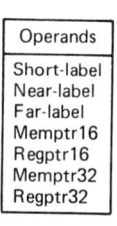

(a)

Operands
Short-label
Near-label
Far-label
Memptr16
Regptr16
Memptr32
Regptr32

(b)

Figure 4.7 (a) Unconditional jump instruction; (b) allowed operands.

An example is the instruction

JMP   LABEL

This means "jump to the point in the program corresponding to the tag LABEL." In this way the programmer does not have to worry about counting the number of bytes from the jump instruction to the location to which program control is to be passed. Moreover, the fact that it is coded as a Short- or Near-label displacement is also determined by the assembler.

The jump to address can also be specified indirectly by the contents of a memory location or the contents of a register. These two types correspond to the Memptr16 and Regptr16 operands, respectively. Just as for the Near-label operand, they both permit a jump of + or − 32K bytes from the address of the jump instruction. But in these cases, the operand is not used as a relative offset for the current value in IP. Instead, it is directly loaded into the instruction pointer register. For example,

JMP   BX

uses the contents of register BX for the new value of IP. That is, the value in BX is copied into IP. Then the physical address of the next instruction is obtained by using the current contents of CS and the new value in IP.

To specify an operand to be used as a pointer, the various addressing modes available with the 80386DX can be used. For instance,

JMP   [BX]

uses the contents of BX as the address of the memory location that contains the offset address. This offset is loaded into IP, where it is used together with the current contents of CS to compute the "jump to" address.

The intersegment unconditional jump instructions correspond to the *Far-label*, *Regptr32*, and *Memptr32 operands* that are shown in Fig. 4.7(b). Far-label uses a 32-bit immediate operand to specify the jump to address. The first 16 bits of this 32-bit pointer are loaded into IP and are an offset address relative to the contents of the code-segment register. The next 16 bits are loaded into the CS register and define the new 64K-byte code segment.

An indirect way to specify the offset and code segment address for an intersegment jump is by using the Memptr32 operand. In this case, four consecutive memory bytes starting at the specified address contain the offset address and the new code segment address. Just like the Memptr16 operand, the Memptr32 operand may be specified using any one of the various addressing modes of the 80386DX.

An example is the instruction

```
JMP    DWORD PTR [DI]
```

It uses the contents of DS and DI to calculate the address of the memory location that contains the double word pointer that identifies the location to which the jump will take place. The two-word pointer starting at this address is read into IP and CS to pass control to the new point in the program.

Conditional Jump Instruction

The second type of jump instruction is that which performs conditional jump operations. Figure 4.8(a) shows a general form of this instruction; Fig. 4.8(b) is a list of each of the conditional jump instructions in the 80386DX's instruction set. Notice that each of these instructions tests for the presence or absence of certain status conditions.

For instance, the *jump on carry* (JC) instruction makes a test to determine if carry flag (CF) is set. Depending on the result of the test, the jump to the location specified by its operand either takes place or does not. If CF equals zero, the test fails and execution continues with the instruction at the address following the JC instruction. On the other hand, if CF is set to 1, the test condition is satisfied and the jump is performed.

Notice that for some of the instructions in Fig. 4.8(b) two different mnemonics can be used. This feature can be used to improve program readability. That is, for each occurrence of the instruction in the program, it can be identified with the mnemonic that best describes its function.

For instance, the instruction *jump on parity* (JP)/*jump on parity even* (JPE) can be used to test parity flag PF for logic 1. Since PF is set to 1 if the result from a computation has even parity, this instruction can initiate a jump based on the occurrence of even parity. The reverse instruction JNP/JPO is also provided. It can be used to initiate a jump based on the occurrence of a result with odd parity instead of even parity.

In a similar manner, the instructions *jump if equal* (JE) and *jump if zero* (JZ) have the same function. Either notation can be used in a program to determine

Mnemonic	Meaning	Format	Operation	Flags Affected
Jcc	Conditional jump	Jcc Operand	If the specific condition cc is true, the jump to the address specified by the Operand is initiated; otherwise, the next instruction is executed	None

(a)

Mnemonic	Meaning	Condition
JA	above	CF = 0 and ZF = 0
JAE	above or equal	CF = 0
JB	below	CF = 1
JBE	below or equal	CF = 1 or ZF = 1
JC	carry	CF = 1
JCXZ	CX register is zero	CX = 0000H
JECXZ	ECX register is zero	ECX = 00000000H
JE	equal	ZF = 1
JG	greater	ZF = 0 and SF = OF
JGE	greater or equal	SF = OF
JL	less	(SF xor OF) = 1
JLE	less or equal	((SF xor OF) or ZF) = 1
JNA	not above	CF = 1 or ZF = 1
JNAE	not above nor equal	CF = 1
JNB	not below	CF = 0
JNBE	not below nor equal	CF = 0 and ZF = 0
JNC	not carry	CF = 0
JNE	not equal	ZF = 0
JNG	not greater	((SF xor OF) or ZF) = 1
JNGE	not greater nor equal	(SF xor OF) = 1
JNL	not less	SF = OF
JNLE	not less nor equal	ZF = 0 and SF = OF
JNO	not overflow	OF = 0
JNP	not parity	PF = 0
JNS	not sign	SF = 0
JNZ	not zero	ZF = 0
JO	overflow	OF = 1
JP	parity	PF = 1
JPE	parity even	PF = 1
JPO	parity odd	PF = 0
JS	sign	SF = 1
JZ	zero	ZF = 1

(b)

Figure 4.8 (a) Conditional jump instruction; (b) types of conditional jump instructions.

if the result of a computation was zero. All other conditional jump instructions work in a similar way except that they test different conditions to decide whether or not the jump is to take place. Examples of these conditions are that the contents of CX are zero, an overflow has occurred, or the result is negative.

To distinguish between comparisons of signed and unsigned numbers by jump instructions, two different names, which seem to be the same, have been devised. They are *above* and *below* for comparison of unsigned numbers and *less* and *greater* for comparison of signed numbers. For instance, the number $ABCD_{16}$ is above the number 1234_{16} if considered as an unsigned number. On the other hand, if they are considered as signed numbers, $ABCD_{16}$ is negative and 1234_{16} is positive. Therefore, $ABCD_{16}$ is less than 1234_{16}.

EXAMPLE 4.4

Write a program to move a block of N bytes of data starting at offset address BLK1ADDR to another block starting at offset address BLK2ADDR. Assume that both blocks are in the same data segment, whose starting point is defined by the data segment address DATASEGADDR.

SOLUTION The steps to be implemented to solve this problem are outlined in the flowchart in Fig. 4.9(a). It has four basic operations. The first operation is initialization. Initialization involves establishing the initial address of the data segment. This is done by loading the DS register with the value DATASEGADDR. Furthermore, source index register SI and destination index register DI are initialized with addresses BLK1ADDR and BLK2ADDR, respectively. In this way they point to the beginning of the source block and the beginning of the destination block, respectively. To keep track of the count, register CX is initialized with N, the number of points to be moved. This leads us to the following assembly language statements.

```
MOV   AX, DATASEGADDR
MOV   DS, AX
MOV   SI, BLK1ADDR
MOV   DI, BLK2ADDR
MOV   CX, N
```

Notice that DS cannot be directly loaded by immediate data with a MOV instruction. Therefore, the segment address was first loaded into AX and then moved to DS. SI, DI, and CX can be loaded directly with immediate data.

The next operation that must be performed is the actual movement of data from the source block of memory to the destination block. The offset addresses are already loaded into SI and DI; therefore, move instructions that employ indirect addressing can be used to accomplish the data transfer operation. Remember that the 80386DX does not allow direct memory-to-memory moves. For this reason AX will be used as an intermediate storage location for data. The source byte is moved into AX with one instruction and then another instruction is needed to move it from AX to the destination location. Thus the data move is accomplished by the following instructions.

```
        MOV    AX, DATASEGADDR
        MOV    DS, AX
        MOV    SI, BLK1ADDR
        MOV    DI, BLK2ADDR
        MOV    CX, N
NXTPT:  MOV    AH, [SI]
        MOV    [DI], AH
        INC    SI
        INC    DI
        DEC    CX
        JNZ    NXTPT
        HLT
              (b)
```

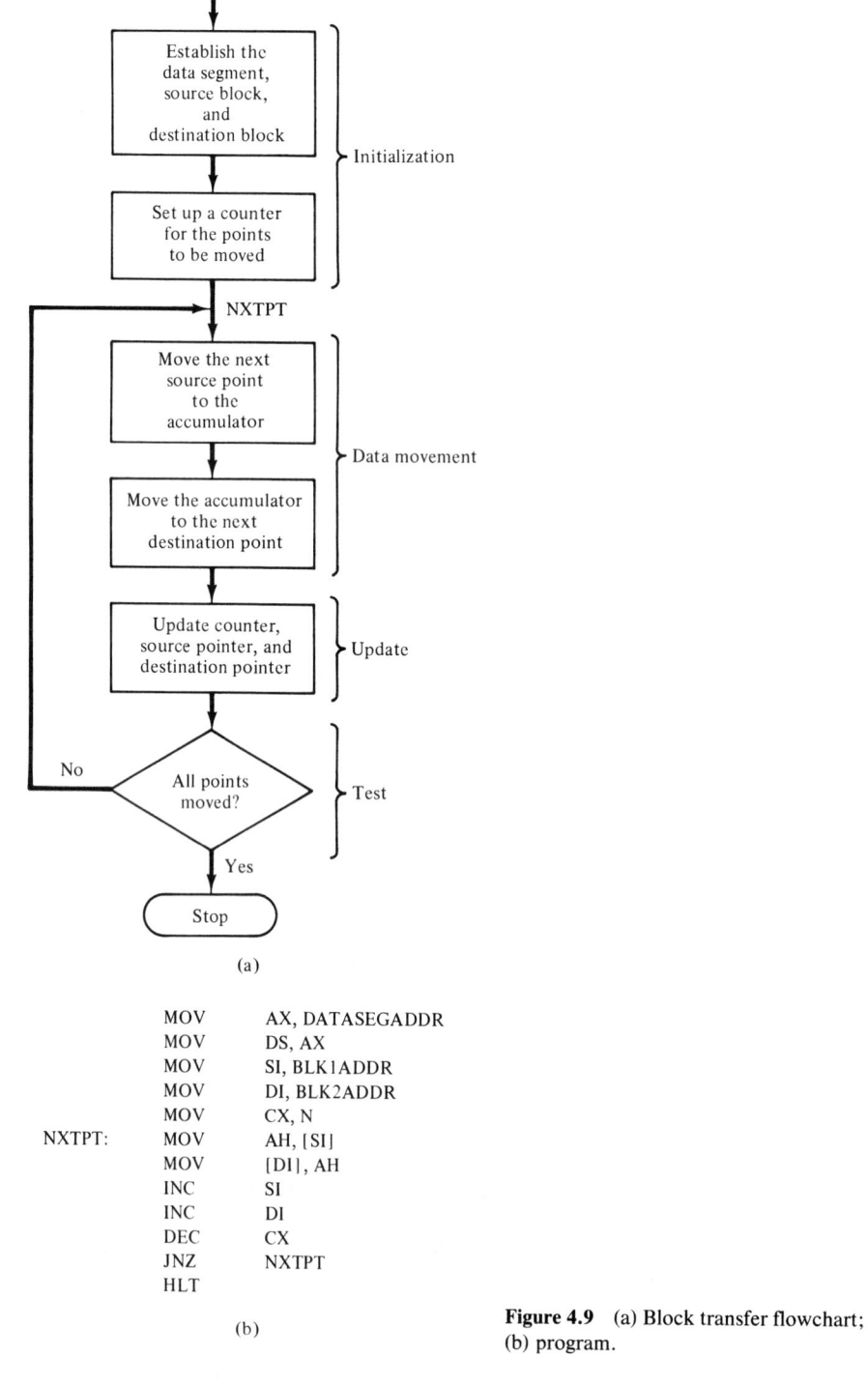

Figure 4.9 (a) Block transfer flowchart; (b) program.

```
NXTPT:  MOV   AH, [SI]
        MOV   [DI], AH
```

Notice that for a byte move only the higher eight bits of AX are used. Therefore, the operand is specified as AH instead of AX.

The next operation is to update the pointers in SI and DI so that they are ready for the next byte move. Also, the counter must be decremented so that it corresponds to the number of bytes that remain to be moved. These updates can be done by the following sequence of instructions:

```
        INC   SI
        INC   DI
        DEC   CX
```

The test operation determines whether or not all the data bytes have been moved. The contents of CX represents this condition. When its value is not zero, there still are points to be moved; whereas a value of zero indicates that the block move is complete. This zero condition is reflected by 1 in ZF. The instruction needed to perform this test is

```
        JNZ   NXTPT
```

Here NXTPT is a label that corresponds to the first instruction in the data move operation. The last instruction in the program can be a *halt* (HLT) instruction to indicate the end of the block move operation. The entire program is shown in Fig. 4.9(b).

4.5 SUBROUTINES AND THE SUBROUTINE HANDLING INSTRUCTIONS

A *subroutine* is a special segment of program that can be called for execution from any point in a program. Figure 4.10(a) illustrates the concept of a subroutine. Here we see a program structure where one part of the program is called the *main program*. In addition to this, we find a smaller segment attached to the main program, known as a subroutine. The subroutine is written to provide a function that must be performed at various points in the main program. Instead of including this piece of code in the main program each time the function is needed, it is put into the program just once as a subroutine.

Wherever the function must be performed, a single instruction is inserted into the main body of the program to "call" the subroutine. Remember that the contents of CS and IP always identifies the next instruction to be fetched for execution. Thus, to branch to a subroutine that starts elsewhere in memory, the value in either IP or CS and IP must be modified. After executing the subroutine, we want to return control to the instruction that follows the one that called the subroutine. In this way, program execution resumes in the main program at the

Main program

Call subroutine A
Next instruction
⋮
Call subroutine A
Next instruction
⋮

Subroutine A
First instruction
⋮
Return

(a)

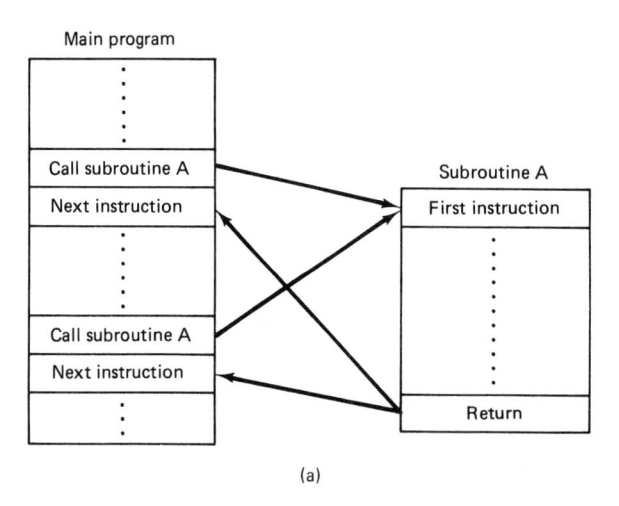

Mnemonic	Meaning	Format	Operation	Affected flags
CALL	Subroutine call	CALL Operand	Execution continues from the address of the subroutine specified by the operand. Information required to return back to the main program such as IP and CS are saved on the stack.	None

(b)

Operands
Near-proc
Far-proc
Memptr16
Regptr16
Memptr32
Regptr32

(c)

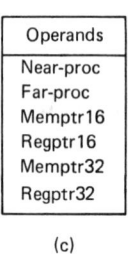

Figure 4.10 (a) Subroutine concept; (b) subroutine call instruction; (c) allowed operands.

point where it left off due to the subroutine call. A return instruction must be included at the end of the subroutine to initiate the *return sequence* to the main program environment.

The instructions provided to transfer control from the main program to a subroutine and return control back to the main program are called *subroutine handling instructions*. Let us now examine the instructions provided for this purpose.

CALL and RET Instructions

There are two basic instructions in the instruction set of the 80386DX for subroutine handling. They are the *call* (CALL) and *return* (RET) instructions. Together they provide the mechanism for calling a subroutine into operation and returning control back to the main program at its completion. We first discuss these two instructions and later introduce other instructions that can be used in conjunction with subroutines.

Just like the JMP instruction, CALL allows implementation of two types of operations, the *intrasegment call* and the *intersegment call*. The CALL instruction is shown in Fig. 4.10(b); its allowed operand variations are shown in Fig. 4.10(c).

It is the operand that initiates either an intersegment or intrasegment call. The operands Near-proc, Memptr16, and Regptr16 all specify intrasegment calls to a subroutine. In all three cases, execution of the instruction causes the contents of IP to be saved on the stack. Then the stack pointer (SP) is decremented by two. The saved value of IP is the address of the instruction that follows the CALL instruction. After saving the return address, a new 16-bit value, which points to the storage location of the first instruction in the subroutine, is loaded into IP.

The three types of intrasegment operands represent different ways of specifying this new value of IP. In a Near-proc operand, the displacement of the first instruction of the subroutine from the current value of IP is supplied directly by the instruction. An example is

```
                    CALL    NPROC
```

Here the label NPROC determines the 16-bit displacement and is coded as an immediate operand following the opcode for the call instruction. This form of call is actually a relative addressing-mode instruction; that is, the offset address is calculated relative to the address of the call instruction itself. With 16 bits, the displacement is limited to + or − 32K bytes.

The Memptr16 and Regptr16 operands provide indirect subroutine addressing by specifying a memory location or an internal register, respectively, as the source of a new value for IP. The value specified in this way is not a displacement. It is the actual offset that is to be loaded into IP. An example of the Regptr16 operand is

```
                    CALL    BX
```

When this instruction is executed, the contents of BX are loaded into IP and execution continues, with the subroutine starting at a physical address derived from CS and the new value in IP.

By using one of the various addressing modes of the 80386DX, an internal register can be used as a pointer to an operand that resides in memory. This represents a Memptr16 type of operand. In this case, the value of the physical address of the offset is obtained from the current contents of the data segment register DS and the address or addresses held in the specified registers. For instance, the instruction

                        CALL   WORD   PTR   [BX]

has its subroutine offset address at the memory location whose physical address is derived from the contents of DS and BX. The value stored at this memory location is loaded into IP. Again, the current contents of CS and the new value in IP point to the first instruction of the subroutine.

Notice that in both intrasegment call examples the subroutine was located within the same code segment as the call instruction. The other type of CALL instruction, the intersegment call, permits the subroutine to reside in another code segment. It corresponds to the Far-proc, Regptr32, and Memptr32 operands. These operands specify both a new offset address for IP and a new segment address for CS. In both cases, execution of the call instruction causes the contents of the CS and IP registers to be saved on the stack, and then new values are loaded into IP and CS. The saved values of CS and IP permit return to the main program from a different code segment.

Far-proc represents a 32-bit immediate operand that is stored in the four bytes that follow the opcode of the call instruction in program memory. These two words are loaded directly from code segment memory into IP and CS with execution of the CALL instruction. An example is the instruction

                            CALL   FPROC

On the other hand, when the operand is Memptr32, the pointer for the subroutine is stored as four bytes in data memory. The location of the double word pointer can be specified indirectly by one of the 80386DX's registers. An example is

                        CALL   DWORD PTR [DI]

Here the physical address of the four-byte pointer in memory is derived from the contents of DS and DI.

Every subroutine must end by executing an instruction that returns control to the main program. This is the return (RET) instruction. It is described in Fig. 4.11(a) and (b). Notice that its execution causes the value of IP or both the values of IP and CS that were saved on the stack to be returned back to their corresponding registers. In general, an intrasegment return results from an intrasegment call and an intersegment return results from an intersegment call.

There is an additional option with the return instruction: A two-byte offset can be included following the return instruction. This offset is added to the stack pointer after restoring the return address into IP or IP and CS. The stack pointer displacement provides a simple means for discarding the parameters saved on the stack before initiating the call to the subroutine.

Mnemonic	Meaning	Format	Operation	Affected flags
RET	Return	RET or RET Operand	Return to the main program by restoring IP (and CS for far-procedure). If operand is present it is added to the content of SP.	None

(a)

Operands
None
Disp16

(b)

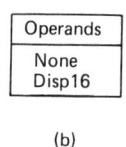

Figure 4.11 (a) Return instruction; (b) allowed operands.

PUSH and POP Instructions

After the context switch to a subroutine, we find that it is usually necessary to save the contents of certain registers or some other main program parameters. These values are saved by pushing them onto the stack. Typically, these data correspond to registers and memory locations that are used by the subroutine. In this way, their original values are kept intact in the stack segment of memory during the execution of the subroutine. Before a return to the main program takes place, the saved registers and main program parameters are restored. This is done by popping the saved values from the stack back into their original locations. Thus a typical structure of a subroutine is that shown in Fig. 4.12.

The instruction that is used to save parameters on the stack is the *push* (PUSH) instruction, and that used to retrieve them back is the *pop* (POP) instruc-

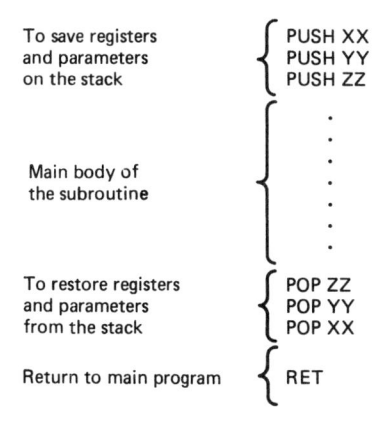

```
To save registers        ┌ PUSH XX
and parameters          ┤ PUSH YY
on the stack             └ PUSH ZZ
                         ┌ .
                         │ .
Main body of             │ .
the subroutine           ┤
                         │ .
                         │ .
                         └ .
To restore registers     ┌ POP ZZ
and parameters          ┤ POP YY
from the stack           └ POP XX

Return to main program   ┤ RET
```

Figure 4.12 Structure of a subroutine.

Sec. 4.5   Subroutines and the Subroutine Handling Instructions

Mnemonic	Meaning	Format	Operation	Flags affected
PUSH	Push word onto stack	PUSH S	((SP − 2)) ← (S) (SP) ← (SP) − 2	None
POP	Pop word off stack	POP D	(D) ← ((SP)) (SP) ← (SP) + 2	None

(a)

Operands (S or D)
Immediate (PUSH only)
Register
Seg-Reg
Memory

(b)

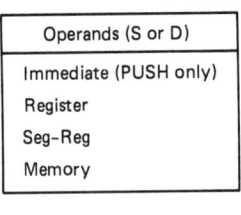

Figure 4.13 (a) Push and pop instructions; (b) allowed operands.

tion. These instructions are shown in Fig. 4.13(a). Notice in Fig. 4.13(b) that the standard PUSH and POP instructions can be written with a general-purpose register, a segment register, or a storage location in memory as their operand. The PUSH instruction also allows for use of a double-word, word-wide, or sign-extended byte immediate source operand.

Execution of a PUSH instruction causes the data corresponding to the operand to be pushed onto the top of the stack. For instance, if the instruction is

$$\text{PUSH} \quad \text{AX}$$

the result is as follows:

$$((SP) - 1) \leftarrow (AH)$$
$$((SP) - 2) \leftarrow (AL)$$
$$(SP) \leftarrow (SP) - 2$$

This shows that the two bytes of AX are saved in the stack part of memory and the stack pointer is decremented by two such that it points to the new top of the stack. On the other hand, if the instruction is

$$\text{POP} \quad \text{AX}$$

its execution results in

$$(AL) \leftarrow ((SP))$$
$$(AH) \leftarrow ((SP) + 1)$$
$$(SP) \leftarrow (SP) + 2$$

In this manner, the saved contents of AX are restored back into the register. When a 32-bit operand, such as EAX, is pushed to or popped from the stack, the value in the stack pointer register is decremented or incremented by four, respectively.

EXAMPLE 4.5

Write a procedure named SUM that adds 31_{16} to the memory location TOTAL. Their sum is to be formed in DX. Assume that this procedure is to be called from another procedure in the same code segment and that at the time it is to be called, DX contains the value $ABCD_{16}$ and this value must be saved at entry of the procedure and restored at its completion.

SOLUTION The beginning of the procedure is defined with the pseudo-op statement

```
SUM    PROC NEAR
```

At entry of the procedure, we must save the value currently held in DX. This is done by pushing its contents to the stack with the instruction

```
PUSH   DX
```

Now we load DX with the value at TOTAL using the instruction

```
MOV    DX, TOTAL
```

Add 31_{16} to it with the instruction

```
ADD    DX, 31H
```

and then place the result back in TOTAL.

```
MOVE   TOTAL, DX
```

This completes the addition operation; but before we return to the main part of the program, the original contents of DX that were saved on the stack are restored with the pop instruction

```
POP    DX
```

Then a return instruction is used to pass control back to the main program.

```
RET
```

The procedure must be terminated with the end procedure pseudo-op statement that follows:

```
SUM    ENDP
```

The complete instruction sequence is shown in Fig. 4.14.

```
SUM     PROC    NEAR
        PUSH    DX
        MOV     DX,TOTAL
        ADD     DX,31H
        MOV     TOTAL,DX
        POP     DX
        RET
SUM     ENDP
```

Figure 4.14 Program for Example 4.5.

At times, we also want to save the contents of the flag register; if saved, we can later restore them. These operations can be accomplished with the *push flags* (PUSHF) and *pop flags* (POPF) instructions, respectively. These instructions are shown in Fig. 4.15. Notice that PUSHF saves the contents of the flag register on the top of the stack. On the other hand, POPF restores the flags from the top of the stack.

Mnemonic	Meaning	Operation	Flags affected
PUSHF	Push flags onto stack	((SP−2)) ← (Flags) (SP) ← (SP)−2	None
POPF	Pop flags from stack	(Flags) ← ((SP)) (SP) ← (SP)+2	OF, DF, IF, TF, SF, ZF, AF, PF, CF

Figure 4.15 Push flags and pop flags instructions.

When writing the compiler for a high-level language like C, it is very common to push the contents of all of the general registers of the 80386DX to the stack before calling a subroutine. If we use the PUSH instruction to perform this operation, many instructions would need to be coded. To simplify this operation, special instructions are provided in the instruction set of the 80386DX. They are called *push all* (PUSHA) and *pop all* (POPA).

Looking at Fig. 4.16 we see that execution of PUSHA causes the values in AX, CX, DX, BX, OLD SP, BP, SI, and DI to be pushed in that order onto the top of the stack. Figure 4.17 shows the state of the stack before and after execution of the instruction. As shown in Fig. 4.18, executing a POPA instruction at the end of the subroutine restores the old state of the 80386DX.

It is also possible to push the contents of all of the 80386DX's 32-bit registers to the stack with a single instruction. Looking at Fig. 4.16, we see that this is done with the PUSHAD instruction. The contents of these registers can be restored from the stack by executing a POPAD instruction.

Mnemonic	Meaning	Operation	Flags affected
PUSHA	Push all 16-bit general registers onto stack	((SP − 2)) ← (AX) ((SP − 4)) ← (CX) ((SP − 6)) ← (DX) ((SP − 8)) ← (BX) ((SP − A)) ← (OLD SP) ((SP − C)) ← (BP) ((SP − E)) ← (SI) (SP − 10)) ← (DI)	None
PUSHAD	Push all 32-bit general registers onto stack	((SP − 4)) ← (EAX) ((SP − 8)) ← (ECX) ((SP − C)) ← (EDX) ((SP − 10)) ← (EBX) ((SP − 14)) ← (OLD ESP) ((SP − 18)) ← (EBP) ((SP − 1C)) ← (ESI) ((SP − 20)) ← (EDI)	None
POPA	Pop all 16-bit general registers from stack	(DI) ← ((SP)) (SI) ← ((SP + 2)) (BP) ← ((SP + 4)) (OLD SP) ← ((SP + 6)) (BX) ← ((SP + 8)) (DX) ← ((SP + A)) (CX) ← ((SP + C)) (AX) ← ((SP + E))	None
POPAD	Pop all 32-bit general registers from stack	(EDI) ← ((SP)) (ESI) ← ((SP + 4)) (EBP) ← ((SP + 8)) (OLD ESP) ← ((SP + C)) (EBX) ← ((SP + 10)) (EDX) ← ((SP + 14)) (ECX) ← ((SP + 18)) (EAX) ← ((SP + 1C))	

Figure 4.16 Push all and pop all instructions.

Stack Frame Instructions: ENTER and LEAVE

Before the main program calls a subroutine, it is often necessary for the calling program to pass the values of some parameters to the subroutine. It is a common practice to push these parameters onto the stack before calling the routine. Then during the execution of the subroutine, they are accessed by reading them from the stack and used in computations. Two instructions are provided in the instruction set of the 80386DX to allocate and deallocate a data area called a *stack frame*. This data area, which is located in the stack part of memory, is used for local storage of parameters and other data.

Normally, high-level languages allocate a stack frame for each procedure in a program. The stack frame provides a dynamically allocated local storage space for the procedure and contains data such as variables, pointers to the stack frames of the previous procedures from which the current procedure was called, and a return address for linkage to the stack frame of the calling procedure. This mechanism also permits access to the data in stack frames of the calling procedures.

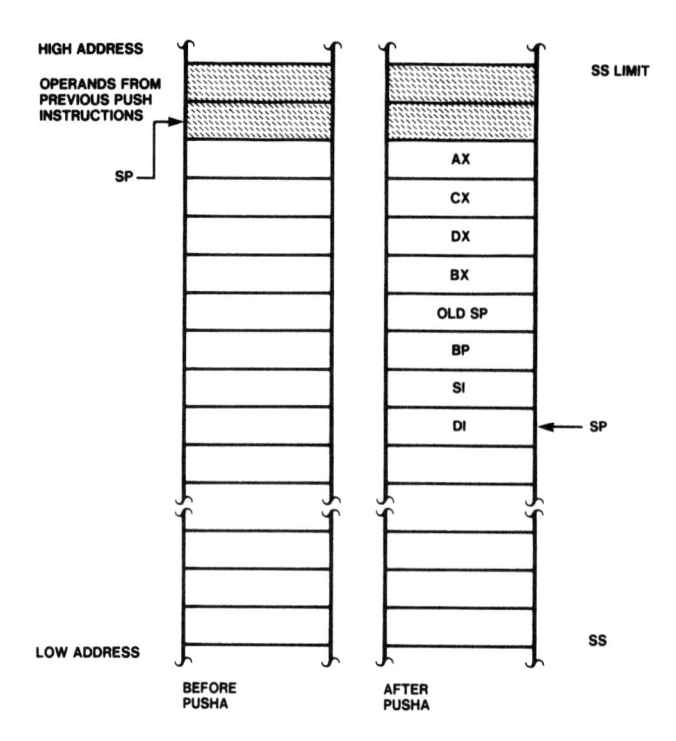

Figure 4.17 State of the stack before and after executing PUSHA. (Reprinted by permission of Intel Corp. Copyright Intel Corp. 1987.)

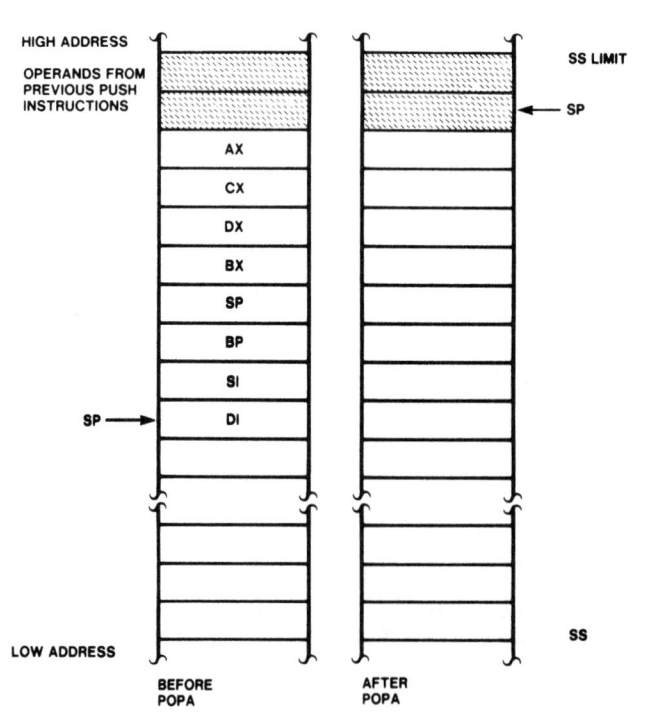

Figure 4.18 State of the stack before and after executing POPA. (Reprinted by permission of Intel Corp. Copyright Intel Corp. 1987.)

Mnemonic	Meaning	Format	Operation

ENTER	Make stack frame	ENTER Imm16,Imm8	Push BP
			TEMP ← (SP)
			If Imm8 > 0 then
			    Repeat (Imm8 − 1) times
			(BP) ← (BP) − 2
			Push ((BP))
			End repeat
			PUSH TEMP
			End if
			(BP) ← TEMP
			(SP) ← (SP) − Imm16

LEAVE	Release stack frame	LEAVE	(SP) ← (BP) Pop to BP

Figure 4.19 Enter and leave instructions.

The instructions used for allocation and deallocation of stack frames are given in Fig. 4.19 as *enter* (ENTER) and *leave* (LEAVE). Execution of an ENTER instruction allocates a stack frame and it is deallocated by executing LEAVE. For this reason, as shown in Fig. 4.20, the ENTER instruction is used at the beginning of a subroutine and LEAVE at the end just before the return instruction.

Looking at Fig. 4.19, we find that the ENTER instruction has two operands. The first operand, identified as Imm16, is a word-size immediate operand. This operand specifies the number of bytes to be allocated on the stack for local data storage of the procedure. The second operand, Imm8, which is a byte-size immediate operand, specifies what is called the *lexical nesting level* of the routine. This lexical level defines how many pointers to previous stack frames can be stored in the current stack frame. The value of the lexical level byte must be limited to 0 through 31 in 80386DX programs. This list of previous stack frame pointers is called a *display*.

An example of an ENTER instruction is

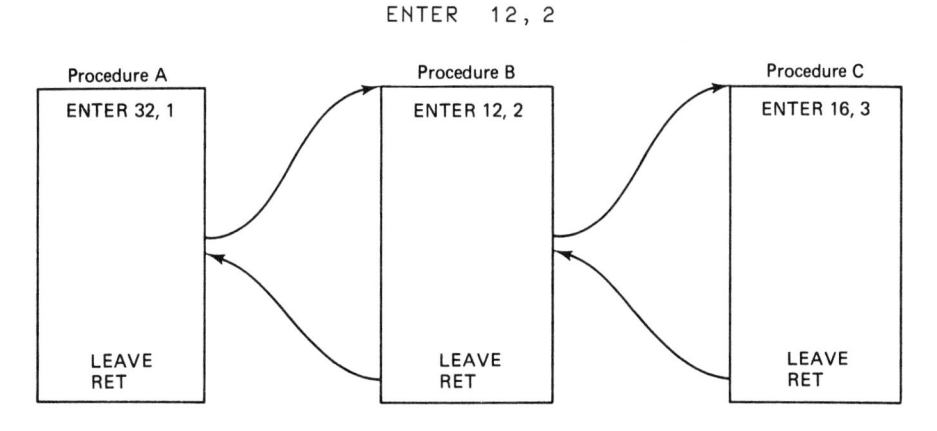

ENTER 12, 2

Procedure A	Procedure B	Procedure C
ENTER 32, 1	ENTER 12, 2	ENTER 16, 3
LEAVE RET	LEAVE RET	LEAVE RET

Figure 4.20 Enter/leave example.

Execution of this instruction allocates 12 bytes of local storage on the stack for use as a stack frame. It does this by decrementing SP by 0CH. This defines a new top of stack at address SS:SP-0CH. Also the base pointer (BP) that identifies the beginning of the previous stack frame is copied into the stack frame created by the ENTER instruction. This value is called the *dynamic link* and is held in the first storage location of the stack frame. The number of stack frame pointers that can be saved in a stack frame is equal to the value of the byte that specifies the lexical level of the procedure. Therefore, in our example, just two levels of nesting are permitted.

The BP register is used as a pointer into the stack segment of memory. When a procedure is called, the value in BP points to the stack frame location that contains the previous stack frame pointer (dynamic link). Therefore, based-indexed addressing can be used to access variables in the stack frame by referencing the BP register.

The LEAVE instruction reverses the process of an ENTER instruction. That is, its execution deallocates the stack frame. This is done by first automatically loading SP from the BP register. This returns the storage locations of the current stack frame to the stack. Now SP points to the location where the dynamic link (pointer to the previous stack frame) is stored. Next, popping the contents of the stack into BP returns the pointer to the stack frame of the previous procedure.

To illustrate the operation of the stack frame instructions, let us consider the example of Fig. 4.20. Here we find three procedures. Procedure A is used to call procedure B, which in turn calls procedure C. It is assumed that the lexical levels for these procedures are 1, 2, and 3, respectively. The ENTER, LEAVE, and RET instructions for each procedure are shown in the diagram. Notice that the ENTER instructions specify the lexical levels for the procedures.

The stack frames created by executing the ENTER instructions in the three procedures are shown in Fig. 4.21. As the ENTER instruction in procedure A is executed, the old BP from the procedure that called procedure A is pushed onto the stack. BP is loaded from SP to point to the location of the old BP. Since the second operand is 1, only the current BP, that is the BP for procedure A, is pushed onto the stack. Finally, to allocate 32 bytes for the stack frame, 20H is subtracted from the current value in SP.

After entering procedure B, a second ENTER instruction is encountered. This time the lexical level is 2. The instruction first pushes the old BP, that is the BP for procedure A, onto the stack, then pushes the BP previously stored on the stack frame for A to the stack, and last it pushes the current BP for procedure B to the stack. This mechanism provides access to the stack frame for procedure A from procedure B. Next 12 bytes, as specified by the instruction, are allocated for local storage.

The ENTER instruction in procedure C is next executed. This instruction has the lexical level of 3 and therefore it pushes the BPs for the two previous procedures, that is, B and A to the stack, in addition to the BP for C and the BP for the calling procedure B.

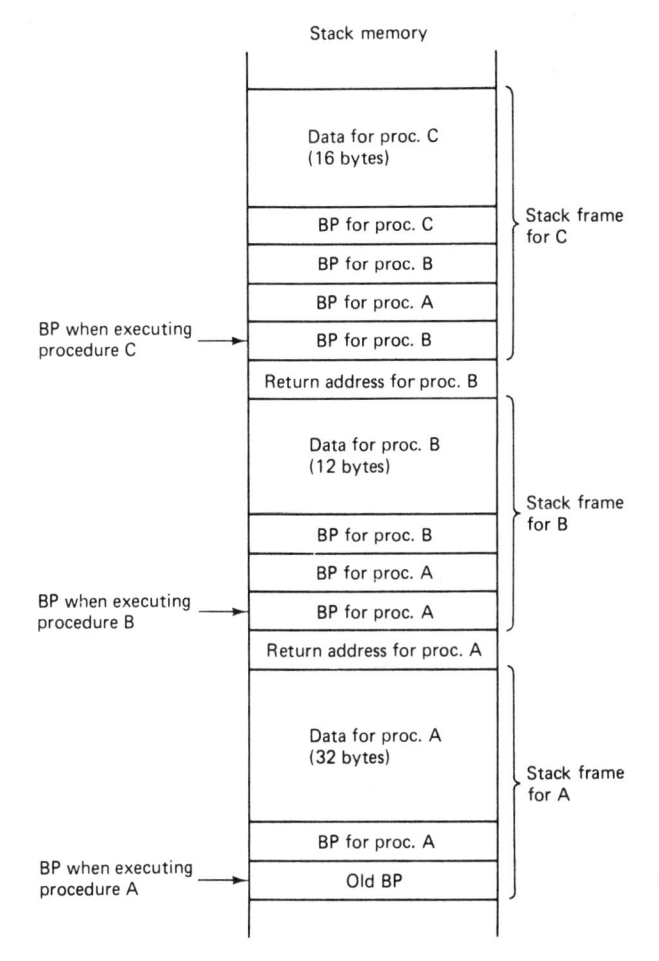

Figure 4.21 Stack after execution of ENTER instructions for procedures A, B, and C.

4.6 THE LOOP AND THE LOOP HANDLING INSTRUCTIONS

The 80386DX microprocessor has three instructions specifically designed for implementing *loop operations*. These instructions can be used in place of certain conditional jump instructions and give the programmer a simpler way of writing loop sequences. The loop instructions are listed in Fig. 4.22.

The first instruction, *loop* (LOOP), works with respect to the contents of the CX register. CX must be preloaded with a count that represents the number of times the loop is to be repeated. Whenever LOOP is executed, the contents of CX are first decremented by one and then checked to determine if they are equal to zero. If equal to zero, the loop is complete and the instruction following LOOP is executed; otherwise, control is returned to the instruction at the label

Mnemonic	Meaning	Format	Operation
LOOP	Loop	LOOP Short-label	$(CX) \leftarrow (CX) - 1$ Jump is initiated to location defined by Short-label if $(CX) \neq 0$; otherwise, execute next sequential instruction.
LOOPE/LOOPZ	Loop while equal/loop while zero	LOOPE/LOOPZ Short-label	$(CX) \leftarrow (CX) - 1$ Jump to location defined by Short-label if $(CX) \neq 0$ and $(ZF) = 1$; otherwise, execute next sequential instruction.
LOOPNE/LOOPNZ	Loop while not equal/loop while not zero	LOOPNE/LOOPNZ Short-label	$(CX) \leftarrow (CX) - 1$ Jump to location defined by Short-label if $(CX) \neq 0$ and $(ZF) = 0$; otherwise, execute next sequential instruction.

Figure 4.22 Loop instructions.

specified in the loop instruction. In this way, we see that LOOP is a single instruction that functions the same as a decrement CX instruction followed by a JNZ instruction.

For example, the LOOP instruction sequence shown in Fig. 4.23(a) will cause the part of the program from the label NEXT through the instruction LOOP to be repeated a number of times equal to the value of count stored in CX. For

```
NEXT:   MOV   CX,COUNT        Load count for the number of repeats
          .
          .
          .                   Body of routine that is repeated
          .
          .
        LOOP  NEXT             Loop back to label NEXT if count not zero
```

(a)

```
                MOV   AX,DATASEGADDR
                MOV   DS,AX
                MOV   SI,BLK1ADDR
                MOV   DI,BLK2ADDR
                MOV   CX,N
       NXTPT:   MOV   AH,[SI]
                MOV   [DI],AH
                INC   SI
                INC   DI
                LOOP  NXTPT
                HLT
```

(b)

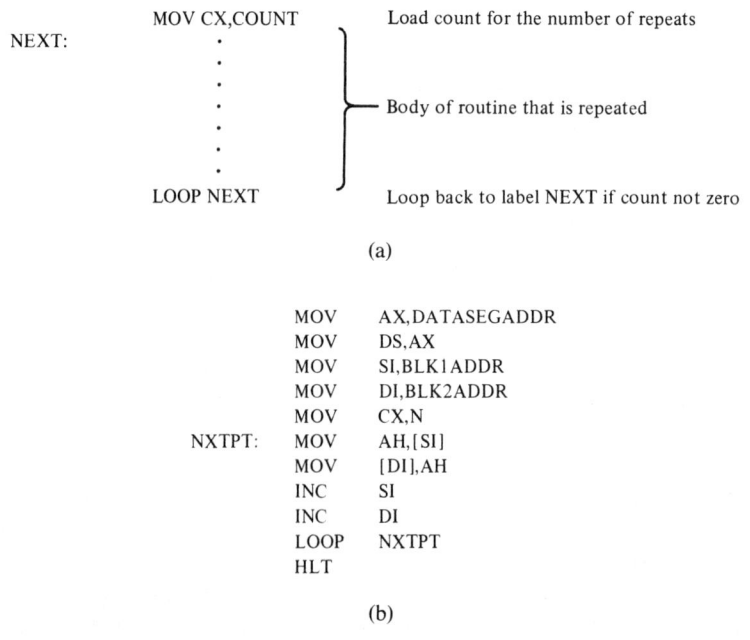

Figure 4.23 (a) Typical loop routine structure; (b) block move program employing the LOOP instruction.

example, if CX contains 000A$_{16}$, the sequence of instructions included in the loop is executed 10 times.

The other two instructions in Fig. 4.22 operate in a similar way except that they check for two conditions. For instance, the instruction *loop while equal* (LOOPE)/*loop while zero* (LOOPZ) checks the contents of both CX and the ZF flag. Each time the loop instruction is executed, CX decrements by 1 without affecting the flags, its contents are checked for 0, and the state of ZF that results from execution of the previous instruction is tested for 1. If CX is not equal to zero and ZF equals 1, a jump is initiated to the location specified with the Short-label operand and the loop continues. If either CX or ZF is zero, the loop is complete and the instruction following the loop instruction is executed.

Instruction *loop while not equal* (LOOPNE)/*loop while not zero* (LOOPNZ) works in a similar way to the LOOPE/LOOPZ instruction. The difference is that it checks ZF and CX looking for ZF equal to zero together with CX not equal to zero. If these conditions are met, the jump back to the location specified with the Short-label operand is performed and the loop continues.

Figure 4.23(b) shows a practical implementation of a loop. Here we find the block move program that was developed in Example 4.4 rewritten using the LOOP instruction. Comparing this program with the one in Fig. 4.9(b), we see that the instruction LOOP NXTPT has replaced both the DEC and JNZ instructions.

EXAMPLE 4.6

You are given the following sequence of instructions:

```
               MOV    DL, 05H
               MOV    AX, 0A00H
               MOV    DS, AX
               MOV    SI, 0H
               MOV    CX, 0FH
AGAIN:         INC    SI
               CMP    [SI], DL
               LOOPNE AGAIN
```

Explain what happens as they are executed.

SOLUTION The first five instructions are for initializing internal registers. Data register DL is loaded with 05$_{16}$; data segment register DS is loaded via AX with the value 0A00$_{16}$; source index register SI is loaded with 0000$_{16}$; and count register CX is loaded with 0F$_{16}$ (15$_{10}$). After initialization, a data segment is set up at address 0A000$_{16}$ and SI points to the memory location at address 0000$_{16}$ in this data segment. Moreover, DL contains the data 5$_{10}$ and the CX register contains the loop count 15$_{10}$.

The part of the program that starts at the label AGAIN and ends with the LOOPNE instruction is a software loop. The first instruction in the loop increments SI by one. Therefore, the first time through the loop SI points to the memory address A001$_{16}$. The next instruction compares the contents of this memory location with the contents of DL, which are 5$_{10}$. If the data held at A001$_{16}$ are 5$_{10}$, the zero flag is set; otherwise, it is left at logic 0. The LOOPNE instruction decrements CX (mak-

Sec. 4.6   The Loop and the Loop Handling Instructions

ing it E_{16}) and then checks for CX = 0 or ZF = 1. If neither of these two conditions is satisfied, program control is returned to the instruction with the label AGAIN. This causes the comparison to be repeated for the examination of the contents of the next byte in memory. On the other hand, if either condition is satisfied, the loop is complete. In this way, we see that the loop is repeated until either a number 5_{16} is found or all locations in the address range $A001_{16}$ through $A00F_{16}$ have been tested and all are found not to contain 5_{16}.

4.7 STRINGS AND THE STRING HANDLING INSTRUCTIONS

The 80386DX microprocessor is equipped with special instructions to handle *string operations*. By *string* we mean a series of data bytes, words, or double words that reside in consecutive memory locations. The string instructions of the 80386DX permit a programmer to implement operations such as to move data from one block of memory to a block elsewhere in memory. A second type of operation that is easily performed is to scan a string of data elements stored in memory looking for a specific value. Other examples are to compare the elements of two strings to determine whether they are the same or different, and to initialize a group of consecutive memory locations. Complex operations such as these typically require several nonstring instructions to be implemented.

There are five basic string instructions in the instruction set of the 80386DX. These instructions, as listed in Fig. 4.24, are *move string* (MOVS), *compare strings* (CMPS), *scan string* (SCAS), *load string* (LODS), and *store string* (STOS). They are called the *basic string instructions* because each defines an operation for one element of a string. Thus these operations must be repeated to handle a string of more than one element. Let us first look at the basic operations performed by these instructions.

Mnemonic	Meaning	Format	Operation	Flags affected
MOVSB/ MOVSW/ MOVSD	Move string	MOVSB MOVSW MOVSD	((ES)0 + (DI) ← ((DS)0 + (SI)) (SI) ← (SI) ± 1, 2, or 4 (DI) ← (DI) ± 1, 2, or 4	None
CMPSB/ CMPSW/ CMPSD	Compare strings	CMPSB CMPSW CMPSD	Set flags as per ((DS)0 + (SI)) − ((ES)0 + (DI)) (SI) ← (SI) ± 1, 2, or 4 (DI) ← (DI) ± 1, 2, or 4	CF, PF, AF, ZF, SF, OF
SCASB/ SCASW/ SCASD	Scan string	SCASB SCASW SCASD	Set flags as per (AL, AX, or EAX) − ((ES)0 + (DI)) (DI) ← (DI) ± 1, 2, or 4	CF, PF, AF, ZF, SF, OF
LODSB/ LODSW/ LODSD	Load string	LODSB LODSW LODSD	(AL, AX, or EAX) ← ((DS)0 + (SI) (SI) ← (SI) ± 1, 2, or 4	None
STOSB/ STOSW/ STOSD	Store string	STOSB STOSW STOSD	((ES)0 + (DI) ← (AL, AX, or EAX) ± 1, 2, or 4 (DI) ← (DI) ± 1, 2, or 4	None

Figure 4.24 Basic string instructions.

Move String: MOVSB/MOVSW/MOVSD

The instructions MOVSB, MOVSW, and MOVSD all perform the same basic operation. An element of the string specified by the source index (SI) register with respect to the current data segment (DS) register is moved to the location specified by the destination index (DI) register with respect to the current extra segment (ES) register. The move can be performed on a byte, word, or double word of data. After the move is complete, the contents of both SI and DI are automatically incremented or decremented by one for a byte move, by two for a word move, or by four for a double-word move. Remember the fact that the address pointers in SI and DI increment or decrement depending on how the direction flag (DF) is set.

For example, to move a byte, the instruction

```
                    MOVSB
```

is used.

An example of a program that uses MOVSB is shown in Fig. 4.25. This program is a modified version of the block move program of Fig. 4.23(b). Notice that the two MOV instructions that performed the data transfer and INC instructions that update the pointers have been replaced with one move string byte instruction. We have also made DS equal to ES.

```
           MOV    AX,DATASEGADDR
           MOV    DS,AX
           MOV    ES,AX
           MOV    SI,BLK1ADDR
           MOV    DI,BLK2ADDR
           MOV    CX,N
           CLD
  NXTPT:   MOVSB
           LOOP   NXTPT
           HLT
```

Figure 4.25 Block move program using the move string instruction.

Compare Strings and Scan Strings: CMPSB/CMPSW/CMPSD and SCASB/SCASW/SCASD

The compare strings instruction can be used to compare two elements in the same or different strings. It subtracts the destination operand from the source operand and adjusts flags CF, PF, AF, ZF, SF, and OF accordingly. The result of subtraction is not saved; therefore, the operation does not affect the operands in any way.

An example of a compare strings instruction for bytes of data is

```
                    CMPSB
```

Again, the source element is pointed to by the address in SI with respect to the current value in DS and the destination element is specified by the contents of DI relative to the contents of ES. When executed, the operands are compared, the flags are adjusted, and both SI and DI are updated such that they point to the next elements in their respective strings.

The scan string instruction is similar to compare strings; however, it compares the byte, word, or double-word element of the destination string at the physical address derived from DI and ES to the contents of AL, AX, or EAX, respectively. The flags are adjusted based on this result and DI incremented or decremented.

Figure 4.26 shows a program that reimplements the string scan operation described in Example 4.6 using the SCASB instruction. Again, we have made DS equal to ES.

```
                MOV     AX, 0H
                MOV     DS,AX
                MOV     ES,AX
                MOV     AL,05H
                MOV     DI,A000H
                MOV     CX,0FH
                CLD
        AGAIN:  SCASB
                LOOPNE  AGAIN
        NEXT:
```

Figure 4.26 Block scan operation using the SCAS instruction.

Load and Store Strings: LODSB/LODSW/LODSD and STOSB/STOSW/STOSD

The last two instructions in Fig. 4.24, load string and store string, are specifically provided to move string elements between the accumulator and memory. LODSB loads a byte from a string in memory into AL. The address in SI is used relative to DS to determine the address of the memory location of the string element. Similiarly, the instruction

```
                LODSW
```

indicates that the word string element at the physical address derived from DS and SI is to be loaded into AX. Then the index in SI is automatically incremented by 2.

On the other hand, STOSB stores a byte from AL into a string location in memory. This time the contents of ES and DI are used to form the address of the storage location in memory. For example, the program in Fig. 4.27 will load the block of memory locations from $0A000_{16}$ through $0A00F_{16}$ with number 5.

STOSD would be used to store the double word from EAX into a 32-bit string location in memory.

```
                    MOV     AX,0H
                    MOV     DS,AX
                    MOV     ES,AX
                    MOV     AL,05H
                    MOV     DI,A000H
                    MOV     CX,0FH
                    CLD
           AGAIN:   STOSB
                    LOOPNE  AGAIN
           NEXT:
```

Figure 4.27 Initializing a block of memory with a store string operation.

Repeat String: REP

In most applications the basic string operations must be repeated in order to process arrays of data. This is done by inserting a repeat prefix before the instruction that is to be repeated. The *repeat prefixes* of the 80386DX are shown in Fig. 4.28.

Prefix	Used with	Meaning
REP	MOVS STOS	Repeat while not end of string $CX \neq 0$
REPE/REPZ	CMPS SCAS	Repeat while not end of string and strings are equal $CX \neq 0$ and $ZF = 1$
REPNE/REPNZ	CMPS SCAS	Repeat while not end of string and strings are not equal $CX \neq 0$ and $ZF = 0$

Figure 4.28 Prefixes for use with the basic string instructions.

The first prefix, REP, causes the basic string operation to be repeated until the contents of register CX become equal to zero. Each time the instruction is executed, it causes CX to be tested for 0. If CX is found not to be zero, it is decremented by one and the basic string operation is repeated. On the other hand, if it is zero, the repeat string operation is finished and the next instruction in the program is executed. The repeat count must be loaded into CX prior to executing the repeat string instruction. Figure 4.29 is the memory initialization routine of Fig. 4.27 modified by using the REP prefix.

```
                    MOV     AX,0H
                    MOV     DS,AX
                    MOV     ES,AX
                    MOV     AL,05H
                    MOV     DI,A000H
                    MOV     CX,0FH
                    CLD
                    REPSTOSB
           NEXT:
```

Figure 4.29 Initializing a block of memory by repeating the STOSB instruction.

The prefixes REPE and REPZ stand for the same function. They are meant for use with the compare and scan string instructions. With REPE/REPZ, the basic compare or scan operation can be repeated as long as both the contents of CX are not equal to zero and the zero flag is one. The first condition, CX not equal to zero, indicates that the end of the string has not yet been reached; the second condition, ZF = 1, indicates that the elements that were compared are equal.

The last prefix, REPNE/REPNZ, works similarly to REPE/REPZ except that now the operation is repeated as long as CX is not equal to zero and ZF is zero. That is, the comparison or scanning is to be performed as long as the string elements are unequal and the end of the string is not yet found.

Autoindexing for String Instructions

Earlier we pointed out that during the execution of a string instruction the address indices in SI and DI are either automatically incremented or decremented. Moreover, we indicated that the decision to increment or decrement is made based on the setting of the direction flag DF. The 80386DX provides two instructions, clear direction flag (CLD) and set direction flag (STD), to permit selection between *autoincrement* and *autodecrement modes* of operation. These instructions are shown in Fig. 4.30. When CLD is executed, DF is set to zero. This selects autoincrement mode, and each time a string operation is performed SI and/or DI are incremented by one if byte data are processed, by two if word data are processed, and by four if double-word data are processed.

Mnemonic	Meaning	Format	Operation	Affected flags
CLD	Clear DF	CLD	(DF) $\leftarrow$ 0	DF
STD	Set DF	STD	(DF) $\leftarrow$ 1	DF

Figure 4.30 Instructions for selection of autoincrementing and autodecrementing in string instructions.

EXAMPLE 4.7

Describe what happens as the following sequence of instructions is executed.

```
CLD
MOV   AX, DATA_SEGMENT
MOV   DS, AX
MOV   AX, EXTRA_SEGMENT
MOV   ES, AX
MOV   CX, 20H
MOV   SI, OFFSET MASTER
MOV   DI, OFFSET COPY
REPMOVSB
```

SOLUTION The first instruction clears the direction flag and selects autoincrement mode of operation for string addressing. The next two instructions initialize DS with the value DATA_SEGMENT. It is followed by two instructions that load ES with the value EXTRA_SEGMENT. Then the number of repeats, 20_{16}, is loaded into CX. The next two instructions load SI and DI with beginning offset addresses MASTER and COPY for the source and destination strings. Now we are ready to perform the string operation. Execution of REPMOVSB moves a block of 32 consecutive bytes from the block of memory locations starting at offset address MASTER with respect to the current data segment (DS) to a block of locations starting at offset address COPY with respect to the current extra segment (ES).

ASSIGNMENT

Section 4.2

1. Explain what happens when the instruction sequence

    ```
    LAHF
    MOV    [BX + DI], AH
    ```

   is executed.

2. What operation is performed by the following instruction sequence?

    ```
    MOV    AH, [BX + SI]
    SAHF
    ```

3. Which instruction should be executed to assure that the carry flag is in the CY state? The NC state?
4. Which instruction when executed disables the interrupt interface?
5. Write an instruction sequence to configure the 80386DX as follows: interrupts not accepted; save the original contents of flags SF, ZF, AF, PF, and CF at the address $A000_{16}$; and then clear CF.

Section 4.3

6. Describe the difference in operation and the effect on status flags due to the execution of the subtract words and compare words instructions.
7. Describe the operation performed by each of the following instructions.
    (a) CMP   [0100H], AL
    (b) CMP   AX, [SI]
    (c) CMP   DWORD PTR [DI], 1234H
8. What is the state of the 80386DX's flags after executing the instructions in Problem 7(a) through (c)? Assume that the following initial state exists before executing the instructions.

$$EAX = 00008001H$$
$$ESI = 00000200H$$
$$EDI = 00000300H$$
$$DS:100H = F0H$$
$$DS:200H = F0H$$
$$DS:201H = 01H$$
$$DS:300H = 34H$$
$$DS:301H = 12H$$
$$DS:302H = 00H$$
$$DS:303H = 00H$$

9. What happens to the ZF and CF status flags as the following sequence of instructions is executed? Assume that they are both initially cleared.

```
MOV   BX,1111H
MOV   AX,BBBBH
CMP   BX,AX
```

10. What does the mnemonic SETNC stand for?
11. What flag condition does the instruction SETNZ test for?
12. Which byte set on condition instruction tests for the conditional relationship CF = 1 + ZF = 1?
13. Describe the operation performed by the instruction

```
SETO   OVERFLOW
```

14. Write an instruction that will load FF_{16} into memory location NEGATIVE if the result of the preceding arithmetic instruction was a negative number.

Section 4.4

15. What is the primary difference between the unconditional jump instruction and conditional jump instruction?
16. Which registers have their contents changed during an intrasegment jump? An intersegment jump?
17. How large is a Short-label displacement? A Near-label displacement? A Memptr16 operand? A Regptr32 operand?
18. Is a Far-label used to initiate an intrasegment jump or an intersegment jump?
19. Identify the type of jump, the type of operand, and operation performed by each of the following instructions.
    (a) JMP   10H
    (b) JMP   1000H
    (c) JMP   WORD PTR [SI]

20. If the following state of the 80386DX exists before executing each of the instructions in Problem 19:

$$CS = 1075H$$
$$IP = 0300H$$
$$SI = 0100H$$
$$DS:100H = 00H$$
$$DS:101H = 10H$$

to what address is program control passed?
21. Which flags are tested by the various conditional jump instructions?
22. What flag condition is tested for by the instruction JNS?
23. What flag conditions are tested for by the instruction JA?
24. Identify the type of jump, type of operand, and operation performed by each of the following instructions:
    (a) JNC   10H
    (b) JNP   PARITY_ERROR
    (c) JO   DWORD PTR [BX]

25. What value must be loaded into BX such that execution of the instruction JMP BX transfers control to the memory location offset from the beginning of the current code segment by 256_{10}?
26. The following program implements what is known as a *delay loop:*

```
        MOV  CX, 1000H
DLY:    DEC  CX
        JNZ  DLY
NXT:    ---  ---
```

   (a) How many times does the JNZ DLY instruction get executed?
   (b) Change the program so that JNZ DLY is executed just 17 times.
   (c) Change the program so that JNZ DLY is executed 2^{32} times.
27. Given a number N in the range $0 < N \leq 5$, write a program that computes its factorial and saves the result in memory location FACT.
28. Write a program that compares the elements of two arrays, A(I) and B(I). Each array contains 100 16-bit signed numbers. The comparison is to be done by comparing the corresponding elements of the two arrays until either two elements are found to be unequal or all elements of the two arrays have been compared and found to be equal. Assume that the arrays start at addresses $A000_{16}$ and $B000_{16}$, respectively. If the two arrays are found to be unequal, save the address of the first unequal element of A(I) in memory location FOUND; otherwise, write all zeros into this location.
29. Given an array A(I) of 100 16-bit signed numbers that are stored in memory starting at address $A000_{16}$, write a program to generate two arrays from the given array such that one P(J) consists of all the positive numbers and the other array N(K) contains all the negative numbers. Store the array of positive numbers in memory starting at address $B000_{16}$ and the array of negative numbers starting at address $C000_{16}$.

30. Given a 16-bit binary number in DX, write a program that converts it to its equivalent BCD number in DX. If the result is bigger than 16 bits, place all ones in DX.
31. Given an array A(I) with 100 16-bit signed numbers, write a program to generate a new array B(I) as follows:

    B(I) = A(I), for I = 1, 2, 99, and 100

and

    B(I) = median value of A(I − 2), A(I − 1), A(I), A(I + 1), and A(I + 2),

    for all other Is

Section 4.5

32. Describe the difference between a jump and call instruction.
33. Why are intersegment and intrasegment call instructions provided in the 80386DX?
34. What is saved on the stack when a call instruction with a Memptr16 operand is executed? A Memptr32 operand?
35. Identify the type of call, the type of operand, and operation performed by each of the following instructions.
    (a) CALL   1000H
    (b) CALL   [0100H]
    (c) CALL   DWORD PTR [BX + SI]
    (d) CALL   EDX
36. The following state of the 80386DX exists before executing the instructions in Problem 35(a), (b), and (c):

    CS = 1075H
    IP = 0300H
    EBX = 00000100H
    ESI = 00000100H
    DS:100H = 00H
    DS:101H = 10H
    DS:200H = 00H
    DS:201H = 01H
    DS:202H = 00H
    DS:203H = 10H

    to what address is program control passed?
37. What function is performed by the RET instruction?
38. Describe the operation performed by each of the following instructions.
    (a) PUSH   1000H
    (b) PUSH   DS
    (c) PUSH   [SI]
    (d) POP   EDI

(e) POP   [BX + DI]
(f) POPF

39. At what addresses will the bytes of the immediate operand in Problem 38(a) be stored after the instruction is executed?
40. Write a subroutine that converts a given 16-bit BCD number to its equivalent binary number. The BCD number is to be passed to a subroutine through register DX and the routine returns the equivalent binary number in DX.
41. When is it required to include PUSHF and POPF instructions in a subroutine?
42. Given an array A(I) of 100 16-bit signed integer numbers, write a subroutine to generate a new array B(I) such that

$$B(I) = A(I) \quad \text{for I = 1 and 100}$$

and

$$B(I) = \tfrac{1}{4}[A(I-1) - 5A(I) + 9A(I+1)] \quad \text{for all other Is}$$

The values of $A(I-1)$, $A(I)$, and $A(I+1)$ are to be passed to the subroutine in registers AX, BX, and CX and the subroutine returns the result B(I) in register AX.

43. Write a segment of main program and show its subroutine structure to perform the following operations. The program is to check continuously the three most significant bits in register DX, and depending on their setting, execute one of three subroutines: SUBA, SUBB, or SUBC. The subroutines are selected as follows:

(a) If bit 15 of DX is set, initiate SUBA.
(b) If bit 14 of DX is set and bit 15 is not set, initiate SUBB.
(c) If bit 13 of DX is set and bits 14 and 15 are not set, initiate SUBC.

If the subroutine is executed, the corresponding bits of DX are to be cleared and then control returned to the main program. After returning from the subroutine, the main program is repeated.

Section 4.6

44. Which flag is tested by the various conditional loop instructions?
45. What two conditions can terminate the operation performed by the instruction LOOPNE?
46. How large a jump can be employed in a loop instruction?
47. What is the maximum number of repeats that can be run with a loop instruction?
48. Using loop instructions, implement the program in Problem 27.
49. Using loop instructions, implement the program in Problem 28.

Section 4.7

50. What determines whether the SI and DI registers increment or decrement during a string operation?
51. Which segment register is used to form the destination address for a string move instruction?
52. Write equivalent string instruction sequences for each of the following.
    (a) MOV   AH, [SI]
        MOV   [DI], AH

```
            INC   SI
            INC   DI
    (b) MOV   AX, [SI]
            INC   SI
            INC   SI
    (c) MOV   AL, [DI]
            CMP   AL, [SI]
            DEC   SI
            DEC   DI
```

53. Use string instructions to implement the program in Problem 28.

54. Write a program to convert a table of 100 ASCII characters stored starting at offset address ASCII_CHAR into their equivalent table of EBCDIC characters and store them at offset address EBCDIC_CHAR. The translation is to be done using an ASCII-to-EBCDIC conversion table starting at offset address ASCII_TO_EBCDIC. Assume that all three tables are located in different segments of memory.

5

Protected-Mode Software Architecture of the 80386DX

5.1 INTRODUCTION

Having completed our study of the real-mode operation and instruction set of the 80386DX microprocessor, we are now ready to turn our attention to its protected-address mode (protected mode) of operation. Earlier we indicated that whenever the 80386DX microprocessor is reset, it comes up in real mode. Moreover, we identified that the PE bit of control register zero (CR_0) can be used to switch the 80386DX into the protected mode under software control. When configured for protected mode, the 80386DX provides an advanced software architecture that supports memory management, virtual addressing, paging, protection, and multitasking. This is the mode of operation used by operating systems such as UNIX™ and OS/2. In this chapter we examine the 80386DX's protected-mode software architecture and advanced system concepts. The topics covered are as follows:

1. Protected-mode register model
2. Protected-mode memory management and address translation
3. Descriptor and page table entries
4. Protected-mode system control instruction set
5. Multitasking and protection
6. Virtual 8086 mode

5.2 PROTECTED-MODE REGISTER MODEL

We will begin our study of the 80386DX's protected mode software architecture with its register model. The protected-mode register set of the 80386DX microprocessor is illustrated in Fig. 5.1. Looking at this diagram, we see that its application register model is a superset of the real-mode register set shown in Fig. 2.2. Comparing these two diagrams, we find four new registers in the protected-mode model: the *global descriptor table register* (GDTR), *interrupt descriptor table register* (IDTR), *local descriptor table register* (LDTR), and *task register* (TR). Furthermore, the function of a few registers have been extended. For example, the instruction pointer, which is now called EIP, is 32 bits in length; more bits of the flag register (EFLAGS) are active; and all four control registers, CR_0 through CR_3, are functional. Let us next discuss the purpose of each of the new registers and extended registers and how they are used in the segmented memory protected-mode operation of the microprocessor.

Global Descriptor Table Register

As shown in Fig. 5.2, the contents of the global descriptor table register define a table in the 80386DX's physical memory address space called the *global descriptor table* (GDT). This global descriptor table is one important element of the 80386DX's memory management system.

GDTR is a 48-bit register that is located inside the 80386DX. The lower two bytes of this register, which are identified as LIMIT in Fig. 5.2, specify the size in bytes of the GDT. The decimal value of LIMIT is one less than the actual size of the table. For instance, if LIMIT equals $00FF_{16}$, the table is 256 bytes in length. Since LIMIT has 16 bits, the GDT can be up to 65,536 bytes long. The upper four bytes of the GDTR, which are labeled BASE in Fig. 5.2, locate the beginning of the GDT in physical memory. This 32-bit base address allows the table to be positioned anywhere in the 80386DX's address space.

EXAMPLE 5.1

If the limit and base in the global descriptor table register are $0FFF_{16}$ and 00100000_{16}, respectively, what is the beginning address of the descriptor table, size of the table in bytes, and ending address of the table?

SOLUTION The starting address of the global descriptor table in physical memory is given by the base. Therefore,

$$GDT_{START} = 00100000_{16}$$

The limit is the offset to the end of the table. This gives

$$GDT_{END} = 00100000_{16} + 0FFF_{16} = 00100FFF_{16}$$

Finally, the size of the table is equal to the decimal value of the limit plus one.

$$GDT_{SIZE} = FFF_{16} + 1_2 = 4096 \text{ bytes}$$

The GDT provides a mechanism for defining the characteristics of the 80386DX's *global memory* address space. Global memory is a general system resource that is shared by many or all software tasks. That is, storage locations in global memory are accessible by any task that runs on the microprocessor.

This table contains what are called *system segment descriptors*. It is these descriptors that identify the characteristics of the segments of global memory. For instance, a segment descriptor provides information about the size, starting point, and access rights of a global memory segment. Each descriptor is eight bytes long; thus our earlier example of a 256-byte table provides enough storage space for just 32 descriptors. Remember that the size of the global descriptor table can be expanded simply by changing the value of LIMIT in the GDTR under software control. If the table is increased to its maximum size of 65,536 bytes, it can hold up to 8,192 descriptors.

EXAMPLE 5.2

How many descriptors can be stored in the global descriptor table defined in Example 5.1?

SOLUTION Each descriptor takes up eight bytes; therefore, a 4096-byte table can hold

$$\text{DESCRIPTORS} = \frac{4096}{8} = 512$$

The value of the BASE and LIMIT must be loaded into the GDTR before the 80386DX is switched from real mode of operation to the protected mode. Special instructions are provided for this purpose in the system control instruction set of the 80386DX. These instructions will be introduced later in this chapter. Once the 80386DX is in protected mode, the location of the table is typically not changed.

Interrupt Descriptor Table Register

Just like the global descriptor table register, the interrupt descriptor table register (IDTR) defines a table in physical memory. However, this table contains what are called *interrupt descriptors*, not segment descriptors. For this reason it is known as the *interrupt descriptor table* (IDT). This register and table of descriptors provide the mechanism by which the microprocessor passes program control to interrupt and exception service routines.

As shown in Fig. 5.3, just like the GDTR, the IDTR is 48 bits in length. Again, the lower two bytes of the register (LIMIT) define the table size. That is, the size of the table equals LIMIT + 1 bytes. Since two bytes define the size, the IDT can also be up to 65,536 bytes long. But the 80386DX only supports up to 256 interrupts and exceptions; therefore, the size of the IDT should not be set to support more than 256 interrupts. The upper three bytes of IDTR (BASE) identify the starting address of the IDT in physical memory.

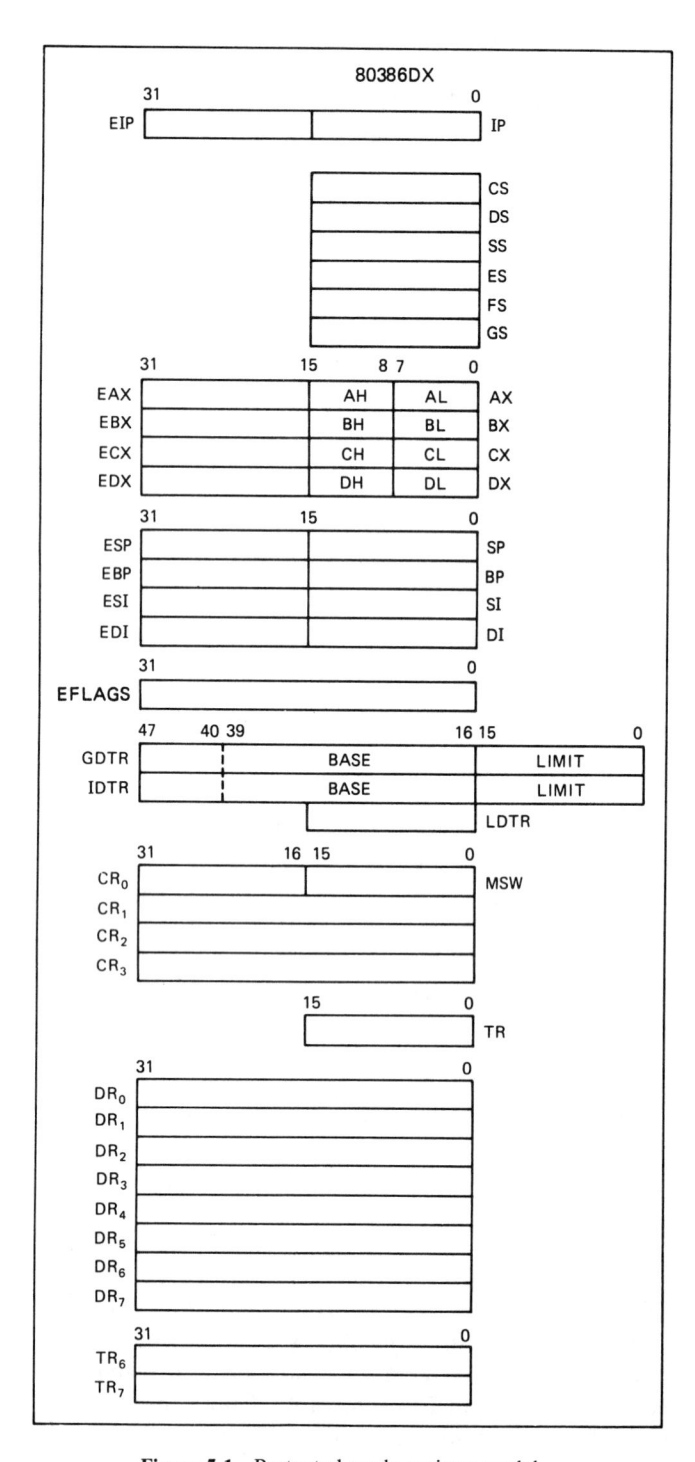

Figure 5.1 Protected-mode register model.

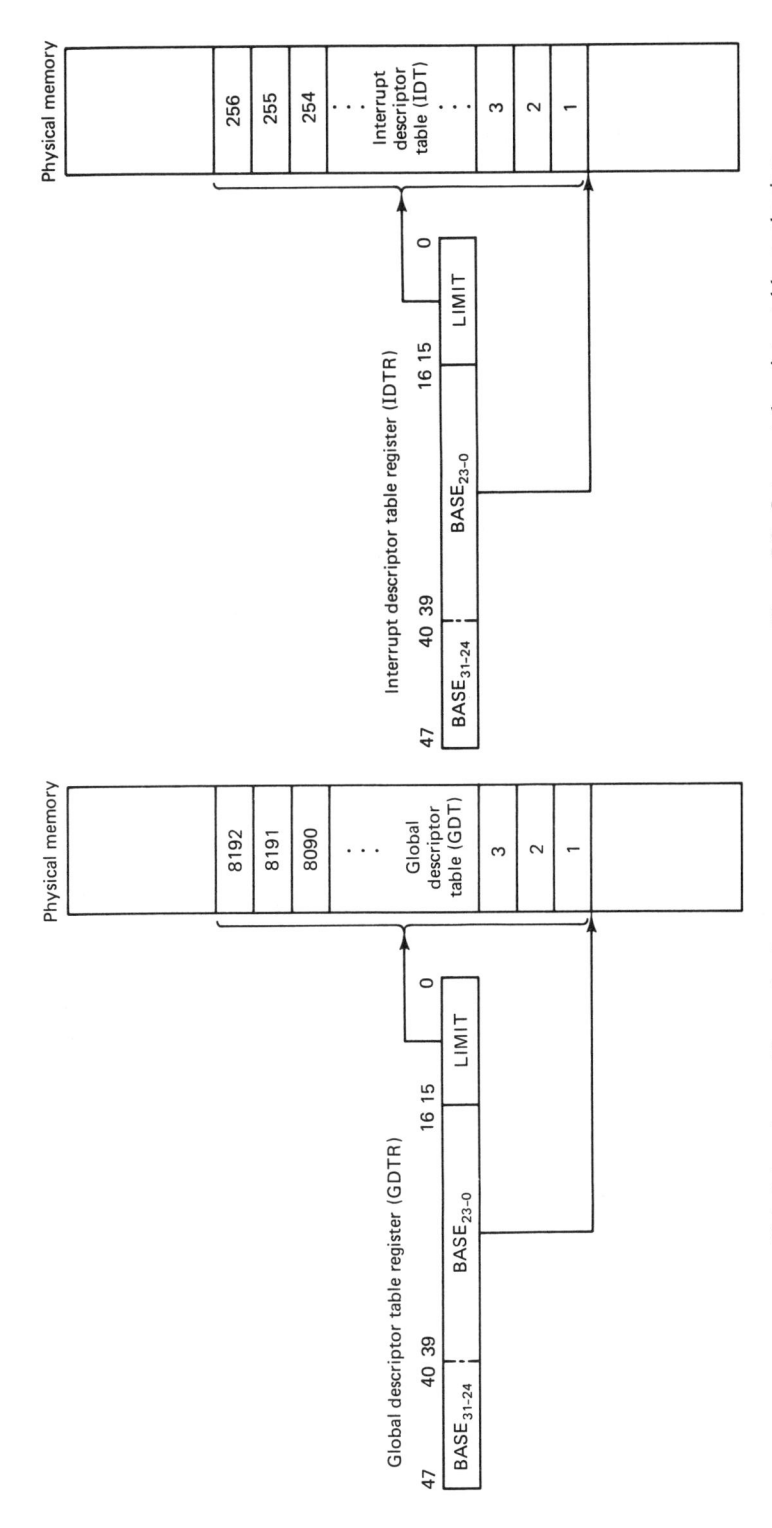

Figure 5.2 Global descriptor table mechanism.

Figure 5.3 Interrupt descriptor table mechanism.

161

The type of descriptor used in the IDT are what are called *interrupt gates*. These gates provide a means for passing program control to the beginning of an interrupt service routine. Each gate is eight bytes long and contains both attributes and a starting address for the service routine.

EXAMPLE 5.3

What is the maximum value that should be assigned to the limit in the IDTR?

SOLUTION The maximum number of interrupt descriptors that can be used in an 80386DX microcomputer system is 256. Therefore, the maximum table size in bytes is

$$IDT_{SIZE} = 8 \times 256 = 4096 = 1000_{16} \text{ bytes}$$

$$LIMIT = 1000_{16} - 1 = 0FFF_{16}$$

This table can also be located anywhere in the linear address space addressable with the 80386DX's 32-bit address. Just like the GDTR, the IDTR needs to be loaded before the 80386DX is switched from the real mode to the protected mode. Special instructions are provided for loading and saving the contents of the IDTR. Once the location of the table is set, it is typically not changed after entering the protected mode.

EXAMPLE 5.4

What is the address range of the last descriptor in the interrupt descriptor table defined by base address 00011000_{16} and limit $01FF_{16}$?

SOLUTION From the values of the base and limit, we find that the table is located in the address range

$$IDT_{START} = 00011000_{16}$$

$$IDT_{END} = 000111FF_{16}$$

The last descriptor in this table takes up the eight bytes of memory from address $000111F8_{16}$ through $000111FF_{16}$.

Local Descriptor Table Register

The *local descriptor table register* (LDTR) is also part of the 80386DX's memory management support mechanism. As shown in Fig. 5.4(a), each task can have access to its own private descriptor table in addition to the global descriptor table. This private table is called the *local descriptor table* (LDT) and defines a *local memory* address space for use by the task. The LDT holds segment descriptors that provide access to code and data in segments of memory that are reserved for the current task. Since each task can have its own segment of local memory,

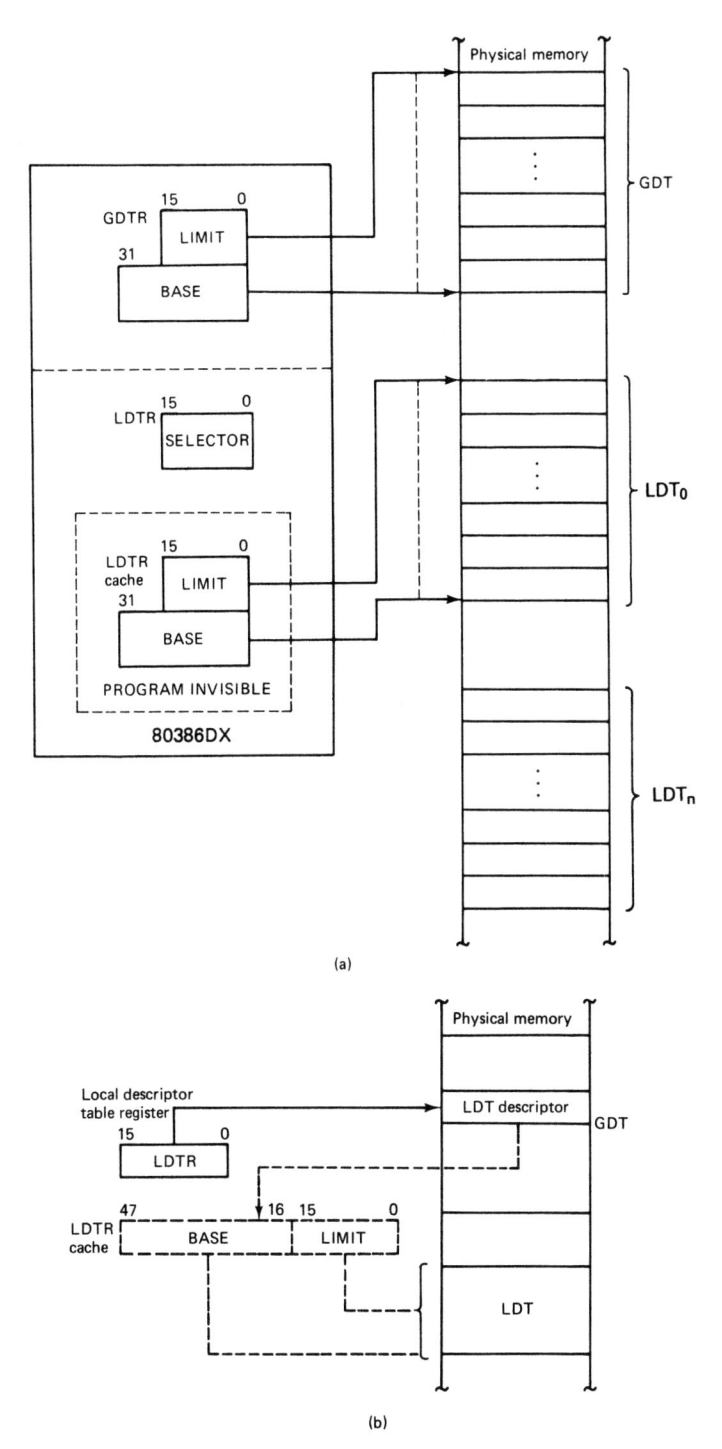

Figure 5.4 (a) Task with global and local descriptor tables; (b) loading the local descriptor table register to define a local descriptor table.

Sec. 5.2 Protected-Mode Register Model

the protected-mode software system may contain many local descriptor tables. For this reason we have identified LDT_0 through LDT_N in Fig. 5.4(a).

Figure 5.4(b) shows us that the contents of the 16-bit LDTR does not directly define the local descriptor table. Instead, it holds a selector that points to an *LDTdescriptor* in the GDT. Whenever a selector is loaded into the LDTR, the corresponding descriptor is transparently read from global memory and loaded into the *local descriptor table cache* within the 80386DX. It is this descriptor that defines the local descriptor table. As shown in Fig. 5.4(b), the 32-bit base value identifies the starting point of the table in physical memory and the value of the 16-bit limit determines the size of the table. Loading of this descriptor into the cache creates the LDT for the current task. That is, every time a selector is loaded into the LDTR, a local descriptor table descriptor is cached and a new LDT is activated.

Control Registers

The protected-mode model includes the four system control registers, identified as CR_0 through CR_3 in Fig. 5.1. Figure 5.5 shows these registers in more detail.

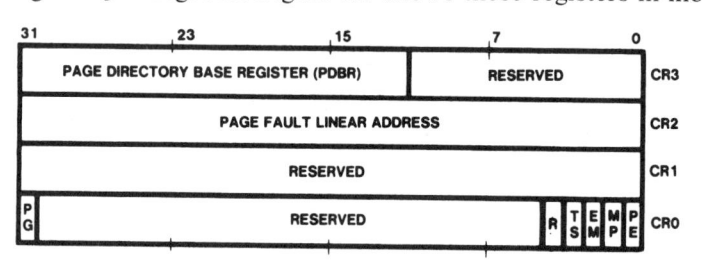

Figure 5.5 Control registers. (Reprinted by permission of Intel Corp. Copyright Intel Corp. 1986.)

Notice that the lower five bits of CR_0 are system control flags. These bits make up what are known as the *machine status word* (MSW). The most significant bit of CR_0 and registers CR_2 and CR_3 are used by the 80386DX's paging mechanism.

Let us continue by examining the machine status word bits of CR_0. They contain information about the 80386DX's protected-mode configuration and status. The four bits labeled PE, MP, EM, and R are control bits that define the protected mode system configuration. The fifth bit, TS, is a status bit. These bits can be examined or modified through software.

The protected-mode enable (PE) bit determines if the 80386DX is in the real or protected mode. At reset, PE is cleared. This enables the real mode of operation. To enter the protected mode, we simply switch PE to 1 through software. Once in the protected mode, the 80386DX can be switched back to real mode under software control by clearing the PE bit. It can also be returned to real mode by a hardware reset.

The *math present* (MP) bit is set to 1 to indicate that a numerics coprocessor is present in the microcomputer system. On the other hand, if the system is to be configured so that a software emulator is used to perform numeric operations instead of a coprocessor, the *emulate* (EM) bit is set to 1. Only one of these two bits can be set at a time. Finally, the *extension type* (R) bit is used to indicate whether an 80287 or 80387DX numerics coprocessor is in use. Logic 1 in R indicates that an 80387DX is installed. The last bit in the MSW, *task switched* (TS), automatically gets set whenever the 80386DX switches from one task to another. It can be cleared under software control.

The protected mode software architecture of the 80386DX also supports paged memory operation. Paging is turned on by switching the PG bit in CR_0 to logic 1. Now addressing of physical memory is implemented with an address translation mechanism that consists of a page directory and page table that are both held in physical memory. Looking at Fig. 5.5, we see that CR_3 contains the *page directory base register* (PDBR). This register holds a 20-bit *page directory base address* that points to the beginning of the page directory. A page fault error occurs during the page translation process if the page is not present in memory. In this case, the 80386DX saves the address at which the page fault occurred in register CR_2. This address is denoted as *page fault linear address* in Fig. 5.5.

Task Register

The *task register* (TR) is one of the key elements in the protected mode task switching mechanism of the 80386DX microprocessor. This register holds a 16-bit index value called a *selector*. The initial selector must be loaded into TR under software control. This starts the initial task. After this is done, the selector is changed automatically whenever the 80386DX executes an instruction that performs a task switch.

As shown in Fig. 5.6, the selector in TR is used to locate a descriptor in the global descriptor table. Notice that when a selector is loaded into the TR, the corresponding *task state segment (TSS) descriptor* automatically gets read from

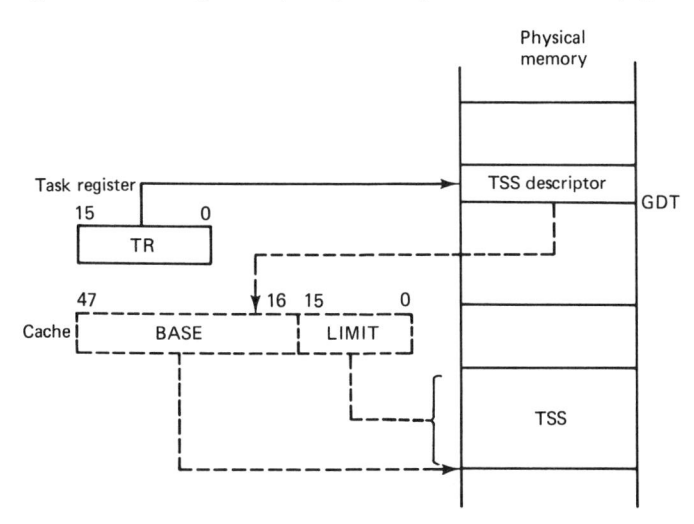

Figure 5.6 Task register and the task-switching mechanism.

memory and loaded into the on-chip *task descriptor cache*. This descriptor defines a block of memory called the *task state segment* (TSS). It does this by providing the starting address (BASE) and the size (LIMIT) of the segment. Every task has its own TSS. The TSS holds the information needed to initiate the task, such as initial values for the user-accessible registers.

> **EXAMPLE 5.5**
>
> What is the maximum size of a TSS? Where can it be located in the linear address space?
>
> **SOLUTION** Since the value of LIMIT is 16 bytes in length, the TSS can be as long as 64K bytes. Moreover, the base is 32 bits in length. Therefore, the TSS can be located anywhere in the 80386DX's 4G-byte address space.

> **EXAMPLE 5.6**
>
> Assume that the base address of the global descriptor table is 00011000_{16} and the selector in the task register is 2108_{16}. What is the address range of the TSS descriptor?
>
> **SOLUTION** The beginning address of the TSS descriptor is
>
> $$\text{TSS\_DESCRIPTOR}_{\text{START}} = 00011000_{16} + 2108_{16}$$
>
> $$= 00013108_{16}$$
>
> Since the descriptor is eight bytes long, it ends at
>
> $$\text{TSS\_DESCRIPTOR}_{\text{END}} = 0001310F_{16}$$

Registers with Changed Functionality

Earlier we pointed out that the function of a few of the registers that are common to both the real- and protected-mode register models changes as the 80386DX is switched into the protected mode of operation. For instance, the segment registers are now called the *segment selector registers*, and instead of holding a base address they are loaded with what is known as a *selector*. The selector does not directly specify a storage location in memory. Instead, it selects a descriptor that defines the size and characteristics of a segment of memory.

The format of a selector is shown in Fig. 5.7. Here we see that the two least significant bits are labeled RPL, which stands for *requested privilege level*. These bits contain either $00 = 0$, $01 = 1$, $10 = 2$, or $11 = 3$ and assign a request protection level to the selector. The next bit, which is identified as *table indicator* (TI) in Fig. 5.7, selects the table to be used when accessing a segment descriptor. Remember that in the protected mode, two descriptor tables are active at a time, the global descriptor table and a local descriptor table. Looking at Fig. 5.7, we find that if TI is 0, the selector corresponds to a descriptor in the global descriptor table. Finally, the 13 most significant bits contain an *index* that is used as a pointer to a specific descriptor entry in the table selected by the TI bit.

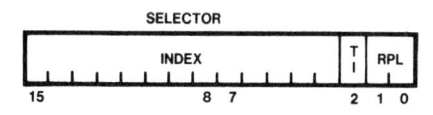

BITS	NAME	FUNCTION
1-0	REQUESTED PRIVILEGE LEVEL (RPL)	INDICATES SELECTOR PRIVILEGE LEVEL DESIRED
2	TABLE INDICATOR (TI)	TI = 0 USE GLOBAL DESCRIPTOR TABLE (GDT) TI = 1 USE LOCAL DESCRIPTOR TABLE (LDT)
15-3	INDEX	SELECT DESCRIPTOR ENTRY IN TABLE

Figure 5.7 Selector format. (Reprinted by permission of Intel Corp. Copyright Intel Corp. 1986.)

EXAMPLE 5.7

Assume that the base address of the LDT is 00120000_{16} and the GDT base address is 00100000_{16}. If the value of the selector loaded into the CS register is 1007_{16}, what is the request privilege level? Is the segment descriptor in the GDT or LDT? What is the address of the segment descriptor?

SOLUTION Expressing the selector in binary form, we get

$$CS = 0001000000000111_2$$

Since the two least significant bits are both 1,

$$RPL = 3$$

The next bit, bit 2, is also 1. This means that the segment descriptor is in the LDT. Finally, the value in the 13 most significant bits must be scaled by 8 to give the offset of the descriptor from the base address of the table. Therefore,

$$OFFSET = 0001000000000_2 \times 8 = 512 \times 8 = 4096$$
$$= 1000_{16}$$

and the address of the segment descriptor is

$$DESCRIPTOR_{ADDRESS} = 00120000_{16} + 1000_{16}$$
$$= 00121000_{16}$$

Another register whose function changes when the 80386DX is switched to protected mode is the flag register. As shown in Fig. 5.1, the flag register is now identified as EFLAGS and expands to 32 bits in length. The functions of the bits in EFLAGS are given in Fig. 5.8. Comparing this illustration to the real-mode 80386DX flag register in Fig. 2.16, we see that five additional bits are active when the 80386DX is in protected mode. They are the two-bit *input/output privilege level* (IOPL) code, the *nested task* (NT) flag, the *resume* (RF) flag, and the *virtual 8086 mode* (VM) flag.

Notice in Fig. 5.8 that each of these flags is identified as a system flag. That is, they represent protected-mode system operations. For example, the IOPL bits

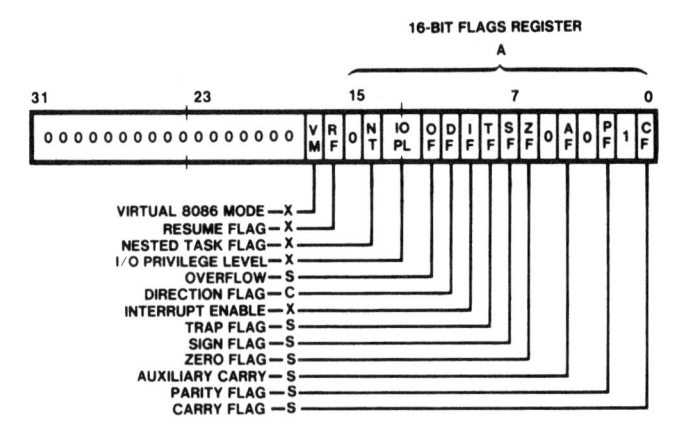

Figure 5.8 Protected-mode flag register. (Reprinted by permission of Intel Corp. Copyright Intel Corp. 1986.)

are used to assign a maximum privilege level to input/output. For instance, if 00 is loaded into IOPL, I/O can be performed only when the 80386DX is in the highest privilege level, which is called *level 0*. On the other hand, if IOPL is 11, I/O is assigned to the least privileged level, level 3.

The NT flag identifies whether or not the current task is a nested task, that is, if it was called from another task. This bit is automatically set whenever a nested task is initiated and can only be reset through software.

5.3 PROTECTED-MODE MEMORY MANAGEMENT AND ADDRESS TRANSLATION

Up to this point in the chapter, we have introduced the register set of the protected-mode software model for the 80386DX microprocessor. However, the software model of a microprocessor also includes its memory structure. Because of the memory management capability of the 80386DX, the organization of protected mode memory appears quite complex. Here we will examine how the *memory management unit* (MMU) of the 80386DX implements the address space and how it translates virtual (logical) addresses to physical addresses. We begin here with what are called the *segmented* and *paged models* of memory.

The Virtual Address and Virtual Address Space

The protected-mode memory management unit employs memory pointers that are 48 bits in length and consists of two parts, the *selector* and the *offset*. This 48-bit memory pointer is called a *virtual address* and is used by the program to specify the memory location of instructions or data. As shown in Fig. 5.9, the selector

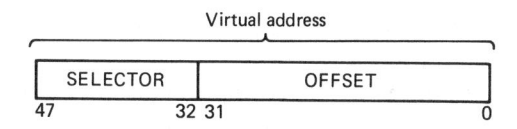

```
                    Virtual address
        ┌─────────────────────────────────┐
        ┌──────────┬──────────────────────┐
        │ SELECTOR │       OFFSET         │     **Figure 5.9**  Protected-mode memory
        └──────────┴──────────────────────┘     pointer.
        47         32 31                  0
```

is 16 bits in length and the offset is 32 bits long. Earlier we pointed out that one source of selectors is the segment selector registers within the 80386DX. For instance, if code is being accessed in memory, the active segment selector will be that held in CS. This part of the pointer selects a unique segment of the 80386DX's *virtual address space.*

The offset is normally held in one of the 80386DX's other user-accessible registers. For our example of a code access, the offset would be in the EIP register. This part of the pointer is the displacement of the memory location that is to be accessed within the selected segment of memory. In our example it points to the first byte of the double word of instruction code that is to be fetched for execution. Since the offset is 32 bits in length, the segment size can be as large as 4G bytes. We say "as large as 4G bytes" because segment size is actually variable and can be defined to be as small as one byte to as large as 4G bytes.

Figure 5.10 shows that the 16-bit selector breaks down into a *13-bit index, table select bit*, and two bits used for a *request privilege level*. The two RPL bits

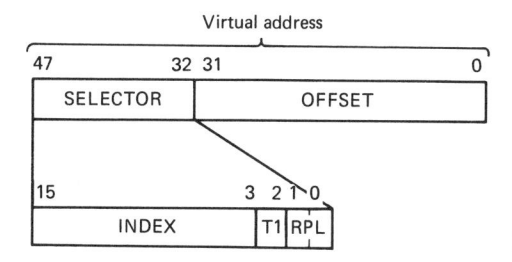

```
                 Virtual address
            47       32 31              0
         ┌──────────┬──────────────────┐
         │ SELECTOR │     OFFSET       │
         └──────────┴──────────────────┘
         15                3  2  0
         ┌──────────────────┬──┬──┐
         │      INDEX       │T1│RPL│             **Figure 5.10**  Segment selector format.
         └──────────────────┴──┴──┘
```

are not used in the selection of the memory segment. That is, just 14 of its 16 bits are employed in addressing memory. Therefore, the virtual address space can consist of 2^{14} (16,384 = 16K) unique segments of memory, each of which has a maximum size of 4G bytes. These segments are the basic element into which the memory management unit of the 80386DX organizes the virtual address space.

A benefit of the segmented organization of memory is the variable segment size. Segments can be made as small or as large as needed for an application. This assures efficient use of memory.

Another way of looking at the size of the virtual address space is that by combining the 14-bit segment selector with the 32-bit offset, we get a 46-bit virtual address. Therefore, the 80386DX's virtual address space can contain 2^{46} equals 64T bytes.

Segmented Partitioning of the Virtual Address Space

The memory management unit of the 80386DX implements both a segmented and paged model of virtual memory. In the segmented model the 80386DX's 64T-byte virtual address space is partitioned into a 32T-byte *global memory address space* and a 32T-byte *local memory address space*. This partitioning is illustrated in general by Fig. 5.11. The TI bit of the selector shown in Fig. 5.10 is used to select

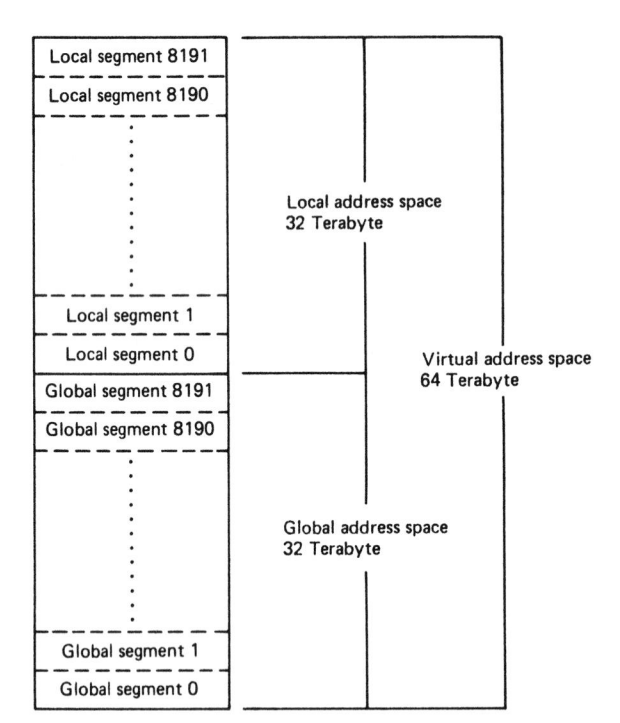

Figure 5.11 Partitioning the virtual address space.

between the global or local descriptor tables that define the virtual address space. Within each of these address spaces, as many as 8192 segments of memory may exist. This assumes that every descriptor in both the global descriptor table and local descriptor table is used and set for maximum size. These descriptors define the attributes of the corresponding segment. However, in practical system applications not all of the descriptors are normally in use. Let us now look briefly at how global and local segments of memory are used by software.

In the multiprocessing software environment of the 80386DX, an application is expressed as a collection of tasks. By *task* we mean a group of program routines that together perform a specific function. When the 80386DX initiates a task, it can have both global and local segments of memory active. This idea is illustrated in Fig. 5.12. Notice that tasks 1, 2, and 3 each have a reserved segment of the local address space. This part of memory stores data or code that can only be accessed by the corresponding task. That is, task 2 cannot access any of the

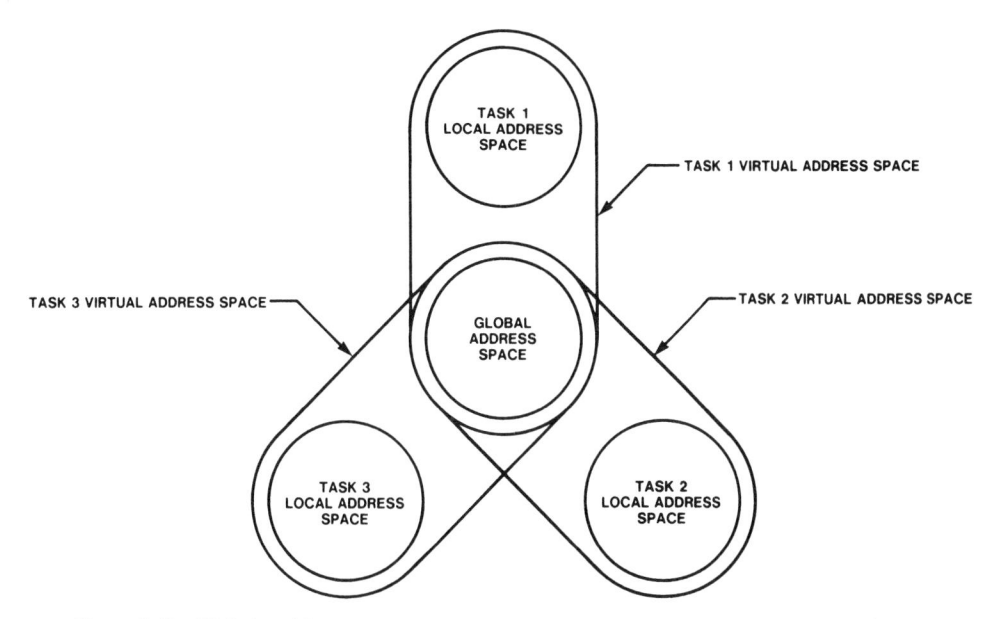

Figure 5.12 Global and local memory for a task. (Reprinted by permission of Intel Corp. Copyright Intel Corp. 1987.)

information in the local address space of task 1. On the other hand, all of the tasks are shown to share the same segment of the global address space. This segment typically contains operating system resources and data that are to be shared by all or many tasks.

Physical Address Space and Virtual to Physical Address Translation

We just found that the virtual address space available to the programmer is 64T bytes in length. However, the 32-bit protected-mode address bus of the 80386DX supports just a 4G-byte *physical address space*. Just a small amount of the information in virtual memory can reside in physical memory at a time. For this reason systems that employ a virtual address space that is larger than the implemented physical memory are equipped with a secondary storage device such as a hard disk.

Information that is not currently in use is stored on disk. If a segment of memory that is not present in physical memory is accessed by a program and space is available in physical memory the segment is simply read from the hard disk and copied in physical memory. On the other hand, if the physical memory address space is full, another segment must first be sent out to the hard disk to make room for the new information. The memory manager part of the operating system controls the allocation and deallocation of physical memory and the swapping of data between the hard disk and physical memory of the computer. In this way, the memory address space of the computer appears much larger than the physical memory in the computer.

The segmentation and paging memory management units of the 80386DX provide the mechanism by which 48-bit virtual addresses are mapped into the 32-bit physical addresses needed by hardware. They employ a memory-based lookup table address translation process. This address translation is illustrated in general by the diagram in Fig. 5.13. Notice that first a *segment translation* is performed on the virtual (logical) address. Then if paging is disabled, the *linear address* produced is equal to the physical address. However, if paging is enabled, the linear address goes through a second translation step, known as *page translation*, to produce the physical address.

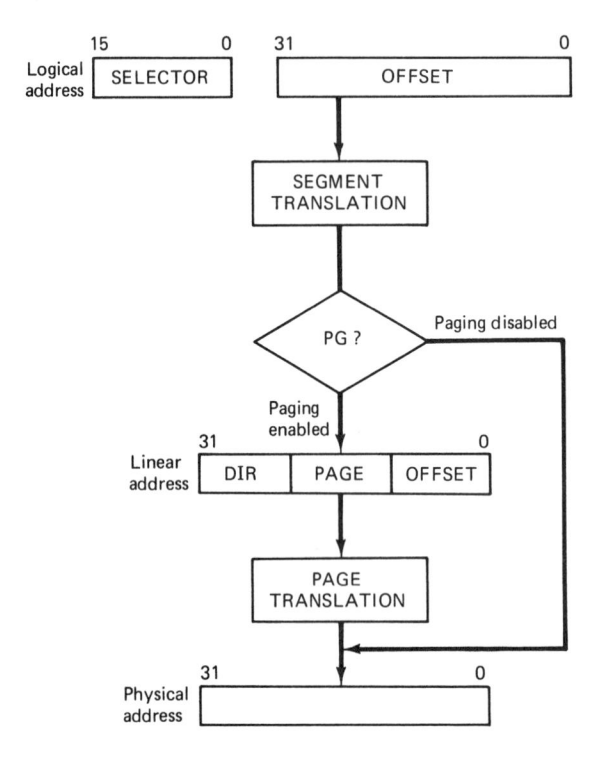

Figure 5.13 Virtual to physical address translation. (Reprinted by permission of Intel Corp. Copyright Intel Corp. 1986.)

As part of the translation process, the MMU determines whether or not the corresponding segment or page of the virtual address space currently exists in physical memory. If the segment or page already resides in memory, the operation is performed on the information. However, if the segment is not present, it signals this condition as an error. Once this condition is signaled, the memory manager software initiates loading of the segment or page from the external storage device to physical memory. This operation is called a *swap*. That is, an old segment or page gets swapped out to disk to make room in physical memory, and then the new segment or page is swapped into this space. Even though a swap has taken place, it appears to the program that all segments or pages are available in physical memory. Let us now look more closely at the address translation process.

Segmentation Virtual to Physical Address Translation

We begin by assuming that paging is turned off. In this case, the address translation sequence that takes place is that highlighted in Fig. 5.14(a). Figure 5.14(b) details the operations that take place during the segment translation process. Earlier we found that the 80386DX's segment selector registers, CS, DS, ES, FS, GS, and SS, provide the segment selectors that are used to index into either the global descriptor table or the local descriptor table. Whenever a selector value is loaded into a segment register, the descriptor pointed to by the index in the table selected by the TI bit is automatically fetched from memory and loaded into the corresponding *segment descriptor cache register*. It is the contents of this descriptor, not the selector, that defines the location, size, and characteristics of the segment of memory.

Notice in Fig. 5.15 that the 80386DX has one 64-bit internal segment descriptor cache register for each of the segment selector registers. These cache registers are not accessible by the programmer. Instead, they are transparently loaded with a complete descriptor whenever an instruction is executed that loads a new selector into a segment register. For instance, if an operand is to be accessed from a new data segment, a local memory data segment selector would first be

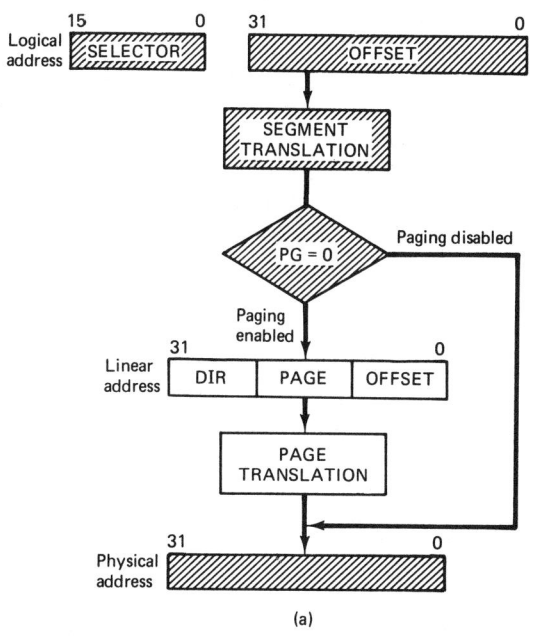

Figure 5.14 (a) Virtual to linear address translation. (Reprinted by permission of Intel Corp. Copyright Intel Corp. 1986.) (b) Translating a virtual address into a physical (linear) address.

Sec. 5.3 Protected-Mode Memory Management and Address Translation 173

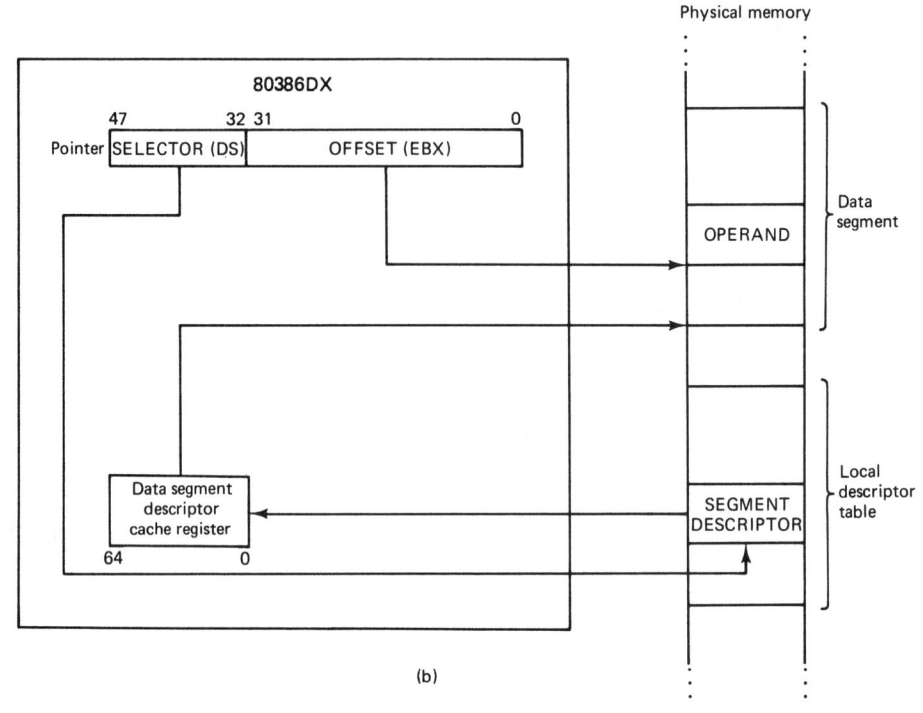

Figure 5.14 (*Continued*)

loaded into DS with the instruction

```
MOV  DS, AX
```

As this instruction is executed, the selector in AX is loaded into DS and then the corresponding descriptor in the local descriptor table is read from memory and loaded into the data segment descriptor cache register. The MMU looks at the information in the descriptor and performs checks to determine whether or not it is valid.

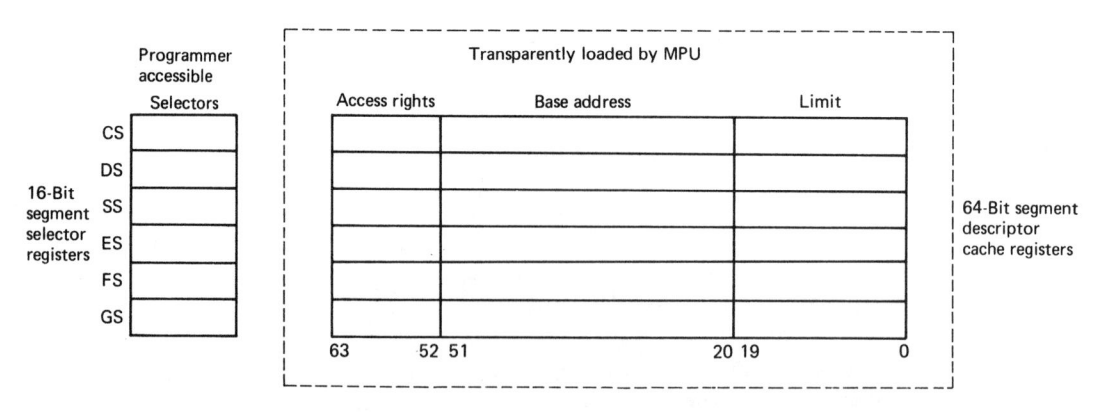

Figure 5.15 Segment selector registers and the segment descriptor cache registers.

In this way, we see that the segment descriptors held in the caches dynamically change as a task is performed. At any one time, the memory management unit permits just six segments of memory to be active. These segments correspond to the six segment selector registers, CS, DS, ES, FS, GS, and SS, and can reside in either local or global memory. Once the descriptors are cached, subsequent references to them are performed without any overhead for loading of the descriptor.

In Fig. 5.15 we find that this data segment descriptor has three parts: 12 bits of *access rights* information, a 32-bit *segment base address*, and a 20-bit *segment limit*. The value of the 32-bit base address identifies the beginning of the data segment that is to be accessed. The loading of the data segment descriptor cache completes the table lookup that maps the 16-bit selector to its equivalent 32-bit data segment base address.

The location of the operand in this data segment is determined by the offset part of the virtual address. For example, let us assume that the next instruction to be executed needs to access an operand in this data segment and that the instruction uses the based addressing mode to specify the operand. Then the EBX register holds the offset of the operand from the base address of the data segment. Figure 5.14(a) shows that the base address is added directly to the offset to produce the 32-bit physical address of the operand. This addition completes the translation of the 48-bit virtual address into the 32-bit linear address. As shown in Fig. 5.14(a), when paging is disabled, PG = 0, the linear address is the physical address of the storage location to be accessed in memory.

EXAMPLE 5.8

Assume that in Fig. 5.14(b), the virtual address is made up of a segment selector equal to 0100_{16} and offset equal to 00002000_{16} and that paging is disabled. If the segment base address read in from the descriptor is 00030000_{16}, what is the physical address of the operand?

SOLUTION The virtual address is

$$\text{VIRTUAL ADDRESS} = 0100:00002000_{16}$$

This virtual address translates to the physical address

$$\begin{aligned}\text{LINEAR ADDRESS} &= \text{BASE ADDRESS} + \text{OFFSET} \\ &= 00030000_{16} + 00002000_{16} \\ &= 00032000_{16}\end{aligned}$$

Paged Partitioning of the Virtual Address Space and Virtual to Physical Address Translation

Earlier we pointed out that the protected mode architecture of the 80386DX also supports paged organization of the memory address space. The paging memory management unit works beneath the segmentation memory management unit, and when enabled it organizes the 80386DX's address space in a different way. We just found that when paging is not in use, the 4G-byte physical address space is organized into segments that can be any size from one byte to 4G bytes. However, when paging is turned on, the paging unit arranges the physical address space into 1,048,496 pages that are each 4096 bytes long. Figure 5.16 shows how the physical address space may be organized in this way. The fixed size blocks of paged memory is a disadvantage in that 4K addresses are allocated by the memory manager even though it may not all be used. This creation of unused sections of memory is called *fragmentation*. Fragmentation results in less efficient use of memory. However, paging greatly simplifies the implementation of the memory manager software. Let us continue by looking at what happens to the address translation process when paging is enabled.

In Fig. 5.17 we see that the linear address produced by the segment translation process is no longer used as the physical address. Instead, it undergoes a second translation called the *page translation*. Figure 5.18 shows the format of

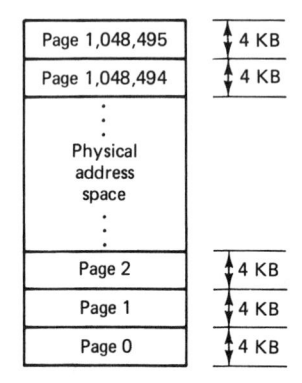

Figure 5.16 Paged organization of the physical address space.

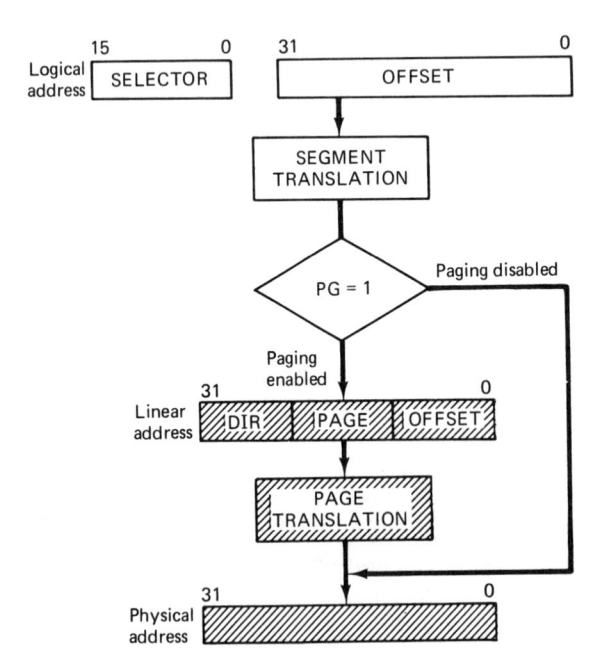

Figure 5.17 Paged translation of a linear address to a physical address. (Reprinted by permission of Intel Corp. Copyright Intel Corp. 1986.)

176 Protected-Mode Software Architecture of the 80386DX   Chap. 5

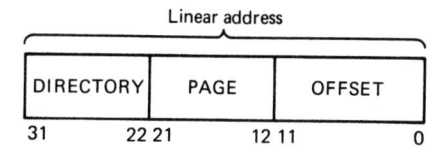

Figure 5.18 Linear address format.

a linear address. Notice that it is composed of three elements: a 20-bit offset field, a 10-bit page field, and a 10-bit directory field.

The diagram in Fig. 5.19 illustrates how a linear address is translated into its equivalent physical address. The location of the *page directory table* in memory is identified by the address in the page directory base register (PDBR) in CR_3. These 20 bits are actually the MSBs of the base address. The 12 lower bits are assumed to start at 000_{16} at the beginning of the directory and range to FFF_{16} at its end. Therefore, the page directory contains 4K-byte memory locations and is organized as 1K, 32-bit addresses. These addresses each point to a separate page table, which is also in physical memory.

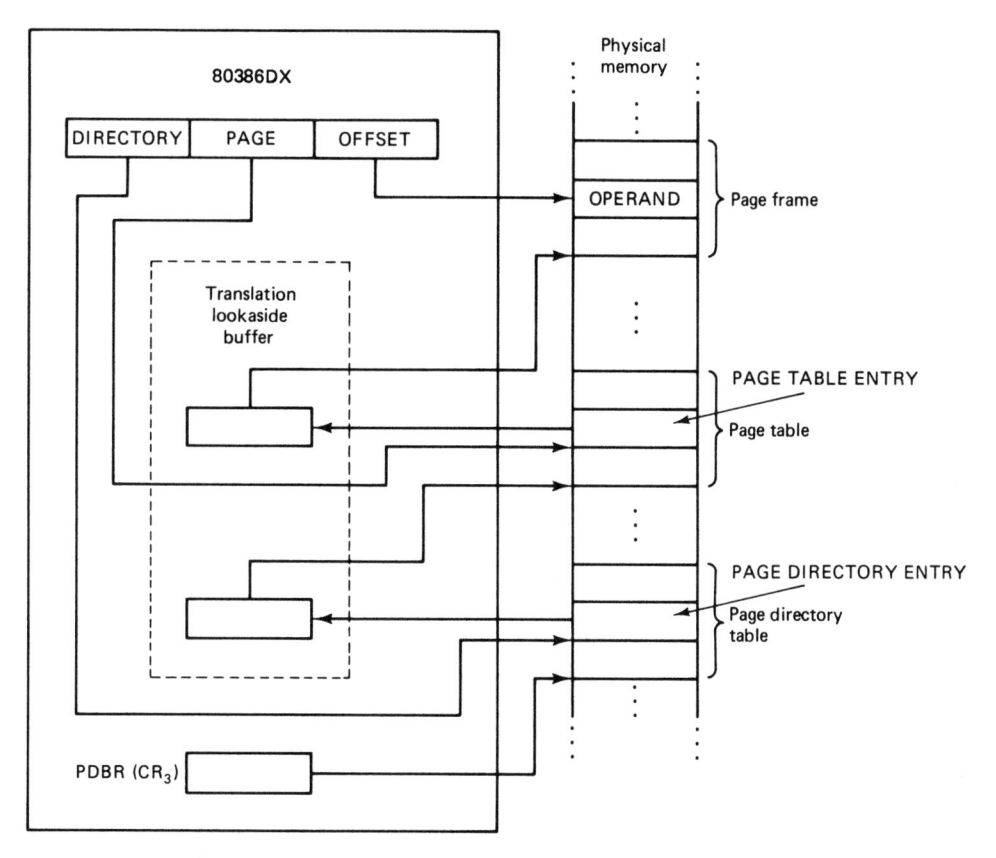

Figure 5.19 Translating a linear address to a physical address.

Sec. 5.3   Protected-Mode Memory Management and Address Translation

Notice that the 10-bit directory field of the linear address is the offset from the value in PDBR that selects one of the 1K, 32-bit *page directory entries* in the page directory table. This pointer is cached inside the 80386DX in what is called the *translation lookaside buffer*. Its value is used as the base address of a *page table* in memory. Just like the page directory, each page table is also 4K bytes long and contains 1K, 32-bit addresses. However, these addresses are called *page frame addresses*. Each page frame address points to a 4K frame of data storage locations in physical memory.

Next the 10-bit page field of the linear address selects one of the 1K, 32-bit *page table entries* from the page table. This table entry is also cached in the translation lookaside buffer. In Fig. 5.19 we see that it is another base address and selects a 4K-byte *page frame* in memory. This frame of memory locations is used for storage of data. The 12-bit offset part of the linear address identifies the location of the operand in the active page frame.

The 80386DX's translation lookaside buffer is actually capable of maintaining 32 sets of table entries. In this way we see that 128K bytes of paged memory are always directly accessible. Operands in these parts of memory can be accessed without first reading new entries from the page tables. If an operand to be accessed is not in one of these pages, overhead is required to first read the page table entry into the translation lookaside buffer.

5.4 DESCRIPTOR AND PAGE TABLE ENTRIES

In the preceding section we frequently used the terms *descriptor* and *page table entry*. We talked about the descriptor as an element of the global descriptor, local descriptor, and interrupt descriptor tables. Actually, there are several kinds of descriptors supported by the 80386DX and they all serve different functions relative to overall system operation. Some examples are the *segment descriptor*, *system segment descriptor*, *local descriptor table descriptor*, *call gate descriptor*, *task state segment descriptor*, and *task gate descriptor*. Moreover, we discussed page table entries in our description of the 80386DX's page translation of linear addresses. There are just two types of page table entries: the *page directory entry* and the *page table entry*. Let us now explore the structure of descriptors and page table entries.

Descriptors are the elements by which the on-chip memory manager hardware manages the segmentation of the 80386DX's 64T-byte virtual memory address space. One descriptor exists for each segment of memory in the virtual address space. Descriptors are assigned to the local descriptor table, global descriptor table, task state segment, call gate, task gate, and interrupts. The contents of a descriptor provides mapping from virtual addresses to linear addresses for code, data, stack, and the task state segments and assigns attributes to the segment.

Each descriptor is eight bytes long and contains three kinds of information. Earlier we identified the 20-bit *LIMIT* field and showed that its value defines the size of the segment or the table. Moreover, we found that the 32-bit *BASE* value provides the beginning address for the segment or the table in the 64G-byte linear address space. The third element of a descriptor, which is called the *access rights byte*, is different for each type of descriptor. Let us now look at the format of just two types of descriptors, the segment descriptor and system segment descriptor.

The segment descriptor is the type of descriptor that is used to describe code, data, and stack segments. Figure 5.20(a) shows the general structure of a segment descriptor. Here we see that the two lowest addressed bytes, bytes 0 and 1, hold the 16 least significant bits of the limit, the next three bytes contain the 24 least significant bits of the base address, byte 5 is the access rights byte, the lower four bits of byte 6 are the four most significant bits of the limit, the upper four bits include the *granularity* (G) and the *programmer available* (AVL) bits, and byte 7 is the eight most significant bits of the 32-bit base. Segment descriptors are found only in the local and global descriptor tables.

Figure 5.21 shows how a descriptor is loaded from the local descriptor table in global memory to define a code segment in local memory. Notice that the LDTR

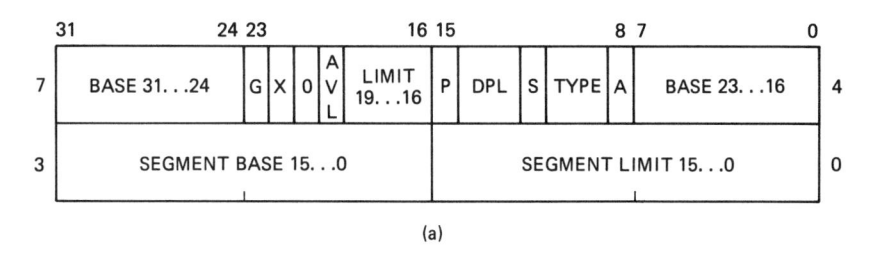

```
  31           24 23        16 15          8 7             0
 ┌──────────────┬───┬─┬─┬────────┬─┬───┬─┬────┬─┬──────────────┐
7│  BASE 31...24│G X 0│A│ LIMIT  │P│DPL│S│TYPE│A│  BASE 23...16│ 4
 │              │     │V│ 19...16│ │   │ │    │ │              │
 │              │     │L│        │ │   │ │    │ │              │
 ├──────────────┴─────┴─┴────────┴─┴───┴─┴────┴─┴──────────────┤
3│        SEGMENT BASE 15...0    │      SEGMENT LIMIT 15...0    │ 0
 └───────────────────────────────┴──────────────────────────────┘
```

(a)

	Bit Position	Name		Function	
	7	Present (P)	P = 1	Segment is mapped into physical memory.	
			P = 0	No mapping to physical memory exists, base and limit are not used.	
	6–5	Descriptor Privilege Level (DPL)		Segment privilege attribute used in privilege tests.	
	4	Segment Descriptor (S)	S = 1	Code or Data (includes stacks) segment descriptor	
			S = 0	System Segment Descriptor or Gate Descriptor	
Type Field Definition	3	Executable (E)	E = 0	Data segment descriptor type is:	If Data Segment (S = 1, E = 0)
	2	Expansion Direction (ED)	ED = 0	Expand up segment, offsets must be ≤ limit.	
			ED = 1	Expand down segment, offsets must be > limit.	
	1	Writeable (W)	W = 0	Data segment may not be written into.	
			W = 1	Data segment may be written into.	
	3	Executable (E)	E = 1	Code Segment Descriptor type is:	If Code Segment (S = 1, E = 1)
	2	Conforming (C)	C = 1	Code segment may only be executed when CPL ≥ DPL and CPL remains unchanged.	
	1	Readable (R)	R = 0	Code segment may not be read.	
			R = 1	Code segment may be read.	
	0	Accessed (A)	A = 0	Segment has not been accessed.	
			A = 1	Segment selector has been loaded into segment register or used by selector test instructions.	

(b)

Figure 5.20 (a) Segment descriptor format; (b) access byte bit definitions. (Reprinted by permission of Intel Corp. Copyright Intel Corp. 1987.)

Sec. 5.4 Descriptor and Page Table Entries

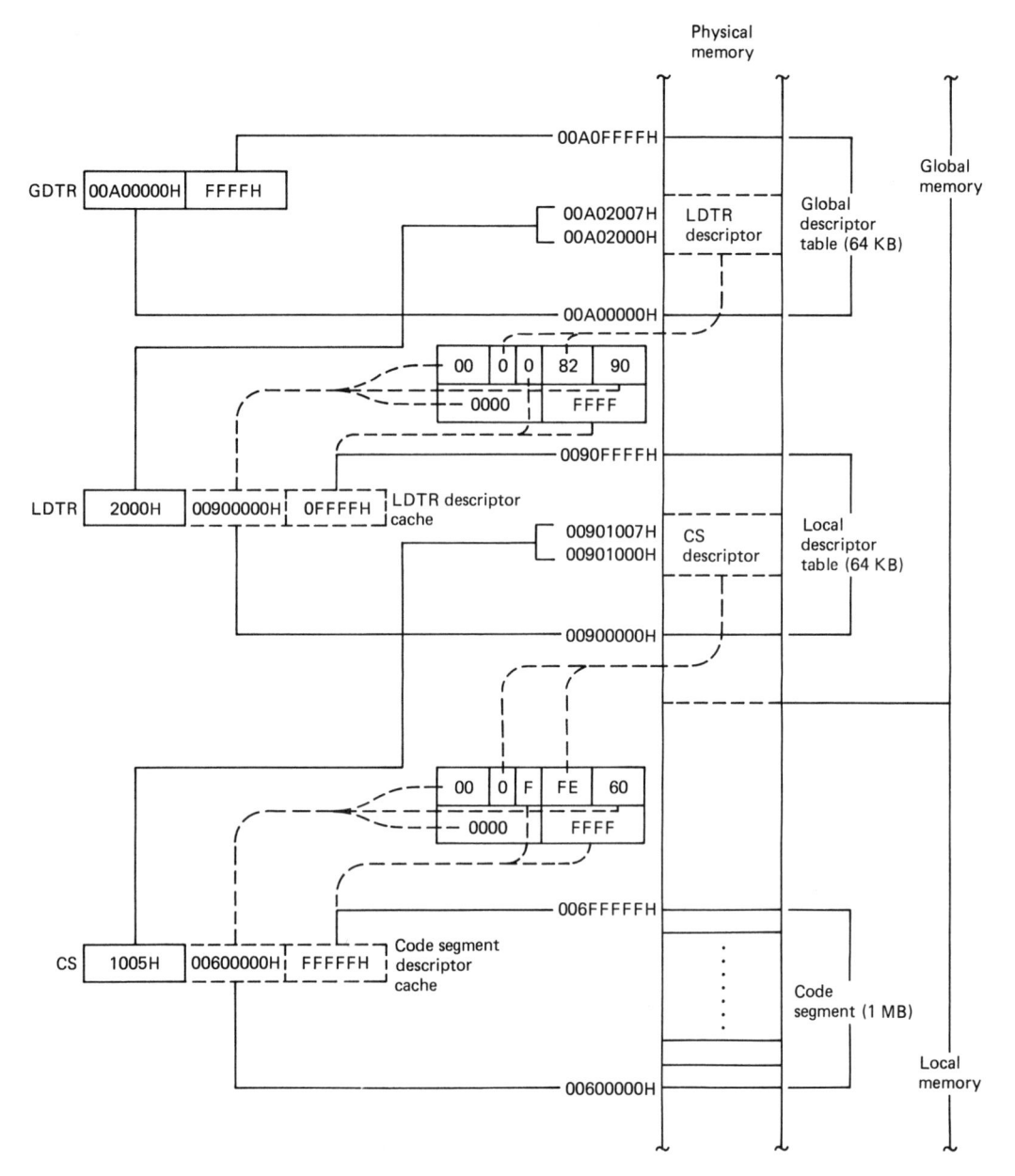

Figure 5.21 Creating a code segment.

descriptor defines a local descriptor table between address 00900000_{16} and $0090FFFF_{16}$. The value 1005_{16}, which is held in the code segment selector register, causes the descriptor at offset 1000_{16} in the local descriptor table to be cached into the code segment descriptor cache. In this way, a 1M-byte code segment is activated starting at address 00600000_{16} in local memory.

The bits of the access rights byte define the operating characteristics of a segment. For example, it contains information about a segment such as whether the descriptor has been accessed, if it is a code or data segment descriptor, its privilege level, if it is readable or writable, and if it is currently loaded into internal memory. Let us next look at the function of each of these bits in detail.

The function of each bit in the access rights byte is listed in Fig. 5.20(b). Notice that if bit 0 is logic 1, the descriptor has been accessed. A descriptor is marked this way to indicate that its value has been cached on the 80386DX. The memory manager software checks this information to find out whether the segment is already in physical memory. Bit 4 identifies whether the descriptor represents a code/data segment or is a control descriptor. Let us assume that this bit is 1 to identify a segment descriptor. Then the type bits, bits 1 through 3, determine whether the descriptor describes a code segment or a data segment. For instance, 000 means that it is a read/write data segment that grows upward from the base to the limit. The DPL bits, bits 5 and 6, assign a descriptor privilege level to the segment. For example, 00 selects the most privileged level, level 0. Finally, the present bit indicates whether or not the segment is currently loaded into physical memory. This bit can be tested by the operating system software to determine if the segment should be loaded from a secondary storage device such as a hard disk. For example, if the access rights byte has logic 1 in bit 7, the segment is already available in physical memory and does not have to be loaded from an external device. Figure 5.22(a) shows the general form of a code segment descriptor and Fig. 5.22(b) a general data/stack segment descriptor.

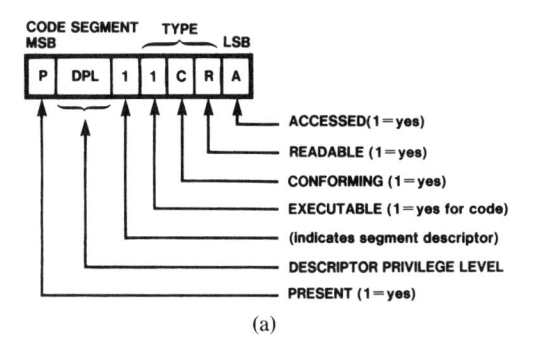

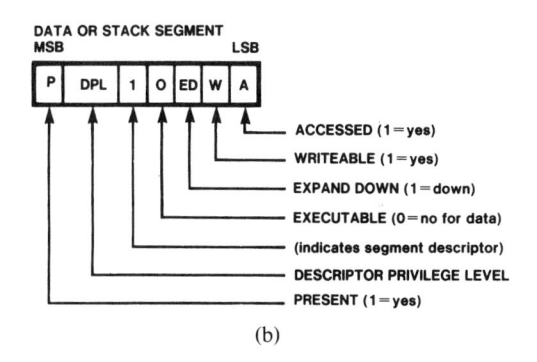

Figure 5.22 (a) Code segment descriptor access byte configuration. (Reprinted by permission of Intel Corp. Copyright Intel Corp. 1987.) (b) Data or stack segment access byte configuration. (Reprinted by permission of Intel Corp. Copyright Intel Corp. 1987.)

EXAMPLE 5.9

The access rights byte of a segment descriptor contains FE_{16}. What type of segment descriptor does it describe, and what are its characteristics?

SOLUTION Expressing the byte in binary form, we get

$$FE_{16} = 11111110_2$$

Since bit 4 is 1, the access rights byte is for a code/data segment descriptor. This segment has the characteristics that follow:

$P = 1 =$ Segment is mapped into physical memory

$DPL = 11 =$ Privilege level 3

$E = 1 =$ Executable code segment

$C = 1 =$ Conforming code segment

$R = 1 =$ Readable code segment

$A = 0 =$ Segment has not been accessed

An example of a system segment descriptor is the descriptor used to define the local descriptor table. This descriptor is located in the GDT. Looking at Fig. 5.23, we find that the format of a system segment descriptor is similar to the segment descriptor we just discussed. However, the type field of the access rights byte takes on new functions.

EXAMPLE 5.10

If a system segment descriptor has an access rights byte equal to 82_{16}, what type of descriptor does it represent? What is its privilege level? Is the descriptor present?

SOLUTION First, we will express the access rights byte in binary form. This gives

$$82_{16} = 10000010_2$$

Now we see that the type of the descriptor is

$TYPE = 0010 =$ Local descriptor table descriptor

The privilege level is

$DPL = 00 =$ Privilege level 0

and since

$P = 1$

the descriptor is present in physical memory.

Now that we have explained the format and use of descriptors, let us continue with page table entries. The format of either a page directory or page table entry is shown in Fig. 5.24. Here we see that the 20 most significant bits are either the base address of the page table if the entry is in the page directory table or the base address of the page frame if the entry is in the page table. Notice that

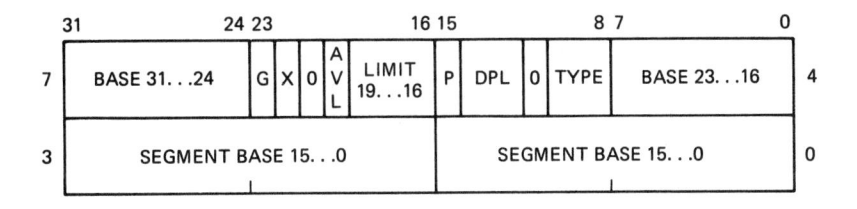

```
 31        24 23        16 15         8 7           0
7 | BASE 31...24 | G|X|0|A| LIMIT |P|DPL|0|TYPE| BASE 23...16 | 4
                        V  19...16
                        L
3 |      SEGMENT BASE 15...0      |     SEGMENT BASE 15...0     | 0
```

Name	Value	Description
TYPE	0	Reserved by Intel
	1	Available 80286 TSS
	2	LDT
	3	Busy 80286 TSS
	4	Call gate
	5	Task gate
	6	80286 interrupt gate
	7	80286 trap gate
	8	Reserved by Intel
	9	Available 80386 TSS
	A	Reserved
	B	Busy 80386 TSS
	C	80386 call gate
	D	Reserved by Intel
	E	80386 interrupt gate
	F	80386 trap gate
P	0	Descriptor contents are not valid
	1	Descriptor contents are valid
DPL	0-3	Descriptor privilege level 0, 1, 2 or 3
BASE	32-bit number	Base address of special system data segment in memory
LIMIT	20-bit number	Offset of last byte in segment from the base

Figure 5.23 System segment descriptor format and field definitions.

only bits 12 through 31 of the base address are supplied by the entry. The 12 least significant bits are assumed to be equal to zero. In this way we see that page tables and page frames are always located on a 4K-byte address boundary. In Fig. 5.19 we found that these entries are cached into the translation lookaside buffer when they are accessed.

The 12 lower bits of the entry supply protection characteristics or statistical information about the use of the page table or page frame. For example, the *user/supervisor* (U/S) and *read/write* (R/W) bits implement a two-level page protection

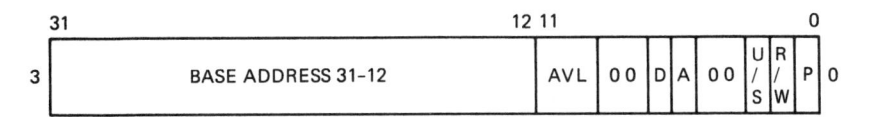

```
  31                                     12 11              0
3 |        BASE ADDRESS 31-12            |AVL|0 0|D|A|0 0|U|R|P| 0
                                                        |/|/|
                                                        |S|W|
```

Figure 5.24 Directory or page table entry format.

Sec. 5.4   Descriptor and Page Table Entries   183

mechanism. Setting U/S to 1 selects user-level protection. User is the low privilege level and is the same as protection level 3 of the segmentation model. That is, user is the protection level assigned to pages of memory that are accessible by application software. On the other hand, making U/S equal to zero assigns supervisor-level protection to the table or frame. Supervisor corresponds to levels 0, 1, and 2 of the segmentation model and is the level assigned to operating system resources. The *read/write* (R/W) bit is used to make a user-level table or frame read-only or read/write. Logic 1 in R/W selects read-only operation. Figure 5.25 summarizes the access characteristics for each setting of U/S and R/W.

U/S	R/W	User	Supervisor
0	0	None	Read/write
0	1	None	Read/write
1	0	Read-only	Read/write
1	1	Read/write	Read/write

Figure 5.25 User- and supervisor-level access rights.

Protection characteristics assigned by a page directory entry are applied to all page frames defined by the entries in the page table. On the other hand, the attributes assigned to a page table entry apply only to the page frame that it defines. Since two sets of protection characteristics exist for all page frames, the page protection mechanism of the 80386DX is designed always to enforce the higher privileged (more restricting) of the two protection rights.

EXAMPLE 5.11

If the page directory entry of the active page frame is $F1000007_{16}$ and its page table entry is 01000005_{16}, is the frame assigned to the user or supervisor? What access is permitted to the frame from user mode and from supervisor mode?

SOLUTION First, the page directory entry is expressed in binary form. This gives

$$F1000007_{16} = 11110001000000000000000000000111_2$$

Therefore, the page protection bits are

$$U/S\ R/W = 11$$

This assigns user-mode and read/write accesses to the complete page frame.

Next, the page table entry for the frame is expressed in binary form as

$$01000005_{16} = 00000001000000000000000000000101_2$$

Here we find that

$$U/S\ R/W = 10$$

This defines the page frame as a user-mode, read-only page. Since the page frame attributes are the more restrictive, they apply. Looking at Fig. 5.25, we see that user software (application software) can only read data in this frame. On the other hand, supervisor software (operating system software) can either read data from or write data into the frame.

The other implemented bits in the directory and page table entry of Fig. 5.24 provide statistical information about the table or frame usage. For instance, the *present* (P) bit identifies whether or not the entry can be used for page address translation. P equal to logic 1 indicates that the entry is valid and is available for use in address translation. On the other hand, if P equals zero, the entry is either undefined or not present in physical memory. If an attempt is made to access a page table or page frame that has its P bit marked zero, a page fault results. This page fault needs to be serviced by the memory manager part of the operating system.

The 80386DX also records the fact that a page table or page frame has been accessed. Just before a read or write is performed to any address in a table or frame, the *accessed* (A) bit of the entry is set to 1. This marks it as having been accessed. For page frame accesses, it also records whether the access was for a read or a write operation. The *dirty* (D) bit is defined only for a page table entry and it gets set if a write is performed to any address in the corresponding page frame. In a virtual demand paged memory system, the operating system can check the state of these bits to determine if a page in physical memory needs to be updated on the virtual storage device (hard disk) when a new page is swapped into its physical memory address space. The last three bits are labeled AVL and are available for use by the programmer.

5.5 PROTECTED-MODE SYSTEM CONTROL INSTRUCTION SET

In Chapters 3 and 4 we studied the real-mode instruction set of the 80386DX microprocessor. The instructions introduced in these chapters represent the base instruction set of the 8086 microprocessor and a number of new instructions called the extended instruction set and 80386DX specific instruction set. In protected mode the 80386DX executes all of the instructions that are available in the real mode. Moreover, it is enhanced with a number of additional instructions that either apply only to protected-mode operation or are used in the real mode to prepare the 80386DX for entry into the protected mode. As shown in Fig. 5.26, these instructions are known as the *system control instruction set*.

The instructions of the system control instruction set are listed in Fig. 5.27. Here we find the format of each instruction along with a description of its op-

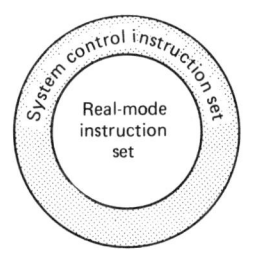

Figure 5.26 Protected-mode instruction set.

Instruction	Description	Mode
LGDT S	Load the global descriptor table register. S specifies the memory location that contains the first byte of the 6 bytes to be loaded into the GDTR.	Both
SGDT D	Store the global descriptor table register. D specifies the memory location that gets the first byte of the 6 bytes to be stored from the GDTR.	Both
LIDT S	Load the interrupt descriptor table register. S specifies the memory location that contains the first byte of the 6 bytes to be loaded into the IDTR.	Both
SIDT D	Store the interrupt descriptor table register. D specifies the memory location that gets the first byte of the 6 bytes to be stored from the IDTR.	Both
LMSW S	Load the machine status word. S is an operand to specify the word to be loaded into the MSW.	Both
SMSW D	Store the machine status word. D is an operand to specify the word location or register where the MSW is to be stored.	Both
LLDT S	Load the local descriptor table register. S specifies the operand to specify a word to be loaded into the LDTR.	Protected
SLDT D	Store the local descriptor table register. D is an operand to specify the word location where the LDTR is to be saved.	Protected
LTR S	Load the task register. S is an operand to specify a word to be loaded into the TR.	Protected
STR D	Store the task register. D is an operand to specify the word location where the TR is to be stored.	Protected
LAR D, S	Load access rights byte. S specifies the selector for the descriptor whose access byte is loaded into the upper byte of the D operand. The low byte specified by D is cleared. The zero flag is set if the loading completes successfully; otherwise it is cleared.	Protected
LSL R16, S	Load segment limit. S specifies the selector for the descriptor whose limit word is loaded into the word register operand R16. The zero flag is set if the loading completes successfully; otherwise it is cleared.	Protected
ARPL D, R16	Adjust RPL field of the selector. D specifies the selector whose RPL field is increased to match the PRL field in the register. The zero flag is set if successful; otherwise it is cleared.	Protected
VERR S	Verify read access. S specifies the selector for the segment to be verified for read operation. If successful the zero flag is set; otherwise it is reset.	Protected
VERW S	Verify write access. S specifies the selector for the segment to be verified for write operation. If successful the zero flag is set; otherwise it is reset.	Protected
CLTS	Clear task switched flag.	Protected

Figure 5.27 Protected-mode system control instruction set.

eration. Moreover, the mode or modes in which the instruction is available are identified. Let us now look at the operation of some of these instructions in detail.

Looking at Fig. 5.27, we see that the first six instructions can be executed in either the real or protected mode. They provide the ability to load (L) or store (S) the contents of the global descriptor table (GDT) register, interrupt descriptor table (IDT) register, and machine status word (MSW) part of CR_0. Notice that the instruction *load global descriptor table register* (LGDT) is used to load the GDTR from memory. Operand S specifies the location of the six bytes of memory that hold the limit and base that specifies the size and beginning address of the GDT. The first word of memory contains the limit and the next four bytes contain the base. For instance, executing the instruction

```
LGDT    [INIT_GDTR]
```

loads the GDTR with the base and limit pointed to by address INIT_GDTR to create a global descriptor table in memory. This instruction is meant to be used during system initialization and before switching the 80386DX to the protected mode.

Once loaded, the current contents of the GDTR can be saved in memory by executing the *store global descriptor table* (SGDT) instruction. An example is the instruction

```
SGDT    [SAVE_GDTR]
```

The instructions LIDT and SIDT perform similar operations for the IDTR. The IDTR is also set up during initialization.

The instructions *load machine status word* (LMSW) and *store machine status word* (SMSW) are provided to load and store the contents of the machine status word (MSW), respectively. These are the instructions that are used to switch the 80386DX from real to protected mode. To do this we must set the least significant bit in the MSW to 1. This can be done by first reading the contents of the machine status word, modifying the LSB (PE), and then writing the modified value back into the MSW part of CR_0. The instruction sequence that follows will switch an 80386DX operating in real mode to the protected mode:

```
SMSW    AX          ;read from the MSW
OR      AX,1        ;modify the PE bit
LMSW    AX          ;write to the MSW
```

The next four instructions in Fig. 5.27 are also used to initialize or save the contents of protected-mode registers. However, they can be used only when the 80386DX is in the protected mode. To load and to save the contents of the LDTR, we have the instructions LLDT and SLDT, respectively. Moreover, for loading and saving the contents of the TR, the equivalent instructions are LTR and STR.

The rest of the instructions in Fig. 5.27 are for accessing the contents of descriptors. For instance, to read a descriptor's access rights byte, the *load access rights byte* (LAR) instruction is executed. An example is the instruction

```
LAR    AX,LDIS_1
```

Execution of this instruction causes the access rights byte of local descriptor 1 to be loaded into AH. To read the segment limit of a descriptor, we use the *load segment limit* (LSL) instruction. For instance, to copy the segment limit for local descriptor 1 into register EBX, the instruction

```
LSL    EBX, LDIS_1
```

is executed. In both cases ZF is set to 1 if the operation is performed correctly.

The instruction *adjust RPL field of selector* (ARPL) can be used to increase the RPL field of a selector in memory or a register, destination (D), to match the protection level of the selector in a register, source (S). If an RPL level increase takes place, ZF is set to 1. Finally, the instructions VERR and VERW are provided to test the accessibility of a segment for a read or write operation, respectively. If the descriptor permits the type of access tested for by executing the instruction, ZF is set to 1.

5.6 MULTITASKING AND PROTECTION

We say that the 80386DX microprocessor implements a *multitasking* software architecture. By this we mean that it contains on-chip hardware that both permits multiple tasks to exist in a software system and allows them to be scheduled for execution in a time-shared manner. That is, program control is switched from one task to another after a fixed interval of time elapses. For instance, the tasks can be executed in a round-robin fashion. This means that the most recently executed task is returned to the end of the list of tasks being executed. Even though the processes are executed in a time-shared fashion, an 80386DX microcomputer has the performance to make it appear to the user that they are all running simultaneously.

Earlier we defined a task as a collection of program routines that performs a specific function. This function is also called a *process*. Software systems typically need to perform many processes. In the protected-mode 80386DX-based microcomputer, each process is identified as an independent task. The 80386DX provides an efficient mechanism, called the *task switching mechanism*, for switching between tasks. For instance, an 80386DX running at 16 MHz can perform a task switch operation in just 19 μs.

We also indicated earlier that when a task is called into operation, it can have both global and local memory resources. The local memory address space is divided between tasks. This means that each task normally has its own private segments of local memory. Segments in global memory can be shared by all tasks. Therefore, a task can have access to any of the segments in global memory. As shown in Fig. 5.28, task A has both a private address space and a global address space available for its use.

Protection and the Protection Model

There are safeguards that can be built into the protected-mode software system to deny unauthorized or incorrect accesses of a task's memory resources. The

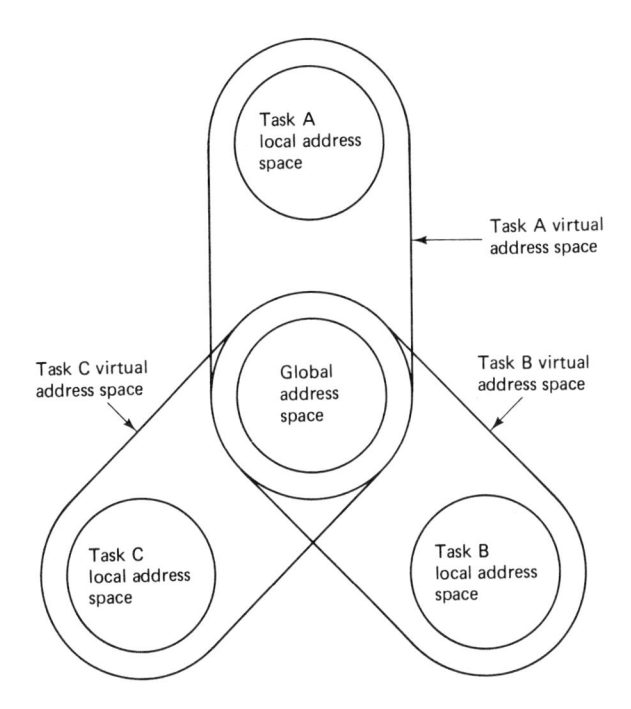

Figure 5.28 Virtual address space of a task. (Reprinted by permission of Intel Corp. Copyright Intel Corp. 1987.)

concept of safeguarding memory is known as *protection*. The 80386DX includes on-chip hardware that implements a *protection mechanism*. This mechanism is designed to put restrictions on the access of local and system resources by a task and to isolate tasks from each other in a multitasking environment.

Segmentation, paging, and descriptors are the key elements of the 80386DX's protection mechanism. We already identified that when using a segmented memory model a segment is the smallest element of the virtual memory address space that has unique protection attributes. These attributes are defined by the access rights information and limit fields in the segment's descriptor. As shown in Fig. 5.29(a), the on-chip protection hardware performs a number of checks during all memory accesses. Figure 5.29(b) is a list of the protection checks and restrictions imposed on software by the 80386DX. For example, when a data storage location in memory is written to, the type field in the access rights byte of the segment is tested to assure that its attributes are consistent with the register cache being loaded and the offset is checked to verify that it is within the limit of the segment.

Let us just review the attributes that can be assigned to a segment with the access rights information in its descriptor. Figure 5.30 shows the format of a data segment descriptor and a executable (code) segment descriptor. The P bit defines whether a segment of memory is present in physical memory. Assuming that a segment is present, bit 4 of the type field makes it either a code segment or data segment. Notice that this bit is 0 if the descriptor is for a data segment and it is 1 for code segments. Segment attributes such as readable, writable, conforming, expand up or expand down, and accessed are assigned by other bits in the type field. Finally, a privilege level is assigned with the DPL field.

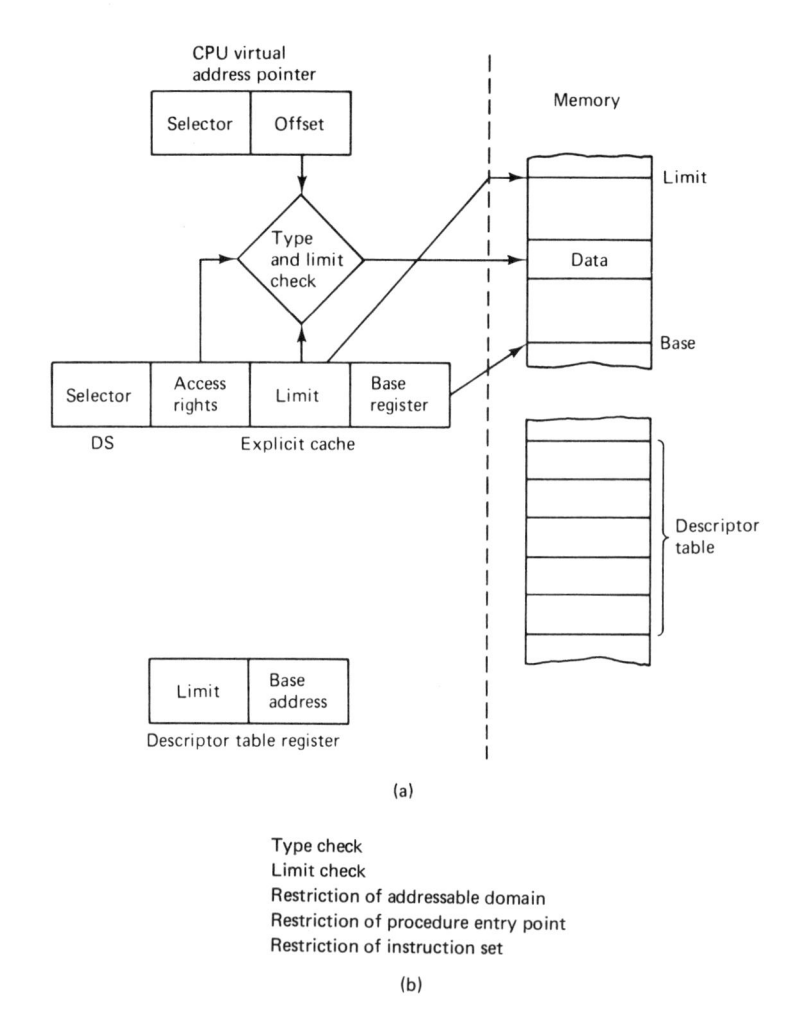

```
                    CPU virtual
                 address pointer
                ┌────────┬────────┐
                │Selector│ Offset │                    Memory
                └────────┴────────┘
                                                              Limit
                       ┌─────┐
                       │Type │
                  ┌───▶│ and │───▶              Data
                  │    │limit│
                  │    │check│
                  │    └─────┘
                  │       ▲                              Base
                  │       │
        ┌────────┬──────┬─────┬────────┐
        │Selector│Access│Limit│ Base   │
        │        │rights│     │register│
        └────────┴──────┴─────┴────────┘
           DS         Explicit cache
                                                        Descriptor
                                                          table


        ┌─────┬────────┐
        │Limit│  Base  │
        │     │address │
        └─────┴────────┘
        Descriptor table register

                            (a)

                  Type check
                  Limit check
                  Restriction of addressable domain
                  Restriction of procedure entry point
                  Restriction of instruction set

                            (b)
```

Figure 5.29 (a) Testing the access rights of a descriptor. (Reprinted by permission of Intel Corp. Copyright Intel Corp. 1982.) (b) Protection checks and restrictions.

Earlier we showed that whenever a segment is accessed, the base address and limit are cached inside the 80386DX. In Fig. 5.29(a) we find that the access rights information is also loaded into the cache register. However, before loading the descriptor the MMU verifies that the selected segment is currently present in physical memory, that the segment is at a privilege level that is accessible from the privilege level of the current program, that the type is consistent with the target segment selector register (CS = code segment, DS, ES, FS, GS, or SS = data segment), and that the reference into the segment does not exceed the address limit of the segment. If a violation is detected, an error condition is signaled. The memory manager software can determine the cause of the error, correct the problem, and then reinitiate the operation.

Let us now look at some examples of memory accesses that result in protection violations. For example, if the selector loaded into the CS register points

DATA SEGMENT DESCRIPTOR

```
 32        23         15              7              0
┌──────────┬───┬──┬──────┬──┬────┬────────┬──────────┐
│ BASE 31..24│G B│AV│LIMIT │P │DPL │ TYPE   │BASE 23..16│ 4
│          │  0│ L│19..16│  │    │1 0 E W A│          │
├──────────┴───┴──┴──────┴──┴────┴────────┴──────────┤
│       SEGMENT BASE 15..0    │   SEGMENT LIMIT 15..0 │ 0
└─────────────────────────────┴───────────────────────┘
```

EXECUTABLE SEGMENT DESCRIPTOR

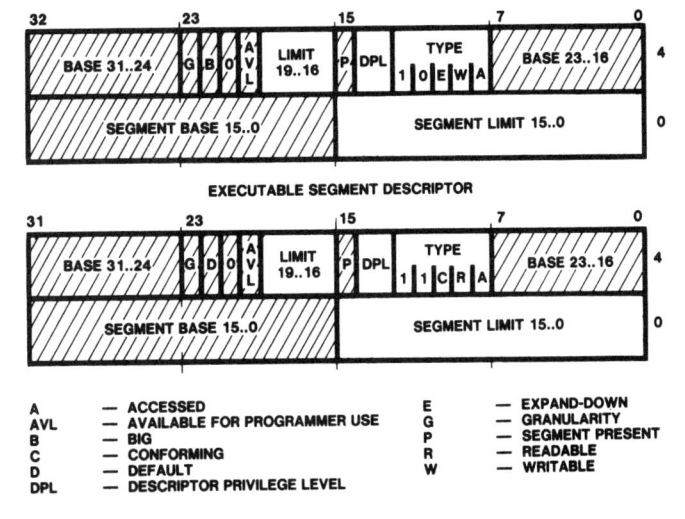

```
 31        23         15              7              0
┌──────────┬───┬──┬──────┬──┬────┬────────┬──────────┐
│ BASE 31..24│G D│AV│LIMIT │P │DPL │ TYPE   │BASE 23..16│ 4
│          │  0│ L│19..16│  │    │1 1 C R A│          │
├──────────┴───┴──┴──────┴──┴────┴────────┴──────────┤
│       SEGMENT BASE 15..0    │   SEGMENT LIMIT 15..0 │ 0
└─────────────────────────────┴───────────────────────┘
```

A	— ACCESSED		E	— EXPAND-DOWN
AVL	— AVAILABLE FOR PROGRAMMER USE		G	— GRANULARITY
B	— BIG		P	— SEGMENT PRESENT
C	— CONFORMING		R	— READABLE
D	— DEFAULT		W	— WRITABLE
DPL	— DESCRIPTOR PRIVILEGE LEVEL			

Figure 5.30 Data segment and executable (code) segment descriptors. (Reprinted by permission of Intel Corp. Copyright Intel Corp. 1986.)

to a descriptor that defines a data segment, the type check leads to a protection violation. Another example of an invalid memory access is an attempt to read an operand from a code segment that is not marked as readable. Finally, any attempt to access a byte of data at an offset greater than LIMIT, a word at an offset equal to or greater than LIMIT, or a double word at an offset equal to or greater than LIMIT-2 extends beyond the end of the data segment and results in a protection violation.

The 80386DX's protection model provides four possible privilege levels for each task. They are called *levels 0, 1, 2, and 3* and can be illustrated by concentric circles as in Fig. 5.31. Here level 0 is the most privileged level and level 3 is the least privileged level.

System and application software are typically partitioned in the manner shown in Fig. 5.31. The kernel represents application-independent software that provides microprocessor-oriented functions such as I/O control, task sequencing, and memory management. For this reason it is kept at the most privileged level, level 0. Level 1 contains processes that provide system services such as file accessing. Level 2 is used to implement custom routines to support special-purpose system operations. Finally, the least privileged level, level 3, is the level at which user applications are run. This example also demonstrates how privilege levels are used to isolate system-level software (operating system software in levels 0 through 2) from the user's application software (level 3). Tasks at a level can use programs from the more privileged levels but cannot modify the contents of these routines in any way. In this way, applications are permitted to use system software routines from the three higher privilege levels without affecting their integrity.

Earlier we indicated that protection restrictions are put on the instruction

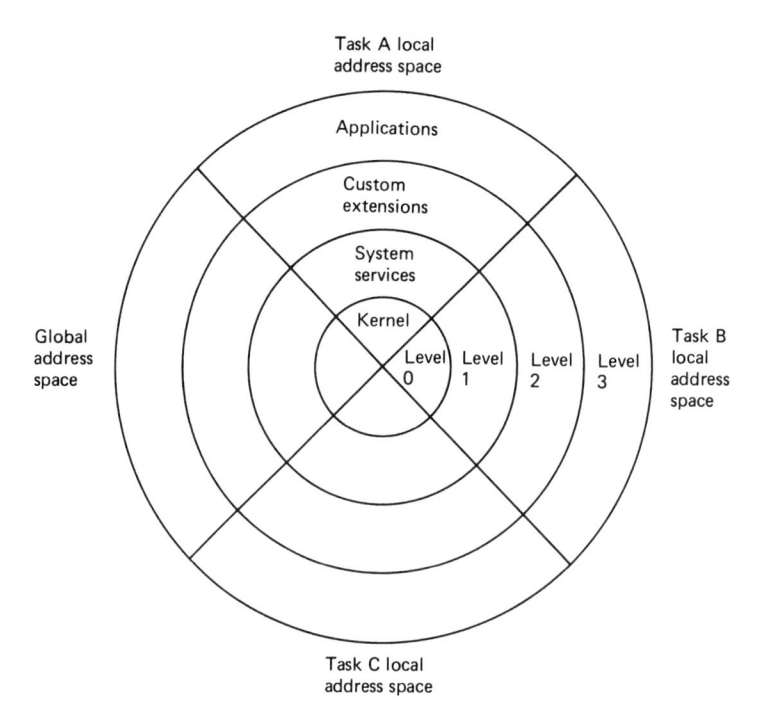

Figure 5.31 Protection model.

set. One example of this is that the system control instructions can only be executed in a code segment that is at protection level 0. We also pointed out that each task is assigned its own local descriptor table. Therefore, as long as none of the descriptors in a task's local descriptor table reference code or data available to another task, it is isolated from all other tasks. That is, it has been assigned a unique part of the virtual address space. For example, in Fig. 5.31 multiple applications running at level 3 are isolated from each other by assigning them different local resources. This shows that segments, privilege levels, and the local descriptor table provide protection for both code and data within a task. These types of protection result in improved software reliability because errors in one application will not affect the operating system or other applications.

Let us now look more closely at how the privilege level is assigned to a code or data segment. Remember that when a task is running it has access to both local and global code segments, local and global data segments, and stack segments. A privilege level is assigned to each of these segments through the access rights information in its descriptor. A segment may be assigned to any privilege level simply by entering the number for the level into the DPL bits.

To provide more flexibility, input/output has two levels of privilege. First, the I/O drivers, which are normally system resources, are assigned to a privilege level. For the software system of Fig. 5.31, we indicated that the I/O control routines were part of the kernel and are at level 0.

The IN, INS, OUT, OUTS, CLI, and STI instructions are called *trusted instructions*. This is because the protection model of the 80386DX puts additional restrictions on their use in protected mode. They can only be executed at a privilege level that is equal to or more privileged than the *input/output privilege level* (IOPL) code. IOPL supplies the second level of I/O privilege. Remember that the IOPL bits exist in the protected-mode flag register. These bits must be loaded with the value of the privilege level that is to be assigned to input/output instructions through software. The value of IOPL may change from task to task. Assigning the I/O instructions to a level higher that 3 restricts applications from directly performing I/O. Instead, to perform an I/O operation, the application must request service by an I/O driver through the operating system.

Accessing Code and Data through the Protection Model

During the running of a task, the 80386DX may either need to pass control to program routines at another privilege level or access data in a segment that is at a different privilege level. Accesses to code or data in segments at a different privilege level are governed by strict rules. These rules are designed to protect the code or data at the more privileged level from contamination by the less privileged routine.

Before looking at how accesses are made for routines or data at the same or different privilege levels, let us first look at some terminology used to identify privilege levels. We have already been using the terms *descriptor privilege level* (DPL) and *I/O privilege level* (IOPL). However, when discussing the protection mechanisms by which processes access data or code, two new terms come into play: *current privilege level* (CPL) and *requested privilege level* (RPL). CPL is defined as the privilege level of the code or data segment that is currently being accessed by a task. For example, the CPL of an executing task is the DPL of the access rights byte in the descriptor cache for the CS register. This value normally equals the DPL of the code segment. RPL is the privilege level of the new selector loaded into a segment register. For instance, in the case of code, it is the privilege level of the code segment that contains the routine that is being called. That is, RPL is the DPL of the code segment to which control is to be passed.

As a task in an application runs, it may require access to program routines that reside in segments at any of the four privilege levels. Therefore, the current privilege level of the task changes dynamically with the programs it executes. This is because CPL of the task is normally switched to the DPL of the code segment currently being accessed.

The protection rules of the 80386DX determine what code or data can be accessed by a program. Before looking at how control is passed to code at different protection levels, let us first look at how data segments are accessed by code at the CPL. Figure 5.32 illustrates the protection-level checks that are made for a data access. The general rule is that code can only access data that are at the same or a less privileged level. For instance, if the current privilege level of a task is 1, it can access operands that are in data segments with DPL equal to 1, 2, or 3. Whenever a new selector is loaded into the DS, ES, FS, or GS register, the DPL of the target data segment is checked to make sure that it is equal to or

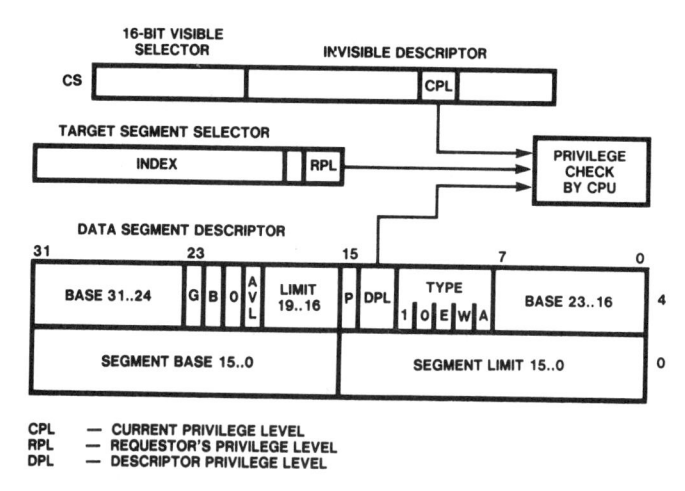

Figure 5.32 Privilege-level checks for a data access. (Reprinted by permission of Intel Corp. Copyright Intel Corp. 1986.)

less privileged than the most privileged of either CPL or RPL. As long as DPL satisfies this condition, the descriptor is cached inside the 80386DX and data access takes place.

One exception to this rule is when the SS register is loaded. In this case, the DPL must always equal the CPL. That is, the active stack (one is required for each privilege level) is always at the CPL.

EXAMPLE 5.12

Assuming that in Fig. 5.32 DPL = 2, CPL = 0, and RPL = 2, will the data access take place?

SOLUTION DPL of the target segment is 2 and this value is less privileged than CPL = 0, which is the more privileged of CPL and RPL. Therefore, the protection criteria are satisfied and the access will take place.

Different rules apply to how control is passed between code at the same privilege level and between code at different privilege levels. To transfer program control to another instruction in the same code segment, one can simply use a near jump or call instruction. In either case, just a limit check is made to assure that the destination of the jump or call does not exceed the limit of the current code segment.

To pass control to code in another segment that is at the same or a different privilege level, a far jump or call instruction is used. For this transfer of program control, both type and limit checks are performed and privilege-level rules are applied. Figure 5.33 shows the privilege checks made by the 80386DX. There are two conditions under which the transfer in program control will take place. First, if CPL equals the DPL, the two segments are at the same protection level and the transfer occurs. Second, if the CPL represents a more privileged level than

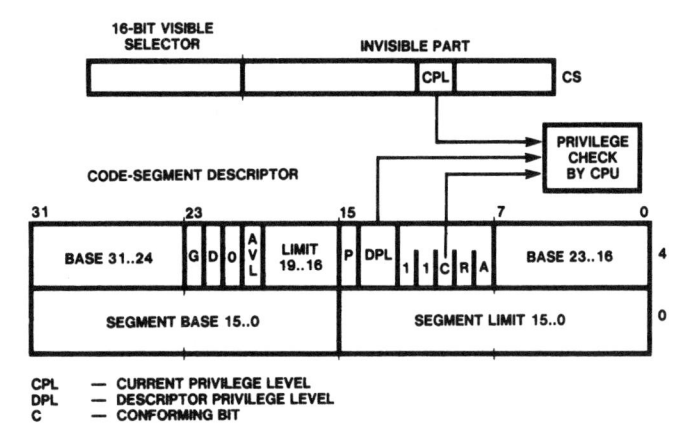

Figure 5.33 Privilege-level checks when passing program control directly to a program in another segment. (Reprinted by permission of Intel Corp. Copyright Intel Corp. 1986.)

DPL, but the conforming code (C) bit in the type field of the new segment is set, the routine is executed at the CPL.

The general rule that applies when control is passed to code in a segment that is at a different privilege level is that the new code segment must be at a more privileged level. A special kind of descriptor called a *gate descriptor* comes into play to implement the change in privilege level. An attempt to transfer control to a routine in a code segment at a higher privilege level is still initiated with either a far call or far jump instruction. This time the instruction does not directly specify the location of the destination code; instead, it references a gate descriptor. In this case the 80386DX goes through a much more complex program control transfer mechanism.

The structure of a gate descriptor is shown in Fig. 5.34. Notice that there are four types of gate descriptors: the *call gate*, *task gate*, *interrupt gate*, and *trap gate*. The call gate implements an indirect transfer of control within a task from code at the CPL to code at a higher privilege level. It does this by defining a valid entry point into the more privileged segment. The contents of a call gate are the virtual address of the entry point: the *destination selector* and the *destination offset*. In Fig. 5.34 we see that the destination selector identifies the code segment that contains the program to which control is to be redirected. The destination offset points to the instruction in this segment where execution is to resume. Call gates can reside in either the GDT or an LDT.

The operation of the call gate mechanism is illustrated in Fig. 5.35. Here we see that the call instruction includes an offset and a selector. When the instruction is executed, this selector is loaded into CS and points to the call gate. In turn, the call gate causes its destination selector to be loaded into CS. This leads to the caching of the descriptor for the called code segment (executable segment descriptor). The executable segment descriptor provides the base address

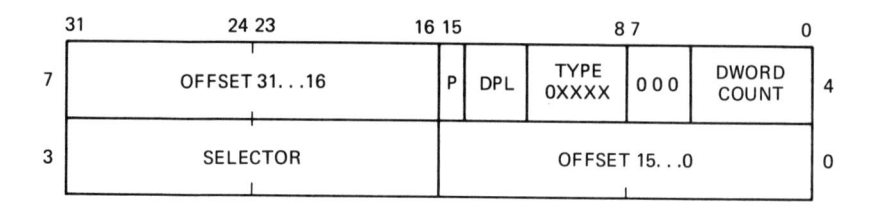

7	31 24 23 OFFSET 31...16	16 15 P	DPL	8 7 TYPE 0XXXX	0 0 0	0 DWORD COUNT	4
3	SELECTOR			OFFSET 15...0			0

Name	Value	Description
TYPE	4 5 6 7	— Call gate — Task gate — Interrupt gate — Trap gate
P	0 1	— Descriptor contents are not valid — Descriptor contents are valid
DPL	0-3	Descriptor privilege level
WORD COUNT	0-31	Number of double words to copy from callers stack to called procedures stack. Only used with call gate
DESTINATION SELECTOR	16-Bit selector	Selector to the target code segment (call interrupt or trap gate) Selector to the target task state segment (task gate)
DESTINATION OFFSET	32-Bit offset	Entry point within the target code segment

Figure 5.34 Gate descriptor format. (Reprinted by permission of Intel Corp. Copyright Intel Corp. 1987.)

for the executable segment (code segment) of memory. Notice that the offset in the call gate descriptor locates the entry point of the procedure in the executable segment.

Whenever the task's current privilege level is changed, a new stack is activated. As part of the program context switching sequence, the old ESP and SS are saved on the new stack along with any parameters and the old EIP and CS. This information is needed to preserve linkage for return to the old program environment.

Now the procedure at the higher privilege level begins to execute. At the end of the routine, a RET instruction must be included to return program control back to the calling program. Execution of RET causes the old values of EIP, CS, the parameters, ESP, and SS to be popped from the stack. This restores the original program environment. Now program execution resumes with the instruction following the call instruction in the lower privileged code segment.

Figure 5.36 shows the privilege checks that are performed when program control transfer is initiated through a call gate. For the call to be successful, the

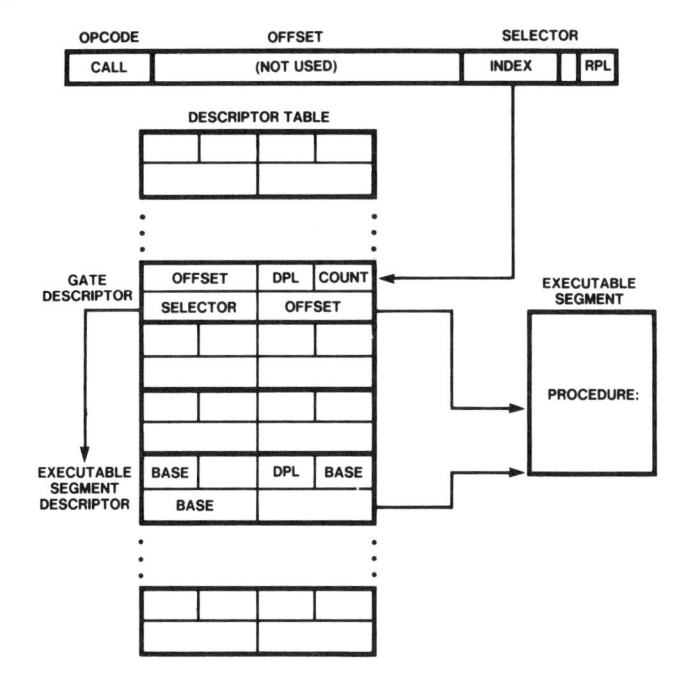

Figure 5.35 Call gate operation. (Reprinted by permission of Intel Corp. Copyright Intel Corp. 1986.)

DPL of the gate must be the same as the CPL, and the RPL of the called code must be higher than the CPL.

Task Switching and the Task State Segment Table

Earlier we identified the task as the key program element of the 80386DX's multitasking software architecture and that another important feature of this architecture is the high-performance task switching mechanism. A task can be invoked either directly or indirectly. This is done by executing either the intersegment

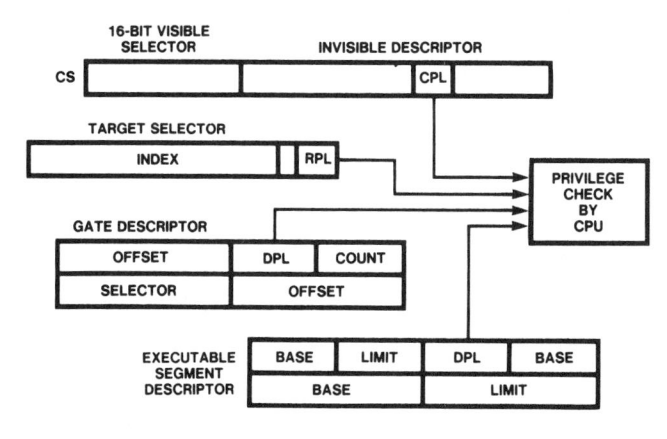

Figure 5.36 Privilege-level checks for program control transfer with a call gate. (Reprinted by permission of Intel Corp. Copyright Intel Corp. 1986.)

CPL — CURRENT PRIVILEGE LEVEL
RPL — REQUESTOR'S PRIVILEGE LEVEL
DPL — DESCRIPTOR PRIVILEGE LEVEL

Sec. 5.6 Multitasking and Protection

jump or intersegment call instruction. When a jump instruction is used to initiate a task switch, no return linkage to the prior task is supported. On the other hand, if a call is used to switch to the new task instead of a jump, back linkage information is saved automatically. This information permits a return to be performed to the instruction that follows the calling instruction in the old task at completion of the new task.

Each task that is to be performed by the 80386DX is assigned a unique selector called a *task state selector*. This selector is an index to a corresponding *task state segment descriptor* in the global descriptor table. The format of a task state segment descriptor is given in Fig. 5.37.

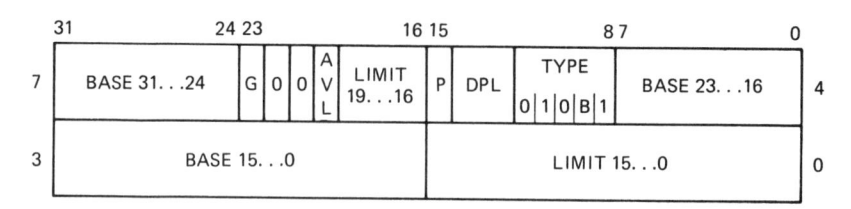

31	24 23			A V L	16 15			8 7		0	
7	BASE 31...24	G	0 0		LIMIT 19...16	P	DPL	TYPE 0 1 0 B 1	BASE 23...16		4
3	BASE 15...0					LIMIT 15...0					0

Figure 5.37 TSS descriptor format. (Reprinted by permission of Intel Corp. Copyright Intel Corp. 1986.)

If a jump or call instruction has a task state selector as its operand, a direct entry is performed to the task. As shown in Fig. 5.38, when a call instruction is executed, the selector is loaded into the 80386DX's task register (TR). Then the corresponding task state segment descriptor is read from the GDT and loaded into the task register cache. Of course, this happens only if the criteria specified by the access rights information of the descriptor are satisfied. That is, the descriptor is present (P = 1); the task is not busy (B = 0); and protection is not violated (CPL must be equal to DPL). Looking at Fig. 5.38, we see that once loaded, the base address and limit specified in the descriptor define the starting point and size of the task's *task state segment* (TSS). This TSS contains all of the information that is needed to either start or stop a task.

Before explaining the rest of the task switch sequence, let us first look more closely at what is contained in the task state segment. A typical TSS is shown in Fig. 5.38. Its minimum size is 103 bytes. For this reason, the minimum limit that can be specified in a TSS descriptor is 00067_{16}. Notice that the segment contains information such as the state of the microprocessor (general registers, segment selectors, instruction pointer, and flags) needed to initiate the task, a back link selector to the TSS of the task that was active when this task was called, the local descriptor table register selector, a stack selector and pointer for privilege levels 0, 1, and 2, and an I/O permission bit map.

Now we will continue with the procedure by which a task is invoked. Let us assume that a task was already active when a new task was called. Then the new task is what is called a *nested task* and causes the NT bit of the flags register to be set to 1. In this case the current task is first suspended and then the state of the 80386DX's user-accessible registers is saved in the old TSS. Next, the B

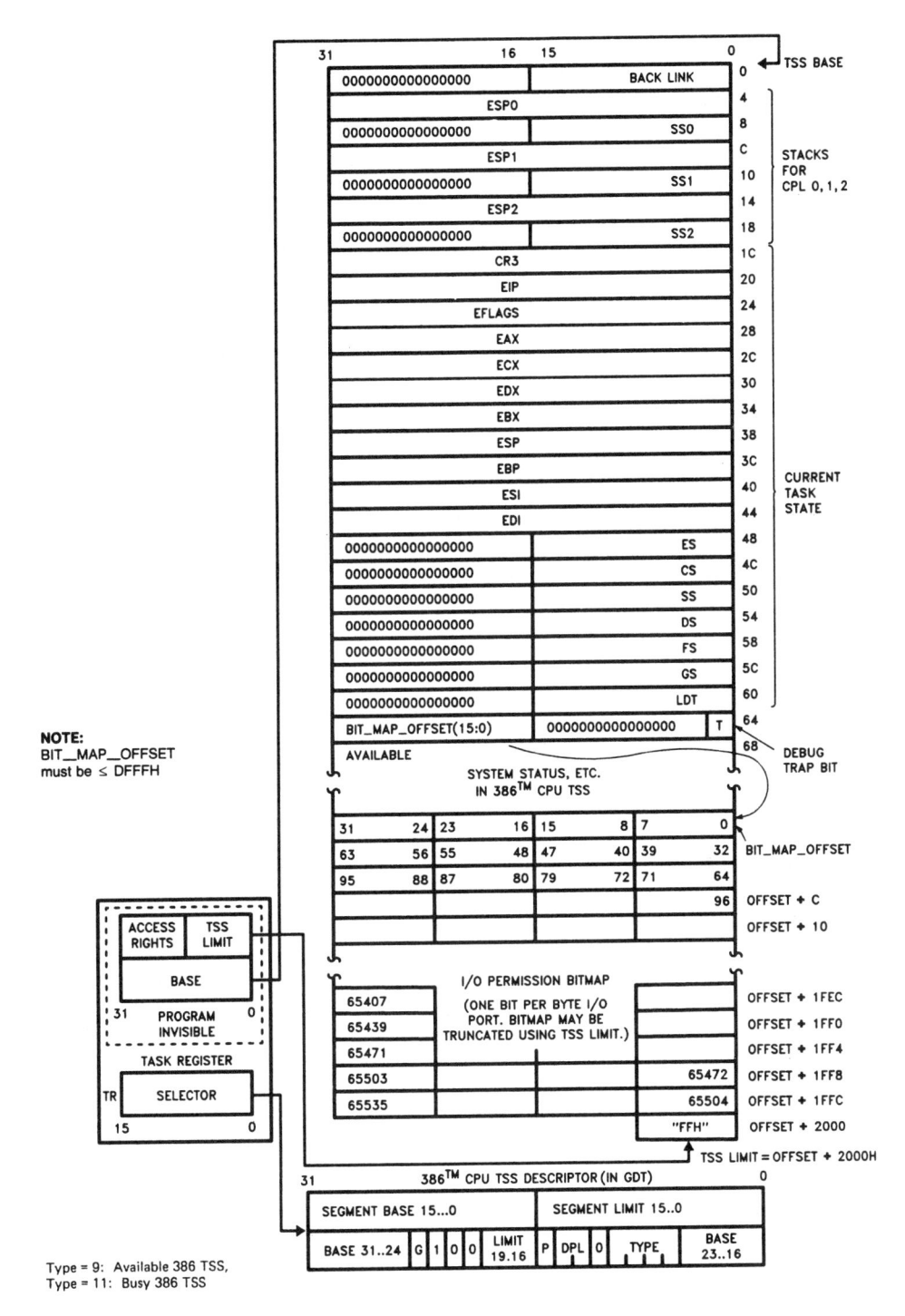

Figure 5.38 Task state segment table. (Reprinted by permission of Intel Corp. Copyright Intel Corp. 1987.)

Sec. 5.6 Multitasking and Protection 199

bit in the new task's descriptor is marked busy; the TS bit in the machine status word is set to indicate that a task is active; the state information from the new task's TSS is loaded into the MPU; and the selector for the old TSS is saved as the back-link selector in the new task state segment. The task switch operation is now complete and execution resumes with the instruction identified by the new contents of the code segment selector (CS) and instruction pointer (EIP).

The old program context is preserved by saving the selector for the old TSS as the back-link selector in the new TSS. By executing a return instruction at the end of the new task, the back-link selector for the old TSS is automatically reloaded into TR. This activates the old TSS and restores the prior program environment. Now program execution resumes at the point where it left off in the old task.

The indirect method of invoking a task is by jumping to or calling a *task gate*. This is the method used to transfer control to a task at an RPL that is higher than the CPL. Figure 5.39 shows the format of a task gate. This time the instruction

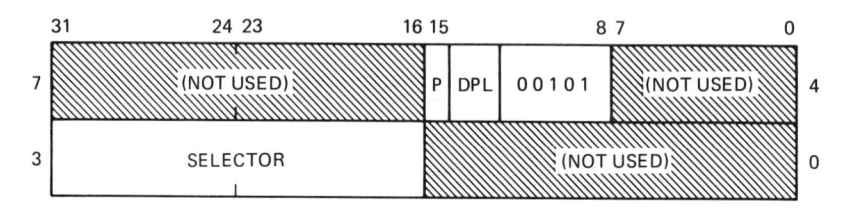

31	24 23	16 15	8 7	0
7	(NOT USED)	P DPL 0 0 1 0 1	(NOT USED)	4
3	SELECTOR	(NOT USED)		0

Figure 5.39 Task gate descriptor format. (Reprinted by permission of Intel Corp. Copyright Intel Corp. 1986.)

includes a selector that points to the task gate, which is in either the LDT or GDT, instead of a task state selector. The TSS selector held in this gate is loaded into TR to select the TSS and initiate the task. Figure 5.40 illustrates a task initiated through a task gate.

Let us consider an example to illustrate the principle of task switching. In Fig. 5.41 we have a table that contains TSS descriptors SELECT0 through SELECT3. These descriptors contain access rights and a selector for tasks 0 through 3, respectively. To invoke the task corresponding to selector SELECT2 in the data segment where these selectors are stored, we can use the following procedure. First, the data segment register is loaded with the address SELECTOR_DATA_SEG_START to point to the segment that contains the selectors. This is done with the instructions

```
        MOV     AX, SELECTOR_DATA_SEG_START
        MOV     DS, AX
```

Since each selector is eight bytes long, SELECT2 is offset from the beginning of the segment by 16 bytes. Let us load this offset into register EBX.

```
        MOV     EBX, 10H
```

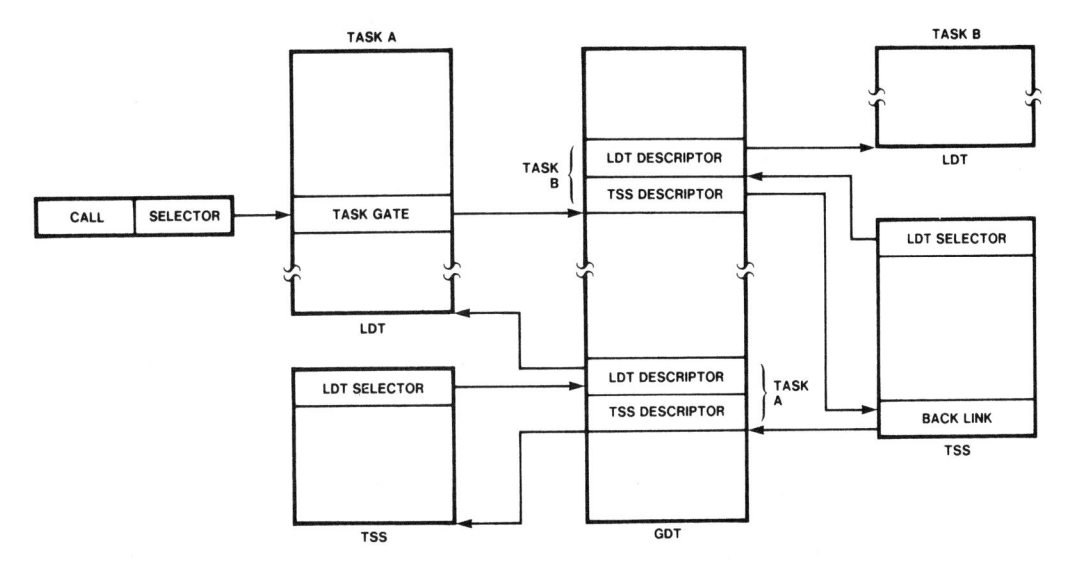

Figure 5.40 Task switch through a task gate. (Reprinted by permission of Intel Corp. Copyright Intel Corp. 1987.)

At this point we can use SELECT2 to implement an intersegment jump with the instruction

```
JMP    DWORD PTR [EBX]
```

Execution of this instruction switches program control to the task specified by the selector in descriptor SELECT2. In this case, no program linkage is preserved.

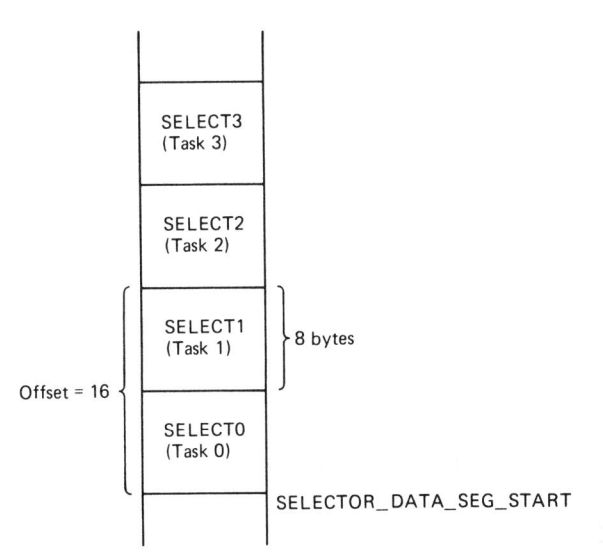

Figure 5.41 Task gate selectors.

Sec. 5.6   Multitasking and Protection

On the other hand, by calling the task with the instruction

```
CALL    DWORD PTR [EBX]
```

linkage is maintained.

5.7 VIRTUAL 8086 MODE

8086 and 8088 application programs, such as those written for the PC DOS operating system, can be run directly on the 80386DX in real mode. A protected-mode operating system, such as UNIX™, can also run DOS applications without change. This is done through what is called *virtual 8086 mode*. When in this mode, the 80386DX supports an 8086 microprocessor programming model and can directly run programs written for the 8086. That is, it creates a virtual 8086 machine for executing programs.

In this kind of application, the 80386DX is switched back and forth between protected mode and virtual 8086 mode. The UNIX operating system and UNIX applications are run in protected mode, and when the DOS operating system and a DOS application are to be run, the 80386DX is switched to the virtual 8086 mode. This mode switching is controlled by a program known as a *virtual 8086 monitor*.

Virtual 8086 mode of operation is selected by the bit called *virtual mode* (VM) in the extended flag register. VM must be switched to 1 to enable the virtual 8086 mode of operation. Actually, the VM bit in EFLAGS is not directly switched to 1 by software. This is because virtual 8086 mode is normally entered as a protected-mode task. Therefore, the copy of EFLAGS in the TSS for the task would include VM equal 1. These EFLAGS are loaded as part of the task switching process. In turn, the virtual 8086 mode of operation is initiated. The virtual 8086 program is run at privilege level 3. The virtual 8086 monitor is responsible for setting and resetting the VM bit in the task's copy of EFLAGS and permits both protected-mode tasks and virtual 8086 mode tasks to coexist in a multitasking program environment.

Another way of initiating virtual 8086 mode is through an interrupt return. In this case, the EFLAGS are reloaded from the stack. Again the copy of EFLAGS must have the VM bit set to 1 to enter the virtual 8086 mode of operation.

ASSIGNMENT

Section 5.2

1. List the protected-mode registers that are not part of the real-mode model.
2. What are the two parts of the GDTR called?
3. What is the function of the GDTR?

4. If the contents of the GDTR are $0021000001FF_{16}$, what are the starting and ending addresses of the table? How large is the table? How many descriptors can be stored in the table?
5. What is stored in the GDT?
6. What do *IDTR* and *IDT* stand for?
7. What is the maximum limit that should be used in the IDTR?
8. What is stored in the IDT?
9. What descriptor table defines the local memory address space?
10. What gets loaded into the LDTR? What happens when it gets loaded?
11. Which control register contains the MSW?
12. Which bit is used to switch the 80386DX from real-address mode to protected-address mode?
13. What MSW bit settings identify that floating-point operations are to be performed by an 80387DX coprocessor?
14. What does *TS* stand for?
15. What must be done to turn on paging?
16. Where is the page directory base register located?
17. How large is the page directory?
18. What is held in the page table?
19. What gets loaded into TR? What is its function?
20. What is the function of the task descriptor cache?
21. What determines the location and size of a task state segment?
22. What is the name of the CS register in the protected mode? The DS register?
23. What are the names and sizes of the three fields in a selector?
24. What does "TI equal 1" mean?
25. If the GDT register contains $0013000000FF_{16}$ and the selector loaded into the LDTR is 0040_{16}, what is the starting address of the LDT descriptor that is to be loaded into the cache?
26. What does *NT* stand for? *RF*?
27. If the IOPL bits of the flag register contain 10, what is the privilege level of the I/O instructions?

Section 5.3

28. How large is the 80386DX's virtual address?
29. What are the two parts of a virtual address called?
30. How large can a data segment be? How small?
31. How large is the 80386DX's virtual address space? What is the maximum number of segments that can exist in the virtual address space?
32. How large is the global memory address space? How many segments can it contain?
33. In Fig. 5.12, which segments of memory does task 3 have access to? Which segments does it not have access to?
34. What part of the 80386DX is used to translate virtual addresses to physical addresses?

35. What happens when the following instruction sequence is executed?

    ```
    MOV   AX, [SI]
    MOV   CS, AX
    ```

36. If the descriptor accessed in Problem 35 has the value $00200000FFFF_{16}$ and IP contains 0100_{16}, what is the physical address of the next instruction to be fetched?
37. Into how many pages is the 80386DX's address space mapped when paging is enabled? What is the size of a page?
38. What are the three elements of the linear address that is produced by page translation? Give the size of each element.
39. What is the purpose of the translation lookaside buffer?
40. How large is a page frame? What selects the specific storage location in the page frame?

Section 5.4

41. How many bytes are in a descriptor? Name each of its fields and give their size.
42. Which registers are segment descriptors associated with? System segment descriptors?
43. The selector 0224_{16} is loaded into a segment register. This value points to a segment descriptor starting at address 00100220_{16} in the local descriptor table. If the words of the descriptor are

    $$00100220H = 0110H$$
    $$00100222H = 0000H$$
    $$00100224H = 1A20H$$
    $$00100226H = 0000H$$

    what are the LIMIT and BASE?
44. Is the segment of memory identified by the descriptor in Problem 43 already loaded into physical memory? A code segment or data segment?
45. If the current value of EIP is 00000226_{16}, what is the physical address of the next instruction to be fetched from the code segment of Problem 43?
46. What do the 20 most significant bits of a page directory or page table entry stand for?
47. The page-mode protection of a page frame is to provide no access from the user protection level and read/write operation at the supervisor protection level. What are the settings of R/W and U/S?
48. What happens when an attempt is made to access a page frame that has P = 0 in its page table entry?
49. What does "D bit" in a page directory entry stand for?

Section 5.5

50. If the instruction LGDT [INIT_GDTR] is to load the limit $FFFF_{16}$ and base 00300000_{16}, show how the descriptor must be stored in memory.
51. Write an instruction sequence that can be used to clear the task-switched bit of the MSW.

52. Write an instruction sequence that will load the local descriptor table register with the selector $02F0_{16}$ from register BX.

Section 5.6

53. Define the term *multitasking*.
54. What is a task?
55. What two safeguards are implemented by the 80386DX's protection mechanism?
56. What happens if either the segment limit check or segment attributes check fails?
57. What is the highest privilege level of the 80386DX protection model called? What is the lowest level called?
58. At what protection level are applications run?
59. What is the protection mechanism used to isolate local and global resources?
60. What protection mechanism is used to isolate tasks?
61. What is the privilege level of the segment defined by the descriptor in Problem 43?
62. What does CPL stand for? RPL?
63. Overview the data access protection rule.
64. Which privilege-level data segments can be accessed by an application running at level 3?
65. Summarize the code access protection rules.
66. If an application is running at privilege level 3, what privilege-level operating system software is available to it?
67. What purpose does a call gate serve?
68. Explain what happens when the instruction CALL [NEW_ROUTINE] is executed within a task. Assume that NEW_ROUTINE is at a privilege level that is higher than the CPL.
69. What is the purpose of the task state descriptor?
70. What is the function of a task state segment?
71. Where is the state of the prior task saved? Where is the linkage to the prior task saved?
72. Into which register is the TSS selector loaded to initiate a task?
73. Give an overview of the task switch sequence illustrated in Fig. 5.40.

Section 5.7

74. Which bit position in EFLAGS is VM?
75. Is 80386DX protection active or inactive in the virtual 8086 mode? If so, what is the privilege level of a virtual 8086 program?
76. Can both protected-mode and virtual 8086 tasks coexist is an 80386DX multitasking environment?
77. Can multiple virtual 8086 tasks be active in an 80386DX multitasking environment?

6

The 80386DX Microprocessor and Its Memory Interface

6.1 INTRODUCTION

Up to this point in the book, we have studied the 80386DX microprocessor from a software point of view. We have covered its software architecture, instruction set, and how to write programs in assembly language. Now we continue by examining the hardware architecture of the 80386DX-based microcomputer system. In this chapter we begin with the 80386DX microprocessor's signal interfaces, memory interface, and external memory subsystem design. In the chapters that follow we cover input/output interfaces, interrupts and exception processing, and the design of the microcomputer in a PC/AT compatible personal computer. For this purpose we have included the following topics in the chapter:

1. The 80386DX microprocessor
2. Signal interfaces of the 80386DX
3. System clock
4. Bus states
5. Nonpipelined and pipelined bus cycles
6. Read and write bus cycle timing
7. Hardware organization of the memory address space
8. Memory interface
9. Program storage memory: ROM, PROM, and EPROM
10. Data storage memory: SRAM and DRAM

11. Program and data storage memory circuits
12. Cache memory
13. The 82385DX cache controller and the cache memory subsystem

6.2 THE 80386DX MICROPROCESSOR

The 80386DX, first announced in 1985, was the sixth member of Intel Corporation's 8086 family of microprocessors. We have learned that from the software point of view, the 80386DX offers several modes of operation: real mode, for compatibility with the large existing 8088/8086 software base, protected mode, which offers the user enhanced system-level features such as memory management, multitasking, and protection, and virtual 8086 mode, which provides 8086 real-mode compatibility while in the protected mode.

Hardware compatibility of the 80386DX with either 8086 or 80286 microprocessors is much less of a concern. In fact, a number of changes have been made to the hardware architecture of the 80386DX to improve both its versatility and performance. For example, additional pipelining has been provided within the 80386DX and the address and data buses have both been made 32 bits in length. These two changes in the hardware result in increased performance for 80386DX-based microcomputers. Another feature, *dynamic bus sizing* for the data bus, provides more versatility in system hardware design.

The 80386DX is manufactured using Intel's complementary high-performance metal-oxide-semiconductor III (CHMOSIII) process. Its circuitry is equivalent to approximately 275,000 transistors, more than twice those used in the design of the 80286 MPU and almost 10 times that of the 8086.

The 80386DX is available in a 132-pin ceramic *pin grid array* (PGA) package. The signal pinned out to each lead is shown in Fig. 6.1(a). Notice that all of the 80386DX's signals are supplied at separate pins on the package. This is intended to simplify the microcomputer circuit design.

Looking at Fig. 6.1(a), we see that the rows of pins on the package are identified by rows 1 through 14 and the columns of pins are labeled A through P. Therefore, the location of the pin for each signal is uniquely defined by a column and row coordinate. For example, in Fig. 6.1(a), address line A_{31} is at the junction of column N and row 2. That is, it is at pin N2. The pin locations for all of the 80386DX's signals are listed in Fig. 6.1(b).

EXAMPLE 6.1

At what pin location is the signal D_0?

SOLUTION Looking at Fig. 6.1(a), we find that the pin for D_0 is located in column H at row 12. Therefore, its pin is identified as H12.

6.3 SIGNAL INTERFACES OF THE 80386DX

A block diagram of the 80386DX microprocessor is shown in Fig. 6.2. Here we have grouped its signal lines into four interfaces: the *memory/IO interface, interrupt interface, DMA interface,* and *coprocessor interface.* Figure 6.3 lists each

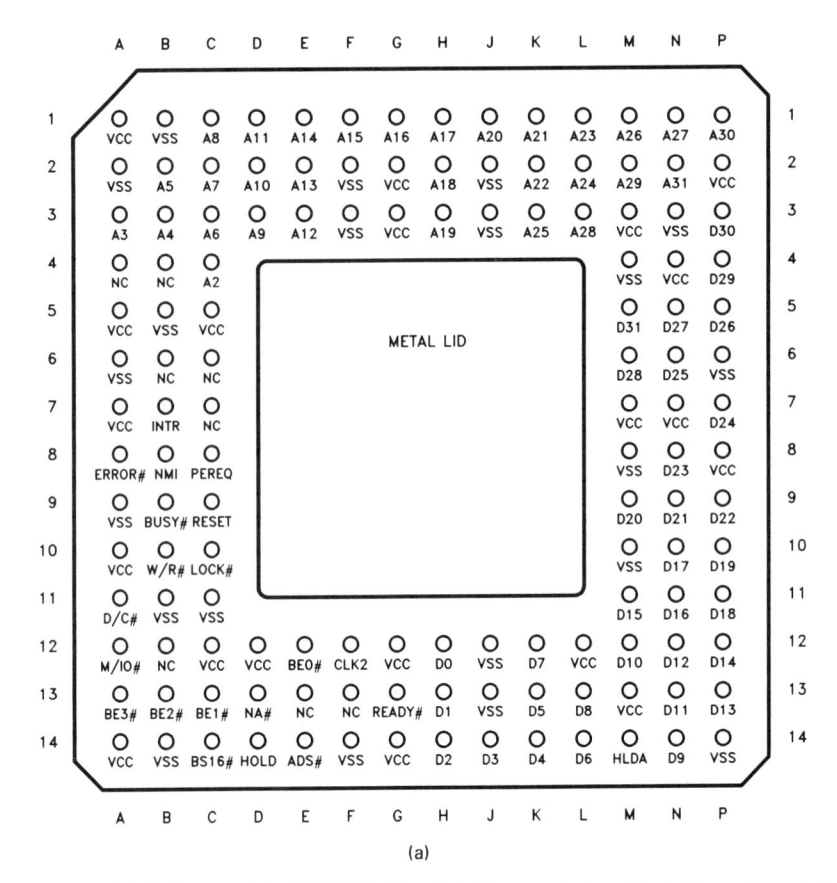

Figure 6.1 (a) Pin layout of the 80386DX. (Reprinted by permission of Intel Corp. Copyright Intel Corp. 1987.) (b) Signal pin numbering. (Reprinted by permission of Intel Corp. Copyright Intel Corp. 1987.)

of the signals at the 80386DX's interfaces. Included in this table are a mnemonic, function, type, and active level for each signal. For instance, we see that the memory/IO interface signal with the mnemonic M/$\overline{\text{IO}}$ stands for memory/IO indication. This signal is an output produced by the 80386DX that is used to signal external circuitry whether the current address available on the address bus is for memory or an I/O device. Its active level is identified as 1/0, which means that logic 1 on this line identifies a memory bus cycle and logic 0 an I/O bus cycle. On the other hand, the signal INTR at the interrupt interface is the maskable interrupt request input of the 80386DX. This input is active when at logic 1. By using this input, external devices can signal the 80386DX that they need to be serviced.

Memory/IO Interface

In a microcomputer system, the address bus and data bus signal lines form a parallel path over which the MPU talks with its memory and I/O subsystems. Like the 80286 microprocessor, but unlike the older 8086 and 8088, the 80386DX

Pin / Signal	Pin / Signal	Pin / Signal	Pin / Signal
N2 A31	M5 D31	A1 V_{CC}	A2 V_{SS}
P1 A30	P3 D30	A5 V_{CC}	A6 V_{SS}
M2 A29	P4 D29	A7 V_{CC}	A9 V_{SS}
L3 A28	M6 D28	A10 V_{CC}	B1 V_{SS}
N1 A27	N5 D27	A14 V_{CC}	B5 V_{SS}
M1 A26	P5 D26	C5 V_{CC}	B11 V_{SS}
K3 A25	N6 D25	C12 V_{CC}	B14 V_{SS}
L2 A24	P7 D24	D12 V_{CC}	C11 V_{SS}
L1 A23	N8 D23	G2 V_{CC}	F2 V_{SS}
K2 A22	P9 D22	G3 V_{CC}	F3 V_{SS}
K1 A21	N9 D21	G12 V_{CC}	F14 V_{SS}
J1 A20	M9 D20	G14 V_{CC}	J2 V_{SS}
H3 A19	P10 D19	L12 V_{CC}	J3 V_{SS}
H2 A18	P11 D18	M3 V_{CC}	J12 V_{SS}
H1 A17	N10 D17	M7 V_{CC}	J13 V_{SS}
G1 A16	N11 D16	M13 V_{CC}	M4 V_{SS}
F1 A15	M11 D15	N4 V_{CC}	M8 V_{SS}
E1 A14	P12 D14	N7 V_{CC}	M10 V_{SS}
E2 A13	P13 D13	P2 V_{CC}	N3 V_{SS}
E3 A12	N12 D12	P8 V_{CC}	P6 V_{SS}
D1 A11	N13 D11		P14 V_{SS}
D2 A10	M12 D10		
D3 A9	N14 D9	F12 CLK2	A4 N.C.
C1 A8	L13 D8		B4 N.C.
C2 A7	K12 D7	E14 ADS#	B6 N.C.
C3 A6	L14 D6		B12 N.C.
B2 A5	K13 D5	B10 W/R#	C6 N.C.
B3 A4	K14 D4	A11 D/C#	C7 N.C.
A3 A3	J14 D3	A12 M/IO#	E13 N.C.
C4 A2	H14 D2	C10 LOCK#	F13 N.C.
A13 BE3#	H13 D1		
B13 BE2#	H12 D0	D13 NA#	C8 PEREQ
C13 BE1#		C14 BS16#	B9 BUSY#
E12 BE0#		G13 READY#	A8 ERROR#
	D14 HOLD		
C9 RESET	M14 HLDA	B7 INTR	B8 NMI

(b)

Figure 6.1 (*Continued*)

has a demultiplexed address/data bus. Notice in Fig. 6.1(b) that the address bus and data bus lines are located at different pins of the package.

From a hardware point of view, there is only one difference between an 80386DX configured for the real-address mode or protected virtual-address mode. This difference is the size of the address bus. When in real mode, just the lower 18 address lines, A_2 through A_{19}, are active, while in the protected mode all 30 lines, A_2 through A_{31}, are functional. Of these, A_{19} and A_{31} are the most significant address bits, respectively. Actually, real-mode addresses are 20 bits long and protected-mode addresses are 32 bits long. The other two bits, A_0 and A_1, are decoded internal to the 80386DX, along with information about the size

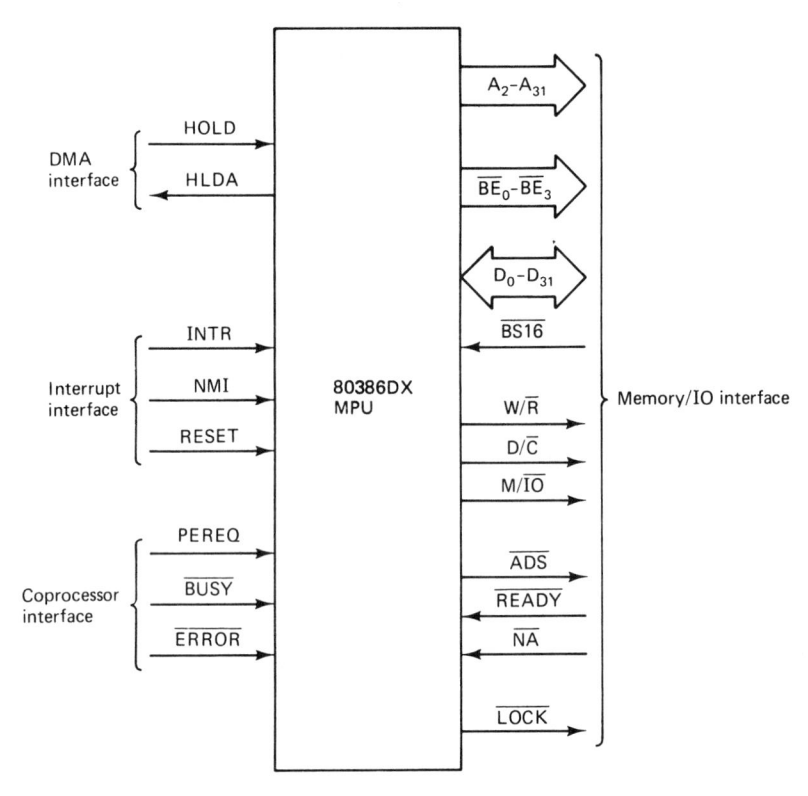

Figure 6.2 Block diagram of the 80386DX.

of the data to be transferred, to produce *byte enable* outputs, $\overline{BE}_0$, $\overline{BE}_1$, $\overline{BE}_2$, and $\overline{BE}_3$.

As shown in Fig. 6.3, the address lines are outputs. They are used to carry address information from the 80386DX to memory and I/O ports. In real-address mode, the 20-bit address gives the 80386DX the ability to address a 1M-byte physical memory address space. On the other hand, in protected mode the extended 32-bit address results in a 4G-byte physical memory address space. Moreover, when in protected mode, virtual addressing is provided through software. This results in a 64T-byte virtual memory address space.

In both the real and protected modes, the 80386DX microcomputer has an independent I/O address space. This I/O address space is 64K bytes in length. Therefore, just address lines A_2 through A_{15} and the $\overline{BE}$ outputs are used when addressing I/O devices.

Since the 80386DX is a 32-bit microprocessor, its data bus is formed from the 32 data lines D_0 through D_{31}. D_{31} is the most significant bit and D_0 the least significant bit. These lines are identified as bidirectional in Fig. 6.3. This is because they have the ability to carry data either in or out of the MPU. The kinds of data transferred over these lines are read/write data for memory, input/output data for I/O devices, and interrupt type codes from an interrupt controller.

Name	Function	Type	Level
CLK2	System clock	I	—
A_{31}-A_2	Address bus	O	1
BE_3-BE_0	Byte enables	O	0
D_{31}-D_0	Data bus	I/O	1
$\overline{BS}_{16}$	Bus size 16	I	0
W/$\overline{R}$	Write/read indication	O	1/0
D/$\overline{C}$	Data/control indication	O	1/0
M/$\overline{IO}$	Memory I/O indication	O	1/0
$\overline{ADS}$	Address status	O	0
$\overline{READY}$	Transfer acknowledge	I	0
$\overline{NA}$	Next address request	I	0
$\overline{LOCK}$	Bus lock indication	O	0
INTR	Interrupt request	I	1
NMI	Nonmaskable interrupt request	I	1
RESET	System reset	I	1
HOLD	Bus hold request	I	1
HLDA	Bus hold acknowledge	O	1
PEREQ	Coprocessor request	I	1
$\overline{BUSY}$	Coprocessor busy	I	0
$\overline{ERROR}$	Coprocessor error	I	0

Figure 6.3 Signals of the 80386DX.

Earlier we indicated that the 80386DX supports dynamic bus sizing. Even though the 80386DX has 32 data lines, the size of the bus can be dynamically switched to 16 bits. This is done simply by switching the *bus size 16* ($\overline{BS16}$) input to logic 0. When in this mode, 32-bit data transfers are performed as two successive 16-bit transfers over bus lines D_0 through D_{15}.

Remember that the 80386DX supports byte, word, and double-word data transfers over its data bus. Therefore, it must signal external circuitry what type of data transfer is taking place and over which part of the data bus the data will be carried. The MPU does this by activating the appropriate byte enable ($\overline{BE}$) output signals.

Figure 6.4 lists each byte enable output and the part of the data bus it is intended to enable. For instance, here we see that $\overline{BE}_0$ corresponds to data bus lines D_0 through D_7. If a byte of data is being read from memory, just one of the $\overline{BE}$ outputs is made active. For example, if the most significant byte of an aligned double word is read from memory, $\overline{BE}_3$ is switched to logic 0. On the other hand, if a word of data is being read, two outputs become active. An example would be to read the most significant word of an aligned double word from memory. In this case, $\overline{BE}_2$ and $\overline{BE}_3$ are both switched to logic 0. Finally, if an aligned double-word read is taking place, all four $\overline{BE}$ outputs are made active.

Sec. 6.3 Signal Interfaces of the 80386DX

Byte Enable	Data Bus Lines
$\overline{BE}_0$	D_0–D_7
$\overline{BE}_1$	D_8–D_{15}
$\overline{BE}_2$	D_{16}–D_{23}
$\overline{BE}_3$	D_{24}–D_{31}

Figure 6.4 Byte enable outputs and data bus lines.

EXAMPLE 6.2

What code is output on the $\overline{BE}$ lines whenever the address on the bus is for an instruction acquisition bus cycle?

SOLUTION Since code is always fetched as 32-bit words (aligned double words), all of the byte enable outputs are made active. Therefore,

$$\overline{BE}_3\overline{BE}_2\overline{BE}_1\overline{BE}_0 = 0000_2$$

The byte enable lines work exactly the same when write data transfers are performed over the bus. Figure 6.5(a) identifies what type of data transfer takes

$\overline{BE}_3$	$\overline{BE}_2$	$\overline{BE}_1$	$\overline{BE}_0$	D_{31}–D_{24}	D_{23}–D_{16}	D_{15}–D_8	D_7–D_0
1	1	1	0				XXXXXXXX
1	1	0	1			XXXXXXXX	
1	0	1	1		XXXXXXXX		
0	1	1	1	XXXXXXXX			
1	1	0	0			XXXXXXXX	XXXXXXXX
1	0	0	1		XXXXXXXX	XXXXXXXX	
0	0	1	1	XXXXXXXX	XXXXXXXX		
1	0	0	0		XXXXXXXX	XXXXXXXX	XXXXXXXX
0	0	0	1	XXXXXXXX	XXXXXXXX	XXXXXXXX	
0	0	0	0	XXXXXXXX	XXXXXXXX	XXXXXXXX	XXXXXXXX

(a)

$\overline{BE}_3$	$\overline{BE}_2$	$\overline{BE}_1$	$\overline{BE}_0$	D_{31}–D_{24}	D_{23}–D_{16}	D_{15}–D_8	D_7–D_0
1	1	1	0				XXXXXXXX
1	1	0	1			XXXXXXXX	
1	0	1	1		XXXXXXXX		DDDDDDDD
0	1	1	1	XXXXXXXX		DDDDDDDD	
1	1	0	0			XXXXXXXX	XXXXXXXX
1	0	0	1		XXXXXXXX	XXXXXXXX	
0	0	1	1	XXXXXXXX	XXXXXXXX	DDDDDDDD	DDDDDDDD
1	0	0	0		XXXXXXXX	XXXXXXXX	XXXXXXXX
0	0	0	1	XXXXXXXX	XXXXXXXX	XXXXXXXX	
0	0	0	0	XXXXXXXX	XXXXXXXX	XXXXXXXX	XXXXXXXX

(b)

Figure 6.5 (a) Types of data transfers for the various byte enable combinations; (b) data transfers that include duplication.

place for all the possible variations of the byte enable outputs. Here we find that $\overline{BE}_3\ \overline{BE}_2\ \overline{BE}_1\ \overline{BE}_0 = 1110_2$ means that a byte of data is written over data bus lines D_0 through D_7.

EXAMPLE 6.3

What type of data transfer takes place and over which data bus lines are data transferred if the $\overline{BE}$ code output is

$$\overline{BE}_3\overline{BE}_2\overline{BE}_1\overline{BE}_0 = 1100_2$$

SOLUTION Looking at the table in Fig. 6.5(a), we see that a word of data is transferred over data bus lines D_0 through D_{15}.

The 80386DX performs what is called *data duplication* during certain types of write cycles. Data duplication is provided in the 80386DX to optimize the performance of the data bus when it is set for 16-bit mode. Notice that whenever a write cycle is performed in which data are transferred only over the upper part of the 32-bit data bus, the data are duplicated on the corresponding lines of the lower part of the bus. For example, looking at Fig 6.5(b), we see that when $\overline{BE}_3\overline{BE}_2\overline{BE}_1\overline{BE}_0 = 1011_2$, data (denoted as XXXXXXXX) are actually being written over data bus lines D_{16} through D_{23}. However, at the same time, the data [denoted as DDDDDDDD in Fig. 6.5(b)] are automatically duplicated on data bus lines D_0 through D_7. Despite the fact that the byte is available on the lower eight data bus lines, $\overline{BE}_0$ stays inactive. The same thing happens when a word of data is transferred over D_{16} through D_{31}. In this example, $\overline{BE}_3\overline{BE}_2\overline{BE}_1\overline{BE}_0 = 0011_2$ and Fig. 6.5(b) shows that the word is duplicated on data lines D_0 through D_{15}.

EXAMPLE 6.4

If a word of data that is being written to memory is accompanied by the byte enable code 1001_2, over which data bus lines are the data carried? Is data duplication performed for this data transfer?

SOLUTION In the tables of Fig. 6.5, we find that for the byte enable code 1001_2 the word of data is transferred over data bus lines D_8 through D_{23}. For this transfer, data duplication does not occur.

Control signals are required to support data transfers over the 80386DX's address and data buses. They are needed to signal when a valid address is on the address bus, in which direction data are to be transferred over the data bus, when valid write data are on the data bus, and when an external device can put read data on the data bus. The 80386DX does not directly produce signals for all of these functions. Instead, it outputs bus cycle definition and control signals at the beginning of each bus cycle. These bus cycle identification signals must be decoded in external circuitry to produce the needed memory and I/O control signals.

Three signals are used to identify the type of 80386DX bus cycle that is in progress. In Figs. 6.2 and 6.3, they are labeled *write/read indication* (W/$\overline{R}$), *data/control indication* (D/$\overline{C}$), and *memory/input–output indication* (M/$\overline{IO}$). The table in Fig. 6.6 lists all possible combinations of the bus cycle definition signals and the corresponding type of bus cycle. Here we find that the logic level of memory/input–output (M/$\overline{\text{IO}}$) tells whether a memory or I/O cycle is to take place over the bus. Logic 1 at this output signals a memory operation, and logic 0 signals an I/O operation. The next signal in Fig. 6.6, data/control indication (D/$\overline{\text{C}}$), identifies whether the current bus cycle is a data or control cycle. In the table we see that it signals control cycle (logic 0) for instruction fetch, interrupt acknowledge, and halt/shutdown operations and data cycle (logic 1) for memory and I/O data read and write operations. Looking more closely at the table in Fig. 6.6, we find that if the code on these two lines, M/$\overline{\text{IO}}$ D/$\overline{\text{C}}$, is 00, an interrupt is to be acknowledged; if it is 01, an input/output operation is in progress; if it is 10, instruction code is being fetched; and finally, if it is 11, a data memory read or write is taking place.

The last signal identified in Fig. 6.6, write/read indication (W/$\overline{\text{R}}$), identifies the specific type of memory or input/output operation that will occur during a bus cycle. For example, when W/$\overline{\text{R}}$ is logic 0 during a bus cycle, data are to be read from memory or an I/O port. On the other hand, logic 1 at W/$\overline{\text{R}}$ says that data are to be written into memory or an I/O device. For example, all bus cycles that read instruction code from memory are accompanied by logic 0 on the W/$\overline{\text{R}}$ line.

EXAMPLE 6.5

If the bus cycle definition code M/$\overline{\text{IO}}$D/$\overline{\text{C}}$W/$\overline{\text{R}}$ equals 010, what type of bus cycle is taking place?

SOLUTION Looking at the table in Fig. 6.6, we see that bus cycle definition code 010 identifies an I/O read (input) bus cycle.

Three bus cycle control signals are produced directly by the 80386DX. They are identified in Figs. 6.2 and 6.3 as *address status* ($\overline{\text{ADS}}$), *transfer acknowledge* ($\overline{\text{READY}}$), and *next address request* ($\overline{\text{NA}}$). The $\overline{\text{ADS}}$ output is switched to logic 0 to indicate that the bus cycle definition (M/$\overline{\text{IO}}$D/$\overline{\text{C}}$W/$\overline{\text{R}}$), byte enable code ($\overline{\text{BE}_3}\overline{\text{BE}_2}\overline{\text{BE}_1}\overline{\text{BE}_0}$), and address ($A_2$ through A_{31}) signals are all stable. Therefore, it is normally applied to an input of the external bus control logic circuit and tell it that a valid bus cycle definition and address are available. In Fig. 6.6 the bus cycle definition code M/$\overline{\text{IO}}$D/$\overline{\text{C}}$W/$\overline{\text{R}}$ = 001 is identified as *idle*. That is, it is the code that is output whenever no bus cycle is being performed.

M/$\overline{\text{IO}}$	D/$\overline{\text{C}}$	W/$\overline{\text{R}}$	Type of Bus Cycle
0	0	0	Interrupt acknowledge
0	0	1	Idle
0	1	0	I/O data read
0	1	1	I/O data write
1	0	0	Memory code read
1	0	1	Halt/shutdown
1	1	0	Memory data read
1	1	1	Memory data write

Figure 6.6 Bus cycle definition signals and types of bus cycles.

$\overline{\text{READY}}$ can be used to insert wait states into the current bus cycle such that it is extended by a number of clock periods. In Fig. 6.3 we find that this signal is an input to the 80386DX. Normally, it is produced by the microcomputer's memory or I/O subsystem and supplied to the 80386DX by way of external bus control logic circuitry. By switching $\overline{\text{READY}}$ to logic 0, slow memory or I/O devices can tell the 80386DX when they are ready to permit a data transfer to be completed.

Earlier we pointed out that the 80386DX supports address pipelining at its bus interface. By address pipelining, we mean that the address and bus cycle definition for the next bus cycle is output before $\overline{\text{READY}}$ becomes active to signal that the prior bus cycle can be completed. This mode of operation is optional. The external bus control logic circuitry activates pipelining by switching the next address request ($\overline{\text{NA}}$) input to logic 0. By using pipelining, the delays introduced by the decode logic can be made transparent and the address to data access time is increased. In this way, the same level of performance can be obtained with slower, lower-cost memory devices.

One other bus interface control output that is supplied by the 80386DX, is *bus lock indication* ($\overline{\text{LOCK}}$). This signal is needed to support multiple processor architectures. In multiprocessor systems that employ shared resources, such as global memory, this signal can be employed to assure that the 80386DX has uninterrupted control of the system bus and the shared resource. That is, by switching its $\overline{\text{LOCK}}$ output to logic 0 the MPU can lock up the shared resource for exclusive use.

Interrupt Interface

Looking at Figs. 6.2 and 6.3, we find that the key interrupt interface signals are *interrupt request* (INTR), *nonmaskable interrupt request* (NMI), and *system reset* (RESET). INTR is an input to the 80386DX that can be used by external devices to signal that they need to be serviced. The 80386DX samples this input at the beginning of each instruction. Logic 1 on INTR represents an active interrupt request.

When an active interrupt request has been recognized by the 80386DX, it signals this fact to external circuitry and initiates an interrupt acknowledge bus cycle sequence. In Fig. 6.6 we see that the occurrence of an interrupt acknowledge bus cycle is signaled to external circuitry with the bus cycle definition M/$\overline{\text{IO}}$D/$\overline{\text{C}}$W/R equals 000. This bus cycle definition code can be decoded in the external bus control logic circuitry to produce an interrupt acknowledge signal. With this interrupt acknowledge signal, the 80386DX tells the external device that its request for service has been granted. This completes the interrupt request/acknowledge handshake. At this point, program control is passed to the interrupt's service routine.

The INTR input is maskable. That is, its operation can be enabled or disabled with the interrupt flag (IF) within the 80386DX's flag register. On the other hand, the NMI input, as its name implies, is a nonmaskable interrupt input. On any 0-

to-1 transition of NMI, a request for service is latched within the 80386DX. Independent of the setting of the IF flag, control is passed to the beginning of the nonmaskable interrupt service routine at the completion of execution of the current instruction.

Finally, the RESET input is used to provide a hardware reset to the 80386DX microcomputer at power-on. Switching RESET to logic 1 initializes the internal registers of the 80386DX. When it is returned to logic 0, program control is passed to the beginning of a reset service routine. This routine is used to initialize the rest of the system's resources, such as I/O ports, the interrupt flag, and data memory. A diagnostic routine that tests the 80386DX microprocessor can also be initiated as part of the reset sequence. This assures an orderly startup of the microcomputer system.

DMA Interface

Now that we have examined the signals of the 80386DX's interrupt interface, let us turn our attention to the *direct memory access* (DMA) interface. From Figs. 6.2 and 6.3 we find that the DMA interface is implemented with just two signals: *bus hold request* (HOLD) and *bus hold acknowledge* (HLDA). When an external device, such as a *DMA controller*, wants to take over control of the local address and data buses, it signals this fact to the 80386DX by switching the HOLD input to logic 1. At completion of the current bus cycle, the 80386DX enters the hold state. When in the hold state, its local bus signals are in the high-impedance state. Next, the 80386DX signals external devices that it has given up control of the bus by switching its HLDA output to the 1 logic level. This completes the hold/hold acknowledge handshake sequence. The 80386DX remains in this state until the hold request is removed.

Coprocessor Interface

In Fig. 6.2 we find that a coprocessor interface is provided on the 80386DX microprocessor to permit it to interface easily to either the *80287* or *80387DX numerics coprocessor*. The 80387DX cannot perform transfers over the data bus by itself. Whenever the 80387DX needs to read or write operands from memory, it must signal the 80386DX to initiate the data transfers. The 80387DX does this by switching the *coprocessor request* (PEREQ) input of the 80386DX to logic 1.

The other two signals included in the external coprocessor interface are $\overline{BUSY}$ and $\overline{ERROR}$. *Coprocessor busy* ($\overline{BUSY}$) is an input of the 80386DX. Whenever the 80387DX is executing a numeric instruction, it signals this fact to the 80386DX by switching the $\overline{BUSY}$ input to logic 0. In this way, the 80386DX knows not to request the numerics coprocessor to perform another calculation until $\overline{BUSY}$ returns to 1. Moreover, if an error occurs in a calculation performed by the numerics coprocessor, this condition is signaled to the 80386DX by switching the *coprocessor error* ($\overline{ERROR}$) input to the 0 logic level.

6.4 SYSTEM CLOCK

The time base for synchronization of the internal and external operations of the 80386DX microprocessor is provided by the *clock* (CLK2) input signal. At present, the 80386DX is available with four different clock speeds. The original 80386DX-16 MPU operates at 16 MHz and its three faster versions, the 80386DX-20, 80386DX-25, and 80386DX-33, operate at 20, 25, and 33 MHz, respectively. The clock signal applied to the CLK2 input of the 80386DX is twice the frequency rating of the microprocessor. Therefore, CLK2 of an 80386DX-16 is driven by a 32-MHz signal. This signal must be generated in external circuitry.

The waveform of the CLK2 input of the 80386DX is given in Fig. 6.7. Here we see that the signal is specified at CMOS-compatible voltage levels and not TTL levels. Its minimum and maximum low logic levels are $V_{ILCmin} = -0.3$ V and $V_{ILCmax} = 0.8$ V, respectively. Moreover, the minimum and maximum high logic levels are $V_{IHCmin} = V_{CC} - 0.8$ V and $V_{IHCmax} = V_{CC} + 0.3$ V, respectively. The minimum period of the 16-MHz clock signal is $t_{Cmin} = 31$ ns (measured at the 2.0-V level); its minimum high time t_{Pmin} and low time t_{lmin} (measured at the 2.0-V level) are both equal to 9 ns; and the maximum rise time t_{rmax} and fall time t_{fmax} of its edges (measured between the $V_{CC} - 0.8$-V and 0.8-V levels) are both equal to 8 ns.

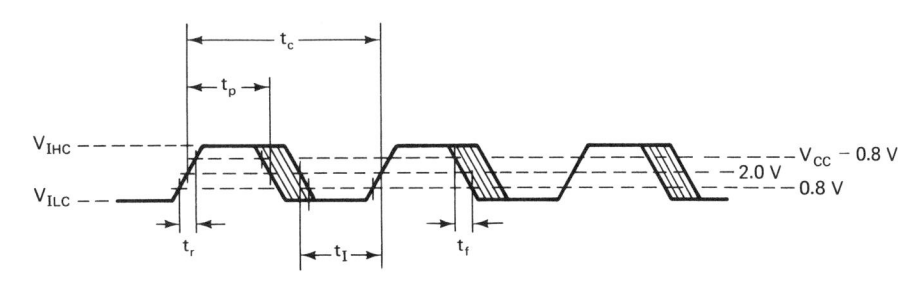

Figure 6.7 System clock (CLK2) waveform.

6.5 BUS STATES AND PIPELINED AND NONPIPELINED BUS CYCLES

Before looking at the bus cycles of the 80386DX, let us first examine the relationship between the timing of the 80386DX's CLK2 input and its bus cycle states. The *internal processor clock* (PCLK) signal is at half the frequency of the external clock input signal. Therefore, as shown in Fig. 6.8, one processor clock cycle corresponds to two CLK2 cycles. Notice that CLK2 cycles are labeled as *phase 1* (ϕ_1) and *phase 2* (ϕ_2). In a 20-MHz 80386DX microcomputer system, CLK2 equals 40 MHz and each clock cycle has a duration of 25 ns. In Fig. 6.8, we see that the two phases (ϕ_1 and ϕ_2) of a processor cycle are identified as one processor clock period. Therefore, a processor clock cycle is a minimum of 50 ns long.

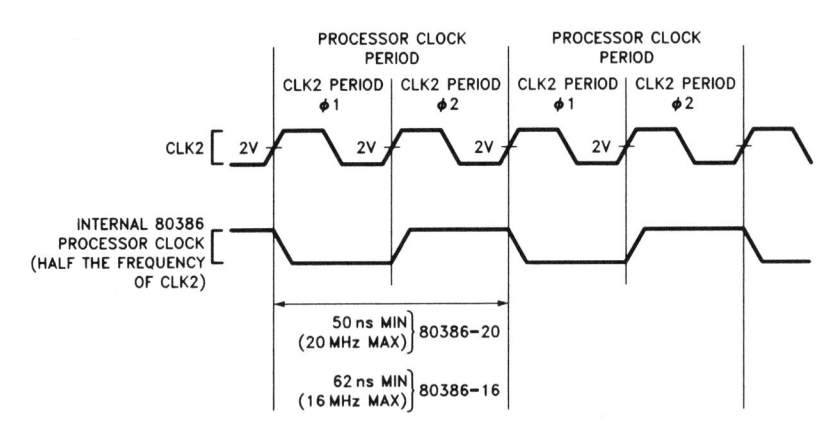

Figure 6.8 Processor clock cycles. (Reprinted by permission of Intel Corp. Copyright Intel Corp. 1987.)

Nonpipelined and Pipelined Bus Cycles

A *bus cycle* is the activity performed whenever a microprocessor accesses information in program memory, data memory, or an I/O device. The 80386DX can perform bus cycles with either of two types of timing: *nonpipelined* and *pipelined*. Here we will examine the difference between these two types of bus cycles.

Figure 6.9 shows a typical nonpipelined microprocessor bus cycle. Notice that the bus cycle contains two T states and that they are called T_1 and T_2. During the T_1 part of the bus cycle, the 80386DX outputs the address of the storage location that is to be accessed on the address bus, a bus cycle definition code, and control signals. In the case of a write cycle, write data are also output on the data bus during T_1. The second state, T_2, is the part of the bus cycle during which external devices are to accept write data from the data bus, or in the case of a read cycle, put data on the data bus.

For instance in Fig. 6.9 we see that the sequence of events start with an address, denoted as n, being output on the address bus in clock state T_1. Later in the bus cycle, while the address is still available on the address bus, a read or write data transfer takes place over the data bus. Notice that the data transfer

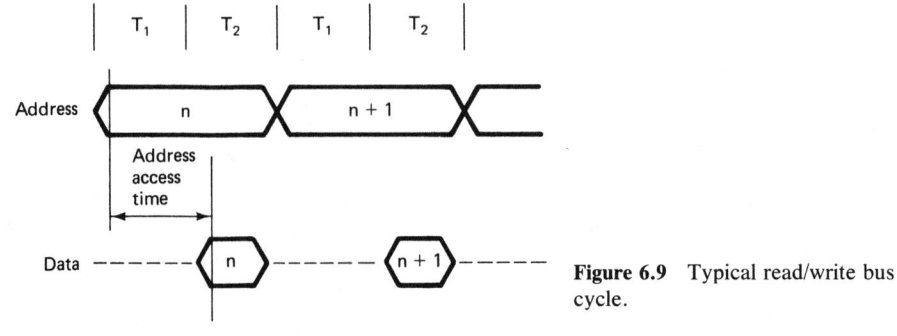

Figure 6.9 Typical read/write bus cycle.

218   The 80386DX Microprocessor and Its Memory Interface   Chap. 6

for address n is shown to occur in clock state T_2. Since each bus cycle has a minimum of two T states (four CLK2 cycles), the minimum bus cycle duration for an 80386DX-20 is 100 ns.

Let us now look at a microprocessor bus cycle that employs *pipelining*. By pipelining we mean that addressing for the next bus cycle is overlapped with the data transfer of the prior bus cycle. When address pipelining is in use, the address, bus cycle definition code, and control signals for the next bus cycle are output during T_2 of the prior cycle, instead of the T_1 that follows.

In Fig. 6.10 we see that address n becomes valid in the T_2 state of the prior bus cycle, and then the data transfer for address n takes place in the next T_2 state. Moreover, notice that at the same time that data transfer n occurs, address n + 1 is output on the address bus. In this way we see that the microprocessor begins addressing the next storage location that it is to access while it is still performing the read or write of data for the previously addressed storage location. Due to the address/data pipelining, the memory or I/O subsystem actually has five CLK2 cycles (125 ns for an 80386DX-20 running at full speed) to perform the data transfer, even though the duration of every bus cycle is just four clock cycles (100 ns).

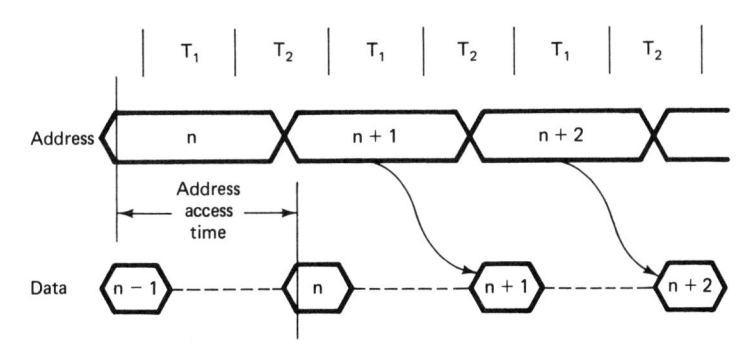

Figure 6.10 Pipelined bus cycle. (Reprinted by permission of Intel Corp. Copyright Intel Corp. 1987.)

The interval denoted as *address access time* in Fig. 6.9 represents the amount of time that the address must be stable prior to the read or write of data actually taking place. Notice that this duration is less than the four CLK2 cycles in a nonpipelined bus cycle. Figure 6.10 shows that in a pipelined bus cycle the *effective address access time* equals the duration of a complete bus cycle. This leads us to the benefit of the 80386DX's pipelined mode of bus operation over the nonpipelined mode of operation. It is that, for a fixed address access time (equal speed memory design), the 80386DX pipelined bus cycle will have a shorter duration than that of its nonpipelined bus cycle. This results in improved bus performance.

Another way of looking at this is to say that when using equal-speed memory designs, an 80386DX that uses a pipelined bus can be operated at a higher clock rate than a design that executes a nonpipelined bus cycle. Once again the result is higher system performance.

In Fig. 6.10 we find that at completion of the bus cycle for address n, another bus cycle is initiated immediately for address n + 1. Sometimes another bus cycle will not be initiated immediately. For instance, if the 80386DX's prefetch queue is already full and the instruction that is currently being executed does not need to access operands in memory, no bus activity will take place. In this case the bus goes into a mode of operation known as an *idle state*, and no bus activity occurs. Figure 6.11 shows a sequence of bus activity in which several idle states exist between the bus cycles for addresses n + 1 and n + 2. The duration of a single idle state is equal to two CLK2 cycles.

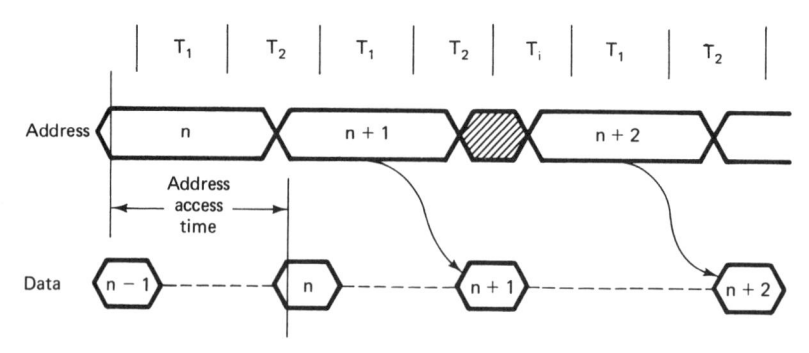

Figure 6.11 Idle states in bus activity.

Wait states can be inserted to extend the duration of the 80386DX's bus cycle. This is done in response to a request by an event in external hardware instead of an internal event such as a full queue. In fact, the $\overline{\text{READY}}$ input of the 80386DX is provided specifically for this purpose. This input is sampled in the later part of the T_2 state of every bus cycle to determine if the data transfer should be completed. Figure 6.12 shows that logic 1 at this input indicates that the current bus cycle should not be completed. As long as $\overline{\text{READY}}$ is held at the 1 level, the read or write data transfer does not take place and the current T_2 state becomes a wait state (T_w) to extend the bus cycle. The bus cycle is not

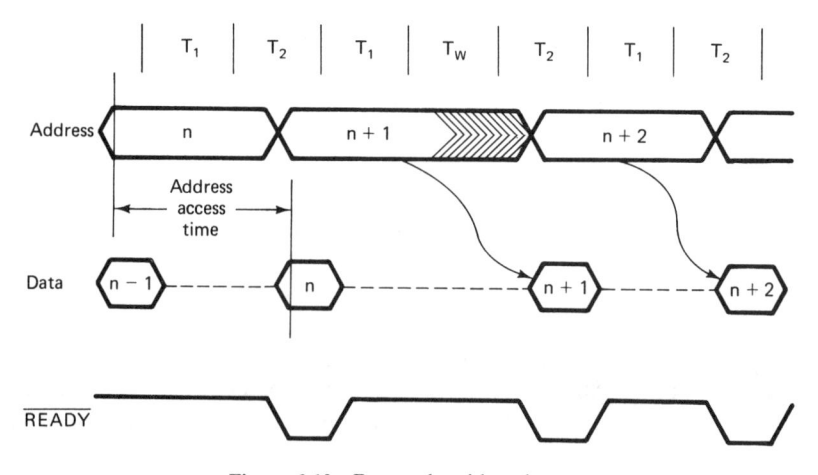

Figure 6.12 Bus cycle with wait states.

completed until external hardware returns $\overline{\text{READY}}$ back to logic 0. This ability to extend the duration of a bus cycle permits the use of slow memory or I/O devices in the microcomputer system.

6.6 READ AND WRITE BUS CYCLE TIMING

In the preceding sections we introduced the 80386DX's memory interface signals, bus cycle states, and the concepts of nonpipelined and pipelined bus cycles. We found that the 80386DX's bus can be dynamically configured to operate in either the 32-bit (standard) mode or 16-bit mode and that data transfers can be performed using either nonpipelined or pipelined bus cycles. In this section we continue by studying in detail the sequence of events that take place during the 80386DX's memory read and write bus cycles.

Nonpipelined Read Cycle Timing

The memory interface signals that occur when the 80386DX reads data from memory are shown in Fig. 6.13. This diagram shows two separate nonpipelined read cycles. They are *cycle 1*, which is performed without wait states, and *cycle 2*, which includes one wait state. Let us now trace through the events that take place in cycle 1 as data or instructions are read from memory.

The occurrence of all signals in the read bus cycle timing diagram are illustrated relative to the two timing states, T_1 and T_2, of the 80386DX's bus cycle. The read operation starts at the beginning of phase 1 (ϕ_1) in the T_1 state of the bus cycle. At this moment, the 80386DX outputs the address of the double-word memory location to be accessed on address bus lines A_2 through A_{31}, the byte enable signals $\overline{BE_0}$ through $\overline{BE_3}$ that identify the bytes of the double word that are to be fetched, and switches address strobe ($\overline{ADS}$) to logic 0 to signal that a valid address is on the address bus.

Notice in Fig. 6.13 that the bus cycle definition signals, $M/\overline{IO}$, $D/\overline{C}$, and $W/\overline{R}$ are also made valid at the beginning of ϕ_1 of state T_1 in the read cycle. The memory read cycle bus cycle definition codes are highlighted in Fig. 6.14. Here we see that if code data is being read from memory $M/\overline{IO}D/\overline{C}W/\overline{R}$ equals 100. That is, signal $M/\overline{IO}$ is set to logic 1 to indicate to the circuitry in the memory interface that a memory bus cycle is in progress, $D/\overline{C}$ is set to 0 to indicate that code memory is to be accessed, and $W/\overline{R}$ is set to 0 to indicate that data are being read from memory. Looking at Fig. 6.13, we see that the address and bus cycle definition signals are maintained stable during the complete bus cycle; however, they must be latched into the external bus control logic circuitry synchronously with the pulse to logic 0 on $\overline{ADS}$. At the end of ϕ_2 of T_1, $\overline{ADS}$ is returned to its inactive 1 logic level.

At the beginning of ϕ_1 in T_2 of the read cycle, external circuitry must signal the 80386DX whether the bus is to operate in the 16- or 32-bit mode. In Fig. 6.13 we see that it does this with the $\overline{BS16}$ signal. The 80386DX samples this input in the middle of the T_2 bus cycle state. The 1 logic level shown in the timing diagram indicates that a 32-bit data transfer is to take place.

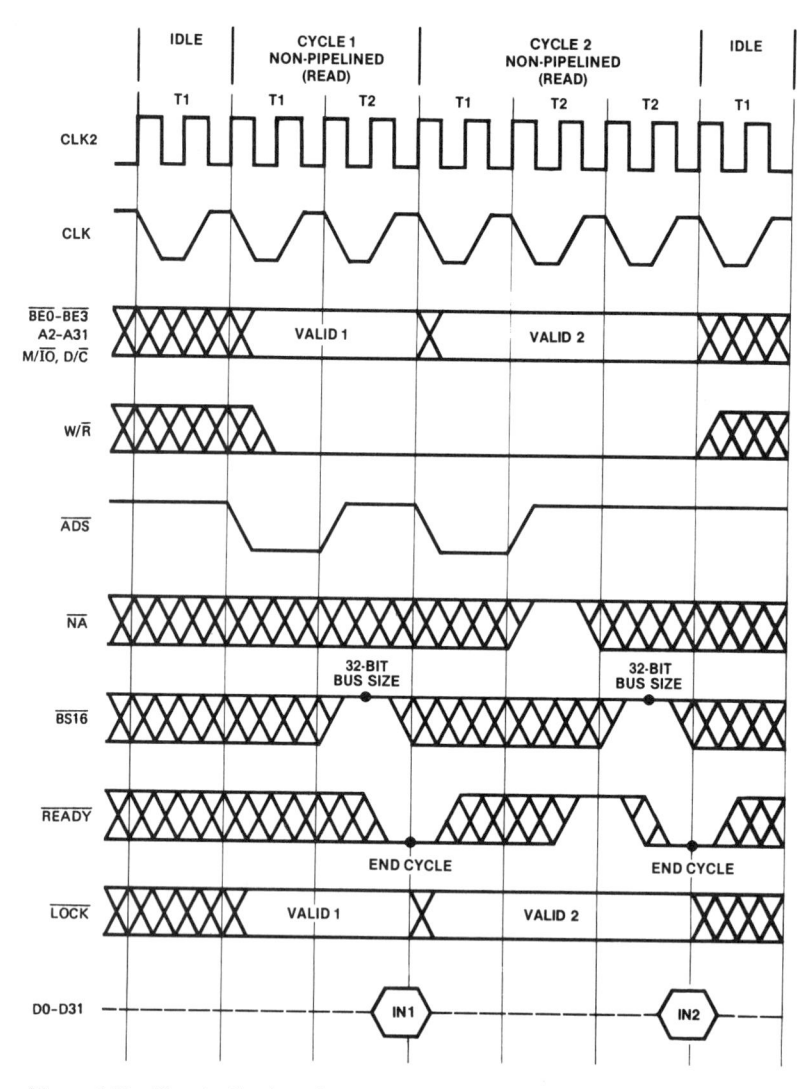

Figure 6.13 Nonpipelined read cycle timing. (Reprinted by permission of Intel Corp. Copyright Intel Corp. 1987.)

M/$\overline{IO}$	D/$\overline{C}$	W/$\overline{R}$	Type of Bus Cycle
0	0	0	Interrupt acknowledge
0	0	1	Idle
0	1	0	I/O data read
0	1	1	I/O data write
1	0	0	Memory code read
1	0	1	Halt/shutdown
1	1	0	Memory data read
1	1	1	Memory data write

Figure 6.14 Memory read bus cycle definition codes.

222   The 80386DX Microprocessor and Its Memory Interface   Chap. 6

Notice in Fig. 6.13 that at the end of T_2 the $\overline{READY}$ input is tested by the 80386DX. The logic level at this input signals whether the current bus cycle is to be completed or extended with wait states. The logic 0 shown at this input means that the bus cycle is to run to completion. For this reason we see that data available on data bus lines D_0 through D_{31} are read into the 80386DX at the end of T_2.

Nonpipelined Write Cycle Timing

The nonpipelined write bus cycle timing diagram, shown in Fig. 6.15, is similar to that given for a nonpipelined read cycle in Fig. 6.13. It includes waveforms for

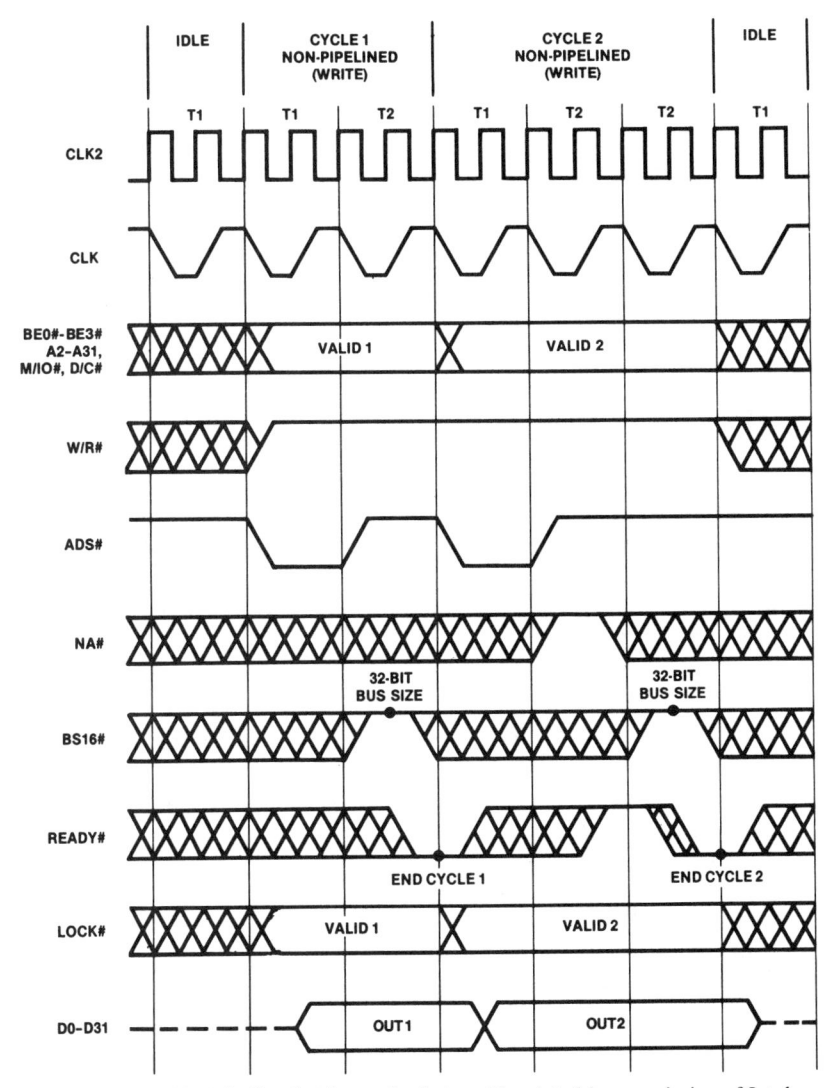

Figure 6.15 Nonpipelined write cycle timing. (Reprinted by permission of Intel Corp. Copyright Intel Corp. 1987.)

Sec. 6.6 Read and Write Bus Cycle Timing

both a no-wait-state write operation (cycle 1) and a one-wait-state write operation (cycle 2). Looking at the write cycle waveforms, we find that the address, byte enable, and bus cycle definition signals are output at the beginning of ϕ_1 of the T_1 state. All these signals are to be latched in external circuitry with the pulse at $\overline{ADS}$. The one difference here is that W/$\overline{R}$ is at the 1 logic level instead of 0. In fact, as shown in Fig. 6.16, the bus cycle definition code for a memory data write is "M/$\overline{IO}$D/$\overline{C}$W/$\overline{R}$ equals 111"; therefore, M/$\overline{IO}$ and D/$\overline{C}$ are also at the logic 1 level.

M/$\overline{IO}$	D/$\overline{C}$	W/$\overline{R}$	Type of Bus Cycle
0	0	0	Interrupt acknowledge
0	0	1	Idle
0	1	0	I/O data read
0	1	1	I/O data write
1	0	0	Memory code read
1	0	1	Halt/shutdown
1	1	0	Memory data read
1	1	1	Memory data write

Figure 6.16 Memory write bus cycle definition code.

Let us now look at what happens on the data bus during a write bus cycle. Notice that the 80386DX outputs the data that are to be written to memory onto the data bus at the beginning of ϕ_2 in the T_1 state. These data are maintained valid until the end of the bus cycle. In the middle of the T_2 state, the logic level of the $\overline{BS16}$ input is tested by the 80386DX and again indicates that the bus is to be used in the 32-bit mode. Finally, at the end of T_2, $\overline{READY}$ is tested and found to be at its active 0 logic level. Since the memory subsystem has made $\overline{READY}$ logic 0, the write cycle is complete and the buses and control signal lines are prepared for the next write cycle.

Wait States in a Nonpipelined Memory Bus Cycle

In the preceding section we showed how wait states are used to lengthen the duration of the memory bus cycle of the 80386DX. Wait states are inserted with the $\overline{READY}$ input signal. Upon request from an event in external hardware, for instance, slow memory, the $\overline{READY}$ input is switched to logic 1. This signals the 80386DX that the current bus cycle should not be completed. Instead, it is extended by repeating the T_2 state. Therefore, the duration of one wait state ($T_w = T_2$) equals 50 ns for 20-MHz clock operation.

Cycle 2 in Fig. 6.13 shows a read cycle extended by one wait state. Notice that the address, byte enable, and bus cycle definition signals are maintained throughout the wait-state period. In this way, the read cycle is not completed until $\overline{READY}$ is switched to logic 0 in the second T_2 state.

EXAMPLE 6.6

If cycle 2 in Fig. 6.15 is for an 80386DX-20 running at full speed, what is the duration of the bus cycle?

SOLUTION Each T state in the bus cycle of an 80386DX running at 20 MHz is 50 ns. Since the write cycle is extended by one wait state, the write cycle takes 150 ns.

Pipelined Read/Write Cycle Timing

Timing diagrams for both nonpipelined and pipelined read and write bus cycles are shown in Fig. 6.17. Here we find that the cycle identified as *cycle 3* is an example of a pipelined write bus cycle. Let us now look more closely at this bus cycle.

Remember that when pipelined addressing is in use, the 80386DX outputs the address information for the next bus cycle during the T_2 state of the current cycle. The signal next address ($\overline{NA}$) is used to signal the 80386DX that a pipelined bus cycle is to be initiated. This input is sampled by the 80386DX during any bus state when $\overline{ADS}$ is not active. In Fig. 6.17 we see that $\overline{NA}$ is first tested as 0

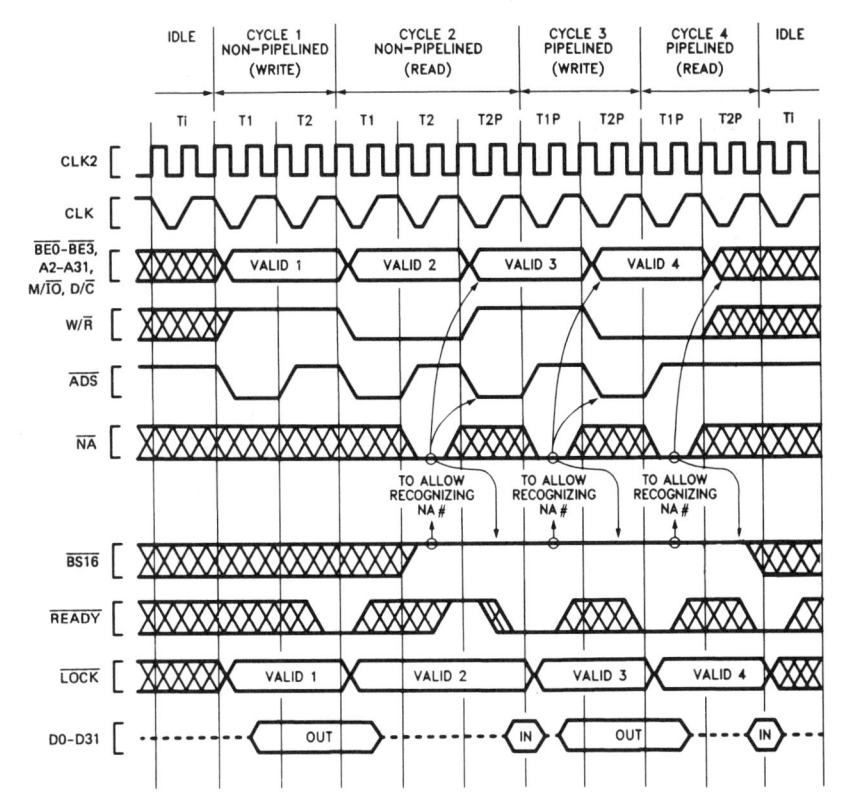

Figure 6.17 Pipelined read and write cycle timing. (Reprinted by permission of Intel Corp. Copyright Intel Corp. 1987.)

(active) during T_2 of cycle 2. This nonpipelined read cycle is also extended with period T_{2P} because $\overline{READY}$ is not active. Notice that the address, byte enable, and bus cycle definition signals for cycle 3 become valid (identified as VALID 3 in Fig. 6.17) during this period and a pulse is produced at $\overline{ADS}$. This information is latched externally synchronously with $\overline{ADS}$ and decoded to produce bus enable and control signals. In this way, the memory access time for a zero-wait-state memory cycle has been increased.

Bus cycle 3 represents a pipelined write cycle. The data to be written to memory are output on D_0 through D_{31} at ϕ_2 of T_{1P} and remains valid for the rest of the cycle. Logic 0 on $\overline{READY}$ at the end of T_{2P} indicates that the write cycle is to be completed without wait states.

Looking at Fig. 6.17, we find that $\overline{NA}$ is also active during T_{1P} of cycle 3. This means that cycle 4 will be performed with pipelined timing. Cycle 4 is an example of a zero-wait-state pipelined read cycle. In this case, the address information, bus cycle definition, and address strobe are output during T_{2P} of cycle 3 (the previous cycle), and memory data are read into the MPU at the end of T_{2P} of cycle 4.

6.7 HARDWARE ORGANIZATION OF THE MEMORY ADDRESS SPACE

Earlier we indicated that in the protected mode the 32-bit address bus of the 80386DX results in a 4G-byte physical memory address space. As shown in Fig. 6.18, from a software point of view, this memory is organized as individual bytes over the address range from 00000000_{16} through $FFFFFFFF_{16}$. The 80386DX can also access data in this memory as words or double words.

From a hardware point of view, the physical address space is implemented as four independent byte-wide banks, and each of these banks is 1G-byte in size.

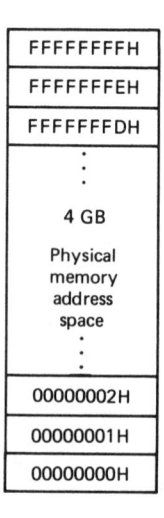

FFFFFFFFH
FFFFFFFEH
FFFFFFFDH
⋮
4 GB Physical memory address space
⋮
00000002H
00000001H
00000000H

Figure 6.18 Physical address space.

In Fig. 6.19 we find that the banks are identified as *bank 0, bank 1, bank 2*, and *bank 3*. Notice that they correspond to addresses that produce byte enable signals $\overline{BE}_0$, $\overline{BE}_1$, $\overline{BE}_2$, and $\overline{BE}_3$, respectively. Logic 0 at a byte enable input selects the bank for operation. Looking at Fig. 6.19, we see that address bits A_2 through A_{31} are applied to all four banks in parallel. On the other hand, each memory bank supplies just eight lines of the 80386DX's 32-bit data bus. For example, byte data transfers for bank 0 take place over data bus lines D_0 through D_7, while byte data transfers for bank 3 are carried over data bus lines D_{24} through D_{31}.

When the 80386DX is operated in real mode, only the value on address lines A_2 through A_{19} and the $\overline{BE}$ signals are used to select the storage location that is to be accessed. For this reason, the physical address space is 1M byte in length, not 4G bytes. The memory subsystem is once again partitioned into four banks, as shown in Fig. 6.19, but this time each bank is 256K bytes in size.

Figure 6.19 shows that in hardware the memory address space is physically organized as a sequence of double words. The address on lines A_2 through A_{31} selects the double-word storage location. Therefore, each aligned double word starts at a physical address that is a multiple of 4. For instance, in Fig. 6.19, we see that aligned double words start at addresses 00000000_{16}, 00000004_{16}, 00000008_{16}, up through $FFFFFFFC_{16}$.

Each of the four bytes of a double word corresponds to one of the byte enable signals. For this reason, they are each stored in a different bank of memory. In Fig. 6.19 we have identified the range of byte addresses that corresponds to the storage locations in each bank of memory. For example, byte data accesses to addresses such as 00000000_{16}, 00000004_{16}, and 00000008_{16} all produce $\overline{BE}_0$, which enables memory bank 0, and the read or write data transfer takes place over data bus lines D_0 through D_7. Figure 6.20(a) illustrates how the byte at double-word aligned memory address X is accessed.

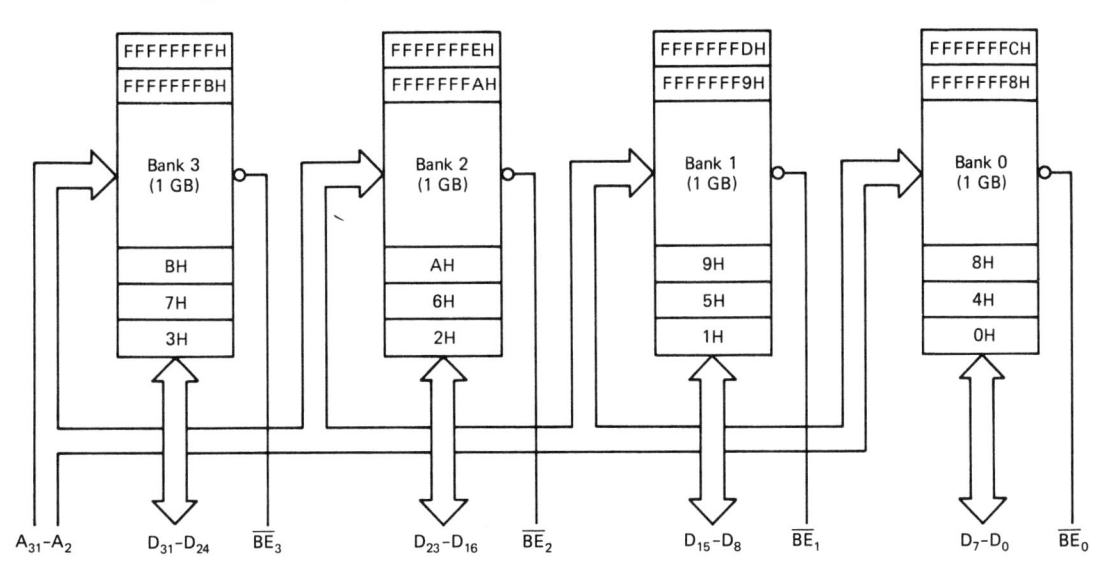

Figure 6.19 Hardware organization of the physical address space.

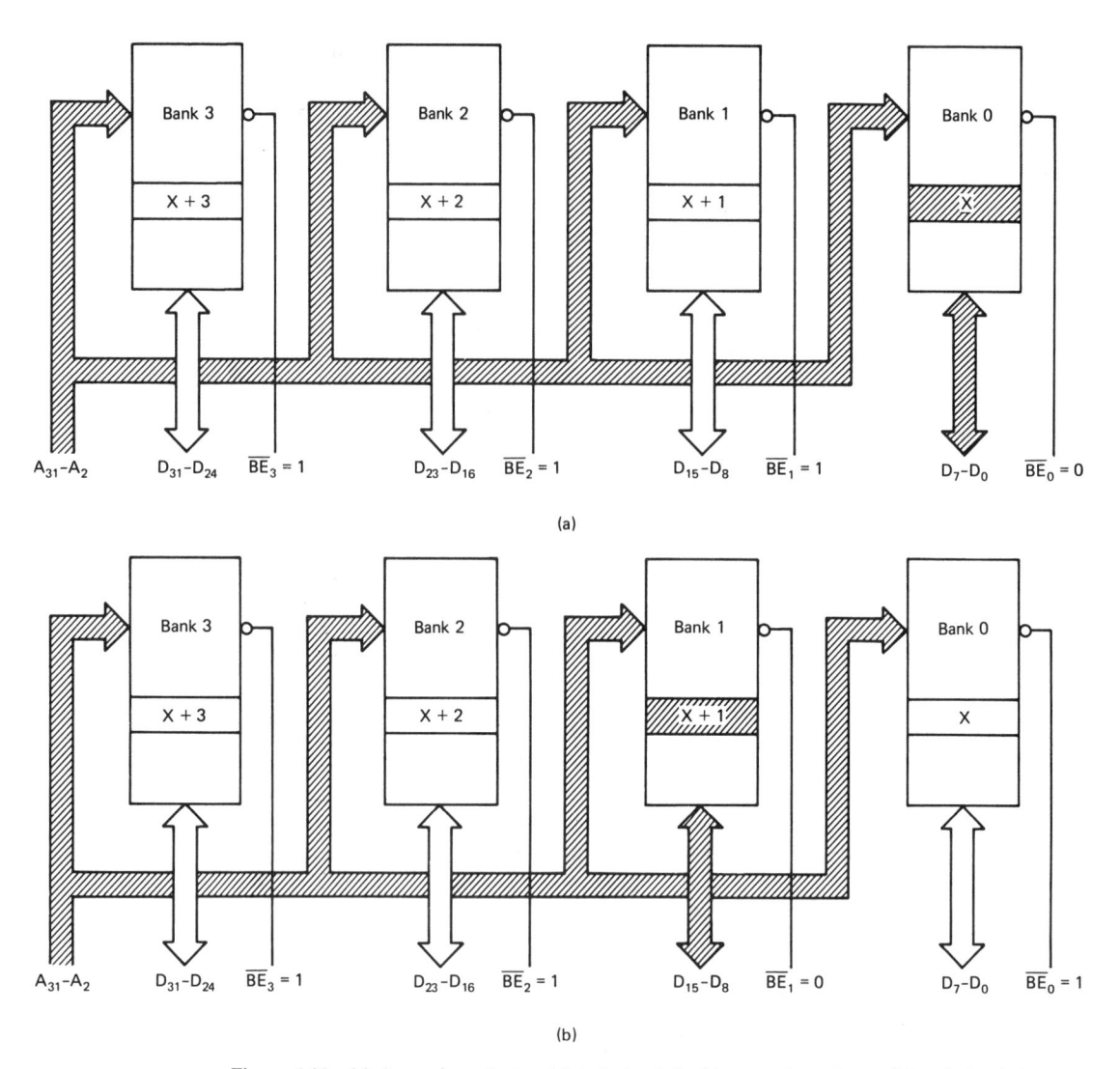

Figure 6.20 (a) Accessing a byte of data in bank 0; (b) accessing a byte of data in bank 1; (c) accessing an aligned word of data in memory; (d) accessing an aligned double word in memory.

On the other hand, in Fig. 6.19 we see that byte addresses 00000001_{16}, 00000005_{16}, and 00000009_{16} correspond to data held in memory bank 1. Figure 6.20(b) shows how the byte of data at address $X + 1$ is accessed. Notice that $\overline{BE}_1$ is made active to enable bank 1 of memory.

Most memory accesses produce more than one bank enable signal. For instance, if the word of data beginning at aligned address X is read from memory, both $\overline{BE}_0$ and $\overline{BE}_1$ are generated. In this way, bank 0 and bank 1 of memory are enabled for operation. As shown in Fig. 6.20(c), the word of data is transferred to the MPU over data bus lines D_0 through D_{15}.

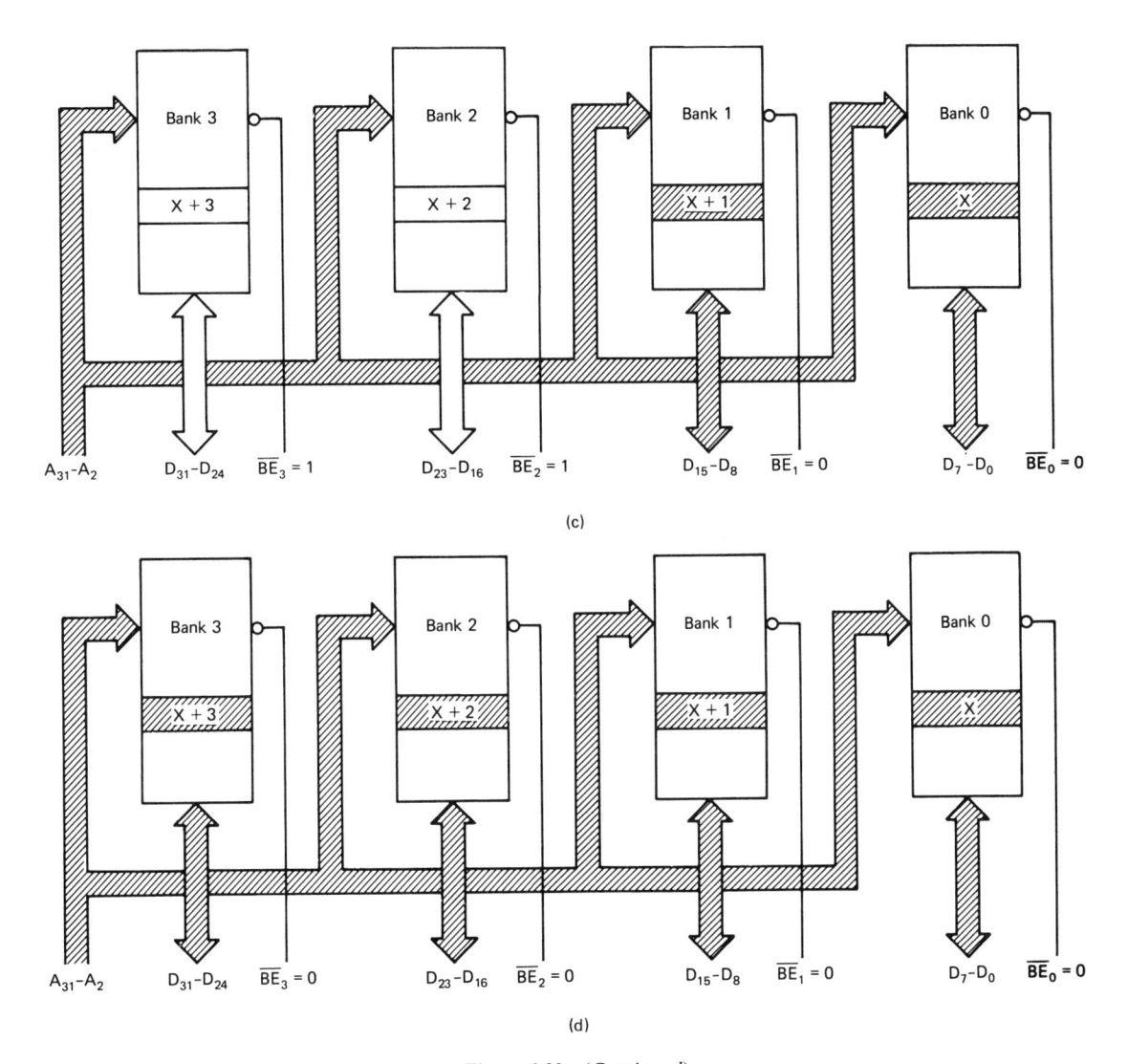

Figure 6.20 (*Continued*)

Let us now look at what happens when a double word of data is written to aligned double-word address X. As shown in Fig. 6.20(d), $\overline{BE}_0$, $\overline{BE}_1$, $\overline{BE}_2$, and $\overline{BE}_3$ are made 0 to enable all four banks of memory and the MPU writes the data to memory over the complete data bus, D_0 through D_{31}.

All the data transfers we have described so far have been for what are called *double-word aligned data*. For each of these pieces of data, all of the bytes existed within the same double word, that is, a double word that is on an address boundary equal to a multiple of 4. Byte, aligned word, and aligned double-word data transfers are all performed by the 80386DX in a single bus cycle.

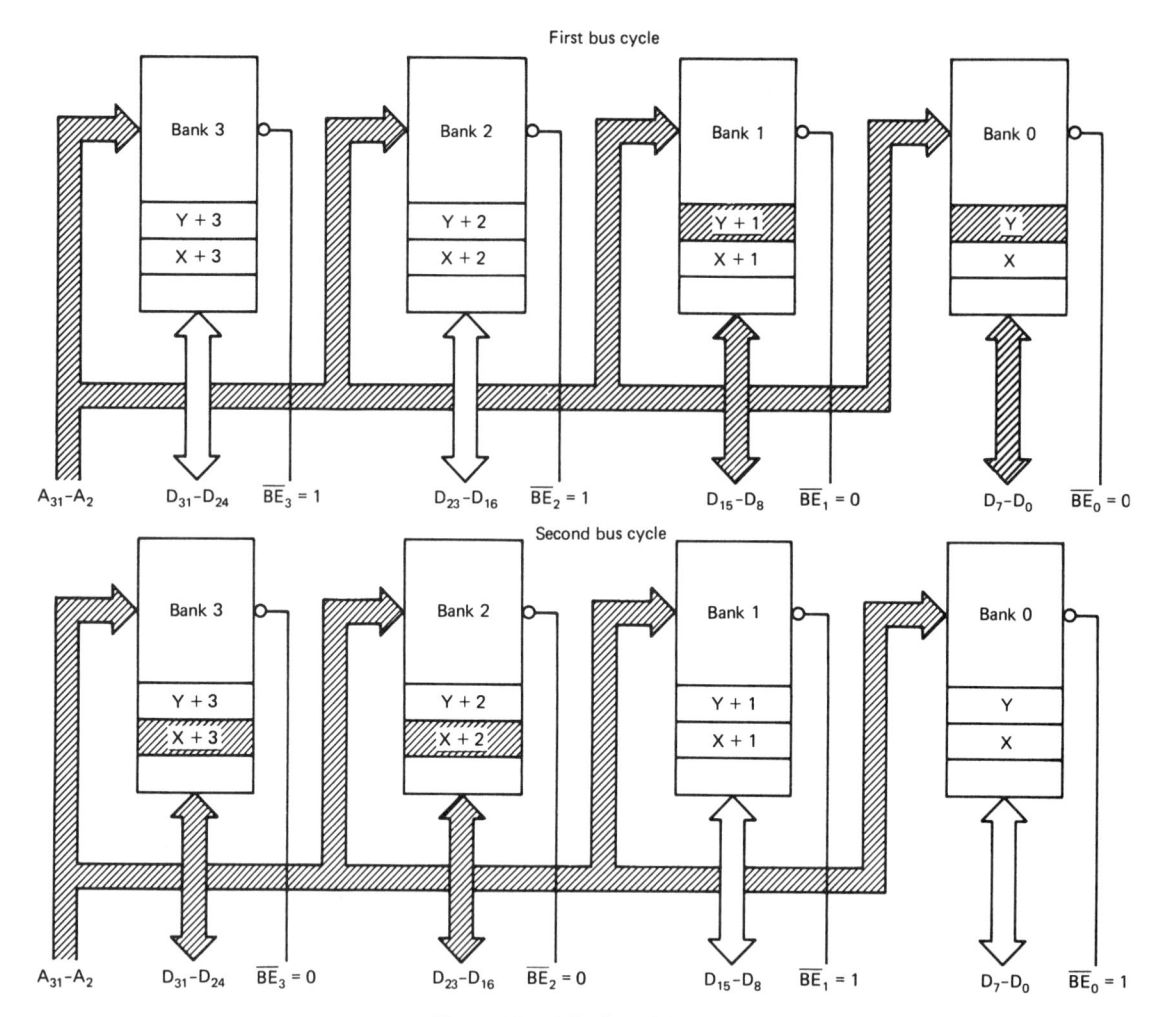

Figure 6.21 Misaligned double-word data transfer.

It is not always possible to have all words or double words of data aligned at double-word boundaries. Let us now look at how misaligned data are transferred over the bus.

The diagram in Fig. 6.21 illustrates a misaligned double-word data transfer. Here the double word of data starting at address X + 2 is to be accessed. However, this word consists of bytes X + 2 and X + 3 of the aligned double word at physical address X and bytes Y and Y + 1 of the aligned double word at physical address Y. Looking at the diagram, we see that $\overline{BE}_0$ and $\overline{BE}_1$ are active during the first bus cycle, and the word at address Y is transferred over D_0 through D_{15}. A second bus cycle automatically follows in which $\overline{BE}_3$ and $\overline{BE}_4$ are active, address X is put on the address bus, and the second word of data, X + 2 and X + 3, is carried over D_{16} through D_{31}. In this way we see that data transfers of misaligned words or double words take two bus cycles.

230   The 80386DX Microprocessor and Its Memory Interface   Chap. 6

> **EXAMPLE 6.7**
>
> Is the word at address $0000123F_{16}$ aligned or misaligned? How many bus cycles are required to read it from memory?
>
> **SOLUTION** The first byte of the word is the fourth byte at aligned double-word address $0000123C_{16}$ and the second byte of the word is the first byte of the aligned double word at address 00001240_{16}. Therefore, the word is misaligned and requires two bus cycles to be read from memory.

6.8 MEMORY INTERFACE CIRCUITS

A memory interface diagram for a protected-mode 80386DX-based microcomputer system is shown in Fig. 6.22. Here we find that the interface includes bus control logic, address bus latches and an address decoder, data bus transceiver/buffers, and bank write control logic. The bus cycle definition signals, $M/\overline{IO}$, $D/\overline{C}$, and $W/\overline{R}$, which are output by the 80386DX, are supplied directly to the bus control logic. Here they are decoded to produce the command and control signals needed to control data transfers over the bus. In Figs. 6.14 and 6.16, the status codes that relate to the memory interface are highlighted. For example, the code $M/\overline{IO}D/\overline{C}W/\overline{R}$ equal 110 indicates that a data memory read bus cycle is in progress. This code makes the $\overline{MRDC}$ command output of the bus control logic switch to logic 0. Notice in Fig. 6.22 that $\overline{MRDC}$ is applied directly to the $\overline{OE}$ input of the memory subsystem.

Next let us look at how the address bus is decoded, buffered, and latched. Looking at Fig. 6.22, we see that address lines A_{29} through A_{31} are decoded to produce chip enable outputs $\overline{CE_0}$ through $\overline{CE_7}$. These chip enable signals are latched along with address bits A_2 through A_{28} and byte enable lines $\overline{BE_0}$ through $\overline{BE_3}$ into the address latches. Notice that the bus control logic produces the address latch enable (ALE) control signal from $\overline{ADS}$ and the bus cycle definition inputs. ALE is applied to the CLK input of the latches and strobes the bits of the address, byte enable, and chip enable signals into the address bus latches. These signals are buffered by the address latch devices and then output directly to the memory subsystem.

This part of the memory interface demonstrates one of the benefits of the 80386DX's pipelined bus mode. When working in the pipelined mode, the 80386DX actually outputs the address in the T_2 state of the prior bus cycle. Therefore, by putting the address decoder before the address latches instead of after, the code at address lines A_{28} through A_{31} can be fully decoded and stable prior to the T_1 state of the next bus cycle. In this way, the access time of the memory subsystem is maximized.

During read bus cycles, the $\overline{MRDC}$ output of the bus control logic enables the data at the outputs of the memory subsystem onto data bus lines D_0 through D_{31}. The 80386DX will read the appropriate byte, word, or double word of data. On the other hand, during write operations to memory, the bank write control logic determines into which of the four memory banks the data are written. This

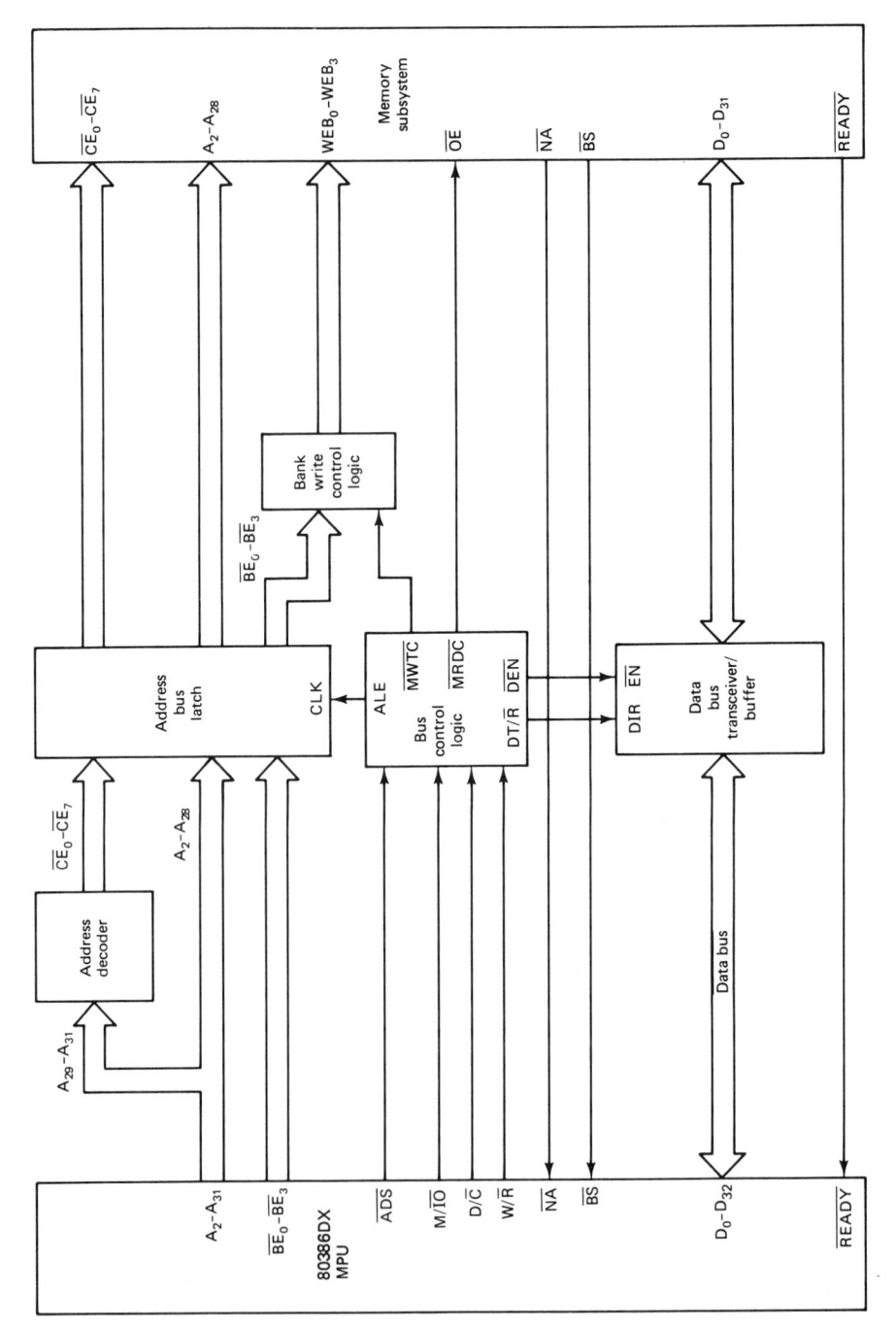

Figure 6.22 Memory interface block diagram.

depends on whether a byte, word, or double-word data transfer is taking place over the bus.

Notice in Fig. 6.22 that the latched byte enable signals $\overline{BE_0}$ through $\overline{BE_3}$ are gated with the memory write command signal $\overline{MWTC}$ to produce a separate write enable signal for each bank. These signals are denoted as $\overline{WEB_0}$ through $\overline{WEB_3}$. For example, if a word of data is to be written to memory over data bus lines D_0 through D_{15}, $\overline{WEB_0}$ and $\overline{WEB_1}$ are switched to their active 0 logic level.

The bus transceivers control the direction of data transfers between the MPU and memory subsystem. In Fig. 6.22 we see that the operation of the transceivers is controlled by the data transmit/receive (DT/$\overline{R}$) and data bus enable ($\overline{DEN}$) outputs of the bus control logic. $\overline{DEN}$ is applied to the enable ($\overline{EN}$) input of the transceivers and enables them for operation. This happens during all read and write bus cycles. DT/$\overline{R}$ selects the direction of data transfer through the transceivers. When a read cycle is in process, DT/$\overline{R}$ is set to 0 and data are passed from the memory subsystem to the MPU. On the other hand, when a write cycle is taking place, DT/$\overline{R}$ is switched to logic 1 and data are carried from the MPU to the memory subsystem.

Address Latches and Buffers

The 74F373 is an example of an octal latch device that can be used to implement the address latch section of the 80386DX's memory interface circuit. A block diagram of this device is shown in Fig. 6.23(a) and its internal circuitry is shown in Fig. 6.23(b). Notice that it accepts eight inputs DI_0 through DI_7. As long as the clock (CLK) input is at logic 1, the outputs of the D-type flip-flops follow the logic level of the data applied to their corresponding inputs. When CLK is switched to logic 0, the current contents of the D-type flip-flops are latched. The

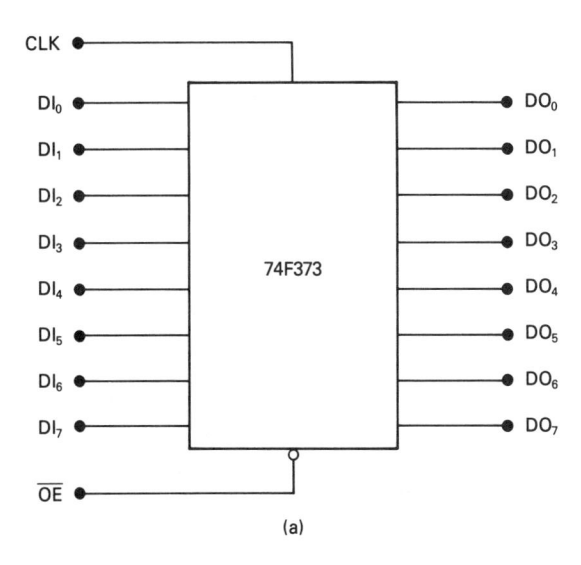

Figure 6.23 (a) Block diagram of an octal D-type latch. (b) Circuit diagram of the 74F373. (Courtesy of Texas Instruments Incorporated.)
(c) Operation of the 74F373. (Courtesy of Texas Instruments Incorporated.)

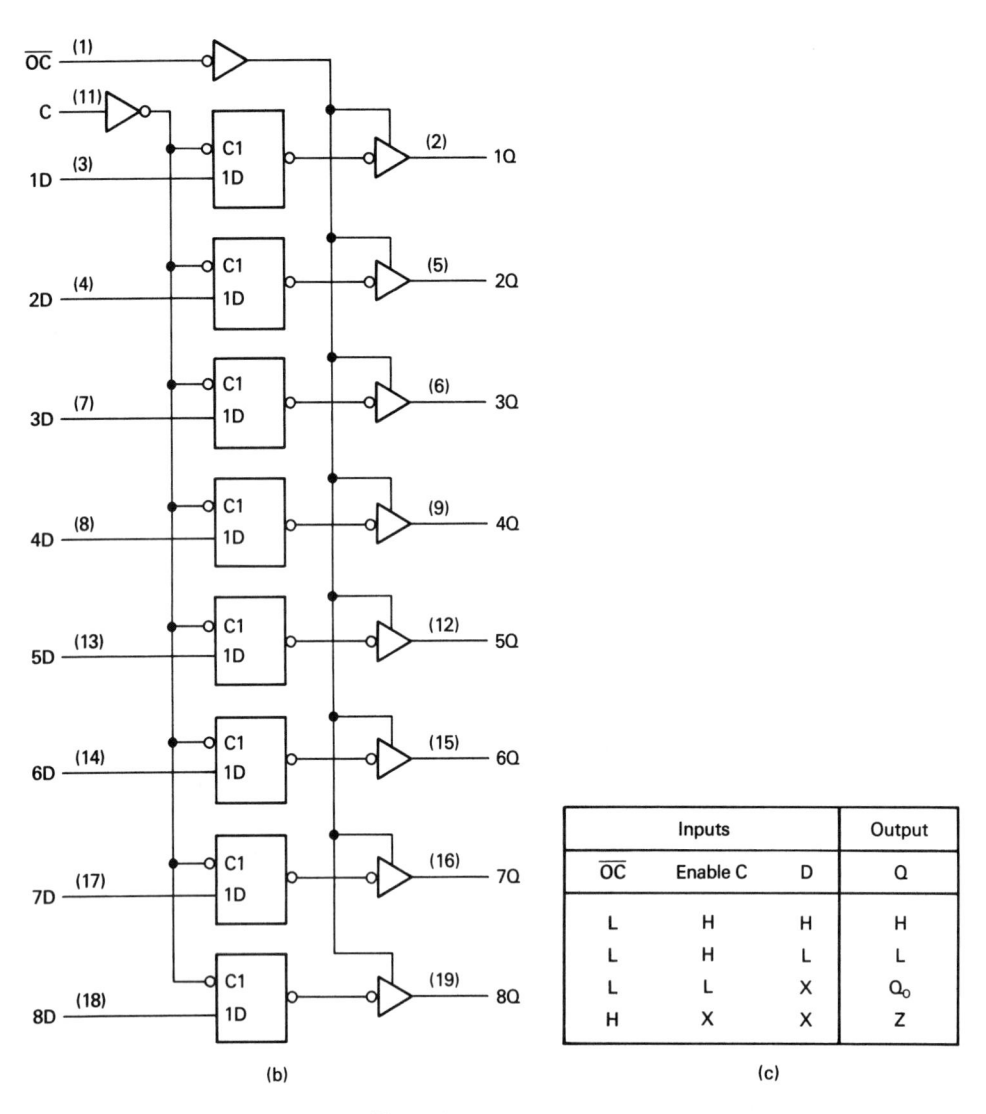

	Inputs		Output
$\overline{OC}$	Enable C	D	Q
L	H	H	H
L	H	L	L
L	L	X	Q_0
H	X	X	Z

(b)   (c)

Figure 6.23   (*Continued*)

latched information in the flip-flops are not output at data outputs DO_0 through DO_7 unless the output enable ($\overline{OE}$) input is at logic 0. If $\overline{OE}$ is at logic 1, the outputs are in the high-impedance state. Figure 6.23(c) summarizes this operation.

In the 80386DX microcomputer system, the 30 address lines A_2 through A_{31} and four byte enable signals $\overline{BE}_0$ through $\overline{BE}_3$ are normally latched in the address latch. The circuit configuration shown in Fig. 6.24 can be used to latch these signals. Notice that the latched outputs AL_2 through AL_{31} and $\overline{BEL}_0$ through $\overline{BEL}_3$ are permanently enabled by fixing $\overline{OE}$ at the 0 logic level. Moreover, the

234   The 80386DX Microprocessor and Its Memory Interface   Chap. 6

address information is latched at the outputs as the $\overline{\text{ALE}}$ signal from the bus control logic returns to logic 1, that is, when the CLK input of all devices is switched to logic 0.

In general, it is important to minimize the propagation delay of the address signals as they go through the bus interface circuit. The switching property of the 74F373 latches that determines this delay for the circuit of Fig. 6.24 is called

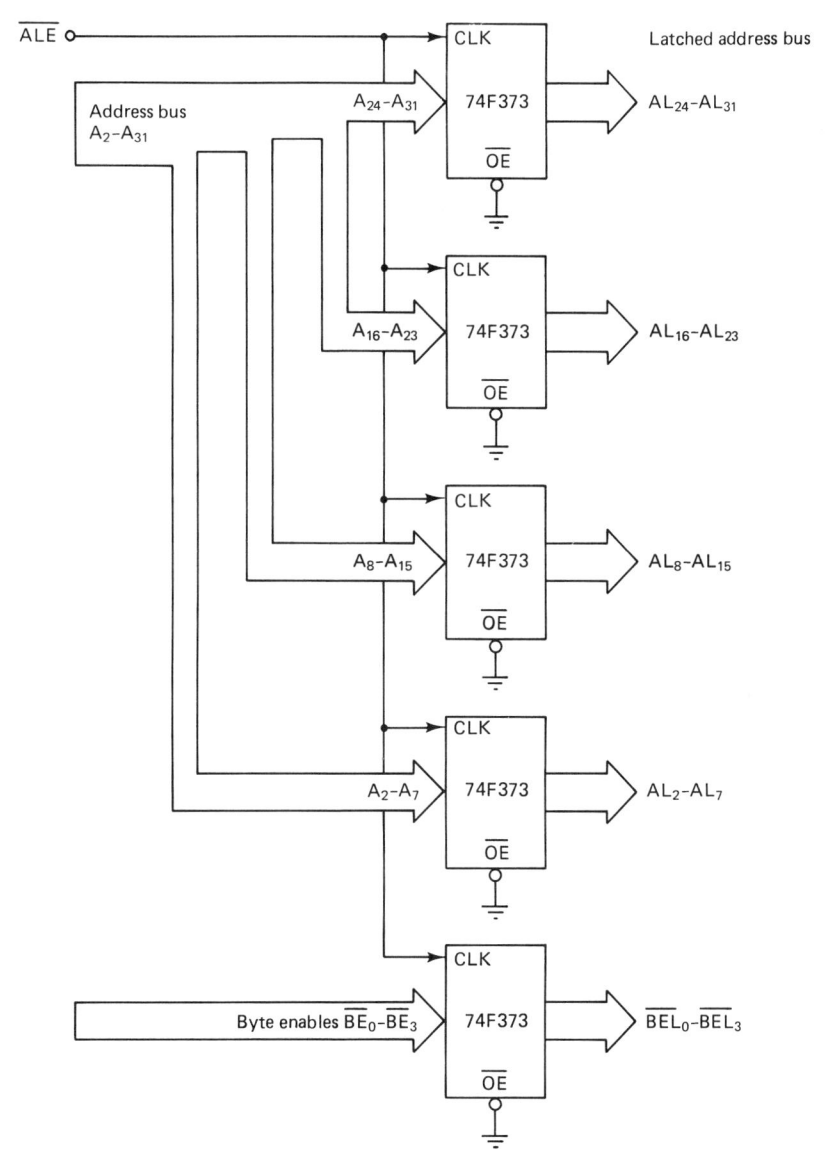

Figure 6.24 Address latch circuit.

enable-to-output propagation delay and has a maximum value of 13 ns. By selecting fast latches, that is, latches with a short propagation delay time, a maximum amount of the 80386DX's bus cycle time is preserved for the access time of the memory devices. In this way, slower, lower-cost memory ICs can be used. These latches also provide buffering for the 80386DX's address lines, which can only sink 4 mA. The outputs of the 74F373 latch can sink a maximum of 24 mA.

The 74F374 is another IC that is frequently used as an address latch in microcomputer systems. The circuit of the 74F374 is similar to the 74F373 we just introduced, in that it is an octal latch device. However, the flip-flops used to implement the latches are edge triggered instead of transparent. For this reason, when the $\overline{OC}$ input is logic 0, the data outputs become equal to the value of the data inputs synchronous with a low-to-high transition at the CLK input.

In some applications, additional buffering is required on the latched address lines. For example, the diagram in Fig. 6.25(a) shows that some of the address lines may be buffered to provide an independent I/O address bus. In this case a simple octal buffer/line driver device such as the 74F244 can be used as the I/O bus buffer. Figure 6.25(b) shows the buffer circuitry provided by the 74F244. The outputs of this device can sink a maximum of 64 mA.

Data Bus Transceivers

The data bus transceiver block of the bus interface circuit can be implemented with 74F245 octal bus transceiver ICs. Figure 6.26(a) shows a block diagram of this device. Notice that its bidirectional input/output lines are called A_0 through A_7 and B_0 through B_7. Looking at the circuit diagram in Fig. 6.26(b), we see that the $\overline{G}$ input is used to enable the buffer for operation. On the other hand, the logic level at the direction (DIR) input selects the direction in which data are transferred through the device. For instance, logic 0 at this input sets the transceiver to pass data from the B lines to the A lines. Switching DIR to logic 1 reverses the direction of data transfer.

(a)

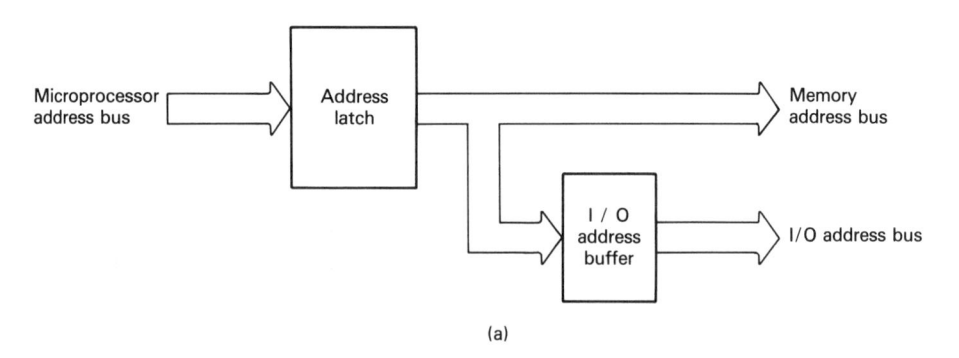

Figure 6.25 (a) Buffering the I/O address; (b) 74F244 circuit diagram. (Courtesy of Texas Instruments Incorporated.)

(b)

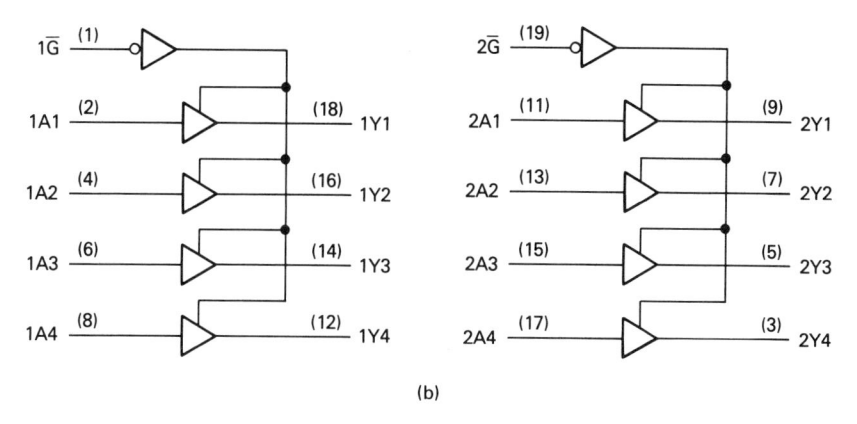

Figure 6.25 (*Continued*)

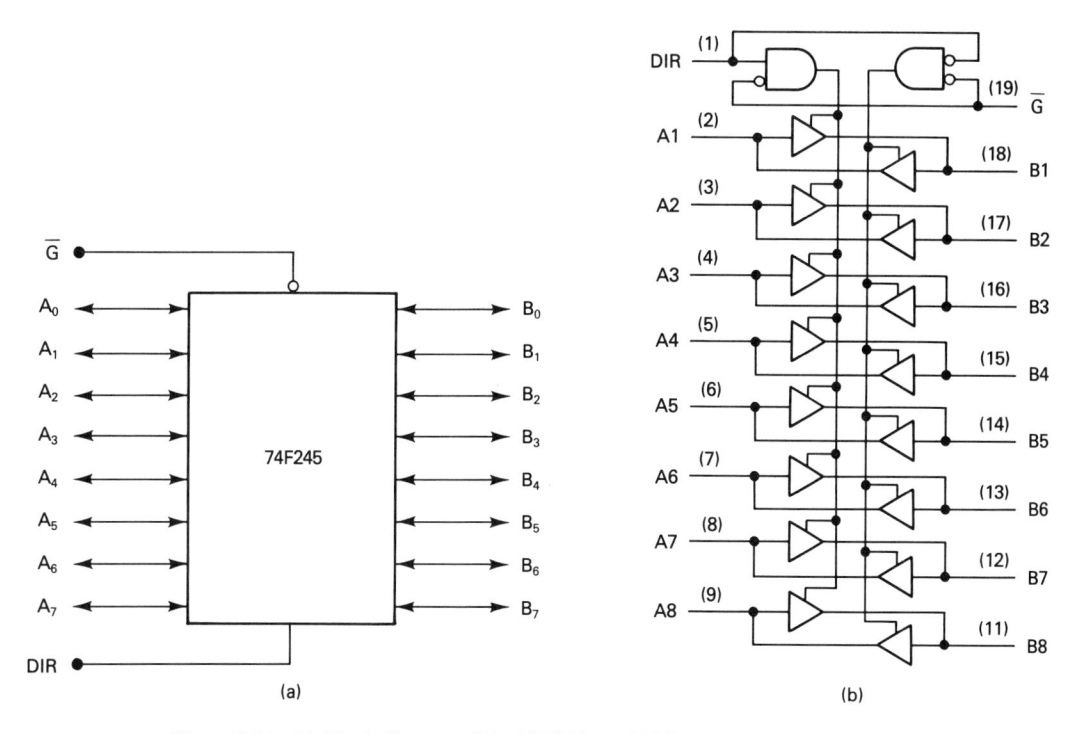

Figure 6.26 (a) Block diagram of the 74F245 octal bidirectional bus transceiver; (b) circuit diagram of the 74F245. (Courtesy of Texas Instruments Incorporated.)

Sec. 6.8 Memory Interface Circuits   237

Figure 6.27 shows a circuit that implements the data bus transceiver block of the bus interface circuit using the 74F245. For the 32-bit data bus of the 80386DX microcomputer, four devices are required. Here the DIR input is driven by the signal data transmit/receive (DT/$\overline{\text{R}}$) and $\overline{\text{G}}$ is supplied by data bus enable ($\overline{\text{DEN}}$). These signals are outputs of the bus control logic.

Another key function of the data bus transceiver circuit is to buffer the data bus lines. This capability is defined by how much current the devices can sink at their outputs. The I_{OL} rating of the 74F245 is 64 mA. The data bus lines of the 80386DX are rated for a maximum of 4 mA.

The 74F646 device is an octal bus transceiver with registers. This device is more versatile than the 74F245 we just described. It can be configured to operate either as a simple bus transceiver or as a registered bus transceiver. Figure 6.28(a) shows the operations that can occur when used as a simple transceiver. Notice that logic 0 at the SAB or SBA input selects the direction of data transfer through the device. Figure 6.28(b) shows that when configured for the registered mode of

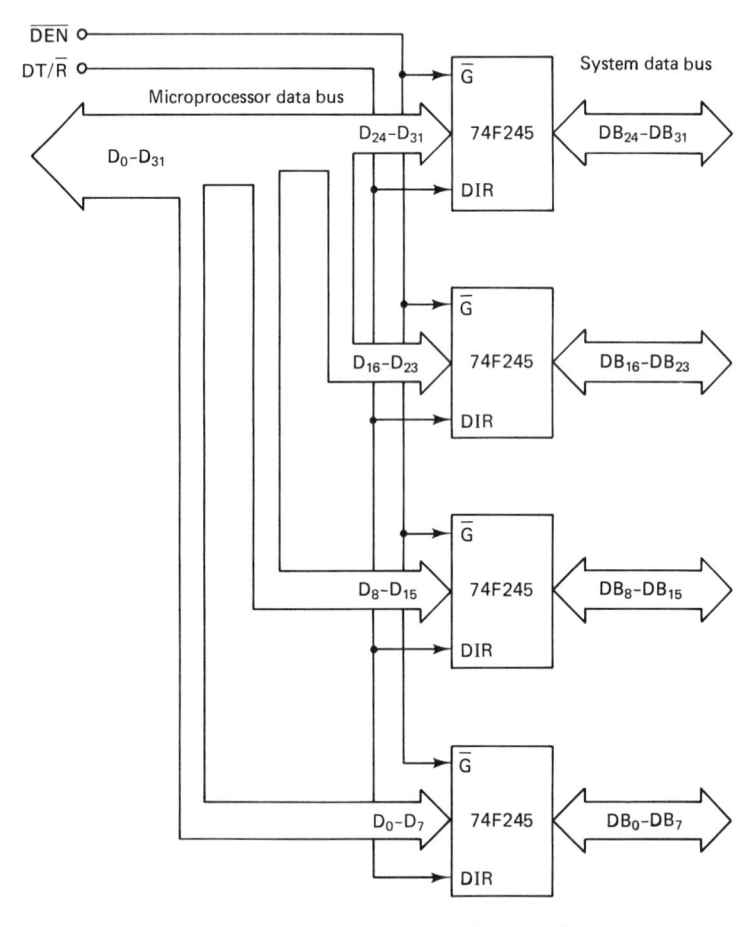

Figure 6.27 Data bus transceiver circuit.

238   The 80386DX Microprocessor and Its Memory Interface   Chap. 6

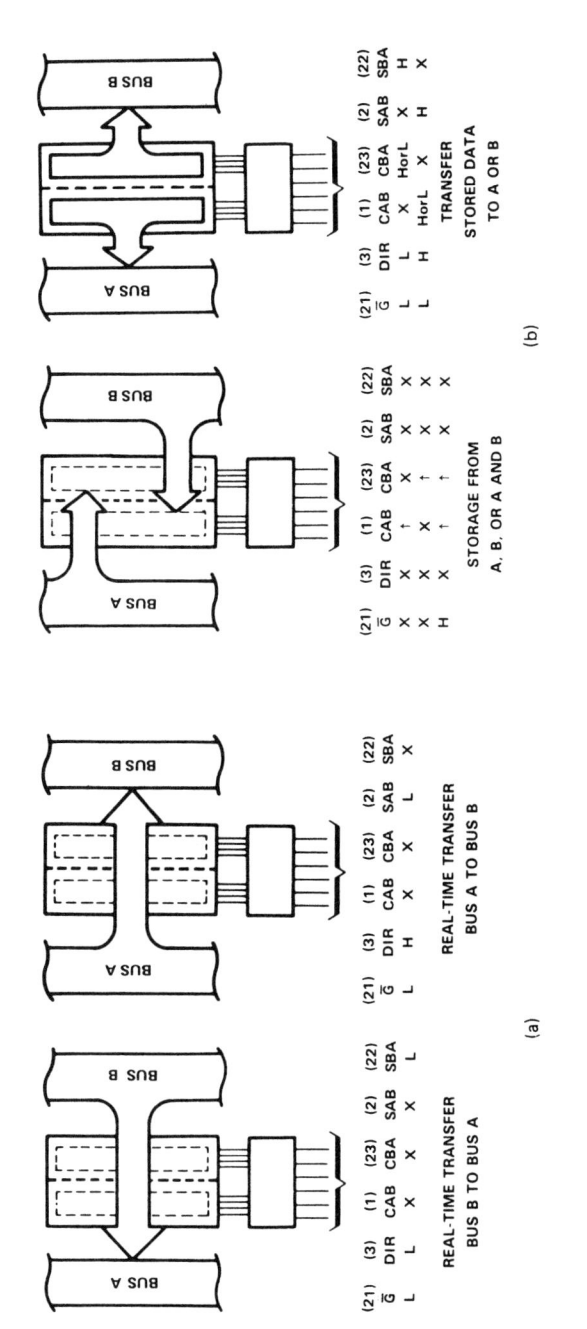

Figure 6.28 (a) Transceiver mode data transfers of the 74F646. (Courtesy of Texas Instruments Incorporated.) (b) Register mode data transfers. (Courtesy of Texas Instruments Incorporated.) (c) Control signals and data transfer operations. (Courtesy of Texas Instruments Incorporated.)

operation, data do not directly transfer between the A and B buses. Instead, data are passed from the bus to an internal register with one control signal sequence and from the internal register to the other bus with another control signal. Notice that lines CAB and CBA control the storage of the data from the A or B bus into the register and SAB and SBA control the transfer of stored data from the registers to the A or B bus. The table in Fig. 6.28(c) summarizes all of the operations performed by the 74F646.

Address Decoders

As shown in Fig. 6.29(a) and (b), the address decoder in the 80386DX microcomputer system can be located on either side of the address latch. A typical device that is used to perform this decode function is the 74F139 dual 2-line to 4-line decoder. Figure 6.30(a) and (b) show a block diagram and circuit diagram for this device, respectively. When the enable ($\overline{G}$) input is at its active 0 logic level, the output corresponding to the code at the BA inputs switches to the 0 logic level. For instance, when BA = 01, output Y_1 is logic 0. The operation of the 74F139 is summarized in the table of Fig. 6.31.

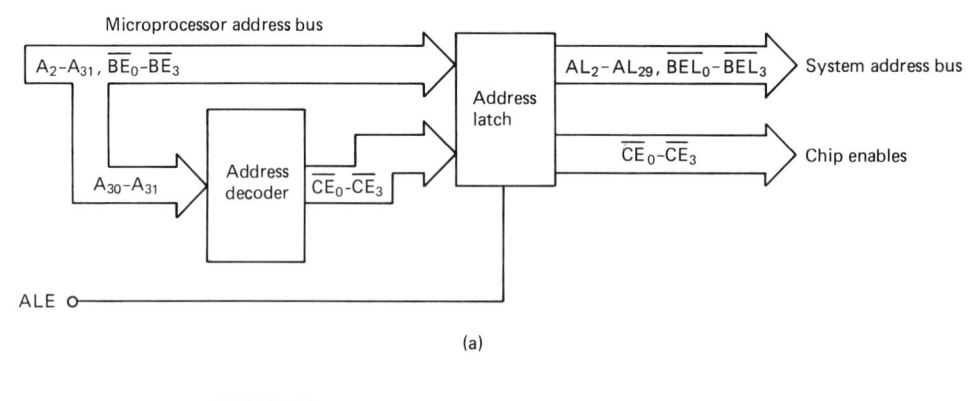

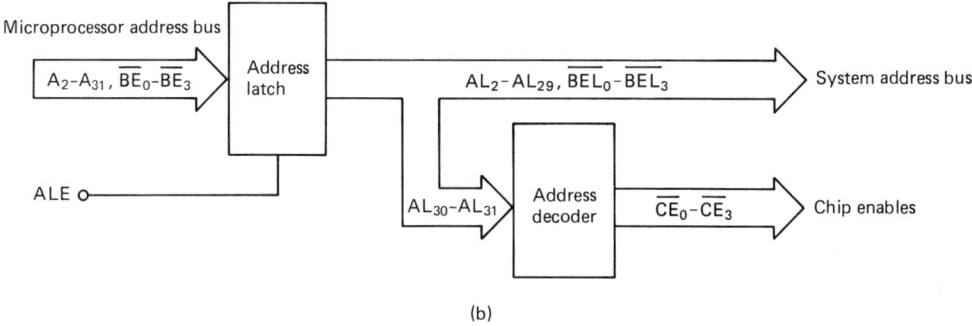

Figure 6.29 (a) Bus configuration with address decoding before the address latch; (b) bus configuration with address decoding after the address latch.

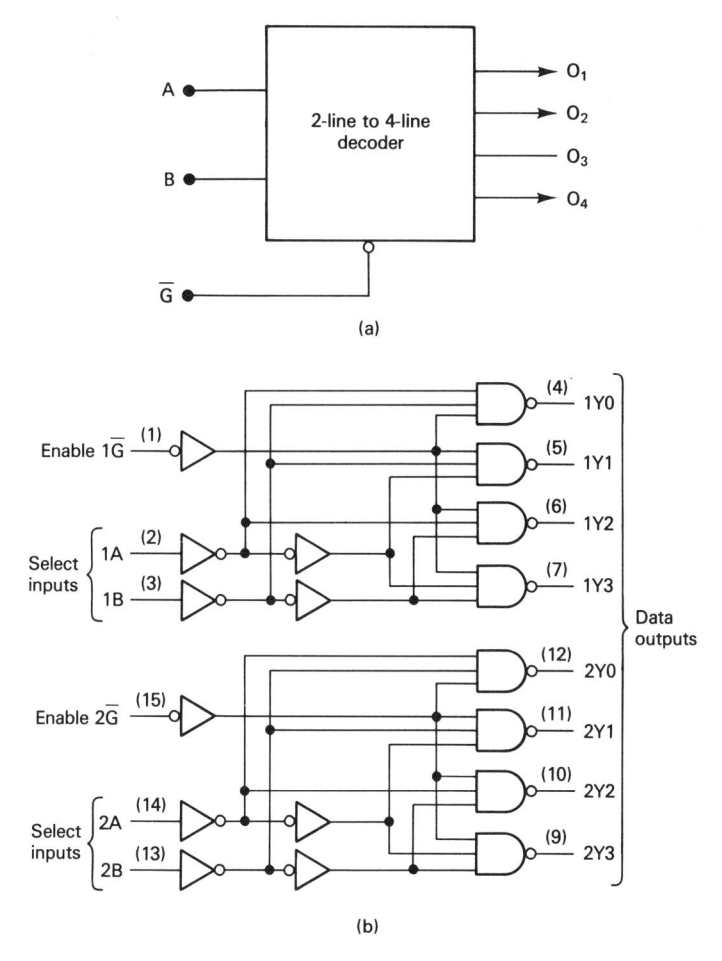

Figure 6.30 (a) Block diagram of the 74F139 2-line to 4-line decoder/demultiplexer; (b) circuit diagram of the 74F139. (Courtesy of Texas Instruments Incorporated.)

INPUTS			OUTPUTS			
ENABLE	SELECT					
$\bar{G}$	B	A	Y0	Y1	Y2	Y3
H	X	X	H	H	H	H
L	L	L	L	H	H	H
L	L	H	H	L	H	H
L	H	L	H	H	L	H
L	H	H	H	H	H	L

Figure 6.31 Operation of the 74F139 decoder. (Courtesy of Texas Instruments Incorporated.)

Sec. 6.8 Memory Interface Circuits

The circuit in Fig. 6.32 employs the address decoder configuration of Fig. 6.29(a). The advantage of this configuration is that the propagation delay involved in decoding the address becomes transparent. That is, since the propagation delay of the decoder takes place prior to the trailing edge of $\overline{\text{ALE}}$, it does not contribute to the total access time of the memory interface. The chip enable outputs $\overline{\text{CE}}_0$ through $\overline{\text{CE}}_3$ of the decoder are loaded into the address latch along with address

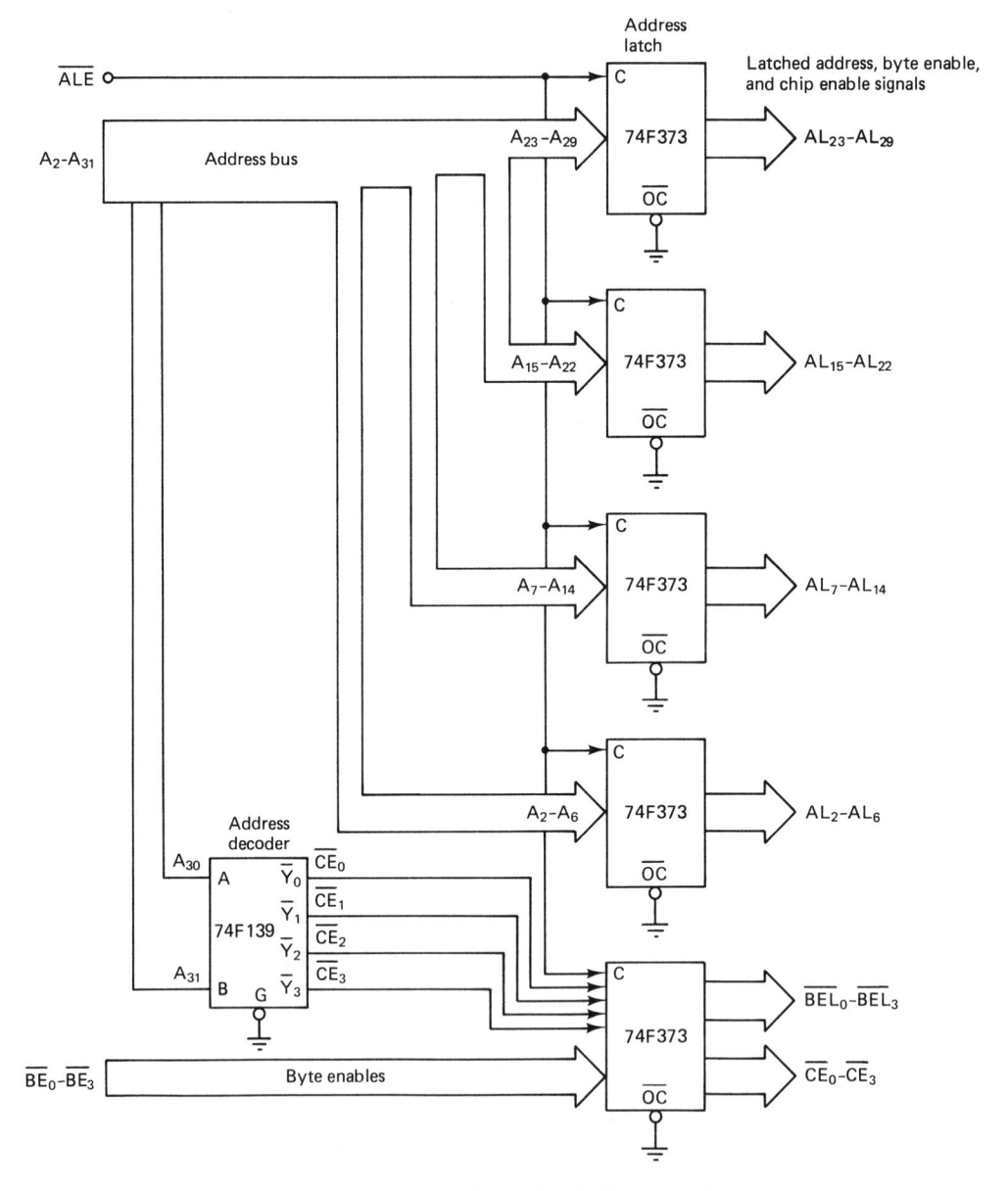

Figure 6.32 Address decoder/latch circuit.

The 80386DX Microprocessor and Its Memory Interface

signals A_2 through A_{29} and byte enable signals $\overline{BE}_0$ through $\overline{BE}_3$. In some applications, independent decoders are used for memory and I/O. In this case, one of the decoders in the 74F139 can be used to produce four memory chip enables and the other decoder to provide 4 I/O device chip selects.

The 74F138 is similar to the 74F139, except that it is a single 3-line to 8-line decoder. The circuit used in this device is shown in Fig. 6.33. Notice that it can be used to produce eight $\overline{CE}$ outputs. When enabled only the output that corresponds to the code at the CBA inputs switches to the active 0 logic level.

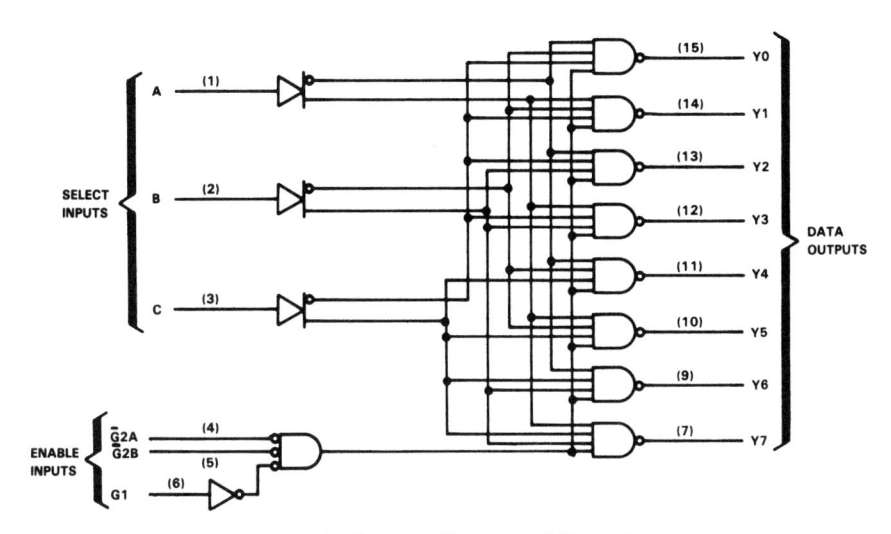

Figure 6.33 74F138 circuit diagram. (Courtesy of Texas Instruments Incorporated.)

6.9 PROGRAMMABLE LOGIC ARRAYS: BUS CONTROL LOGIC

In the preceding section we found that basic logic devices such as latches, transceivers, and decoders are required in the bus interface section of the 80386DX microcomputer system. We showed that these functions were performed with standard logic devices such as the 74F373 octal transparent latch, 74F245 octal bus transceiver, and 74F139 2-line to 4-line decoder, respectively. Today, *programmable logic array* (PLA) devices are becoming very important in the design of high-performance microcomputer systems. For example, the bus control logic section of the memory interface in Fig. 6.22 is normally implemented with PLAs, not with separate logic ICs. Unlike the devices mentioned earlier, PLAs do not implement a specific logic function. Instead, they are general-purpose logic devices that have the ability to perform a wide variety of specialized logic functions. A PLA contains a general purpose AND–OR–NOT array of logic gate circuits. The user has the ability to interconnect the inputs to the AND gates of this array. The definition of these inputs determines the logic function that is implemented.

The process used to connect or disconnect inputs of the AND gate array is known as *programming*. Thus, the name programmable logic array.

Block Diagram of a PLA

The block diagram in Fig. 6.34 represents a typical PLA. Looking at this diagram, we see that it has 16 input leads, marked I_0 through I_{15}. Moreover, there are eight output leads. These leads are labeled F_0 through F_7. This PLA is equipped with three-state outputs. For this reason, it has a chip enable control lead. In the block diagram this control input is marked $\overline{CE}$. The logic level of $\overline{CE}$ determines if the outputs are enabled or disabled.

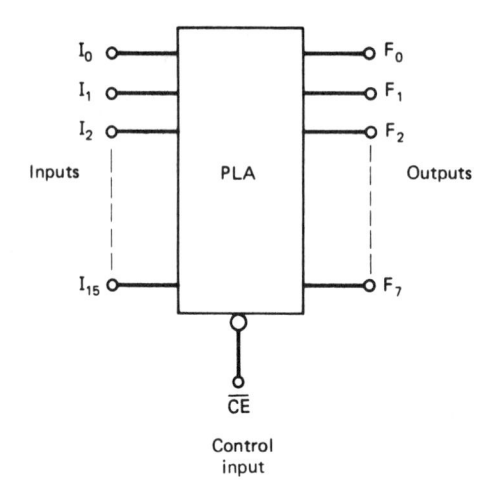

Figure 6.34 Block diagram of a PLA.

When a PLA is used to implement random logic functions, the inputs represent Boolean variables, and the outputs are used to provide eight separate random logic functions. The internal AND–OR–NOT array is programmed to define a sum-of-product equation for each of these outputs in terms of the inputs and their complements. In this way we see that the logic levels applied at inputs I_0 through I_{15} and the programming of the AND array determine what logic levels are produced at outputs F_0 through F_7. Therefore, the capacity of a PLA is measured by three properties: the number of inputs, the number of outputs, and the number of product terms (P-terms).

Architecture of a PLA

We just pointed out that the circuitry of a PLA is a general purpose AND–OR–NOT array. Figure 6.35(a) shows this architecture. Here we see that the input buffers supply input signals A and B and their complements $\overline{A}$ and $\overline{B}$. Programmable connections in the AND array permit any combination of these inputs to be combined to form a P-term. The product term outputs of the AND array are supplied to fixed inputs of the OR array. The output of the OR gate produces a sum-of-products function. Finally, the inverter complements this function.

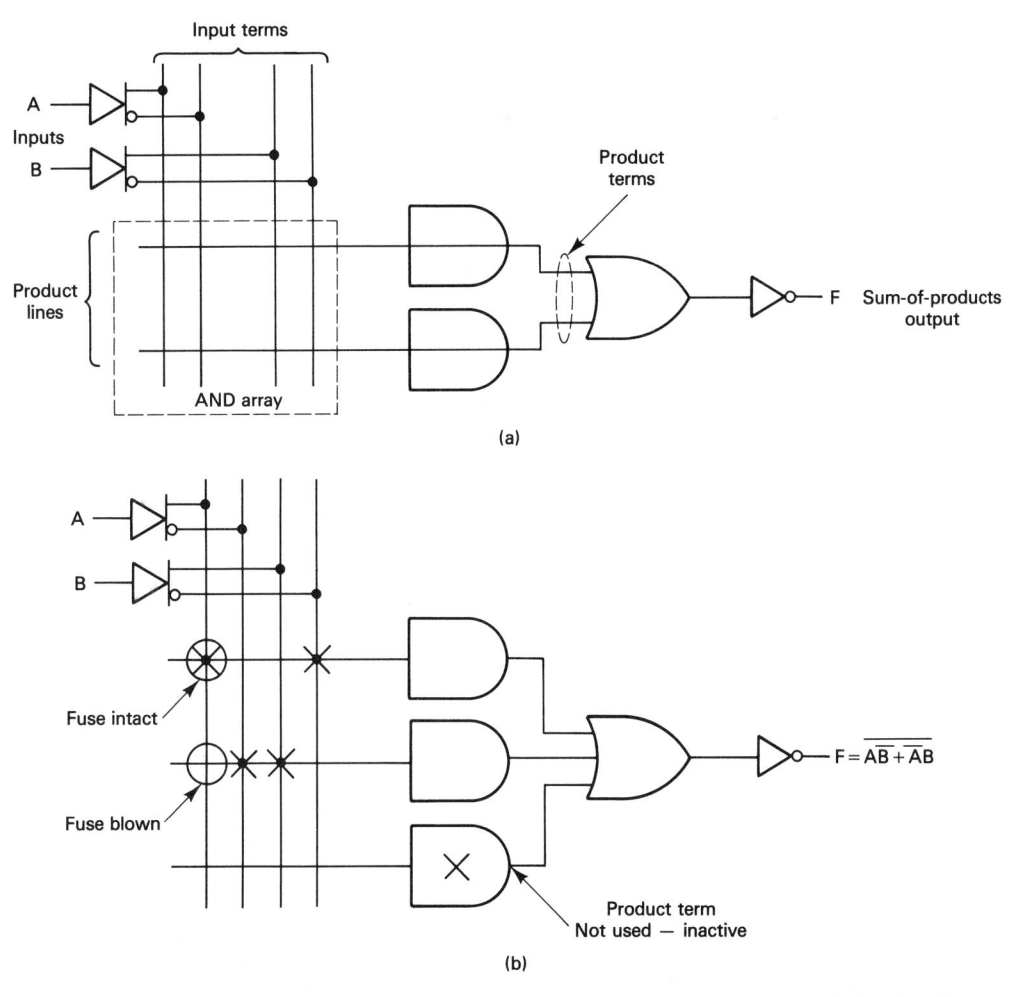

Figure 6.35 (a) Basic PLA architecture; (b) implementing the logic function $F = \overline{(A\overline{B} + \overline{A}B)}$.

The circuit of Fig. 6.35(b) shows how the function $F = \overline{A\overline{B} + \overline{A}B}$ is implemented with the AND–OR–NOT array. Notice that an X marked into the AND array means that the fuse is left intact, and no marking means that it has been blown to form an open circuit. For this reason, the upper AND gate is connected to A and $\overline{B}$ and produces the product term $A\overline{B}$. The second AND gate from the top connects to $\overline{A}$ and B to produce $\overline{A}B$. The bottom AND gate is marked with an X to indicate that it is not in use. Gates like this that are not to be active should have all of their input fuse links left intact.

In Fig. 6.36(a), we have shown the circuit structure that is most widely used in PLAs. It differs from the circuit shown in Fig. 6.35(a) in two ways. First, the inverter has a programmable three-state control and can be used to isolate the logic function from the output. Second, the buffered output is fed back to form

Sec. 6.9 Programmable Logic Arrays: Bus Control Logic 245

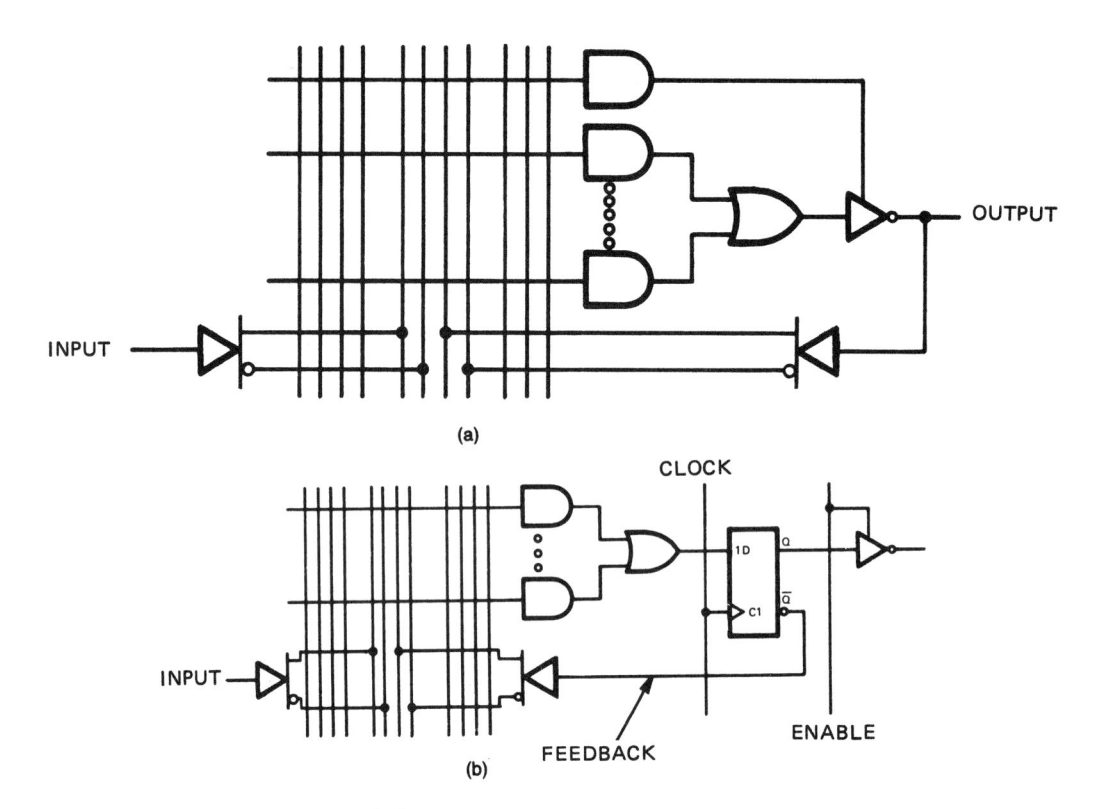

Figure 6.36 (a) Typical PLA architecture. (Courtesy of Texas Instruments Incorporated.) (b) PLA with output latch. (Courtesy of Texas Instruments Incorporated.)

another set of inputs to the AND array. This new output configuration permits the output pin to be programmed to work as a *standard output, standard input,* or *logic-controlled input/output*. For instance, if the upper AND gate, which is the control gate for the output buffer, is set up permanently to enable the inverter, and the fuse links for its inputs that are fed back from the outputs are all blown open, the output functions as a standard output.

PLAs are also available in which the outputs are latched with registers. A circuit for this type of device is shown in Fig. 6.36(b). Here we see that the output of the OR gate is applied to the D input of a clocked D-type flip-flop. In this way, the logic level produced by the AND–OR array is not presented at the output until a pulse is first applied at the CLOCK input. Furthermore, the feedback input is produced from the complemented output of the flip-flop, not the output of the inverter. This configuration is known as a *PLA with registered outputs* and is designed to simplify implementation of *state machine* designs.

Standard PLA Devices

Now that we have introduced the block diagram of the PLA and internal architecture of the PLA, let us continue by examining a few of the widely used PAL™

devices. A programmable array logic (PAL) is a PLA in which the OR array is fixed. That is, only the AND array is programmable.

The 16L8 is one of the more widely used PAL ICs. Its internal circuitry and pin numbering are shown in Fig. 6.37(a). This device is housed in a 20-pin package as shown in Fig. 6.37(b). Looking at this diagram, we see that it employs the PLA

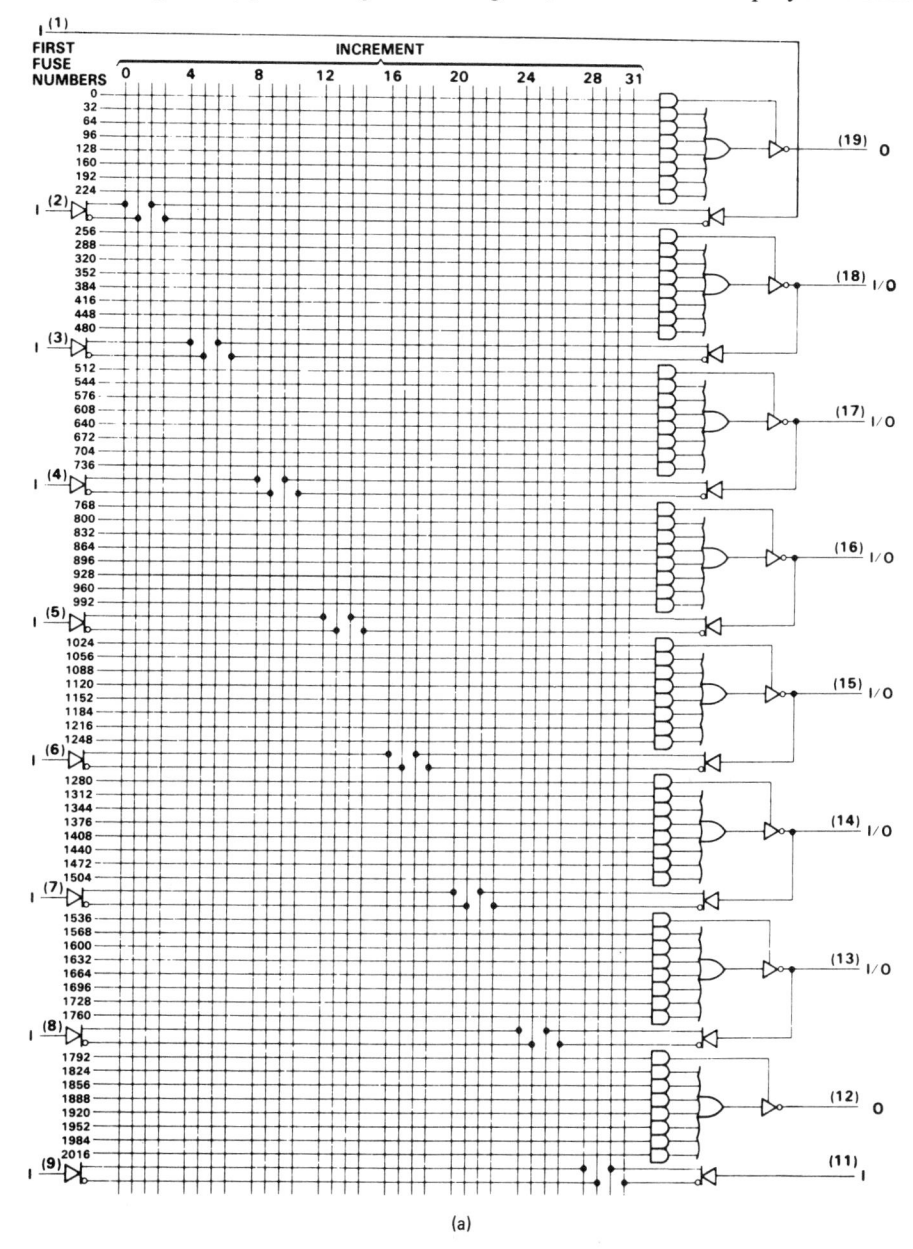

Figure 6.37 (a) 16L8 circuit diagram. (Courtesy of Texas Instruments Incorporated.) (b) 16L8 pin layout. (Courtesy of Texas Instruments Incorporated.)

Sec. 6.9 Programmable Logic Arrays: Bus Control Logic

```
         ┌──┬─┐
       I │1  20│ V_CC
       I │2  19│ O
       I │3  18│ I/O
       I │4  17│ I/O
       I │5  16│ I/O
       I │6  15│ I/O
       I │7  14│ I/O
       I │8  13│ I/O
       I │9  12│ O
     GND │10 11│ I
         └─────┘
```

(b)

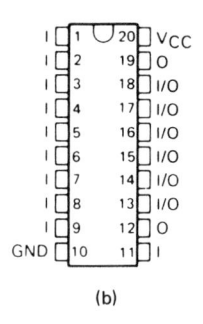

Figure 6.37 (*Continued*)

architecture that was illustrated in Fig. 6.36(a). Notice that it has 10 dedicated input pins. All of these pins are labeled I. There are also two dedicated outputs, which are labeled with the letter O, and six programmable I/O lines, which are labeled I/O. Using the programmable I/O lines, the number of input lines can be expanded to as many as 16 inputs or the number of outputs can be increased to as many as eight lines.

All of the 16L8's inputs are buffered and produce both the original form of the signal and its complement. The outputs of the buffer are applied to the inputs of the AND array. This array is capable of producing 64 product terms. Notice that the AND gates are arranged into eight groups of eight. The outputs of seven gates in each of these groups are used as inputs to an OR gate and the eighth is used to produce an enable signal for the corresponding three-state output buffer. In this way we see that the 16L8 is capable of producing up to seven product terms for each output and the product terms can be formed using any combination of the 16 inputs.

The 16L8 is manufactured with bipolar technology. It operates from a +5-V ±10% dc power supply and draws a maximum of 180 mA. Moreover, all of its inputs and outputs are at TTL-compatible voltage levels. This device exhibits medium-speed input/output propagation delays. In fact, the maximum I-to-O propagation delay is rated as 35 ns.

Another widely used PAL is the 20L8 device. It is similar to that of the 16L8 just described. However, the 20L8 has a maximum of 20 inputs, eight outputs, and 64 P terms.

The 16R8 is also a popular 20-pin PAL. The circuit diagram and pin layout for this device are shown in Fig. 6.38(a) and (b), respectively. From Fig. 6.38(a) we find that its eight fixed I inputs and AND–OR array are essentially the same as those of the 16L8. There is one change. The outputs of eight AND gates, instead of seven, are supplied to the inputs of each OR gate.

On the other hand, a number of changes have been made at the output side. Notice that the outputs of the OR gates are first latched in D-type flip-flops with the CLK signal. They are then buffered and supplied to the eight Q outputs. Another change is that the enable signals for the output inverters are no longer programmable. Now all three-state outputs are enabled by the logic level of the $\overline{OE}$ control input.

The last change is in the part of the circuit that produces the feedback inputs. In the 16R8, these eight input signals are derived from the complementary output of the corresponding latch instead of the output of the buffer. For this reason the

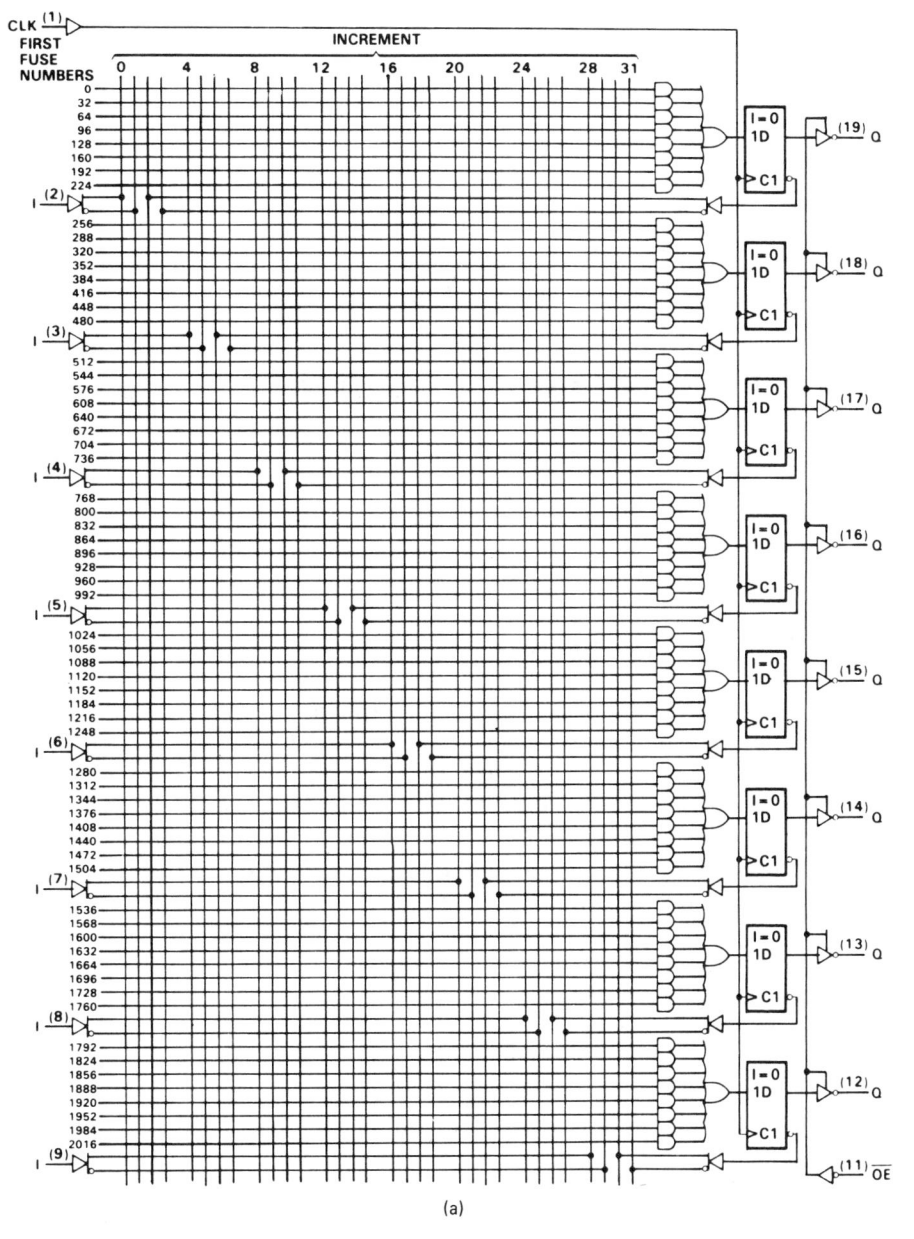

Figure 6.38 (a) 16R8 circuit diagram. (Courtesy of Texas Instruments Incorporated.) (b) 16R8 pin layout. (Courtesy of Texas Instruments Incorporated.)

Sec. 6.9   Programmable Logic Arrays: Bus Control Logic

(b)

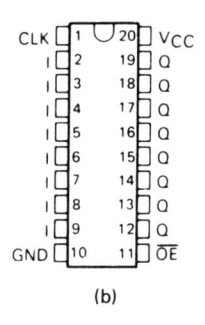

Figure 6.38 (*Continued*)

output leads can no longer be programmed to work as direct inputs. The 20R8 is the registered output version of the 20L8 PAL.

6.10 PROGRAM STORAGE MEMORY: ROM, PROM, AND EPROM

Read-only memory (ROM) is one type of semiconductor memory device. It is most widely used in microcomputer systems for storage of the program that determines overall system operation. The information stored within a ROM integrated circuit is permanent—or *nonvolatile*. This means that when the power supply of the device is turned off, the stored information is not lost.

ROM, PROM, and EPROM

For some ROM devices, information (the microcomputer program) must be built in during manufacturing and for others the data must be electrically entered. The process of entering data into a ROM is called *programming*. As the name ROM implies, once entered into the device this information can be read only. Three types of ROM devices exist. They are known as the *mask-programmable read-only memory* (ROM), the *one-time programmable read-only memory* (PROM), and the *erasable programmable read-only memory* (EPROM).

Let us continue by looking more closely into the first type of device, the mask programmable read-only memory. This device has its data pattern programmed as part of the manufacturing process. This is known as *mask programming*. Once the device is programmed, its contents can never be changed. Because of this and the cost for making the programming masks, ROMs are used mainly in high-volume applications where the data will not change frequently.

The other two types of read-only memories, the PROM and EPROM, differ from the ROM in that the data contents are electrically entered by the user. Programming is usually done with equipment called an *EPROM programmer*. Both the PROM and EPROM are programmed in the same way. Once a PROM is programmed, its contents cannot be changed. This is the reason they are sometimes called one-time programmable EPROMs. On the other hand, the contents of an EPROM can be erased by exposing it to ultraviolet light. In this way, the device can be used over and over again simply by erasing and reprogramming. PROMs and EPROMs are most often used during the design of a product and for early production, when the code of the microcomputer may need to be changed frequently.

Block Diagram of a ROM

A block diagram of a typical ROM is shown in Fig 6.39. Here we see that the device has three sets of signal lines: the address inputs, data outputs, and control inputs. This block diagram is valid for a ROM, PROM, or EPROM. Let us now look at the function of each of these sets of signal lines.

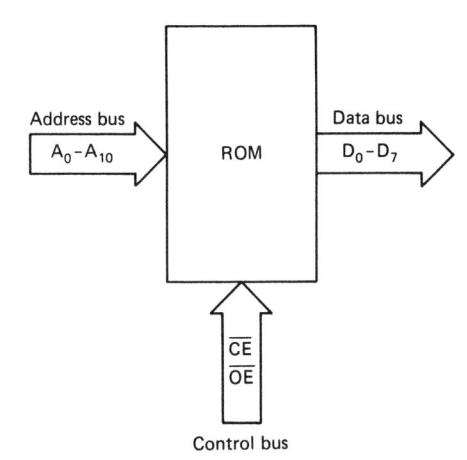

Figure 6.39 Block diagram of a ROM.

The address bus is used to input the signals that select between the data storage locations within the ROM device. In Fig. 6.39 we find that this bus consists of 11 address lines, A_0 through A_{10}. The bits in the address are arranged so that A_{10} is the MSB and A_0 is the LSB. With an 11-bit address, the memory device has $2^{11} = 2048$ unique data storage locations. The individual storage locations correspond to addresses over the range $00000000000_2 = 000_{16}$ through $11111111111_2 = 7FF_{16}$.

Each bit of data is stored inside a ROM, PROM, or EPROM as either a binary 0 or binary 1. Actually, 8 bits of data are stored at every address. Therefore the total storage capacity of the device we are describing is $2048 \times 8 = 16,384$ bits; that is, the device we are describing is really a 16K-bit ROM. By applying the address of a storage location to the address inputs of the ROM, the byte of data held at the addressed location is read out onto the data lines. In the block diagram of Fig. 6.39, we see that the data bus consists of eight lines labeled D_0 through D_7.

The control bus represents the control signals that are required to enable or disable the ROM, PROM, or EPROM device. In the block diagram of Fig. 6.39, two control leads, output enable ($\overline{OE}$) and chip enable ($\overline{CE}$), are identified. For example, logic 0 at $\overline{OE}$ enables the three state outputs, D_0 through D_7, of the device. If $\overline{OE}$ is switched to the 1 logic level, these outputs are disabled (put in the high-Z state). Moreover, $\overline{CE}$ must be at logic 0 for the device to be active. Logic 1 at $\overline{CE}$ puts the device in a low-power standby mode. When in this state, the data outputs are in the high-Z state independent of the logic level of $\overline{OE}$.

Read Operation

For a microprocessor to read a byte of data from the device, it must apply a binary address to inputs A_0 through A_{10}. This address gets decoded inside the device to select the storage location of the byte of data that is to be read. Then the microprocessor must switch $\overline{CE}$ and $\overline{OE}$ to logic 0 to enable the device and outputs. Now the byte of data is available at D_0 through D_7 and the microprocessor can read the data over its data bus.

From our description of the read operation, it appears that after the inputs of the ROM are set up the output appears immediately; however, in practice this is not true. A short delay exists between address inputs and data outputs. This leads us to three important timing properties defined for the read cycle of a ROM. They are called *access time* (t_{ACC}), *chip enable time* (t_{CE}), and *chip deselect time* (t_{DF}).

Access time tells us how long it takes to access data stored in a ROM. Here we assume that both $\overline{CE}$ and $\overline{OE}$ are at their active zero levels, and then an address is applied to the inputs of the ROM. In this case, the delay t_{ACC} occurs before the data stored at the addressed location are stable at the outputs. The microprocessor must wait at least this long before reading the data; otherwise, invalid results may be obtained.

Chip enable time is similar to access time. In fact, for most EPROMs they are equal in value. They differ in how the device is set up initially. This time the address is applied and $\overline{OE}$ is switched to 0, then the read operation is initiated by making $\overline{CE}$ active. Therefore, t_{CE} represents the chip enable to output delay instead of the address to output delay.

Chip deselect time is the opposite of access or chip enable time. It represents the amount of time the device takes for the data outputs to return to the high-Z state after $\overline{OE}$ becomes inactive—that is, the recovery time of the outputs.

Standard EPROM ICs

A large number of standard EPROM ICs are available today. Figure 6.40 lists the part numbers, bit densities, and byte capacities of the seven most popular devices. They range in size from the 2716, which is a 16K-bit density ($2K \times 8$) device, to the 27C010, which is a 1M-bit ($128K \times 8$) device. Higher-density devices, such as the 27C256 through 27C010, are most popular for new system designs. In fact, some of the older devices, such as the 2716 and 2732, have already been discon-

EPROM	Density (bits)	Capacity (bytes)
2716	16K	2K × 8
2732	32K	4K × 8
27C64	64K	8K × 8
27C128	128K	16K × 8
27C256	256K	32K × 8
27C512	512K	64K × 8
27C010	1M	128K × 8

Figure 6.40 Standard EPROM devices.

tinued by many manufacturers. Let us now look at some of these EPROMs in more detail.

The 27C256 is an EPROM IC manufactured with the CMOS technology. Looking at Fig. 6.40, we find that it is a 256K-bit device and its storage array is organized as 32K × 8 bits. Figure 6.41 shows the pin layout of the 27C256. Here we see that it has 15 address inputs, labeled A_0 through A_{14}, eight data outputs, identified as O_0 through O_7, and control signals $\overline{CE}$ and $\overline{OE}$.

The 27C256 is available in three access-time speed selections. In Fig. 6.42 we find that the speed of an EPROM is denoted by a dash and number at the end of the generic part number. For example, the standard 27C256 is a 250-ns-access-time device. On the other hand, the 27C256-1 is faster; it has an access time of 150 ns.

In an erased 27C256, all storage cells hold logic 1. The device is put into the programming mode by switching on the V_{PP} power supply. Once in this mode, the address of the storage location that is to be programmed is applied to the address inputs, and the data that is to be loaded into this location is supplied to the data leads. Now the $\overline{CE}$, input is pulsed to load the data. Actually, a complex series of program and verify operations are performed to program each storage location in an EPROM. The two programming sequences in wide use today are the *Quick-Pulse Programming Algorithm*™ and the *Intelligent Programming Algorithm*™. Flowcharts for these programming algorithms are given in Fig. 6.43(a) and (b), respectively.

Figure 6.41 also shows the pin layouts for the 2716 through 27C512 EPROM devices. In this diagram we find that the 27C256 and 27C512 are both in a 28-pin package. A comparison of the pin configuration of the 27C512 with that of the 27C256 shows that the only differences between the two pinouts are that pin 1 on the 27C512 becomes the new address input A_{15}, and V_{PP}, which was at pin 1 on the 27C256, becomes a second function performed by pin 22 on the 27C512.

Expanding ROM Word Length and Word Capacity

In many applications, the microcomputer system requirements for ROM are greater than what is available in a single device. There are two basic reasons for expanding ROM capacity: first, the byte-wide length is not large enough; second, the total storage capacity is not enough bytes. Both of these expansion needs can be satisfied by interconnecting a number of ICs.

Pin	2716	2732A	27C64	27C128	27C512
28			V_{CC}	V_{CC}	V_{CC}
27			$\overline{PGM}$	$\overline{PGM}$	A_{14}
26	V_{CC}	V_{CC}	N.C.	A_{13}	A_{13}
25	A_8	A_8	A_8	A_8	A_8
24	A_9	A_9	A_9	A_9	A_9
23	V_{PP}	A_{11}	A_{11}	A_{11}	A_{11}
22	$\overline{OE}$	$\overline{OE}/V_{PP}$	$\overline{OE}$	$\overline{OE}$	$\overline{OE}/V_{PP}$
21	A_{10}	A_{10}	A_{10}	A_{10}	A_{10}
20	$\overline{CE}$	$\overline{CE}$	$\overline{CE}$	$\overline{CE}$	$\overline{CE}$
19	O_7	O_7	O_7	O_7	O_7
18	O_6	O_6	O_6	O_6	O_6
17	O_5	O_5	O_5	O_5	O_5
16	O_4	O_4	O_4	O_4	O_4
15	O_3	O_3	O_3	O_3	O_3

```
           ┌──┐
  V_PP  1 ─┤  ├─ 28  V_CC
  A_12  2 ─┤  ├─ 27  A_14
  A_7   3 ─┤  ├─ 26  A_13
  A_6   4 ─┤  ├─ 25  A_8
  A_5   5 ─┤  ├─ 24  A_9
  A_4   6 ─┤  ├─ 23  A_11
  A_3   7 ─┤27C256├─ 22  OE
  A_2   8 ─┤  ├─ 21  A_10
  A_1   9 ─┤  ├─ 20  CE
  A_0  10 ─┤  ├─ 19  O_7
  O_0  11 ─┤  ├─ 18  O_6
  O_1  12 ─┤  ├─ 17  O_5
  O_2  13 ─┤  ├─ 16  O_4
  Gnd  14 ─┤  ├─ 15  O_3
           └──┘
```

27C512	27C128	27C64	2732A	2716	Pin
A_{15}	V_{PP}	V_{PP}			1
A_{12}	A_{12}	A_{12}			2
A_7	A_7	A_7	A_7	A_7	3
A_6	A_6	A_6	A_6	A_6	4
A_5	A_5	A_5	A_5	A_5	5
A_4	A_4	A_4	A_4	A_4	6
A_3	A_3	A_3	A_3	A_3	7
A_2	A_2	A_2	A_2	A_2	8
A_1	A_1	A_1	A_1	A_1	9
A_0	A_0	A_0	A_0	A_0	10
O_0	O_0	O_0	O_0	O_0	11
O_1	O_1	O_1	O_1	O_1	12
O_2	O_2	O_2	O_2	O_2	13
Gnd	Gnd	Gnd	Gnd	Gnd	14

Figure 6.41 Pin layouts of standard EPROMs.

Part number	Access time
27C256	250 ns
27C256-2	200 ns
27C256-1	150 ns

Figure 6.42 Speed selections for the 27C256.

For example, the 80386DX microprocessor has a 32-bit data bus. Therefore, its program memory subsystem needs to be implemented with four 27C256 EPROMs connected as shown in Fig. 6.44(a). Notice that the individual address inputs, chip enable lines, and output enable lines on the four devices are connected in parallel. On the other hand, the eight data outputs of each device are used to

(a)

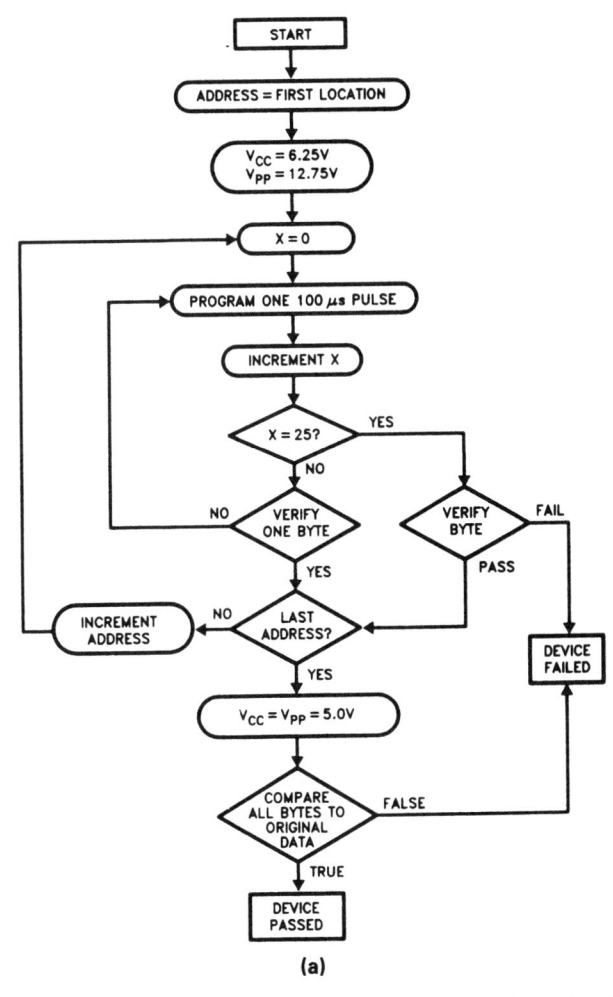

Figure 6.43 (a) Quick-Pulse Programming™ algorithm flowchart. (Reprinted by permission of Intel Corp. Copyright Intel Corp. 1989.) (b) Intelligent Programming™ algorithm flowchart. (Reprinted by permission of Intel Corp. Copyright Intel Corp. 1989.)

Sec. 6.10   Program Storage Memory: ROM, PROM, and EPROM

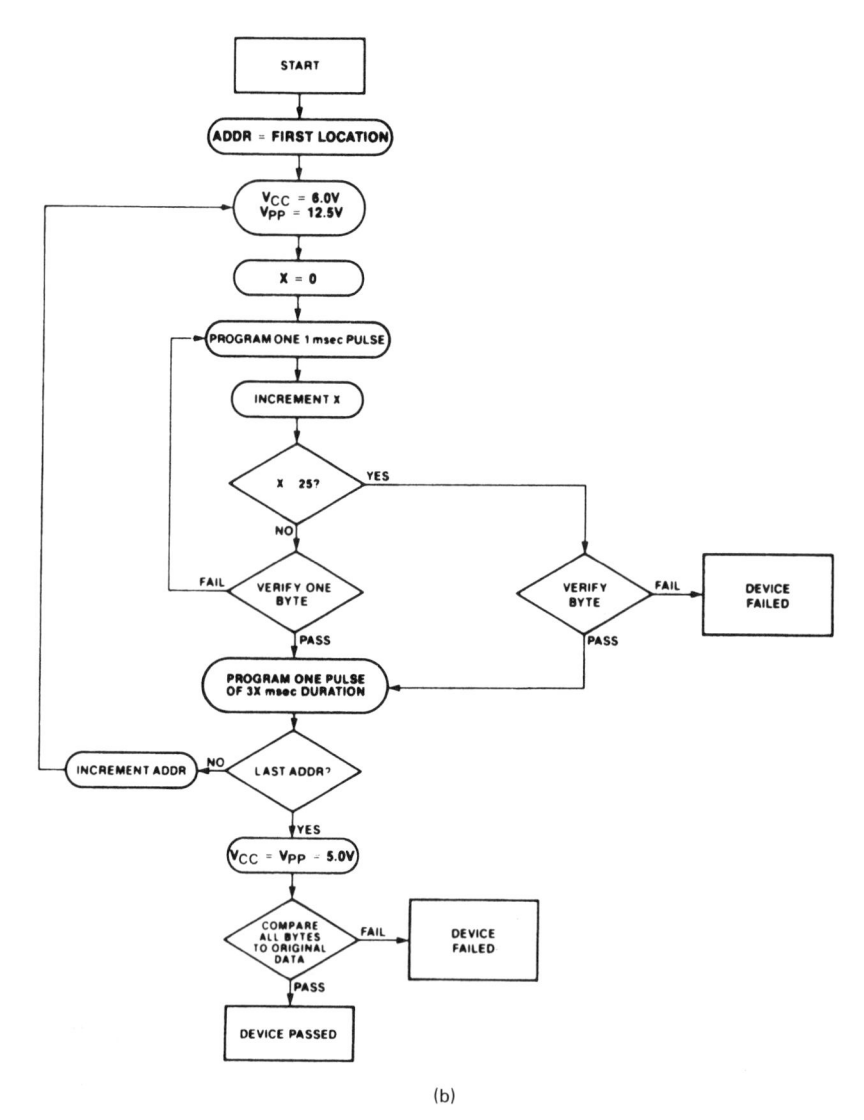

Figure 6.43 (*Continued*)

supply eight lines of the 32-bit data bus. This device configuration has a total storage capacity equal to 32K 32-bit words (1M bits).

Figure 6.44(b) shows how two 27C256s can be interconnected to expand the number of bytes of storage. Here the individual address inputs, data outputs, and output enable lines of the two devices are connected in parallel. However, the $\overline{CE}$ inputs of the individual devices remain independent and can be supplied by different outputs of an address decoder circuit. In this way only one of the two devices will be enabled at a time. This configuration results in a total storage capacity of 64K bytes (512K bits). To expand the word capacity of the circuit in Fig. 6.44(a), this connection must be made for each EPROM.

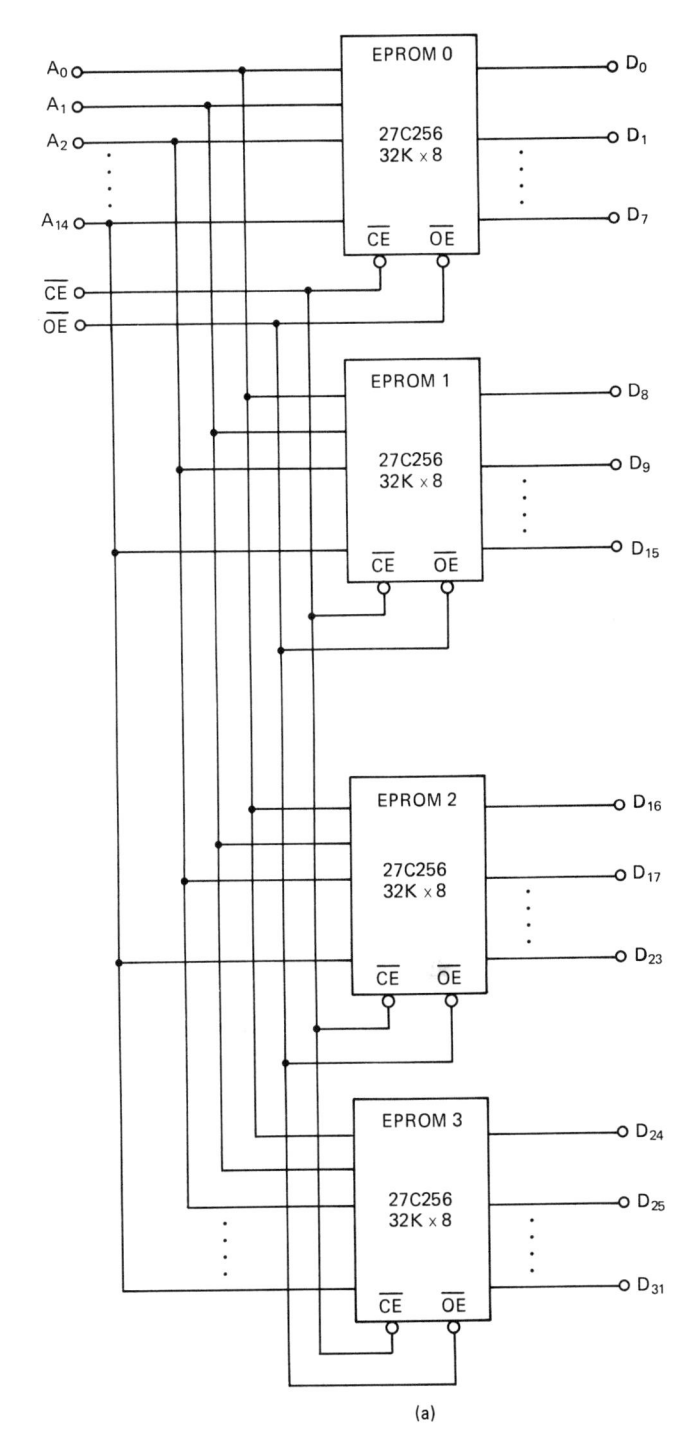

Figure 6.44 (a) Expanding word length; (b) expanding word capacity.

Sec. 6.10 Program Storage Memory: ROM, PROM, and EPROM

(b)

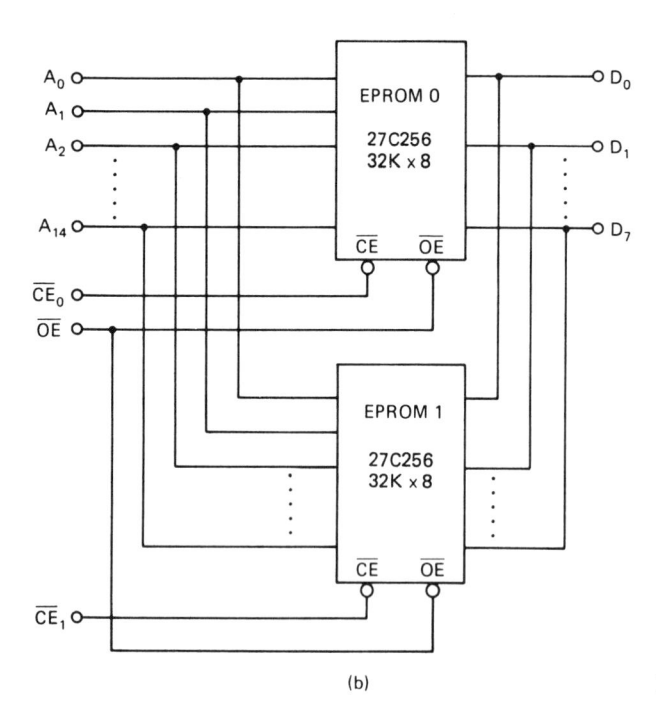

Figure 6.44 (*Continued*)

6.11 DATA STORAGE MEMORY: SRAM AND DRAM

The memory section of a microcomputer system is normally formed from both read-only memories (ROM) and *random access read/write memories* (RAM). Earlier we pointed out that the ROM is used to store permanent information such as the microcomputer's program. RAM is different from ROM in two important ways. First, we are able both to save data by writing it into RAM and to read it back for additional processing. Because of its versatile read and write features, RAM finds wide use where data changes frequently. For this reason, it is normally used to store data such as numerical and character data. The second difference is that RAM is volatile; that is, if power is removed from RAM all data are lost.

Static and Dynamic RAMs

There are two types of RAMs in general use today, the *static RAM* (SRAM) and *dynamic RAM* (DRAM). For static RAMs, data, once entered, remains valid as long as the power supply is not turned off. On the other hand, to retain data in a DRAM, it is not sufficient just to maintain the power supply. For this type of device, we must both keep the power supply turned on and periodically restore the data in each storage location by addressing them. This added requirement is necessary because the storage elements in a DRAM are capacitive nodes. If the storage nodes are not recharged at regular intervals of time, data would be lost. This recharging process is known as *refreshing* the DRAM.

Block Diagram of a Static RAM

A block diagram of a typical static RAM IC is shown in Fig. 6.45. By comparing this diagram with the one shown for a ROM in Fig. 6.39, we see that they are similar in many ways. For example, they both have address lines, data lines, and control lines. The data lines of the RAM, however, act as both inputs and outputs. For this reason, they are identified as a *bidirectional bus*.

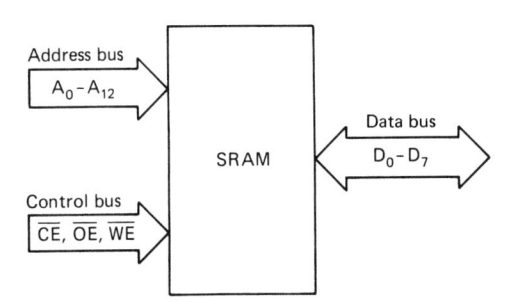

Figure 6.45 Block diagram of a static RAM.

A variety of static RAM ICs are currently available. They differ both in density and organization. The most commonly used densities in new circuit designs are the 64K- and 256K-bit devices. The structure of the data bus determines the organization of the RAM's storage array. In Fig. 6.45, an 8-bit data bus is shown. This type of organization is known as a *byte-wide* RAM. Devices are also manufactured with by 1 and by 4 data I/O organizations. At the 64K-bit density, this results in three standard device organizations: 64K × 1, 16K × 4, and 8K × 8.

The address bus on the RAM in Fig. 6.45 consists of the lines labeled A_0 through A_{12}. This 13-bit address is what is needed to select between the 8K individual storage locations in an 8K × 8-bit RAM IC. The 16K × 4 and 64K × 1 devices have a 14- and 16-bit address bus, respectively.

To either read or write data from the RAM, the device must first be chip enabled. Just like for a ROM, this is done by switching the $\overline{CE}$ input of the RAM to logic 0. Earlier we indicated that data lines D_0 through D_7 in Fig. 6.45 are bidirectional. This means that they will act as inputs when writing data into the RAM or as outputs when reading data from the RAM. The setting of the *write enable* ($\overline{WE}$) control input determines how the data lines operate. During all write operations to a storage location within the RAM, the $\overline{WE}$ input must be switched to the active 0 logic level. This configures the data lines as inputs. On the other hand, if data are to be read from a storage location, $\overline{WE}$ is left at the 1 logic level.

To read data from the RAM, output enable ($\overline{OE}$) must also be active. Logic 0 at this input enables the device's three-state outputs. Three-state data bus lines allow for the parallel busing needed to expand data memory by interconnecting multiple devices. For example, in Fig. 6.46 we see how 8K × 8-bit RAMs are interconnected to form an 8K × 32-bit memory circuit. Here individual chip

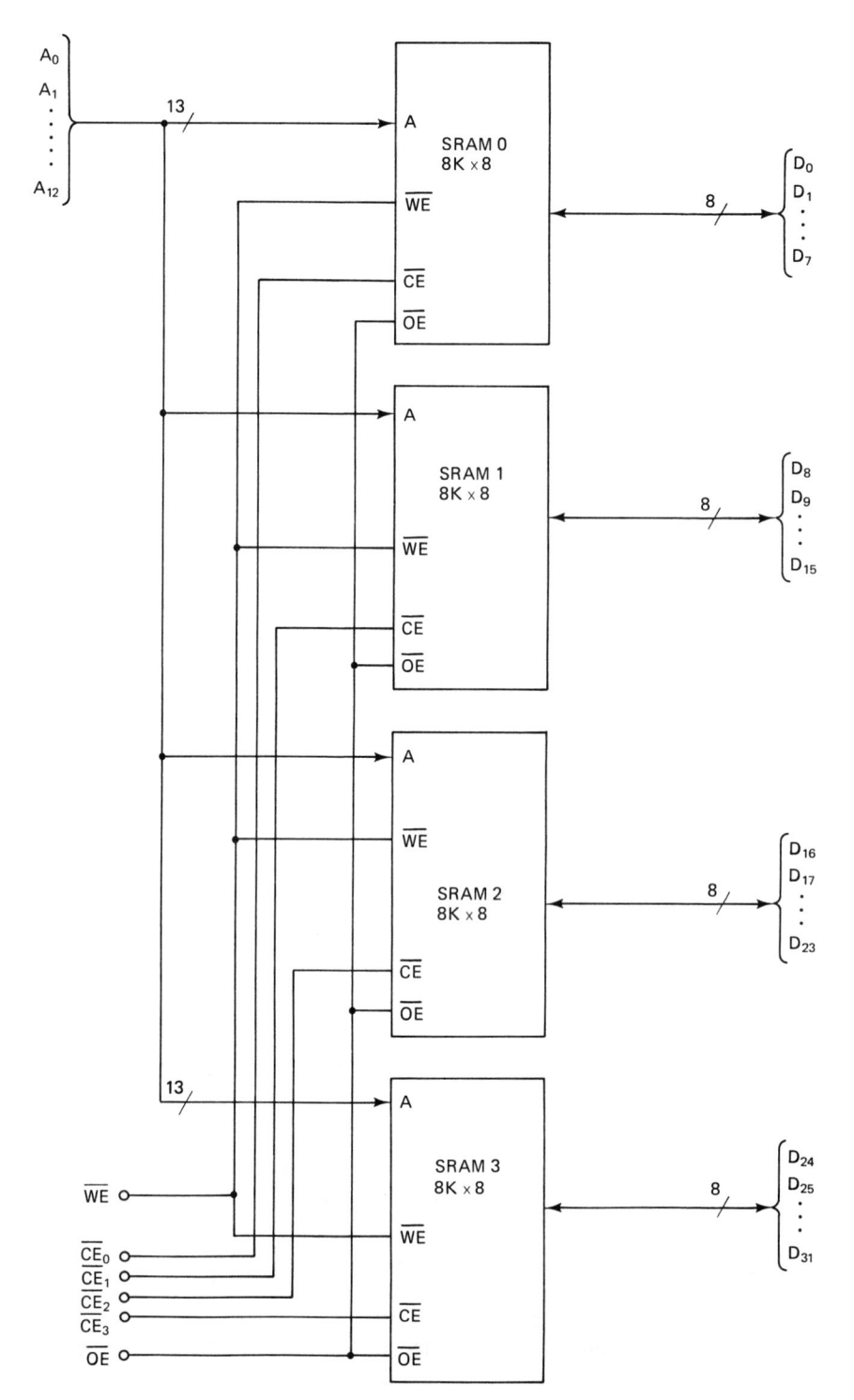

Figure 6.46 8K × 32-bit SRAM circuit.

enables are provided for each RAM. For instance, $\overline{CE_0}$ enables SRAM 0 and $\overline{CE_1}$ SRAM 1. They are required to permit byte and word write operations.

Standard Static RAM ICs

Figure 6.47 is a list of a number of standard static RAM ICs. Here we find their part numbers, densities, and organizations. For example, the 4361, 4363, and 4364 are all 64K-bit density devices; however, they are each organized differently. The 4361 is a 64K × 1-bit device, the 4363 is a 16K × 4-bit device, and the 4364 is an 8K × 8-bit device.

SRAM	Density (bits)	Organization
4361	64K	64K × 1
4363	64K	16K × 4
4364	64K	8K × 8
43254	256K	64K × 4
43256A	256K	32K × 8

Figure 6.47 Standard SRAM devices.

The pin layouts of the 4364 and 43256A ICs are given in Fig. 6.48(a) and (b), respectively. Looking at the 4364 we see that it is almost identical to the block diagram shown in Fig. 6.45. The one difference is that it has two chip enable lines instead of one. They are labeled $\overline{CE_1}$ and CE_2. Notice that one is activated by logic 0 and the other by logic 1.

As shown in Fig. 6.49, the 4364 is available in four speeds. For example, the minimum read cycle and write cycle times for the 4364-10 is 100 ns.

The waveforms for a typical write cycle are illustrated in Fig. 6.50. Let us now trace the events that take place during the write cycle. Here we see that all critical timing is referenced to the point at which the address becomes valid. Notice that the minimum duration of the write cycle is identified as t_{WC}. This is the 100 ns *write cycle time* of the 4364-10. The address must remain stable for this complete interval of time.

Next $\overline{CE_1}$ and CE_2 become active and must remain active until the end of the write cycle. The durations of these pulses are identified as $\overline{CE_1}$ *to end of write* (t_{CW1}) time and CE_2 *to end of write* (t_{CW2}) time. As shown in the waveforms, we are assuming here that they begin at any time after the occurrence of the address but before the leading edge of $\overline{WE}$. The minimum value for both of these times is 80 ns. On the other hand, $\overline{WE}$ is shown to not occur until the interval t_{AS} elapses. This is the *address set-up time* and represents the minimum amount of time the address inputs must be stable before $\overline{WE}$ can be switched to logic 0. For the 4364, however, this parameter is equal to 0 ns. The width of the write enable pulse is identified as t_{WP} and its minimum value equals 60 ns.

Data applied to the D_{IN} data inputs are written into the device synchronous with the trailing edge of $\overline{WE}$. Notice that the data must be valid for an interval equal to t_{DW} before this edge. This interval, which is called *data valid to end of write*, has a minimum value of 40 ns for the 4364-10. Moreover, it is shown to

Sec. 6.11 Data Storage Memory: SRAM and DRAM

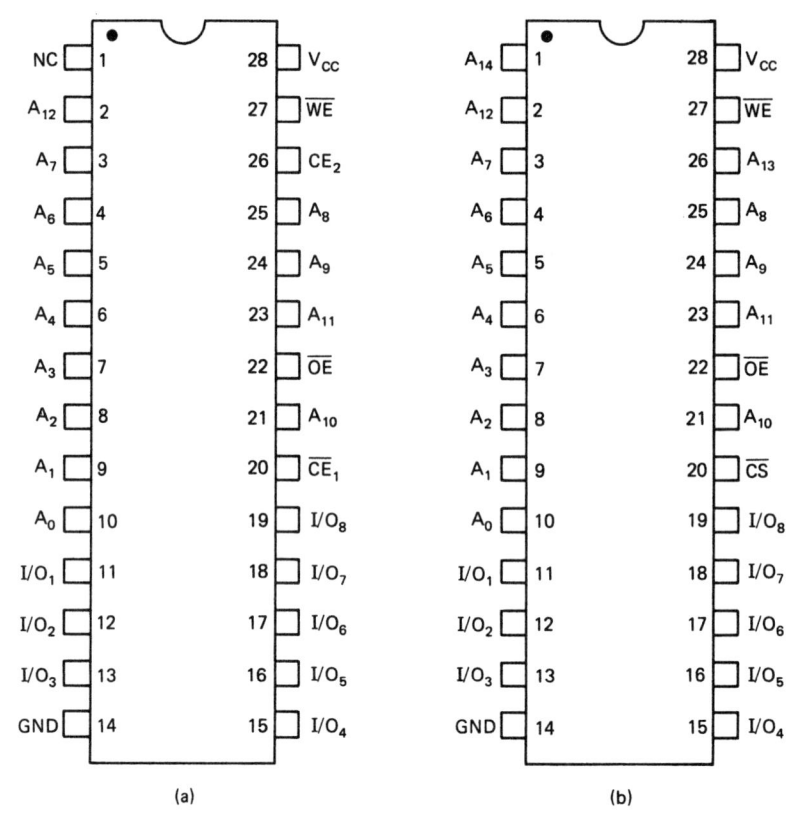

```
     NC ☐ 1        28 ☐ V_CC          A_14 ☐ 1        28 ☐ V_CC
    A_12 ☐ 2       27 ☐ WE            A_12 ☐ 2        27 ☐ WE
     A_7 ☐ 3       26 ☐ CE_2           A_7 ☐ 3        26 ☐ A_13
     A_6 ☐ 4       25 ☐ A_8            A_6 ☐ 4        25 ☐ A_8
     A_5 ☐ 5       24 ☐ A_9            A_5 ☐ 5        24 ☐ A_9
     A_4 ☐ 6       23 ☐ A_11           A_4 ☐ 6        23 ☐ A_11
     A_3 ☐ 7       22 ☐ OE             A_3 ☐ 7        22 ☐ OE
     A_2 ☐ 8       21 ☐ A_10           A_2 ☐ 8        21 ☐ A_10
     A_1 ☐ 9       20 ☐ CE_1           A_1 ☐ 9        20 ☐ CS
     A_0 ☐ 10      19 ☐ I/O_8          A_0 ☐ 10       19 ☐ I/O_8
    I/O_1 ☐ 11     18 ☐ I/O_7         I/O_1 ☐ 11      18 ☐ I/O_7
    I/O_2 ☐ 12     17 ☐ I/O_6         I/O_2 ☐ 12      17 ☐ I/O_6
    I/O_3 ☐ 13     16 ☐ I/O_5         I/O_3 ☐ 13      16 ☐ I/O_5
     GND ☐ 14      15 ☐ I/O_4          GND ☐ 14       15 ☐ I/O_4

            (a)                                  (b)
```

Figure 6.48 (a) 4364 pin layout; (b) 43256A pin layout.

remain valid for an interval of time equal to t_{DH} after this edge. This *data hold time*, however, just like address setup time, equals 0 ns for the 4364. Finally, a short recovery period takes place after $\overline{WE}$ returns to logic 1 before the write cycle is complete. This interval is identified as t_{WR} in the waveforms, and its minimum value equals 5 ns.

The read cycle of a static RAM, such as the 4364, is similar to that of a ROM. Waveforms of a read operation are given in Fig. 6.51.

Standard Dynamic RAM ICs

Dynamic RAMs are available in higher densities than static RAMs. Currently, the most widely used DRAMs are the 64K-, 256K-, and 1M-bit devices. Figure 6.52 is a list of a number of popular DRAM ICs. Here we find the 2164B, which

Part number	Read/write cycle time
4364-10	100 ns
4364-12	120 ns
4364-15	150 ns
4364-20	200 ns

Figure 6.49 Speed selections for the 4364.

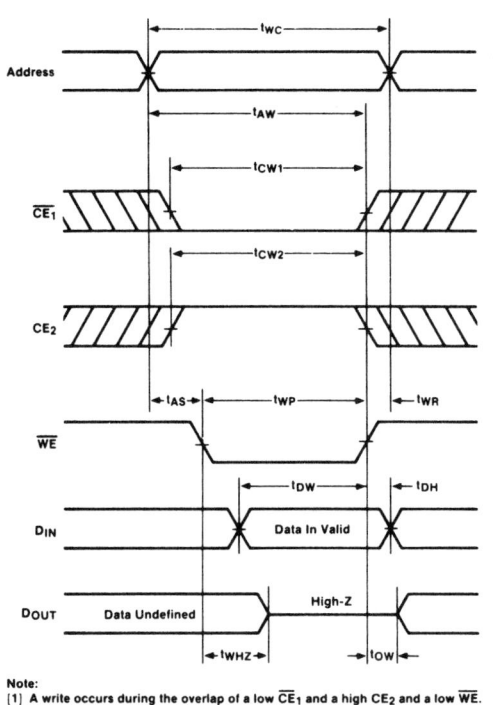

Figure 6.50 Write-cycle timing diagram.

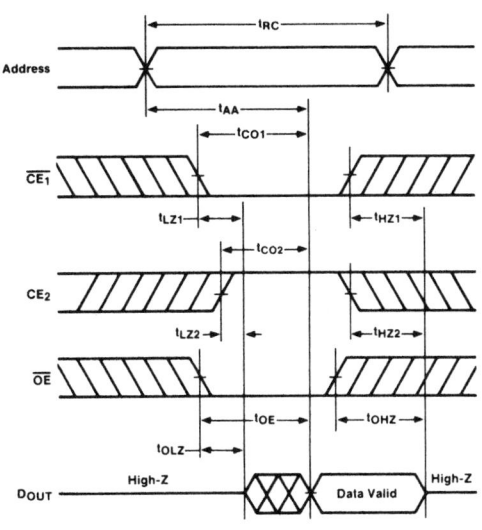

Figure 6.51 Read-cycle timing diagram.

Sec. 6.11   Data Storage Memory: SRAM and DRAM

DRAM	Density (bits)	Organization
2164B	64K	64K × 1
21256	256K	256K × 1
21464	256K	64K × 4
421000	1M	1M × 1
424256	1M	256K × 4

Figure 6.52 Standard DRAM devices.

is organized as 64K × 1 bit, the 21256, which is organized as 256K × 1 bit, the 21464, which is organized as 64K × 4 bits, the 421000, which is organized as 1M × 1 bit, and the 424256, which is organized as 256K × 4 bits. Pin layouts for the 2164B, 21256, and 421000 are shown in Figs. 6.53(a), (b), and (c), respectively.

Some other benefits of using DRAMs over SRAMs are that they cost less, consume less power, and their 16- and 18-pin packages take up less space. For these reasons, DRAMs are normally used in applications that require a large amount of memory. For example, most systems that support at least 1M byte of data memory are designed using DRAMs.

The 2164B is one of the older NMOS DRAM devices. A block diagram of the device is shown in Fig. 6.54. Looking at the block diagram we find that it has eight address inputs, A_0 through A_7, a data input and data output marked D and Q, respectively, and three control inputs, *row address strobe* ($\overline{RAS}$), *column address strobe* ($\overline{CAS}$), and *read/write* ($\overline{W}$).

The storage array within the 2164B is capable of storing 65,536 (64K) individual bits of data. To address this many storage locations, we need a 16-bit address; however, this device's package has just 16 pins. For this reason, the 16-bit address is divided into two separate parts: an 8-bit *row address* and an 8-bit *column address*. These two parts are time-multiplexed into the device over a single set of address lines, A_0 through A_7. First the row address is applied to A_0 through A_7. Then $\overline{RAS}$ is pulsed to logic 0 to latch it into the device. Next, the column address is applied and $\overline{CAS}$ strobed to logic 0. This 16-bit address selects which one of the 64K storage locations is to be accessed.

Data are either written into or read from the addressed storage location in the DRAMs. Write data are applied to the D input and read data are output at Q. The logic levels of control signals $\overline{W}$, $\overline{RAS}$, and $\overline{CAS}$ tell the DRAM whether a read or write data transfer is taking place and control the three-state outputs. For example, during a write operation, the logic level at D is latched into the addressed storage location at the falling edge of either $\overline{CAS}$ or $\overline{W}$. If $\overline{W}$ is switched to logic 0 before $\overline{CAS}$, an early write cycle is performed. During this type of write cycle, the outputs are maintained in the high-Z state throughout the complete bus cycle. The fact that the output is put in the high-Z state during the write operation allows the D input and Q output of the DRAM to be tied together. The Q output is also in high-Z state whenever $\overline{CAS}$ is logic 1. This is the connection and mode of operation normally used when attaching DRAMs to the bidirectional data bus

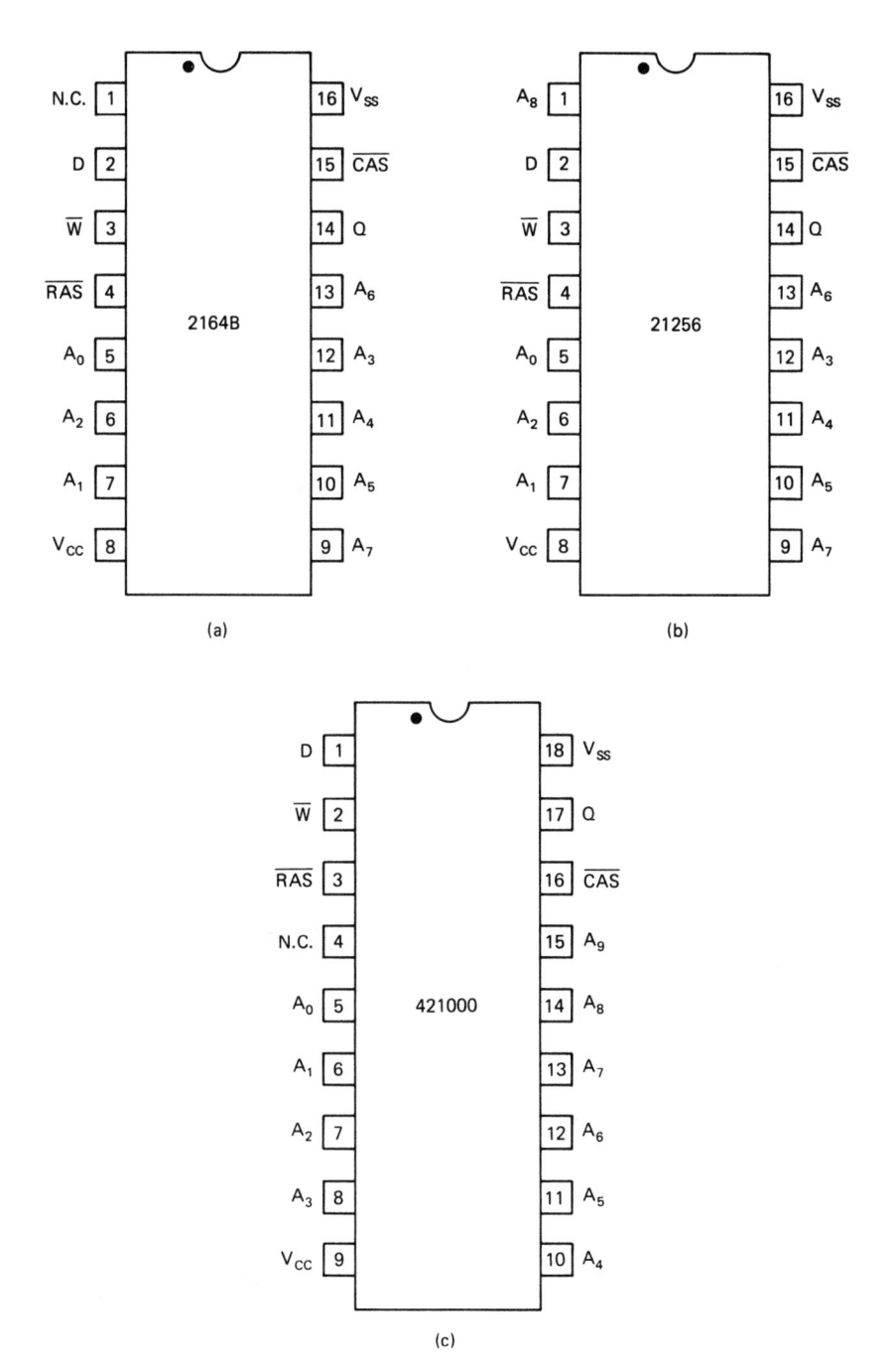

Figure 6.53 (a) 2164B pin layout; (b) 21256 pin layout; (c) 421000 pin layout.

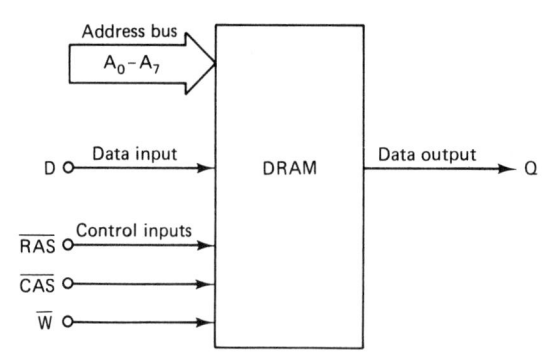

Figure 6.54 Block diagram of the 2164B DRAM.

of a microprocessor. Figure 6.55 shows how 32 2164B devices are connected to make a 64K × 32-bit DRAM array.

The 2164B also has the ability to perform what are called *page mode* accesses. If RAS is left at logic 0 after the row address is latched inside the device, the address is maintained within the device. Then data cells along the selected row can be accessed by simply supplying successive column addresses. This permits faster access of memory by eliminating the time needed to set up and strobe additional row addresses.

Earlier we pointed out that the key difference between the DRAM and SRAM is that the storage cells in the DRAM need to be periodically refreshed; otherwise, they lose their data. To maintain the integrity of the data in a DRAM, each of the rows of the storage array must typically be refreshed every 2 ms. All of the storage cells in an array are refreshed by simply cycling through the row addresses. As long as $\overline{CAS}$ is held at logic 1 during the refresh cycle, no data are output.

External circuitry is required to perform the address multiplexing, $\overline{RAS}/\overline{CAS}$ generation, and refresh operations for a DRAM subsystem. *DRAM refresh controller* ICs are available to permit easy implementation of these functions. An example of such a device is the 82C08 DRAM refresh controller.

6.12 80386DX MICROCOMPUTER SYSTEM MEMORY INTERFACE CIRCUITRY

In earlier sections we introduced the bus cycles, hardware organization of the memory address space, and the memory interface of the 80386DX-based microcomputer system. Here we show how the memory interface can be implemented in a simple microcomputer system. This microcomputer circuit is given in Fig. 6.56. Notice that the memory interface circuits have been highlighted in the circuit diagram.

Let us begin by examining the bus interface circuitry. Looking at the circuit diagram, we see that address line A_{31} and control signal $M/\overline{IO}$ are decoded by the upper 74F139 2-line to 4-line decoder. This device represents the memory

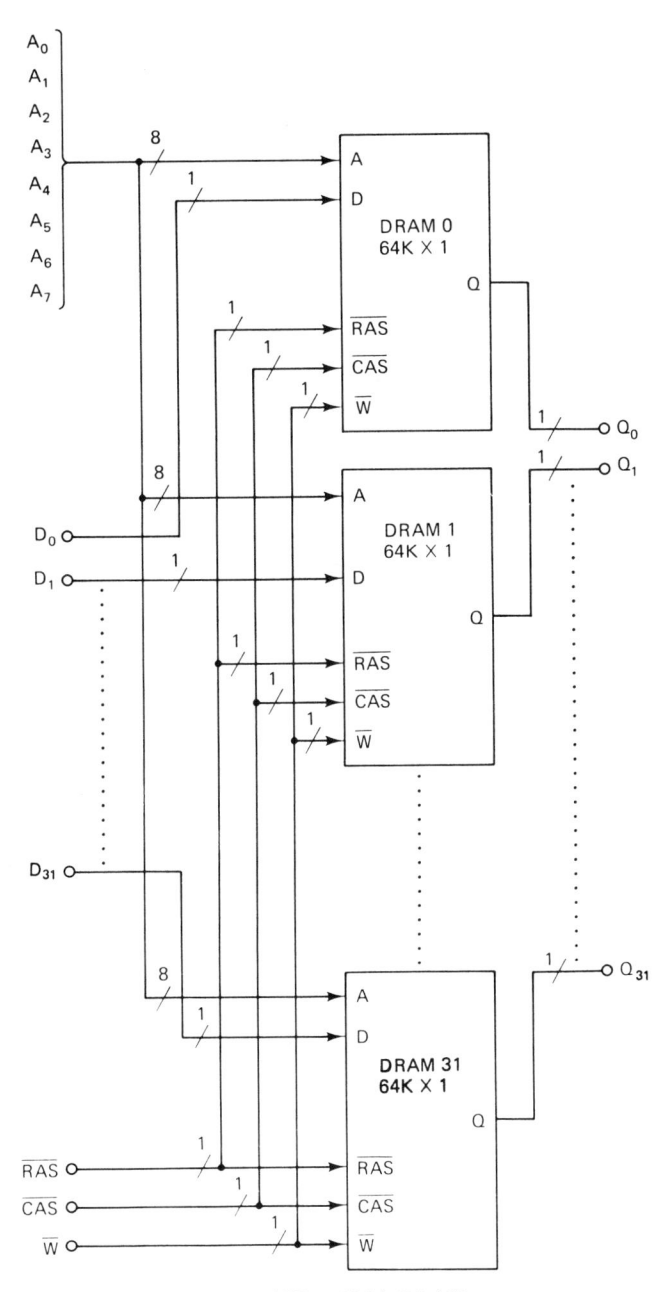

Figure 6.55 64K × 32-bit DRAM array.

address decoder. When M/$\overline{\text{IO}}$ equals 1 and A_{31} equals 0, the $\overline{Y}_2$ output of the decoder is at its active 0 logic level. This signal indicates that a memory address is on the bus. Notice that it is input to PAL 1 of the bus control logic as chip select 1 wait state ($\overline{\text{CS1WS}}$). $\overline{\text{CS1WS}}$ is also latched in the 74AS373 address latch.

Sec. 6.12   80386DX Microcomputer System Memory Interface Circuitry   267

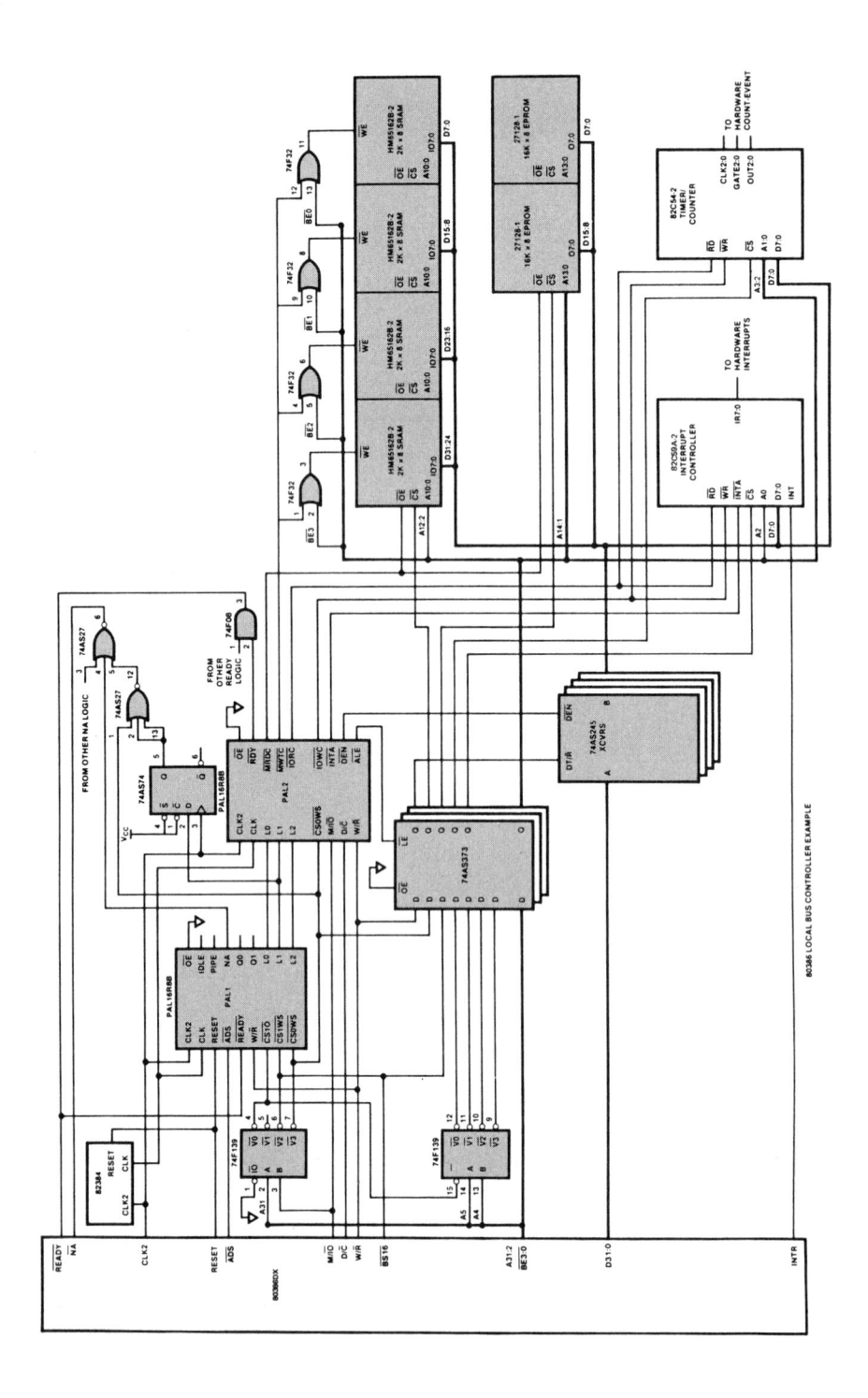

Figure 6.56 Memory circuit for a typical 80386DX microcomputer.

Here it is used as the chip select signal for the 27128 EPROMs. On the other hand, an input of 11 to the decoder produces chip select 0 wait state ($\overline{CS0WS}$). This signal is also used as an input to PAL 1. Moreover, it is latched for use as the chip select signal for the static RAMs.

The address latch is implemented with three 74AS373 octal latch devices. Notice that the outputs of the latches are permanently enabled by applying logic 0 to the $\overline{OE}$ input. The address signals, byte enable signals, chip select outputs of the memory address decoder, and W/$\overline{R}$ are latched into the address latch synchronously with the $\overline{ALE}$ output from PAL 2 of the bus control logic. Actually, just address lines A_2 through A_{14} are used in the memory interface.

The bus control logic section is formed from the devices labeled PAL 1 and PAL 2. In the circuit diagram we find that the inputs to this section include the bus cycle definition signals M/$\overline{IO}$, D/$\overline{C}$, and W/$\overline{R}$, control signal $\overline{ADS}$, and chip selects $\overline{CSIO}$, $\overline{CS1WS}$, and $\overline{CS0WS}$. At the output side the bus control logic produces the control signals $\overline{MRDC}$, $\overline{MWTC}$, $\overline{DEN}$, and $\overline{ALE}$ for the memory interface, along with the $\overline{NA}$ for activating pipelining and $\overline{RDY}$ for introducing wait states. $\overline{NA}$ and $\overline{RDY}$ are supplied through control logic to inputs of the 80386DX.

The bank write control logic section is implemented with the gates of a 74F32 quad OR gate IC. The memory write command ($\overline{MWTC}$) signal is applied to one input of each of the four 2-input OR gates. The other input on each of the four gates is supplied by one of the latched byte enable signals. For instance, $\overline{BE_0}$ enables $\overline{MWTC}$ to the $\overline{WE}$ input of the rightmost static RAM. The signals produced at the outputs are used to activate the write enable ($\overline{WE}$) input of the static RAM is in banks 0 through 3.

The final section in the bus interface section consists of the data bus transceivers/buffers. In the circuit diagram we see that the data bus lines are buffered with four 74AS245 transceivers. The $\overline{DEN}$ output of the bus control logic enables the transceivers for operation, and the latched W/$\overline{R}$ output of the 80386DX is used to control the direction in which data are passed through them.

Let us now look at the program storage memory section. It is implemented with two 27128 EPROMs. These devices attach to the lower 16 data bus lines, D_0 through D_{15}, and provide 16K × 16 bits of code storage. Notice that program memory is 16 bits wide, not 32 bits wide. The storage location to be accessed is selected by the 14-bit address that is applied to the A_0 through A_{13} inputs of both EPROMs. Memory read command ($\overline{MRDC}$) supplies the $\overline{OE}$ input of the EPROMs and enables them whenever a memory read bus cycle is in progress. Finally, chip select signal $\overline{CS1WS}$ is used to enable the EPROMs. $\overline{CS1WS}$ is also returned to the bus size 16 ($\overline{BS16}$) input of the 80386DX and signals it to perform all read cycles from the program memory address space as 16-bit bus cycles.

In Fig. 6.56 we see that the data storage memory section is constructed with four 2K × 8 bit static RAMs. Each of the SRAMs represents a bank and attaches to eight of the data bus lines. For example, the leftmost device is bank 3 and connects to data bus lines D_{24} through D_{31}. This memory subsystem gives a total data storage capacity of 8K bytes, which is organized as 2K × 32 bits. The storage location to be accessed is selected by the 11-bit address applied to address inputs

A_0 through A_{10}. During bus cycles to data memory, all four SRAMs are enabled by the chip select signal $\overline{\text{CS0WS}}$. For this reason, data read and write bus cycles involve no wait states. When data are read from memory, the signal $\overline{\text{MRDC}}$ enables the outputs of all four SRAMs. Therefore, the 32-bit word held at the addressed storage location is output onto D_0 through D_{31} and the 80386DX reads a byte, word, or double word of data off the bus.

On the other hand, during write bus cycles to SRAM, the 80386DX does not always output a 32-bit word; instead, it outputs the appropriate-size element of data: a byte, word, or double word. At the same time, it outputs a code on $\overline{\text{BE}}_0$ through $\overline{\text{BE}}_3$ to identify which part of the bus carries the data. The byte enable signals gate $\overline{\text{MWTC}}$ to the $\overline{\text{WE}}$ input of the appropriate SRAM banks. In this way, we see that data are written only into the appropriate banks of memory, not to all four banks.

6.13 CACHE MEMORY

When an 80386DX microcomputer system employs a large main memory subsystem of several megabytes, it is normally made with high-capacity but relatively slow-speed dynamic RAMs and EPROMs. Even though DRAMs are available with access times as short as 60 ns and EPROMs as fast as 120 ns, these high-speed versions of the devices are expensive and still too slow to work in an 80386DX system that is running with zero wait states. For example, an 80386DX microprocessor running at 25 MHz would require DRAMs with a 40-ns access time to implement a zero-wait-state memory design. For this reason, wait states are introduced in all bus cycles to data and program memory. These wait states degrade the overall performance of the microcomputer system.

Addition of a cache memory subsystem to the 80386DX microcomputer provides a means for improving overall system performance while permitting the use of low-cost, slow-speed memory devices in main memory. In a microcomputer system with cache, a second smaller but very fast memory section is added between the MPU and main memory subsystem. Figure 6.57 illustrates this type of system architecture. This small, high-speed memory section is known as the *cache memory*. The cache is designed with fast, more expensive static RAMs and can be accessed without wait states. During system operation, the cache memory contains recently used instructions and data. The objective is that the MPU accesses code and data in the cache most of the time, instead of from main memory. This results in close to a zero-wait-state memory system operation and higher performance for the microcomputer even though accesses of the main memory require one or more wait states.

Cache memories are widely used in high-performance microcomputer systems today. Notice in Fig. 6.57 that on one side the cache memory subsystem attaches to the local bus of the 80386DX, and at the other side it drives the system bus to the microcomputer's main memory subsystem. Caches typically range in size from 16K bytes to 256K bytes and can be used to cache both data and code.

Let us continue our examination of cache for the 80386DX microcomputer

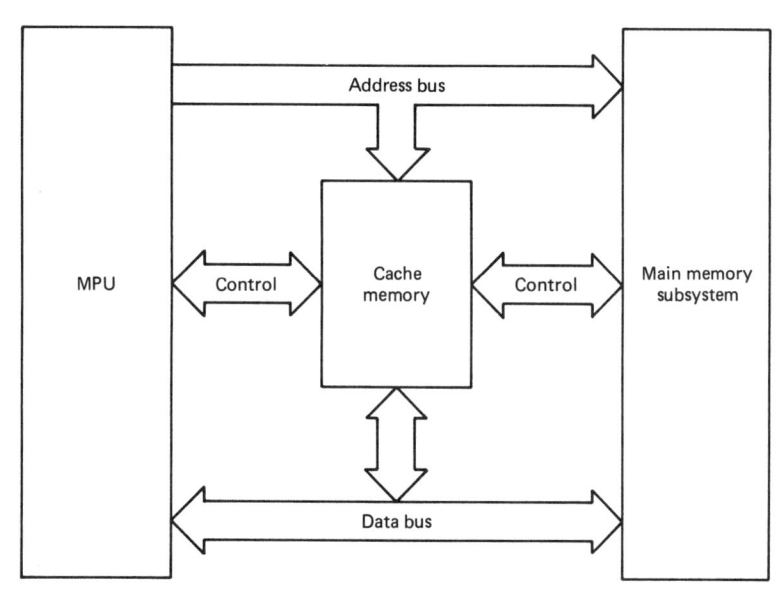

Figure 6.57 Microcomputer system with cache memory.

system by looking at how it affects the execution of a program. The first time the 80386DX executes a segment of program, one instruction after the other is fetched from main memory and executed. The most recently fetched instructions are automatically saved in the cache memory. That is, a copy of these instructions is held within the cache. For example, a segment of program that implements a loop operation could be fetched, executed, and placed in the cache. In this way we find that the cache always holds some of the most recently executed instructions.

Now that we know what the cache holds, let us look at how the cached instructions are used during program execution. Many software operations involve repeated execution of the same sequence of instructions. A loop is a good example of this type of program structure. In Fig. 6.58 we find that the first execution of the loop references code held in the slow main program memory. During this access, the routine is copied into the cache part of memory. When the instructions of the loop are repeated, the MPU reaccesses the routine by using the instructions held in the cache instead of refetching them from main memory. Accesses to code in the cache are performed with no wait states, whereas those of code in main memory normally require wait states. In this way we see that the use of cache has reduced the number of accesses made from the slower main memory. The more frequent instructions held in the cache are used, the closer to zero-wait-state operation is achieved, the more the overall execution time of the program is decreased, and the higher is the performance of the microcomputer system.

During execution of the loop routine, data operands that are accessed can also be cached in the cache memory. If these operands are reaccessed during the

Sec. 6.13   Cache Memory

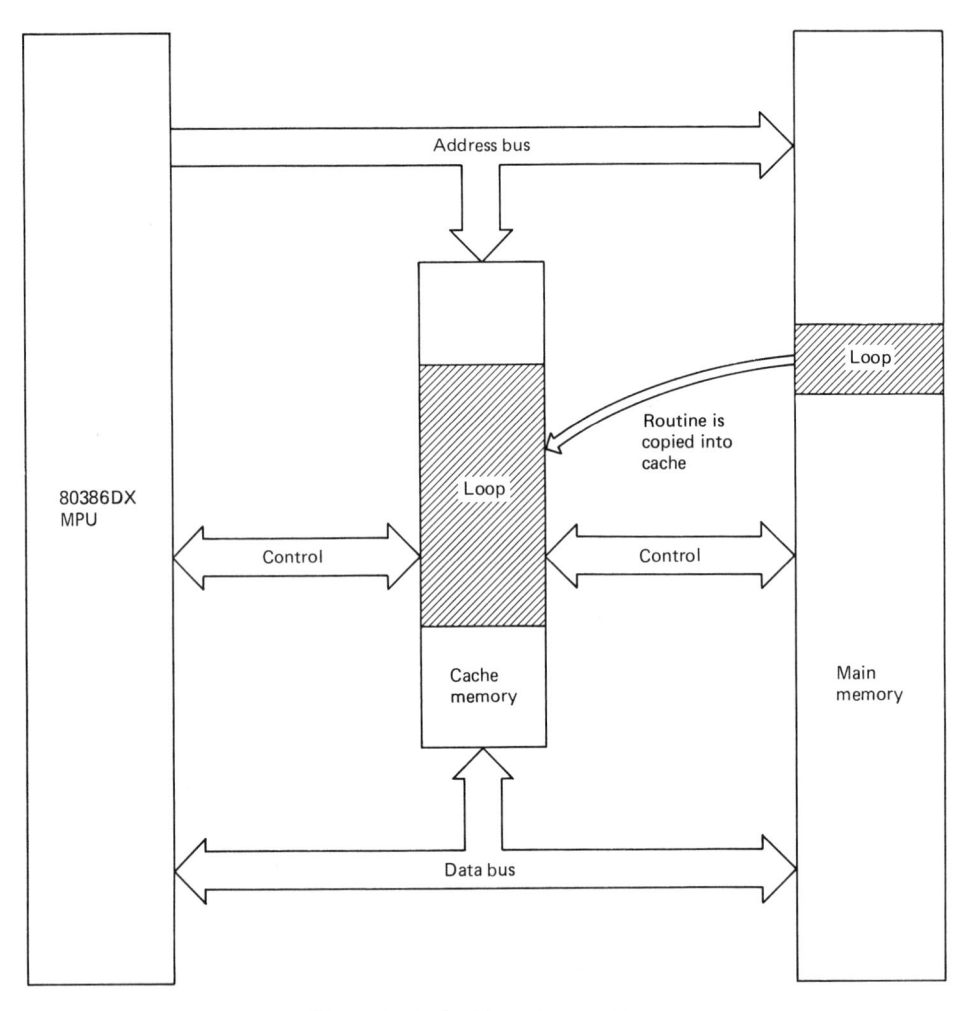

Figure 6.58 Caching a loop routine.

repeated execution of the loop, they are read from the cache instead of from main data memory. This further reduces the execution time of the segment of program.

We just found that the concept behind cache memory is that it stores recently used code and data and that if this information is to be reaccessed, it may be read from the cache with zero-wait-state bus cycles rather than from main memory. When the address of a code or data storage location that is to be read is output on the local bus, the cache subsystem must determine whether or not the information to be accessed resides in both main memory and cache memory. If it does, the memory cycle is considered a *cache hit* condition. In this case a bus cycle is not initiated to the main memory subsystem; instead, the copy of the information in the cache is accessed.

On the other hand, if the address output on the local bus does not correspond to information that is already cached, the condition represents what is called a *cache miss*. This time the MPU reads the code or data from main memory and writes it into a corresponding location in cache.

Hit rate is a measure of how effective the cache subsystem operates. Hit rate is defined as the ratio of the number of cache hits to the total number of memory accesses, expressed as a percentage. That is, hit rate equals

$$\text{hit rate} = \frac{\text{number of hits}}{\text{number of bus cycles}} \times 100\%$$

The higher the value of hit rate, the better the cache memory design. For instance, a cache may have a hit rate of 85%. This means that the MPU reads data from the cache memory for 85% of its memory bus cycles. In other words, just 15% of the memory accesses are from the main memory subsystem. Hit rate is not a fixed value for a cache design. It depends on the code being executed. That is, hit rate may be one value for a specific application program and a totally different value for another.

A number of features of the cache design also affect the hit rate. For instance, the size, organization, and update method of the cache memory subsystem all determine the maximum hit rate that may be achieved by a cache. Earlier we pointed out that practical cache memories for microcomputer systems range in size from as small as 16K bytes to as large as 256K bytes. In general, the larger the size of the cache, the higher the hit rate. This is because larger cache size can contain more data and code, which yields a greater chance that the information to be accessed resides in the cache. However, due to the fact that the improvement in hit rate decreases with increasing cache size and that the cost of the cache subsystem increases substantially with increasing size, most caches used today are either 32K bytes or 64K bytes.

The two most widely used cache memory organizations are those known as the *direct-mapped cache* and the *two-way set associative cache*. The direct-mapped cache is also called a *one-way set associative cache*. The organization of a 64K-byte direct-mapped cache memory is illustrated in Fig. 6.59. Notice that the cache memory array is arranged as a single 64K-byte bank of memory, and the main memory is viewed as a series of 64K-byte pages, denoted page 0 through page n. Notice that the data storage location at the same offset (X) in all pages of main memory, these storage locations are identified as X(0) through X(n) in Fig. 6.59, map to a single storage location, marked X, in the cache memory array. That is, each location in a 64K-byte page of main memory maps to a different location in the cache memory array.

On the other hand, the 64K-byte memory array of a two-way set associative cache memory is organized into two 32K-byte banks. That is, the cache array is divided two ways: *BANK A* and *BANK B*. This cache memory subsystem configuration is shown in Fig. 6.60. Again main memory is mapped into pages equal to the size of a bank in the cache array. But because a bank is now 32K bytes, there are twice as many main memory pages as in direct-mapped organizations. In this case the storage location at a specific offset in every page of main memory can map to the same storage location in either the A or B bank. For example,

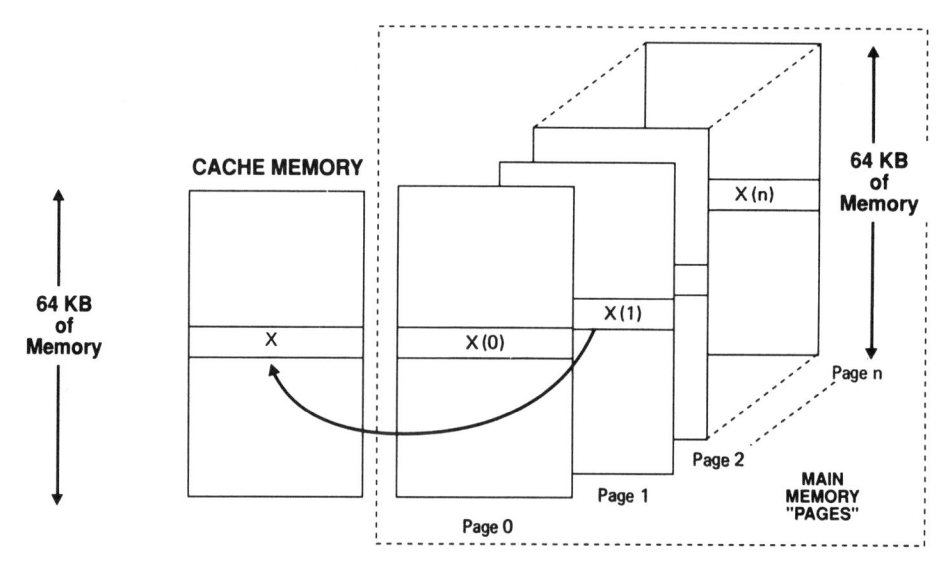

Figure 6.59 Organization of a direct-mapped memory subsystem. (Reprinted by permission of Intel Corp. Copyright Intel Corp. 1990.)

the contents of storage location X(2) can be cached into either X_A or X_B. The two-way set associative organization results in higher hit rate operation.

One example of a memory update method that affects hit rate is the information replacement algorithm. Replacement methods are based on the fact that there is a higher chance that more recently used information will be reused. For

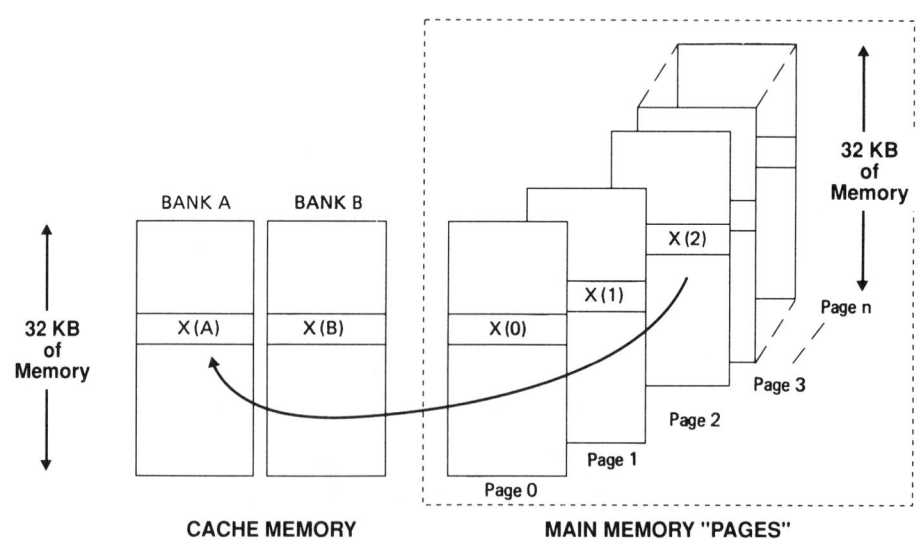

Figure 6.60 Organization of a two-way set associative memory subsystem. (Reprinted by permission of Intel Corp. Copyright Intel Corp. 1990.)

instance, the two-way set associative cache organization permits the use of a *least recently used* (LRU) replacement algorithm. In this method the cache subsystem hardware keeps track of whether information X(A) in the BANK A storage location or X(B) in BANK B is most recently used. For example, let us assume that the value at storage location X(A) in BANK A of Fig. 6.60 is X_0 from page 0 and that it was just loaded into the cache. On the other hand, X(B) in BANK B is from page 1 and has not been accessed for a long time. Therefore, the value of X(B) in BANK B is tagged as the least recently used information. When a new value of code or data, for instance, from offset X(3) in page 3 is accessed, it must replace the value of X(A) in BANK A or X(B) in BANK B. Therefore, the cache replacement algorithm automatically selects the cache storage location corresponding to the least recently accessed bank for storage of X(3). For our example this would be storage location X(B) in BANK B. This shows that the replacement algorithm maintains more recently used information in the cache memory array. The results are a higher maximum hit rate and a higher level of performance for the microcomputer system.

We found earlier that use of a cache reduced the number of accesses of main memory over the system bus. In our example of a loop routine we saw that repeated accesses of the instructions that perform the loop operation were made from the cache, not from main memory. Therefore, fewer code and data accesses are performed across the system bus. That is, availability of the bus has been increased for external devices. This is another advantage of using cache in a microcomputer system. The freed-up bus bandwidth is available to other bus masters, such as DMA controllers or other processors in a multiple-processor system.

Posted write-through is an example of a cache memory update method that affects microcomputer system performance but not hit rate. In this method, all write operations that are used to update addresses of storage locations in main memory go through (*write through*) the cache. When a write bus cycle is initiated, the cache memory checks to determine if the information is cached. If a miss occurs, the write is performed only to the storage location in main memory. However, if the information for the addressed location is cached (a cache hit), the corresponding entry in the cache memory array is also updated. *Posted write* capability allows write operations to main memory to be completed by the cache memory subsystem rather than by the MPU. For this reason, the MPU can terminate the write cycle without incurring any wait states. This capability increases the bus bandwidth of the local bus.

6.14 THE 82385DX CACHE CONTROLLER AND THE CACHE MEMORY SUBSYSTEM

The 82385DX is a VLSI device that is designed to implement the control function for the cache memory subsystem in a 80386DX-based microcomputer system. Use of the 82385DX results in a high-performance, versatile, compact cache design. It is designed to attach directly to the 80386DX microprocessor. For this reason the amount of circuitry needed to construct the cache memory subsystem is minimized. In terms of versatility, the 82385DX controller can be used to implement several different cache configurations. For instance, it can work as either a master or a slave. The master mode of operation is selected when used in a single-processor microcomputer application. The 82385DX supports a 32K-byte cache memory array. This memory array can be set up to operate as either a direct-mapped cache or a two-way set associative cache. When configured for two-way set associative organization, it uses the least recently used replacement algorithm. Finally, memory write cycles are performed in a posted write-through manner. Depending on the application, an 82385DX-based cache memory subsystem can achieve hit rates as high as 99%.

The 82385DX IC is housed in a 132-lead pin grid array package. The pin layout of the 82385DX is shown in Fig. 6.61(a), and Fig. 6.61(b) lists the signal at each lead. For example, address line A_{31} is at pin M2 and byte enable input $\overline{BE_0}$ is at pin G14.

Architecture of an 80386DX Microcomputer with an 82385DX-Based Cache Memory

Let us now look at the architecture of an 80386DX microcomputer that employs an 82385DX-based cache memory. A block diagram of such a microcomputer system is shown in Fig. 6.62. Notice that the 82385DX cache controller and cache memory both connect onto the 80386DX's memory address bus, control bus, and data bus in parallel with the system bus to the main memory subsystem. For this reason, the diagram shows that the cache controller and cache memory attach to the 80386DX's *local bus*. On the other hand, the memory control signals for the main memory subsystem are produced by the 82385DX. These signals, together with the buffered address and data buses, are called the *82385DX local bus*. Therefore, the main memory subsystem, which is attached to the *system bus*, is supplied by the 82385DX's local bus.

Signal Interfaces of the 82385DX

A block diagram of the 82385DX is shown in Fig. 6.63. Here we see that it has *80386DX/82385DX interface signals, 80386DX local bus decode inputs, bus watching support signals, 82385DX bus data transceiver and address latch control signals, bus arbitration signals, 82385DX local bus interface signals, status and control signals, cache memory control signals*, and *configuration inputs*. Let us now look briefly at the function of these signals relative to operation of the cache memory subsystem.

The configuration inputs are strapped to set the mode of operation of the 82385DX. Looking at Fig. 6.63, we find two configuration inputs, *master/slave select* (M/$\overline{S}$) and *two-way/direct mapped select* (2W/$\overline{D}$). To put the 82385DX in master mode, the M/$\overline{S}$ input is strapped to the 1 logic level. This is the normal mode of operation in single-processor microcomputer systems. A multiprocessor system would normally employ some cache controller devices configured as mas-

	A	B	C	D	E	F	G	H	J	K	L	M	N	P
1	A6	VSS	VCC	A9	A12	A15	A18	A19	A22	A24	A27	VCC	VSS	VCC
2	SA2	A3	A7	A8	A11	A14	A17	A21	A23	A25	A29	A31	VSS	VSS
3	SA3	A2	A4	A5	A10	A13	A16	A20	A26	A28	A30	READYO#	NA#	VCC
4	SA7	SA5	SA4			METAL LID						LDSTB	CALEN	VSS
5	SA9	SA10	SA6									CS0#	CT//R#	CS3#
6	SA13	SA11	SA8									CS1#	CS2#	CWEB#
7	SA14	SA15	SA12									COEB#	CWEA#	COEA#
8	SA17	SA16	SA18									WBS	MISS#	BRDYEN#
9	SA20	SA19	SA22									BAOE#	BADS#	BLOCK#
10	SA21	SA24	SA25									DOE#	BT/R#	BACP
11	SA23	SA26	SA27									BHLDA	BHOLD	VCC
12	SA28	SA29	SA31	2W/D#	FLUSH	D/C#	NCA#	BE2#	SEN	BRESET	BBE2#	BBE0#	BBE1#	VSS
13	SA30	M/S#	CLK2	RESET	ADS#	LOCK#	X16#	BE1#	SSTB#	BREADY#	BCLK2	BBE3#	VCC	VCC
14	DEFOE#	VSS	VCC	READYI#	M/IO#	W/R#	BE0#	LBA#	BE3#	BNA#	RESERVED	VSS	VSS	VSS

(a)

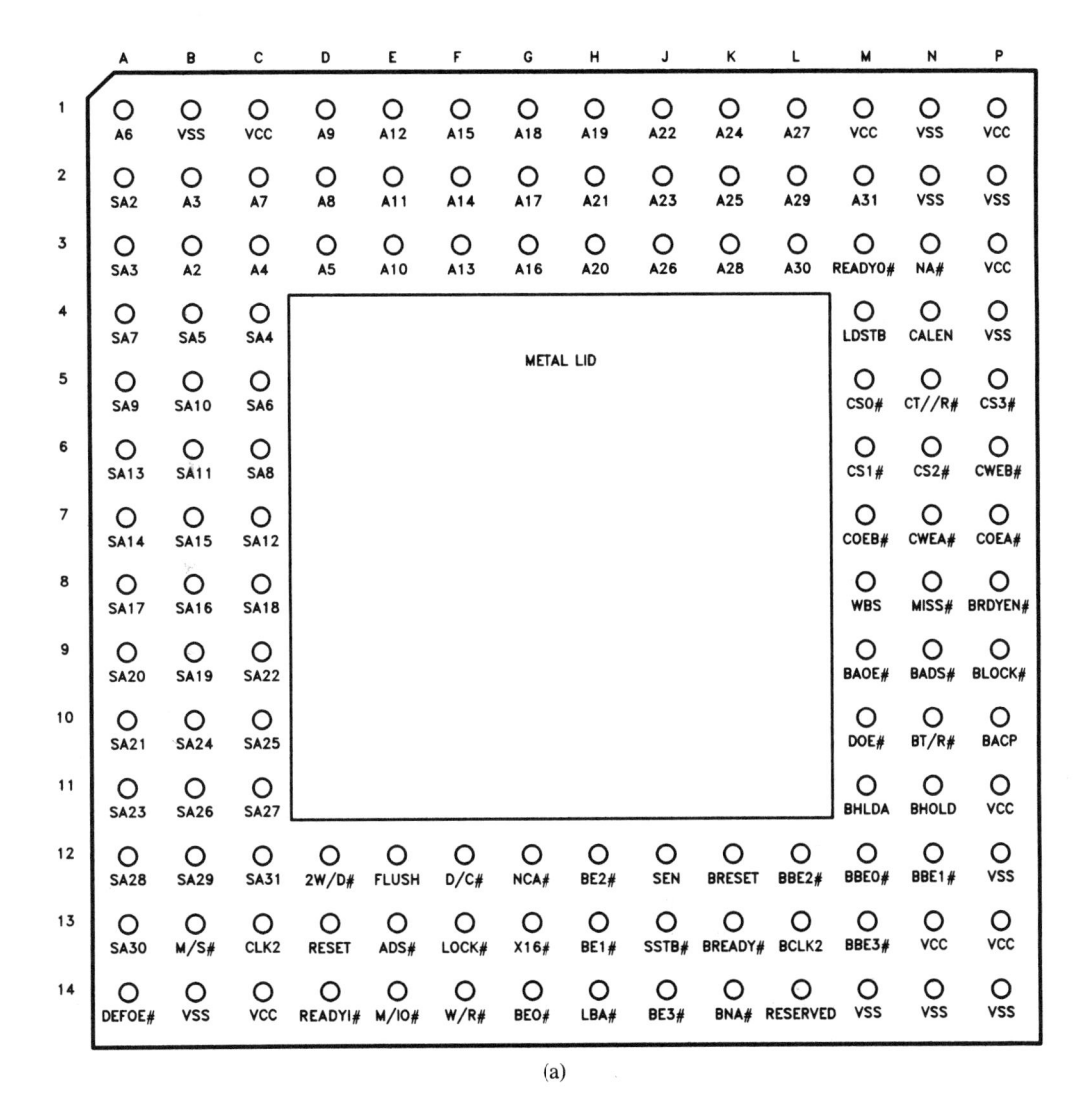

Figure 6.61 (a) Pin layout of the PGA 82385DX cache controller. (Reprinted by permission of Intel Corp. Copyright Intel Corp. 1990.) (b) Signal mnemonics and pin numbers. (Reprinted by permission of Intel Corp. Copyright Intel Corp. 1990.)

ters and others as slaves. Next, the cache organization is picked with the $2W/\overline{D}$ input. Wiring this input to logic 0 selects direct-mapped cache operation, and setting it to 1 selects two-way set associative operation.

The circuit diagram in Fig. 6.64 shows how the 82385DX connects to the 80386DX MPU, the signals of the 82385DX's local bus, which goes to the main memory subsystem, and the signal interface to the cache memory subsystem. Notice that most of the signals identified as 80386DX/82385DX interface signals in Fig. 6.63 connect directly to the corresponding pins of the 80386DX. For instance, the $\overline{ADS}$ output of the 80386DX is attached to the $\overline{ADS}$ input of the

Pin	Signal	Pin	Signal	Pin	Signal	Pin	Signal
M2	A31	C12	SA31	—	V_{CC}	B1	V_{SS}
L3	A30	A13	SA30	C1	V_{CC}	B14	V_{SS}
L2	A29	B12	SA29	C14	V_{CC}	M14	V_{SS}
K3	A28	A12	SA28	M1	V_{CC}	N1	V_{SS}
L1	A27	C11	SA27	N13	V_{CC}	N2	V_{SS}
J3	A26	B11	SA26	P1	V_{CC}	N14	V_{SS}
K2	A25	C10	SA25	P3	V_{CC}	P2	V_{SS}
K1	A24	B10	SA24	P11	V_{CC}	P4	V_{SS}
J2	A23	A11	SA23	P13	V_{CC}	P12	V_{SS}
J1	A22	C9	SA22	E13	$\overline{ADS}$	P14	V_{SS}
H3	A20	A10	SA21	F14	$W/\overline{R}$	N9	$\overline{BADS}$
H1	A19	A9	SA20	F12	$D/\overline{C}$	M12	$\overline{BBE0}$
G1	A18	B9	SA19	E14	$M/\overline{IO}$	N12	$\overline{BBE1}$
G2	A17	C8	SA18	F13	$\overline{LOCK}$	L12	$\overline{BBE2}$
G3	A16	A8	SA17	N3	$\overline{NA}$	M13	$\overline{BBE3}$
F1	A15	B8	SA16	G13	$\overline{X16}$	P9	$\overline{BLOCK}$
F2	A14	B7	SA15	G12	$\overline{NCA}$	K14	$\overline{BNA}$
F3	A13	A7	SA14	H14	$\overline{LBA}$	N4	CALEN
E1	A12	A6	SA13	D14	$\overline{READYI}$	P7	$\overline{COEA}$
E2	A11	C7	SA12	M3	$\overline{READYO}$	M7	$\overline{COEB}$
E3	A10	B6	SA11	E12	FLUSH	N7	$\overline{CWEA}$
D1	A9	B5	SA10	M8	WBS	P6	$\overline{CWEB}$
D2	A8	A5	SA9	N8	$\overline{MISS}$	M5	$\overline{CS0}$
C2	A7	C6	SA8	A14	$\overline{DEFOE}$	M6	$\overline{CS1}$
A1	A6	A4	SA7	D12	$2W/\overline{D}$	N6	$\overline{CS2}$
D3	A5	C5	SA6	B13	$M/\overline{S}$	P5	$\overline{CS3}$
C3	A4	B4	SA5	M10	$\overline{DOE}$	N5	$CT/\overline{R}$
B2	A3	C4	SA4	M4	LDSTB	P8	$\overline{BRDYEN}$
B3	A2	A3	SA3	N11	BHOLD	K13	$\overline{BREADY}$
G14	$\overline{BE0}$	A2	SA2	M11	BHLDA	P10	BACP
H13	$\overline{BE1}$	J12	SEN			M9	$\overline{BAOE}$
H12	$\overline{BE2}$	J13	$\overline{SSTB}$			N10	$BT/\overline{R}$
J14	$\overline{BE3}$	L14	RESERVED				
C13	CLK2						
D13	RESET						
K12	BRESET						
L13	BCLK2						

(b)

Figure 6.61 (*Continued*)

82385DX, the CLK2 input of both devices are tied together and driven by the external clock oscillator, and the $\overline{NA}$ input of the 80386DX is driven by the $\overline{NA}$ output of the 82385DX.

Next we will examine the signals at the cache memory interface. In Fig. 6.64 we find that this interface includes data bus lines D_0 through D_{31}, address lines A_2 through A_{13} or A_2 through A_{14}, and a number of control signals. The address and data bus lines are supplied directly by the 80386DX. On the other hand, the control signals are generated by the 82385DX. Figure 6.63 identifies these control signals as the cache memory control signals.

An example of a direct-mapped cache memory array is shown in Fig. 6.65(a) and a two-way set associative cache array in Fig. 6.65(b). The first cache control signal, as shown in Fig. 6.63, is *cache address latch enable* (CALEN). This output is used to enable the address latch in the cache memory subsystem. In both circuits CALEN is applied to the enable (E) input of the 74F373 address latches. The next control signal, *cache transmit/receive* ($CT/\overline{R}$), is used to control the direction of

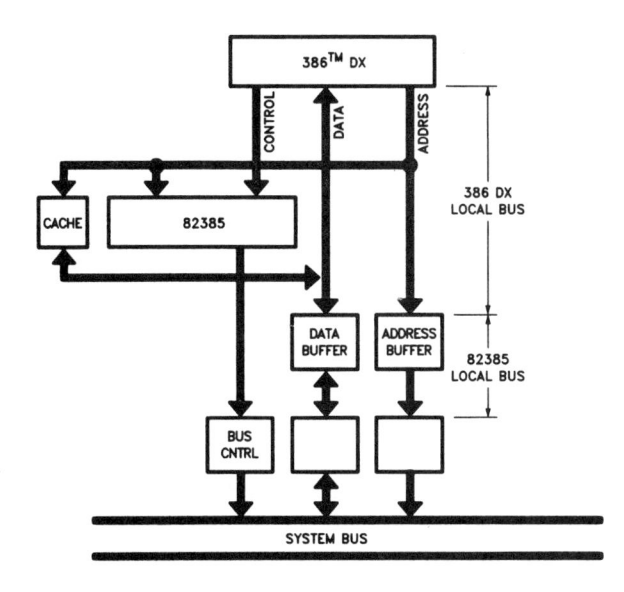

Figure 6.62 82385DX cache memory subsystem architecture. (Reprinted by permission of Intel Corp. Copyright Intel Corp. 1990.)

data transfer through the cache data bus transceivers during read and write bus cycles. Notice in Fig. 6.65(a) and (b) that this signal is applied to the direction (DIR) inputs of the 74F245 data bus transceiver devices.

The 32K-byte memory array in the direct-mapped cache of Fig. 6.65(a) is implemented with four 8K × 8-bit SRAMs. Each RAM is attached to eight of the data bus lines. The *cache chip select* ($\overline{CS_0}$ through $\overline{CS_3}$) outputs of the 82385DX are applied to chip select inputs of the RAMs and are used to enable them for operation. For example, Fig. 6.66 shows that SRAM 0 supplies data bus lines D_0 through D_7 and is enabled by $\overline{CS_0}$. During byte, word, and double-word data transfers, the cache chip select signals are used to enable the appropriate static RAMs for operation.

In Fig. 6.63 we find two more groups of cache control signals, the *cache output enable* ($\overline{COEA}$ and $\overline{COEB}$) signals and *cache write enable* ($\overline{CWEA}$ and $\overline{CWEB}$) signals. They are used to enable the data bus transceivers and the outputs of the SRAMs, respectively. Looking at Fig. 6.65(a), we see that in a direct-mapped cache only $\overline{COEA}$ and $\overline{CWEA}$ are used. $\overline{COEA}$ is applied to the $\overline{OE}$ input of the four 74F245 data bus transceivers and enables them for operation during read and write bus cycles. Moreover, $\overline{CWEA}$ drives the $\overline{WE}$ inputs of the four SRAMs in parallel and signals them whether a read or write data transfer is taking place.

Earlier we found that the 80386DX microcomputer's system bus was formed from the 82385DX's local bus signals. Let us now look at the signals of the *82385DX local bus*. Notice in Fig. 6.64 that the data bus, address bus, and bus cycle indication signals of the 80386DX are used as part of the 82385DX's local bus. Buffered address lines BA_2 through BA_{31} and buffered bus cycle indication signals BM/$\overline{IO}$, BD/$\overline{C}$, and BW/$\overline{R}$ are formed by latching the corresponding signal of the 80386DX into 74F374 latches. The address and bus cycle indication signals

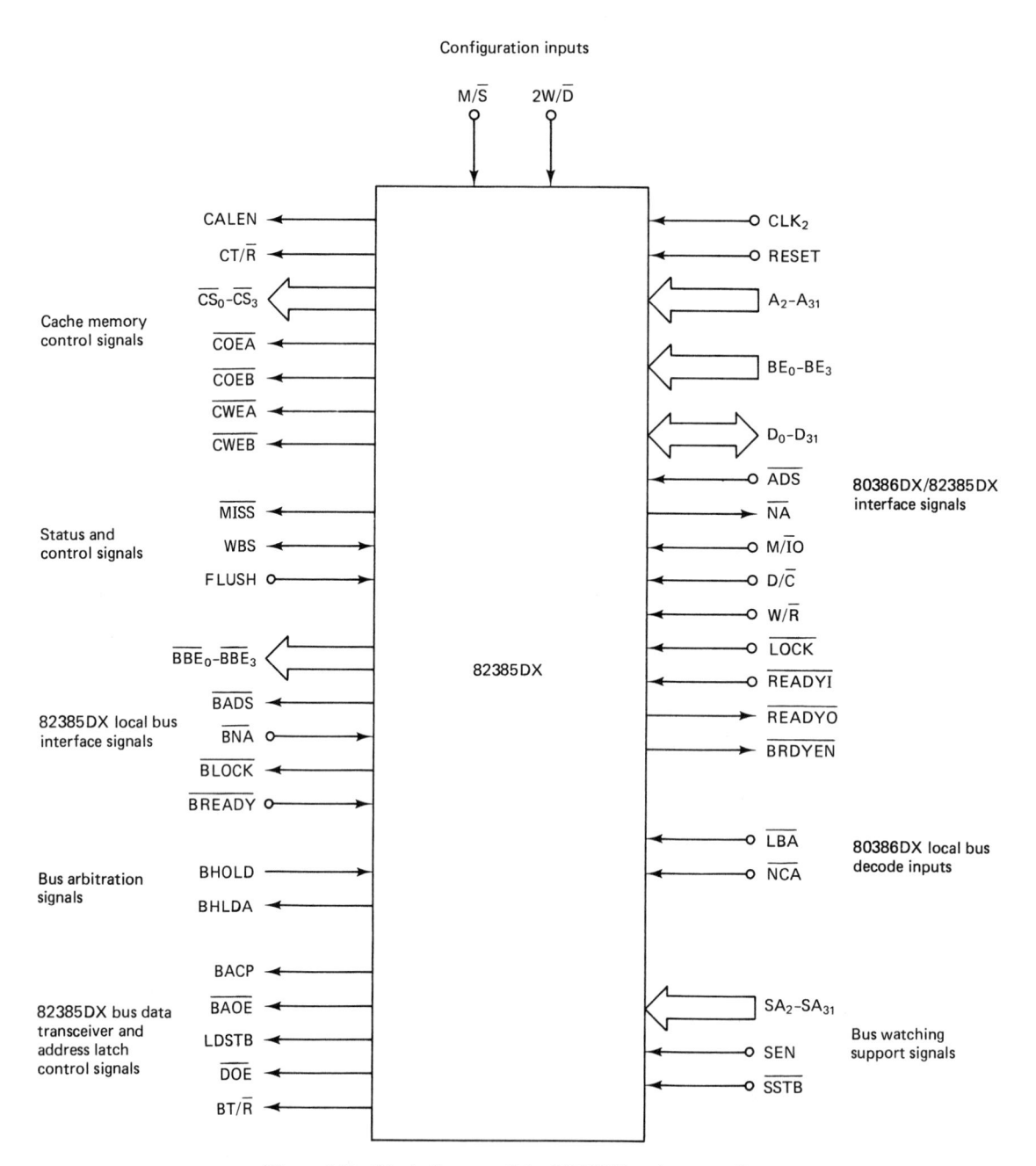

Figure 6.63 Block diagram of the 82385DX cache controller.

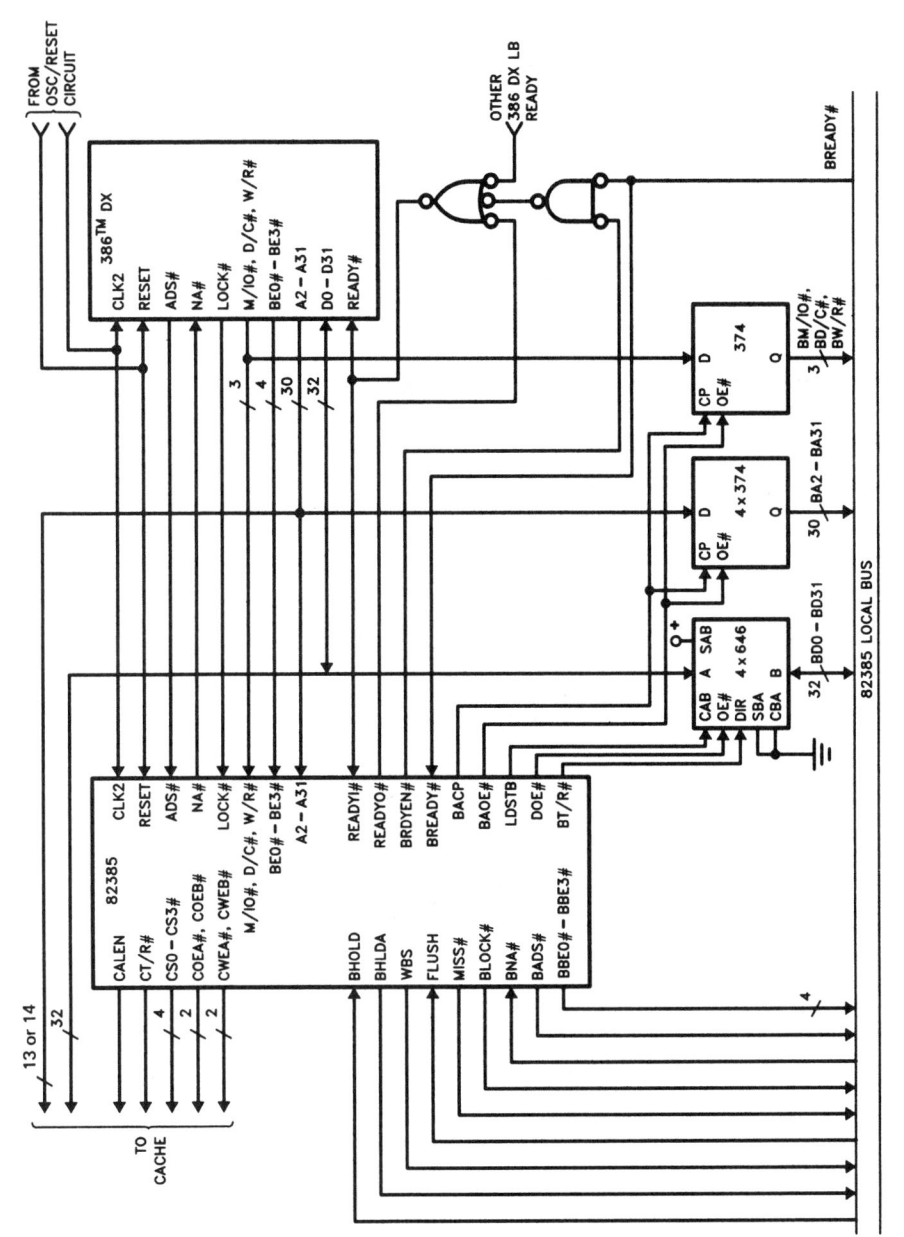

Figure 6.64 Connection of the 82385DX to the 80386DX MPU. (Reprinted by permission of Intel Corp. Copyright Intel Corp. 1990.)

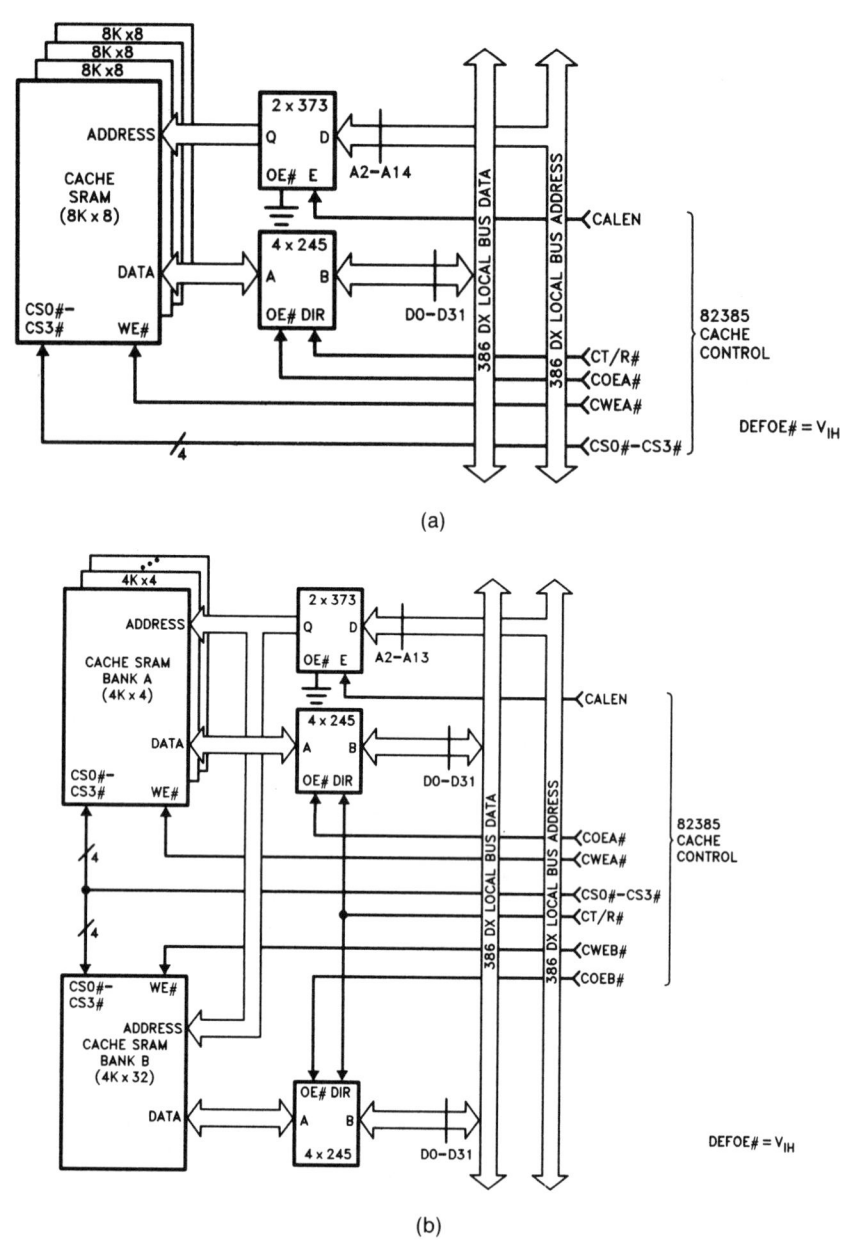

Figure 6.65 (a) Direct-mapped cache memory array. (Reprinted by permission of Intel Corp. Copyright Intel Corp. 1990.) (b) Two-way set associative cache memory array. (Reprinted by permission of Intel Corp. Copyright Intel Corp. 1990.)

SRAM	Data bus lines	Cache chip select
0	D_7–D_0	CS_0
1	D_{15}–D_8	CS_1
2	D_{23}–D_{16}	CS_2
3	D_{31}–D_{24}	CS_3

Figure 6.66 Direct-mapped cache SRAM data bus connections and the cache chip select signals.

are latched into these octal latches with a pulse from the *bus address clock pulse* (BACP) output of the 82385DX and the buffer is enabled by the *bus address output enable* ($\overline{BAOE}$) output. Data bus lines BD_0 through BD_{31} are buffered by 74F646 registered data bus transceivers. Signals *bus transmit/receive* (BT/$\overline{R}$), *data output enable* ($\overline{DOE}$), and *local data strobe* ($\overline{LDSTB}$) are used to set the direction of data transfer through the transceivers, enable the transceivers for operation, and clock the data into the register within the 74F646s during write cycles, respectively. These outputs are identified in Fig. 6.63 as the 82385DX bus data transceiver and address latch control signals.

At the other side of the 82385DX in Fig. 6.63, we find the control signals of the 82385DX local bus. The *bus byte enable* ($\overline{BBE_0}$ through $\overline{BBE_3}$), *bus address status* ($\overline{BADS}$), *bus next address request* ($\overline{BNA}$), *bus lock* ($\overline{BLOCK}$), *bus hold acknowledge* (BHLDA), and *bus hold* (BHOLD) signals perform similar functions as their corresponding signals of the 80386DX. For instance, the $\overline{BADS}$ output indicates that a valid address and bus cycle indication code are available. There are also a few new signals that are unique to the 82385DX local bus. The *cache flush* (FLUSH) input can be used by external circuitry to initiate clearing of the cache memory. If FLUSH is switched to the 1 logic level for eight CLK2 cycles, all of the cache directory valid bits are cleared. In this way, the contents of the cache SRAMs are flushed by invalidating them. *Cache miss* ($\overline{MISS}$) is another output. It signals external circuitry that the current address on the bus does not correspond to data that are already in the cache and that the information must be read from the main memory subsystem. Finally, *write buffer status* (WBS) is an output that signals that the registers within the 74F646 data bus transceivers contain write data that have not yet been written to the main memory subsystem. These control signals correspond to the 82385DX local bus interface signals, bus arbitration signals, and status and control signals groups in Fig. 6.63.

Direct-Mapped Cache Operation

We just examined the signal interfaces of the 82385DX cache controller and the hardware of its cache memory subsystem. Let us look next at the operation of an 82385DX-based direct-mapped cache memory subsystem.

We will begin by describing how the 82385DX keeps track of which main memory locations are cached. The 82385DX accomplishes this through what is called a *cache directory*. The cache directory is located within the 82385DX and contains 1024 26-bit entries. Figure 6.67 illustrates the organization of the direct-mapped cache memory subsystem. Notice that the 80386DX's 4G-byte main memory physical address space is treated as $2^{17} - 1$ (131,071) 32K-byte pages. Since the 80386DX reads code and data as 32-bit double words, the organization of a page in main memory can also be viewed as 8192 (8K) double words.

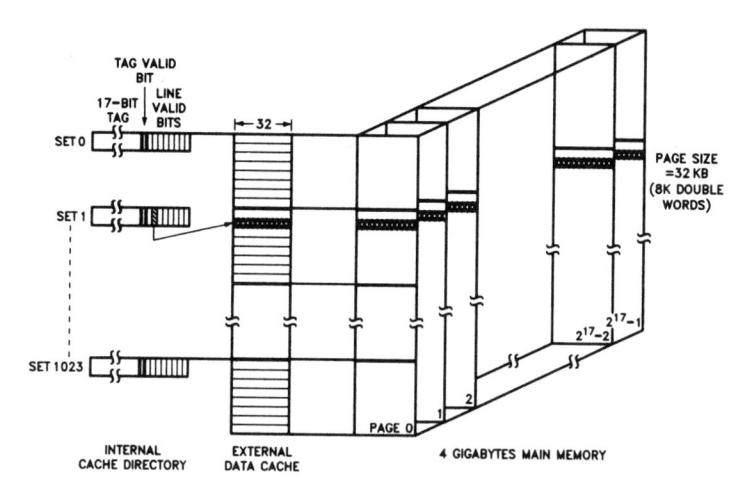

Figure 6.67 Direct-mapped cache organization. (Reprinted by permission of Intel Corp. Copyright Intel Corp. 1990.)

As expected, the page size in main memory matches the size of the cache memory. The 32K-byte cache is organized as double-word storage locations. Figure 6.67 shows that this results in an 8K × 32-bit memory array. Each of these 32-bit double words is called a *line*. Code and data are read from and written into cache one line (a double word) at a time.

The cache directory within the 82385DX is organized in a different way. Within the cache controller, the status of the cache memory locations are tracked with 26-bit cache directory entries. In Fig. 6.67 we see that the entries within the cache directory are called *SET 0* through *SET 1023*. Each of the 1024 SET entries corresponds to eight consecutive lines (double words) of information in the cache memory array. For instance, SET 0 corresponds to lines 0 through 7 of the array.

EXAMPLE 6.8

In which SET entry of the cache directory in Fig. 6.67 is the information for line 9 of the memory array stored?

SOLUTION Looking at Fig. 6.67, we find that line 9 of the cache memory array is in SET 1. In fact, line 9 is the 32-bit storage location that is highlighted in the external data cache section of the drawing.

The format of the cache directory SET entry used by an 82385DX configured for direct-mapped mode is given in Fig. 6.68. Notice that bits 0 through 7 are eight independent *line valid bits*, the eighth bit is the *tag valid bit*, and bits 9 through 25 are a 17-bit *tag*. The 17-bit tag identifies the page number of one of the 131,072 pages in main memory. The tag bit indicates whether the 17-bit tag is valid or invalid. If this bit is 0, all lines of information in the SET are invalid. However, logic 1 in the tag bit does not mean that they are all valid. When the tag bit is logic 1, the eight line valid bits identify whether the individual lines in the SET are valid or invalid. For example, logic 1 in line valid bit 0 of SET 0

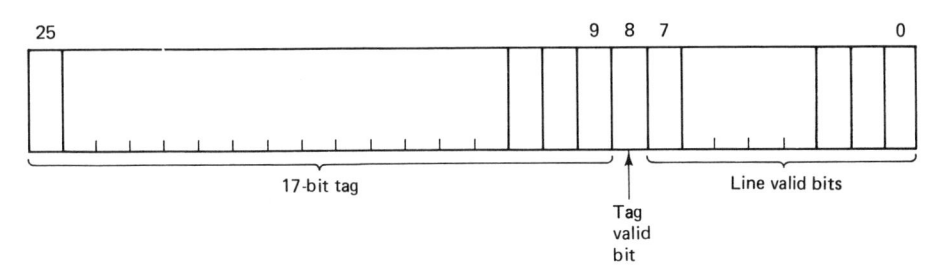

Figure 6.68 Direct-mapped cache directory SET entry format.

means that line 0 in the cache contains valid information, logic 1 in bit 1 means that line 1 is valid, and so on. This information is updated automatically each time a noncached line of information is read from memory.

EXAMPLE 6.9

If the cache directory entry for SET 1 equals $00005FF_{16}$, from which page of main memory is the cached information? Is the cache entry valid? Which lines in SET 1 are valid?

SOLUTION Expressing the SET 1 entry in binary form gives

$$\text{SET 1} = 00000000000000010111111111_2$$

From the binary form of SET 1, we find that

$$\text{TAG} = 00000000000000010_2 = 2_{10}$$

Therefore, the entry is from page 2 of the main memory array. Next the valid bit is

$$\text{TAG VALID} = 1$$

and the tag entry is valid. Finally, the line valid bits are

$$\text{LINE VALID} = 11111111_2$$

This means that all lines of information in SET 1 are valid.

Now that we have looked at what information is used to track entries in the cache memory we will continue by examining the events that take place when data are read from memory. Whenever the 80386DX initiates a read bus cycle, the address output on the address bus is applied in parallel to the 82385DX local bus address latch, the address inputs of the 82385DX cache controller, and the cache memory interface. The cache controller quickly makes a decision whether the information for this address should be read from the cache or from main memory. It does this by interpreting the address and then comparing this information to the corresponding cache directory SET entry.

Figure 6.69 shows how addresses are interpreted by the 82385DX cache controller. Notice that the 17 most significant bits, A_{15} through A_{31}, are the tag that identifies the page of main memory from which information is to be read. The next 10 address bits, A_5 through A_{14}, are called the SET address. This part

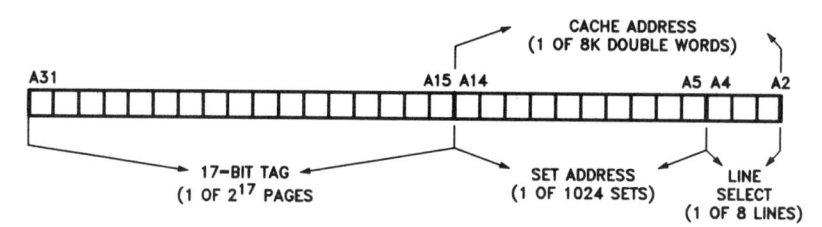

```
                                        CACHE ADDRESS
                                     (1 OF 8K DOUBLE WORDS)
 A31                       A15 A14              A5 A4    A2
 ┌─┬─┬─┬─┬─┬─┬─┬─┬─┬─┬─┬─┬─┬─┬─┬─┬─┬─┬─┬─┬─┬─┬─┬─┬─┬─┬─┬─┬─┐
 └─┴─┴─┴─┴─┴─┴─┴─┴─┴─┴─┴─┴─┴─┴─┴─┴─┴─┴─┴─┴─┴─┴─┴─┴─┴─┴─┴─┴─┘
        17-BIT TAG                SET ADDRESS        LINE
      (1 OF 2¹⁷ PAGES)           (1 OF 1024 SETS)   SELECT
                                                 (1 OF 8 LINES)
```

Figure 6.69 Address bit fields for direct-mapped cache. (Reprinted by permission of Intel Corp. Copyright Intel Corp. 1990.)

of the address identifies which SET of storage locations in the cache memory array are to be accessed. Finally, the three least significant bits, A_2 through A_4, are a line select code and selects one of the eight lines in this SET.

Let us now look at what happens when an address is applied to the address inputs of the 82385DX. The SET part of the address is used to select one of the 1024 SET entries in the cache directory. Then, three checks are made to determine if the information is already in the cache. First, the 82385DX compares the tag field of the address to the value of the tag field in the selected SET entry to verify that they match. Second, the tag valid bit in the SET entry is examined to verify that it is set. Finally, the line valid bit corresponding to the line select code part of the address is tested to verify that it is 1. If all three of these conditions are satisfied, the information held in the storage location to be read is already cached and is valid. That is, a cache hit has occurred. In this case, the 82385DX initiates the read cycle from the cache memory instead of main memory.

We just saw the result of a cache hit. Let us now determine what can cause a cache miss and what happens when one occurs. If during the address interpretation process, the tag is found to match and the tag valid bit is 1, but the line valid bit is 0, a cache miss has occurred. This result represents a situation where the SET entry in the directory corresponds to the correct page of main memory, but the double word (line) to be read has not yet been cached. This is called a *line miss*. For this situation the cache controller initiates a read of the double word from main memory to the MPU. As part of the bus operation, the double word of data is written into the corresponding storage location in the cache memory array and then the line valid bit in its SET entry in the cache directory is set to 1. That is, the information is copied into the cache and marked as valid.

If during the entry verification operation, either the tags do not match or the tag valid bit is found to be 0, a *tag miss* has occurred. This corresponds to an attempt to read information from either a page that has not already been cached or one that is cached but is invalid. When this type of cache miss occurs, the cache controller again initiates a read from main memory and the value transferred over the data bus is written into the cache memory. However, this time the tag in the SET entry of the directory is updated with the value from the tag part of the address, the tag valid bit is set, the line valid bit identified by the address is set, and all other line valid bits are cleared. In this way, a new valid page and line entry have been created and all other entries in the SET, which now correspond to information in another page of main memory, are invalidated.

Two-Way Set Associative Cache Operation

The operation of an 82385DX-based cache memory subsystem configured for two-way set associative mode is somewhat different from that we just described for the direct-mapped subsystem. Let us begin by examining the organization of the two-way set associative cache memory subsystem shown in Fig. 6.70. Here we

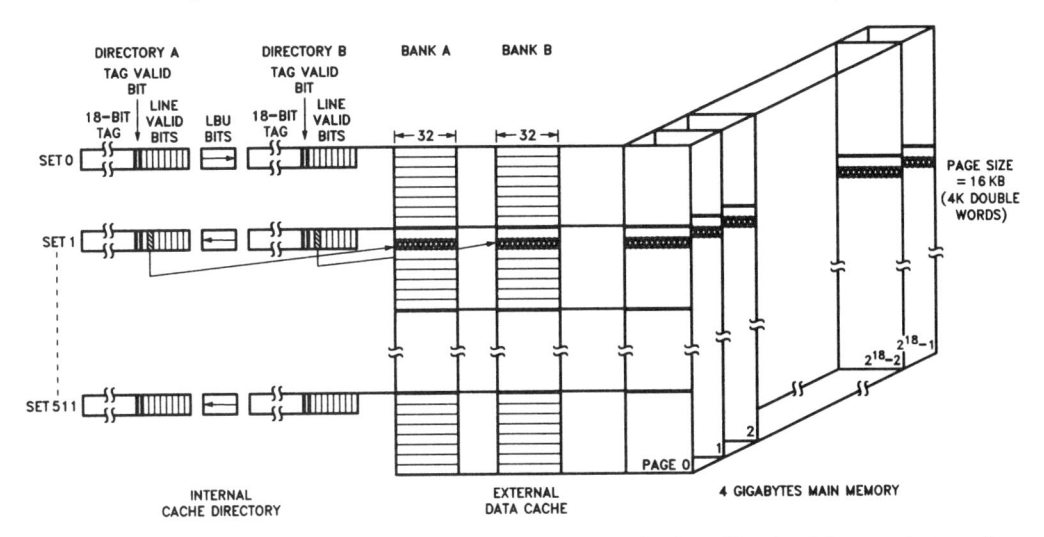

Figure 6.70 Two-way set associative cache organization. (Reprinted by permission of Intel Corp. Copyright Intel Corp. 1990.)

find that the 82385DX's internal cache directory contains two SET entry directories, not one as in the direct-mapped configuration. These directories are identified as *DIRECTORY A* and *DIRECTORY B* and each contains 512 SET entries. The external 32K-byte cache memory array is partitioned into two separate 4K × 32-bit banks called *BANK A* and *BANK B*. This permits information from the same offset in two different pages of main memory to reside in the cache at the same time. The cache directory also includes 512 one-bit *least recently used (LRU) flags*. The setting of these flags keep track of whether BANK A or BANK B holds the least recently used information. These flags are checked by the least recently used replacement algorithm that is employed by the 82385DX. Finally, we find that the 80386DX's 4G-byte physical address space is divided into 2^{18} (262,144) 16K-byte (4K double-word) pages. This page size is half that used in the direct-mapped main memory subsystem.

The SET entries in DIRECTORY A and DIRECTORY B both have the same format. Figure 6.71 shows this format in more detail. Notice that they are 27 bits in length. The difference between this SET entry and that shown in Fig. 6.68 for a direct-mapped cache is that the tag is 18 bits long instead of 17 bits. The extra bit in the tag is needed because making the page size 16K bytes results in twice as many pages.

A read operation from a two-way set associative cache memory is similar

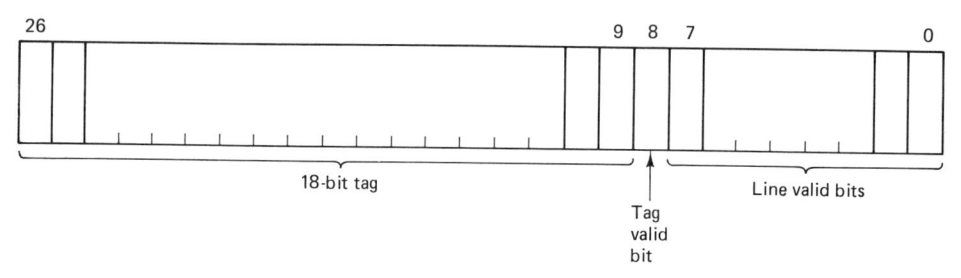

Figure 6.71 Two-way set associative cache directory SET entry format.

to that just described for the direct-mapped cache. The 82385DX interprets the memory address as shown in Fig. 6.72. Notice that the SET address is 9 bits, not 10 bits. The reason the SET address has one less bit than that for a direct mapped cache is that the cache directory contains 512 SET entries rather than 1024. First, the 9-bit SET address is used to select one of the 512 SET entries in both DIRECTORY A and DIRECTORY B. Then the 18-bit tag address is compared to the tag in each of these SET entries, there tag valid bits are checked, and the line valid bit corresponding to the line select code is checked in each SET entry. If these three conditions are satisfied for either of the two SET entries, a cache hit has occurred and the line of information is read from the corresponding bank of cache memory.

On the other hand, a miss is detected if there is no match of either the tags, if both valid bits are cleared, or if the line valid bit is not set in either entry. At this point, operation differs for the two-way set associative and the direct-mapped cache. This is because we are at the point where the least recently used (LRU) algorithm takes over. It checks the LRU flag for the SET selected by the SET address to determine whether the BANK A or BANK B cache entry was least recently used. Then the read operation is initiated from the main memory subsystem and the line of information is read by the MPU and written into the cache location of the least recently used bank. Now the SET entry is updated by replacing the tag, setting the tag valid bit and the line valid bit, and switching the LRU flag. The updated cache memory location now holds the most recently used information.

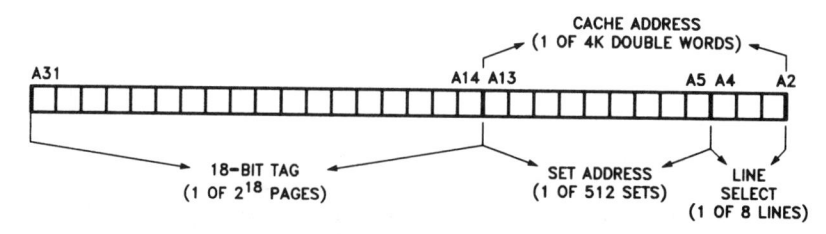

Figure 6.72 Address bit fields for two-way set associative cache. (Reprinted by permission of Intel Corp. Copyright Intel Corp. 1990.)

Cache Coherency and Bus Watching

The ability to maintain the information held at an address in main memory and its copy in a cache the same is called cache *coherency*. In multiprocessor applications, another bus master can take over control of the system bus. This bus master could write data into a main memory storage location whose data are already held in the cache of another processor. When this happens, the data in the cache no longer match those held in main memory. That is, the data are invalid. If this cached value of data were read, processed, and written back to memory, the content of main memory would be contaminated. To protect against this problem, the 82385DX is equipped with a feature known as *bus watching*.

The diagram in Fig. 6.73 shows the architecture of an 80386DX microcomputer system that employs an 82385DX-based cache memory with bus watching.

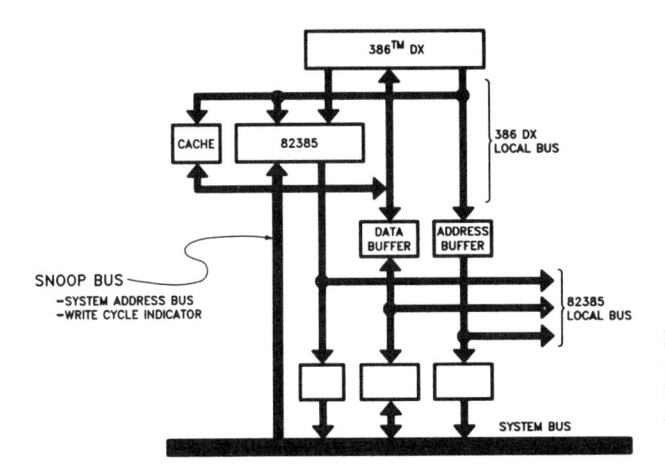

Figure 6.73 82385DX cache memory subsystem architecture that employs bus watching. (Reprinted by permission of Intel Corp. Copyright Intel Corp. 1990.)

Here we find that a *snoop bus* has been added between the microcomputer's system bus and the 82385DX. The snoop bus is implemented with the signals identified as the bus watching support signals in Fig. 6.63. They include the *snoop address bus* (SA_2 through SA_{31}) inputs, the *snoop enable* (SEN) input, and the *snoop strobe* ($\overline{SSTB}$) input.

Whenever a bus master is performing a write bus cycle to the main memory subsystem, the 82385DXs in the system examine the address to determine if they contain valid data for this storage location. The main memory subsystem must signal the 82385DXs that a write cycle is in progress by applying logic 1 at their SEN inputs. The address of the storage location being accessed is applied to the snoop address inputs of the 82385DXs and is strobed into the device with logic 0 at $\overline{SSTB}$. If the cache directory entry corresponding to this address indicates that the cache contains valid data, a *snoop hit* has occurred. The 82385DX automatically clears the tag to invalidate this piece of data in the cache. In this way we see that the 82385DX has the ability to maintain cache coherency by watching what system bus activity takes place.

Noncacheable Memory Address Space

The 82385DX supports partitioning of the main memory subsystem's address space into a cachable and noncacheable address range. This feature is achieved with the *noncacheable access* ($\overline{\text{NCA}}$) input of the 82385DX. By decoding the address in external circuitry and returning logic 0 to the $\overline{\text{NCA}}$ input for those addresses in the noncacheable range, bus cycles to these storage locations are made noncacheable. That is, the cache controller causes their bus cycles to be performed over the 82385DX local bus to the main memory subsystem.

ASSIGNMENT

Section 6.2

1. Name the technology used to fabricate the 80386DX microprocessor.
2. What is the transistor count of the 80386DX?
3. Which signal is located at pin B7?

Section 6.3

4. How large is the real-address mode address and physical address space? How large is the protected-address-mode address and physical address space? How large is the protected-mode virtual address space?
5. If the byte enable code output during a data write bus cycle is $\overline{BE_3}\,\overline{BE_2}\,\overline{BE_1}\,\overline{BE_0}$ = 1110_2, is a byte, word, or double-word data transfer taking place? Over which data bus lines are the data transferred? Does data duplication occur?
6. For which byte enable codes does data duplication take place?
7. What type of bus cycle is in progress when the bus status code M/$\overline{\text{IO}}$ D/$\overline{\text{C}}$ W/$\overline{\text{R}}$ equals 010?
8. Which signals implement the DMA interface?
9. What processor extension is most frequently attached to the processor extension interface?

Section 6.4

10. What speeds of 80386DX ICs are available from Intel Corporation? How are these speeds denoted in the part number?
11. At what pin is the CLK2 input applied?
12. What frequency clock signal must be applied to the CLK2 input of an 80386DX-25 if it is run at full speed?

Section 6.5

13. What is the duration of PCLK for an 80386DX that is driven by CLK2 equal to 50 MHz?

14. What two types of bus cycles can be performed by the 80386DX?
15. How many CLK2 cycles are in an 80386DX bus cycle that has no wait states? How many T states are in this bus cycle? What would be the duration of this bus cycle if the 80386DX were operating at CLK2 equal to 50 MHz?
16. What does T_1 stand for? What happens in this part of the bus cycle?
17. Explain what is meant by *pipelining* of the 80386DX's bus.
18. What is an idle state?
19. What is a wait state?

Section 6.6

20. If an 80386DX-25 is executing a nonpipelined write bus cycle that has no wait states, what would be the duration of this bus cycle if the 80386DX were operating at full speed?
21. If an 80386DX-25 that is running at full speed performs a read bus cycle with two wait states, what is the duration of the bus cycle?

Section 6.7

22. How is memory organized from a hardware point of view in a protected-mode 80386DX microcomputer system? In a real-mode 80386DX microcomputer system?
23. What are the five types of data transfers that can take place over the data bus? How many bus cycles are required for each type of data transfer?
24. If an 80386DX-25 is running at full speed and all memory accesses involve one wait state, how long will it take to fetch the word of data starting at address $0FF1A_{16}$? The word at address $0FF1F_{16}$?
25. During a bus cycle that involves a misaligned word transfer, which byte of data is transferred over the bus during the first bus cycle?

Section 6.8

26. Overview the function of each of the blocks in the memory interface diagram of Fig. 6.22.
27. When the instruction PUSH AX is executed, what bus status code is output by the 80386DX, which byte enable signals are active, and what read/write control signal is produced by the bus control logic?
28. What type of logic function is implemented by the 74F373 and 74F374 ICs?
29. What is the primary difference between the 74F373 and 74F374?
30. Make a drawing like that in Fig. 6.24 for the real-mode 80386DX address bus.
31. Make a drawing to show how the I/O address buffer in Fig. 6.25(a) can be constructed with 74F244 ICs. Assume that the buffer circuits will be permanently enabled.
32. What logic function is implemented by the 74F245 IC?
33. In the circuit of Fig. 6.27, what logic levels must be applied to the $\overline{\text{DEN}}$ and $\text{DT}/\overline{\text{R}}$ inputs to cause data on the system data bus to be transferred to the microprocessor data bus?

34. What are the logic levels of the $\overline{G}$, DIR, CAB, CBA, SAB, and SBA inputs of the 74F646 when stored data in the A register are transferred to the B bus?
35. Name an IC that implements a 2-line to 4-line decoder logic function.
36. If the inputs to a 74F138 decoder are $G_1 = 1$, $\overline{G}_{2A} = 0$, $\overline{G}_{2B} = 0$, and CBA = 101, which output is active?
37. Make a drawing like that in Fig. 6.32 for a real-mode address bus for which a 74F138 decoder is used to decode address lines A_{17} through A_{19} into memory chip selects.

Section 6.9

38. What does *PLA* stand for?
39. List three properties that measure the capacity of a PLA.
40. What is the programming mechanism used in a PAL called?
41. Redraw the circuit in Fig. 6.35(b) to show how it can implement the logic function F = $\overline{AB}$ + AB.
42. How many dedicated inputs, dedicated outputs, programmable input/outputs, and product terms are supported on the 16L8 PAL?
43. What is the maximum number of inputs on a 20L8 PAL? The maximum number of outputs?
44. How do the outputs of the 16R8 differ from those of the 16L8?

Section 6.10

45. What is meant by the term *nonvolatile memory*?
46. What does *PROM* stand for? *EPROM*?
47. What must an EPROM be exposed to in order to erase its stored data?
48. If the block diagram of Fig. 6.39 has address lines A_0 through A_{16} and data lines D_0 through D_7, what are its bit density and byte capacity?
49. Summarize the read cycle of an EPROM. Assume that both $\overline{CE}$ and $\overline{OE}$ are active before the address is applied.
50. Which standard EPROM stores 64K 8-bit words?
51. What is the difference between a 27C64 and a 27C64-1?
52. What are the values of V_{CC} and V_{PP} for the Intelligent Programming algorithm?
53. What is the duration of the programming pulses used for the Intelligent Programming algorithm?

Section 6.11

54. What do *SRAM* and *DRAM* stand for?
55. Are RAM ICs examples of nonvolatile or volatile memory devices?
56. What must be done to maintain the data in a DRAM valid?
57. Find the total storage capacity of the circuit in Fig. 6.46 if the devices are 43256As.
58. List the minimum values of each of the write cycle parameters that follow for the 4364-10 SRAM: t_{WC}, t_{CW1}, t_{CW2}, t_{WP}, t_{DW}, and t_{WR}.
59. Give two advantages of DRAMs over SRAMs.

60. Name the two parts of a DRAM address.
61. Show how the circuit in Fig. 6.55 can be expanded to 128K × 32 bits.
62. Give a disadvantage of the use of DRAMs in an application that does not require a large amount of memory.

Section 6.12

63. The output of what gate in the circuit of Fig. 6.56 supplies the $\overline{\text{READY}}$ input of the 80386DX?
64. In the microcomputer of Fig. 6.56, for what logic levels at M/$\overline{\text{IO}}$ and A_{31} will the $\overline{Y}_3$ output of the memory address decoder become active?
65. The output of which gate drives the $\overline{\text{NA}}$ input of the 80386DX in the circuit of Fig. 6.56? For what input conditions does it become active?
66. What type of PALs are used to implement the bus control logic in Fig. 6.56?
67. For the microcomputer in Fig. 6.56, what input conditions make the output of an OR gate in the bank write control logic active?

Section 6.13

68. What is a cache memory?
69. What is the result obtained by using a cache memory in a microcomputer system?
70. What is the range of cache memory sizes commonly used in microcomputer systems?
71. Define the term *cache hit*.
72. When an application program is tested on a microcomputer system with a code cache, it is found that 1340 instruction acquisition bus cycles are from the cache memory and 97 are from main memory. What is the hit rate?
73. If the cache memory in Problem 72 operates with zero wait states and main memory bus cycles are performed with three wait states, what is the average number of wait states experienced executing the application?
74. Name two widely used cache memory organizations.
75. Name the element into which main memory is organized for use with a cache.
76. How does the organization of a direct-mapped cache memory array differ from that of a two-way set associative cache memory?
77. What does *LRU* stand for?
78. What is *posted write-through*?

Section 6.14

79. At what pin of the 82385DX's package is the signal CALEN located?
80. Which signal is output at pin M10 of the 82385DX's package?
81. What must be done to configure an 82385DX cache controller for two-way set associative operation?
82. Make a list of the mnemonics and signal names for all of the control signals at the cache memory interface.
83. What are the functions of the signals BCAP and $\overline{\text{BAOE}}$?

Chap. 6 Assignment

84. Which signals indicate whether a byte, word, or double word of data is being transferred across the 82385DX's local data bus?
85. Which input, when activated, clears the cache memory?
86. Into what size pages does a direct-mapped 82385DX organize the memory of a real-mode 80386DX microcomputer system? Into how many pages is the real-mode address space organized?
87. What is the size of the cache memory array in an 82385DX-based direct-mapped cache memory subsystem? Express the answer in double words.
88. If the cache directory entry for SET 3 in a direct-mapped cache equals $0001F03_{16}$, from which page of main memory is the cached information? Is the cache entry valid? Which lines in the set are valid?
89. What is the physical address range of the lines of the SET 3 entry in Problem 88?
90. What are the size of BANK A and BANK B in the memory array of an 82385DX-based two-way set associative cache memory? Express the answer in kilobytes.
91. What is the line length in a two-way set associative cache memory that is designed with the 82385DX cache controller?
92. What function is performed by the LRU flag?
93. Which feature of the 82385DX is used to assure cache coherency?

7

Input/Output Interface of the 80386DX Microprocessor

7.1 INTRODUCTION

In Chapter 6 we studied the memory interface of the 80386DX microprocessor. Here we examine another important interface of the 80386DX microcomputer system, the input/output interface. These are the topics in the order in which they are covered:

1. Types of input/output
2. The isolated input/output interface
3. Input and output bus cycle timing
4. I/O instructions
5. 82C55A programmable peripheral interface
6. 82C55A implementation of parallel I/O ports
7. Memory-mapped I/O
8. Input/output polling and handshaking
9. The 82C54 programmable interval timer
10. The 82C37A direct memory access controller
11. I/O circuitry

7.2 TYPES OF INPUT/OUTPUT

The 80386DX microprocessor can employ two different types of input/output (I/O). They are known as *isolated I/O* and *memory mapped I/O*. These I/O methods differ in how I/O ports are mapped into the 80386DX's address spaces. Practical microcomputer systems usually employ both kinds of I/O. That is, some peripheral ICs are treated as isolated I/O devices and others as memory-mapped I/O devices. Let us now look at each of these types of I/O.

Isolated Input/Output

When using isolated input/output in a microcomputer system, the I/O devices are treated separate from memory. This is achieved because the software and hardware architectures of the 80386DX support separate memory and I/O address spaces. Figure 7.1 illustrates the 80386DX's real-mode address spaces.

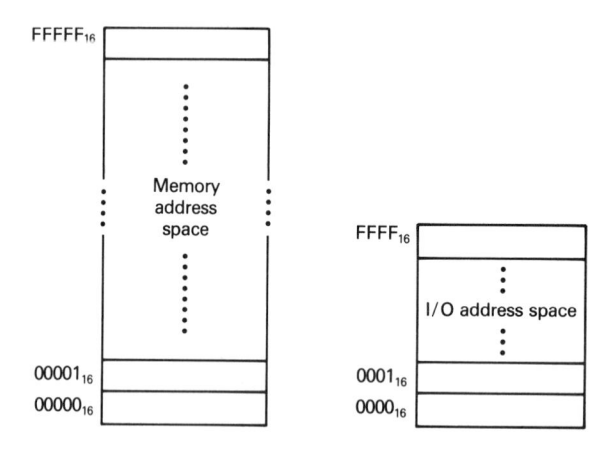

Figure 7.1 Isolated I/O real-mode memory and I/O address spaces.

Let us begin our study of input/output with the I/O address space. Looking at Fig. 7.1, we find that information in memory or at I/O ports is organized as bytes of data, that the real-mode memory address space contains 1M consecutive byte addresses in the range 00000_{16} through $FFFFF_{16}$, and that the I/O address space contains 64K consecutive byte addresses in the range 0000_{16} through $FFFF_{16}$. Bytes of data in two consecutive memory or I/O addresses could be accessed as word-wide data and the contents of four consecutive byte addresses represent a double word of data. For instance, in Fig. 7.2, I/O addresses 0000_{16}, 0001_{16}, 0002_{16}, and 0003_{16} can be treated as independent byte-wide I/O ports, ports 0, 1, 2, and 3; ports 0 and 1 may be considered together as word-wide port 0; and ports 0 through 3 may be accessed as a double-word port.

Notice that the part of the I/O address map in Fig. 7.2 from address 0000_{16} through $00FF_{16}$ is referred to as *page 0*. Certain I/O instructions can only perform operations to ports in this part of the address range. Other I/O instructions can input or output data for ports anywhere in the I/O address space.

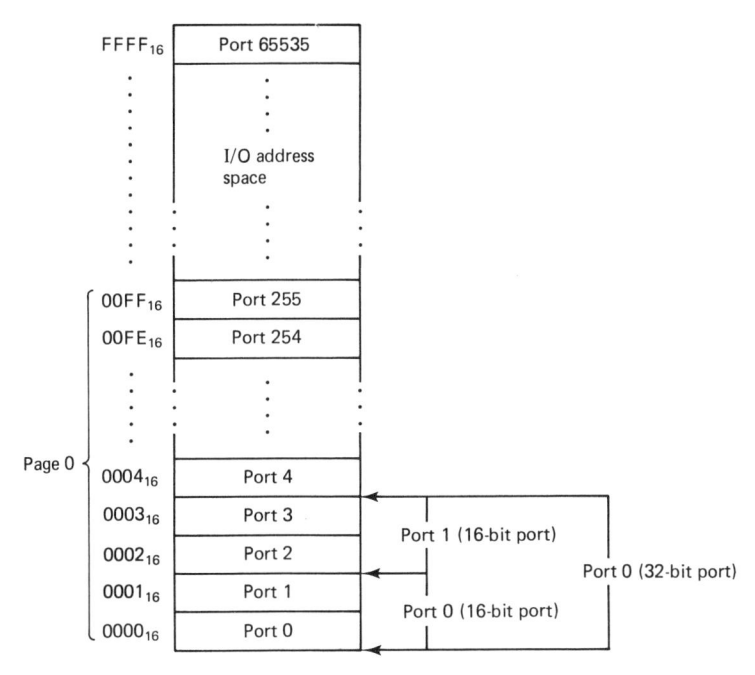

```
                FFFF₁₆  ┌──────────────┐
                        │  Port 65535  │
                   .    │      .       │
                   .    │      .       │
                   .    │  I/O address │
                   .    │    space     │
                   .    │      .       │
                   .    │      .       │
                   .    │      .       │
              ┌ 00FF₁₆  ├──────────────┤
              │         │   Port 255   │
              │ 00FE₁₆  ├──────────────┤
              │         │   Port 254   │
              │    .    │      .       │
              │    .    │      .       │
              │    .    │      .       │
       Page 0 ┤ 0004₁₆  ├──────────────┤
              │         │    Port 4    │
              │ 0003₁₆  ├──────────────┤
              │         │    Port 3    │
              │ 0002₁₆  ├──────────────┤
              │         │    Port 2    │
              │ 0001₁₆  ├──────────────┤
              │         │    Port 1    │
              └ 0000₁₆  ├──────────────┤
                        │    Port 0    │
                        └──────────────┘
```

Figure 7.2 Byte-wide, word-wide, and double-word-wide I/O ports.

The isolated method of input/output offers some advantages. First, the complete 1M-byte memory address space is available for use with memory. Second, special instructions have been provided in the instruction set of the 80386DX to perform isolated I/O input and output operations. These instructions have been tailored to maximize I/O performance. A disadvantage of this type of I/O is that all input and output data transfers must take place between the AL, AX, or EAX register and the I/O port.

Memory-Mapped Input/Output

I/O devices can be placed in the memory address space of the microcomputer as well as in the independent I/O address space. In this case the MPU looks at the I/O port as though it is a storage location in memory. For this reason, the method is known as *memory-mapped I/O*.

In a microcomputer system with memory-mapped I/O, some of the memory address space is dedicated to I/O ports. For example, in Fig. 7.3 the 4096 memory addresses in the range from E0000₁₆ through E0FFF₁₆ are assigned to I/O devices. Here the contents of address E0000₁₆ represents byte-wide I/O port 0; the contents of addresses E0000₁₆ and E0001₁₆ correspond to word-wide port 0; and the contents of addresses E0000₁₆ through E0003₁₆ are double-word-wide port 0.

When I/O is configured in this way, instructions that affect data in memory are used instead of the special input/output instructions. This is an advantage in that many more instructions and addressing modes are available to perform I/O

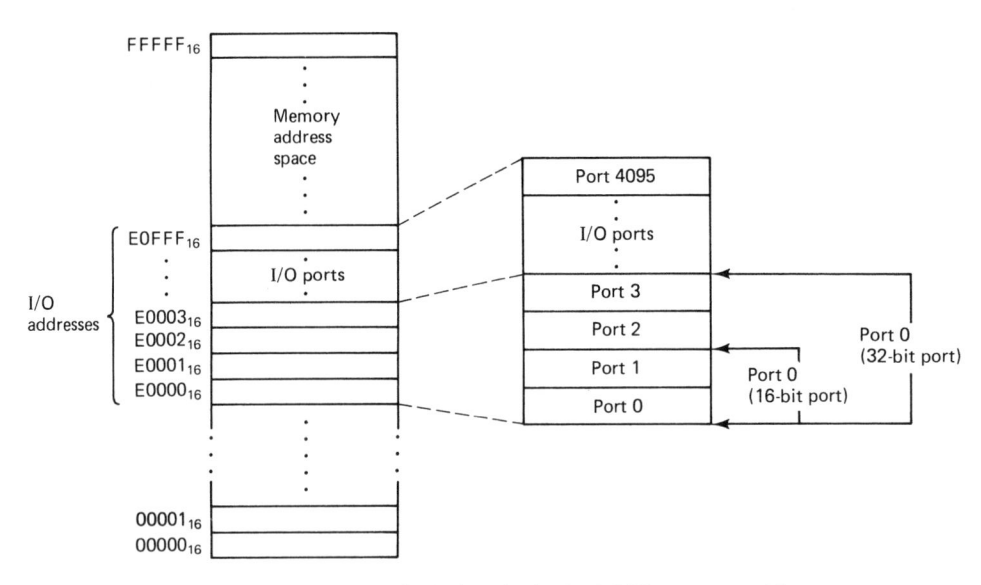

Figure 7.3 Memory-mapped I/O devices in the 80386DX's memory address space.

operations. For instance, the contents of a memory-mapped I/O port can be directly ANDed with a value in an internal register. In addition, I/O transfers can now take place between an I/O port and an internal register other than just AL, AX, or EAX. However, this also leads to a disadvantage. That is, the memory instructions tend to execute more slowly than those specifically designed for isolated I/O. Therefore, a memory-mapped I/O routine may take longer to execute than an equivalent program using the input/output instructions.

Another disadvantage of using this method is that part of the memory address space is lost. For instance, in Fig. 7.3 addresses in the range from $E0000_{16}$ through $E0FFF_{16}$ cannot be used to implement memory.

7.3 THE ISOLATED INPUT/OUTPUT INTERFACE

The isolated input/output interface of the 80386DX microcomputer permits it to communicate with the outside world. The way in which the 80386DX deals with input/output circuitry is similar to the way in which it interfaces with memory circuitry. That is, input/output data transfers also take place over the data bus. This parallel bus permits easy interface to LSI peripheral devices such as parallel I/O expanders, interval timers, and serial communication controllers. Let us continue by looking at how the 80386DX interfaces to an isolated I/O subsystem.

Figure 7.4 shows a typical isolated I/O interface. Here we find that I/O devices 0 through N connect to the 80386DX through an I/O interface circuit. These blocks can represent input devices such as a keyboard, output devices such as a printer, or input/output devices such as an asynchronous serial communications port. An example of a typical I/O device used in the I/O subsystem is a

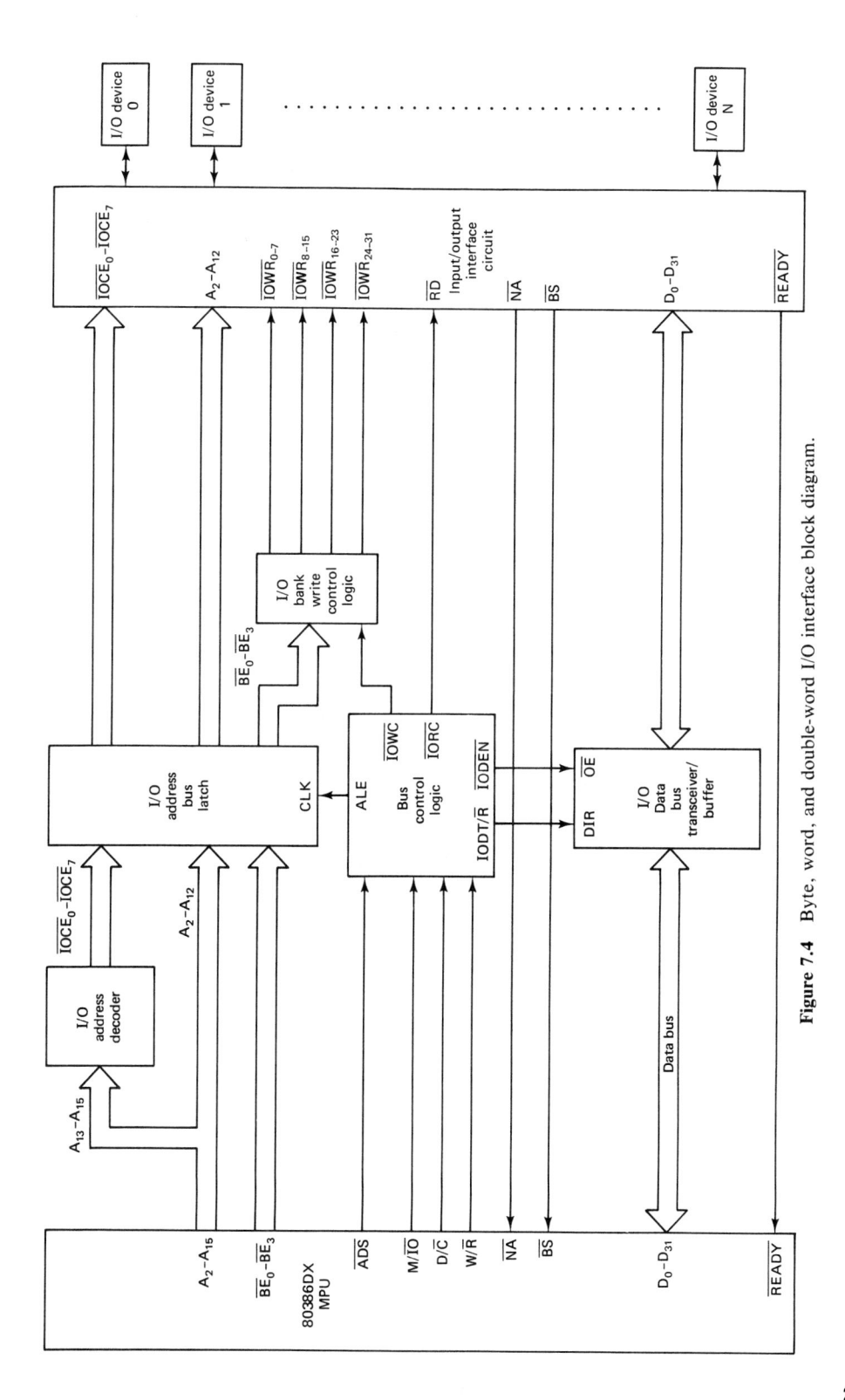

Figure 7.4 Byte, word, and double-word I/O interface block diagram.

programmable peripheral interface (PPI) IC, such as the 82C55A. This type of device is used to implement parallel input and output ports. Let us now look at the function of each of the blocks in this circuit more closely.

Notice that the interface between the microprocessor and I/O subsystem includes the bus controller logic, an I/O address decoder, I/O address latches, I/O data bus transceiver/buffers, and I/O bank write control logic. As in the memory interface, bus control logic produces control signals for the I/O interface. In Fig. 7.4 we see that it decodes the bus cycle indication codes that are output by the 80386DX on M/$\overline{\text{IO}}$, D/$\overline{\text{C}}$, and W/$\overline{\text{R}}$. These 3-bit codes tell which type of bus cycle is in progress. In the table in Fig. 7.5, the bus cycle definition codes that

M/$\overline{\text{IO}}$	D/$\overline{\text{C}}$	W/$\overline{\text{R}}$	Type of bus cycle
0	0	0	Interrupt acknowledge
0	0	1	Idle
0	1	0	I/O data read
0	1	1	I/O data write
1	0	0	Memory code read
1	0	1	Halt/shutdown
1	1	0	Memory data read
1	1	1	Memory data write

Figure 7.5 Input/output bus cycle definition codes.

correspond to input/output bus cycles are highlighted. Notice that the code M/$\overline{\text{IOD}}$/$\overline{\text{CW}}$/$\overline{\text{R}}$ equals 010 identifies an I/O read bus cycle, and M/$\overline{\text{IOD}}$/$\overline{\text{CW}}$/$\overline{\text{R}}$ equals 011 an I/O write bus cycle. In response to these inputs, the bus control logic section produces bus control signals, such as ALE, IODT/$\overline{\text{R}}$, and $\overline{\text{IODEN}}$. These signals are needed to latch the address and set up the data bus for an input or output data transfer.

The control logic must also generate the I/O read command ($\overline{\text{IORC}}$) and I/O write command ($\overline{\text{IOWC}}$) control signals. $\overline{\text{IORC}}$ is applied directly to the I/O devices and tells them when data are to be input to the MPU. In this case, 32-bit data are always put on the data bus. However, the 80386DX only inputs the appropriate byte, word, or double word. On the other hand, $\overline{\text{IOWC}}$ is gated with the $\overline{\text{BE}}$ signals to produce a separate write enable signal for each byte of the data bus. They are labeled $\overline{\text{IOWR}}_{0-7}$, $\overline{\text{IOWR}}_{8-15}$, $\overline{\text{IOWR}}_{16-23}$, and $\overline{\text{IOWR}}_{24-31}$. These signals are needed to support writing of 8-bit, 16-bit, or 32-bit data through the interface.

The I/O device that is accessed for input or output of data is selected by an *I/O address*. This address is specified as part of the instruction that performs the I/O operation. In the 80386DX architecture, all isolated I/O addresses are 16 bits in length. As shown in Fig. 7.4, they are output to the I/O interface over address bus lines A_2 through A_{15} and byte enable lines $\overline{\text{BE}}_0$ through $\overline{\text{BE}}_3$. The more significant address bits, A_{16} through A_{31}, are held at the 0 logic level during the address period of all I/O bus cycles. The 80386DX signals external circuitry that an I/O address is on the bus by switching its M/$\overline{\text{IO}}$ output to logic 0.

The I/O interface shown in Fig. 7.4 is designed to support 8-, 16-, and 32-bit I/O data transfers. The address on lines A_2 through A_{15} is used to specify the double-word I/O port that is to be accessed. When data are output to word-wide or byte-wide output ports, the logic levels of $\overline{BE_0}$ through $\overline{BE_3}$ determine which ports are enabled for operation.

Notice in the circuit diagram that part of the I/O address that is output on address lines A_2 through A_{15} of the 80386DX is decoded by the I/O address decoder. The bits of the address that are decoded produce I/O chip enable signals for the individual I/O devices. For instance, Fig. 7.4 shows that with three address bits, A_{13} through A_{15}, enough chip enable outputs are produced to select up to eight I/O devices. Notice that the outputs of the I/O address decoder are labeled $\overline{IOCE_0}$ through $\overline{IOCE_7}$. These I/O chip enable signals are latched along with the address and byte enable signals in the I/O address latches. Latching of this information is achieved with a pulse at the ALE output of the bus control logic. If a microcomputer employs a very simple I/O subsystem, it may be possible to eliminate the address decoder and simply use some of the latched high-order address bits as I/O enable signals.

In Fig. 7.4 all of the low-order address bits, A_2 through A_{12}, are shown to be latched and sent directly to the I/O devices. Typically, these address bits are used to select the register within the peripheral device that is to be accessed. For example, with just four of these address lines, we can select any one of 16 registers.

EXAMPLE 7.1

If address bits A_7 through A_{15} are used directly as chip enable signals and address lines A_2 through A_6 are used as register select inputs for the I/O devices, how many I/O devices can be used, and what is the maximum number of registers that each device can contain?

SOLUTION When latched into the address latch, the nine address lines produce the nine I/O chip enable signals $\overline{IOCE_0}$ through $\overline{IOCE_8}$ for I/O devices 0 through 8. The lower five address bits are able to select between 2^5 equal 32 registers for each peripheral IC.

Earlier we indicated that all data transfers between the 80386DX and I/O interface take place over data bus lines D_0 through D_{31}. Data written by the MPU to an I/O device are referred to as *output data* and data read in from an I/O device are called *input data*. The 80386DX can input or output data in byte-wide, word-wide, or double-word-wide format. However, many LSI peripherals used as I/O controllers are designed to interface to an 8-bit bus. For this reason I/O operations frequently involve byte-wide data transfers. Just as for the memory interface, the signals $\overline{BE_0}$ through $\overline{BE_3}$ are used to signal which byte of data is being transferred over the bus. Again logic 0 at $\overline{BE_0}$ identifies that a byte of data is input or output over data bus lines D_0 through D_7. On the other hand, logic 0 at $\overline{BE_3}$ means that a byte-data transfer is taking place over bus lines D_{24} through D_{31}. All byte-wide I/O data transfers are performed in just one bus cycle.

During input and output bus cycles, data are passed between the selected register in the enabled I/O device and the 80386DX over data bus lines D_0 through D_{31}. We just mentioned that most LSI peripherals have a byte-wide data bus. For this reason they are normally attached to the lower part of the data bus. That is, they are connected to data bus lines D_0 through D_7. If this is done in the circuit of Fig. 7.4, all I/O addresses must be scaled by four. This is because the first byte I/O address, which corresponds to a byte transfer across the lower eight data bus lines is 0000_{16}; the next byte address, which represents a byte transfer over D_0 through D_7, is 0004_{16}; the third I/O address is 0008_{16}; and so on. In fact, if only 8-bit peripherals are used in the 80386DX microcomputer system and they are all attached to I/O data bus lines D_0 through D_7, the byte enable signals are not needed in the I/O interface. In this case, address bit A_2 is used as the least significant bit of the I/O address and A_{15} as the most significant bit. Therefore, from a hardware point of view, the I/O address space appears as 16K contiguous byte-wide storage locations over the address range from

$$A_{15} \cdots A_3 A_2 = 0000000000000_2$$

to

$$A_{15} \cdots A_3 A_2 = 1111111111111_2$$

This puts the burden on software to assure that bytes of data are input or output only for addresses that are a multiple of 4 and correspond to a data transfer over data bus lines D_0 through D_7.

Earlier we found that the bus control logic section produced the ALE signal that is used to latch the address. It also supplies the control signals needed to set up the data bus for input and output data transfers. The data bus transceiver/buffers control the direction of data transfers between the 80386DX and I/O devices. They are enabled for operation when their output enable ($\overline{OE}$) inputs are switched to logic 0. Notice that the signal *I/O data bus enable* ($\overline{IODEN}$) is applied to the $\overline{OE}$ inputs.

The direction in which data are passed through the transceivers is determined by the logic level of the DIR input. This input is supplied by the *I/O data transmit/receive* (IODT/$\overline{R}$) output of the bus control logic. During all input cycles, IODT/$\overline{R}$ is logic 0 and the transceivers are set to pass data from the selected I/O device to the 80386DX. On the other hand, during output cycles, IODT/$\overline{R}$ is switched to logic 1 and data pass from the 80386DX to the I/O device.

Another input/output interface diagram is shown in Fig. 7.6. This circuit includes an I/O bank select decoder in the data bus interface. The decoder is used to multiplex the 32-bit data bus of the 80386DX to an 8-bit I/O data bus for connection to 8-bit peripheral ICs. By using this circuit configuration, data can be input from or output to all 64K contiguous byte addresses in the I/O address space. In this case, hardware, instead of software, assures that byte data transfers to consecutive byte I/O addresses are performed to contiguous byte-wide I/O ports. The I/O bank select decoder circuit maps bytes of data from the 32-bit data bus to the 8-bit I/O data bus. It does this by assuring that only one byte enable ($\overline{BE}$) output of the 80386DX is active. That is, it checks to assure that a byte input or output operation is in progress. If more than one of the $\overline{BE}$ inputs of the decoder is active, none of the $\overline{OE}$ outputs of the decoder is produced and data transfer does not take place. Now the I/O address 0000_{16} corresponds to an I/O cycle over data bus lines D_0 through D_7 to an 8-bit peripheral attached to I/O data bus lines, IOD_0 through IOD_7, 0001_{16} corresponds to an input or output of a byte of data for the peripheral over lines D_8 through D_{15}, 0002_{16} represents a byte I/O transfer over D_{16} through D_{23}, and finally, 0003_{16} accompanies a byte transfer over data lines D_{24} through D_{31}. That is, even though the bytes of data for addresses 0000_{16} through 0003_{16} are output by the 80386DX on different parts of its data bus, they are all multiplexed in external hardware to the same 8-bit I/O data bus, IOD_0 through IOD_7. In this way the addresses of the peripheral's registers no longer need to be scaled by 4 in software.

7.4 INPUT AND OUTPUT BUS CYCLE TIMING

In Section 7.3 we found that the isolated I/O interface signals of the 80386DX microcomputer are essentially the same as those involved in the memory interface. In fact, the function, logic levels, and timing of all signals other than M/$\overline{\text{IO}}$ are identical to those already described for the memory interface in Chapter 6.

The timing diagram in Fig. 7.7 shows some *nonpipelined input* and *output bus cycles*. Looking at the waveforms for the first input/output bus cycle, which is called cycle 1, we see that it represents a zero-wait-state input bus cycle. Notice that the byte enable signals, $\overline{BE}_0$ through $\overline{BE}_3$, the address A_2 through A_{15}, the bus cycle definition signals M/$\overline{\text{IO}}$, D/$\overline{\text{C}}$, and W/$\overline{\text{R}}$, and address strobe ($\overline{\text{ADS}}$) signal are all output at the beginning of the T_1 state. This time the 80386DX switches M/$\overline{\text{IO}}$ to logic 0, D/$\overline{\text{C}}$ to 1, and W/$\overline{\text{R}}$ to logic 0 to signal external circuitry that an I/O data input bus cycle is in progress.

As shown in the block diagram of Fig. 7.4, the bus cycle definition code is input to the bus control logic. An input of M/$\overline{\text{IO}}$D/$\overline{\text{C}}$W/$\overline{\text{R}}$ equals 010 initiates an I/O input bus control sequence. Let us continue with the sequence of events that take place in external circuitry during the input cycle. First the bus control logic outputs a pulse to the 1 logic level on ALE. As shown in the circuit of Fig. 7.4, this pulse is used to latch the address information into the I/O address latch devices. The decoded part of the latched address ($\overline{\text{IOCE}}_0$ through $\overline{\text{IOCE}}_7$) selects the I/O device to be accessed, and the code on the lower address lines selects the register that is to be accessed. Later in the bus cycle, $\overline{\text{IORC}}$ is switched to logic 0 to signal the enabled I/O device that data are to be input to the MPU. In response to $\overline{\text{IORC}}$, the enabled input device puts the data from the addressed register onto the data bus. A short time later, IODT/$\overline{\text{R}}$ is switched to logic 0 to set the data bus transceivers to the input direction, and then the transceivers are enabled as $\overline{\text{IODEN}}$ is switched to logic 0. At this point the data from the I/O device are available on the 80386DX's data bus.

In the waveforms of Fig. 7.7 we see that at the end of the T_2 state the 80386DX tests the logic level at its ready input to determine if the I/O bus cycle should be completed or extended with wait states. As shown in Fig. 7.7, $\overline{\text{READY}}$ is at its active 0 logic level when sampled. Therefore, the 80386DX inputs the data off the bus. Finally, the bus control logic returns $\overline{\text{IORC}}$, $\overline{\text{IODEN}}$, and IODT/$\overline{\text{R}}$ to their inactive logic levels and the input bus cycle is finished.

Figure 7.6 Byte-wide I/O interface block diagram.

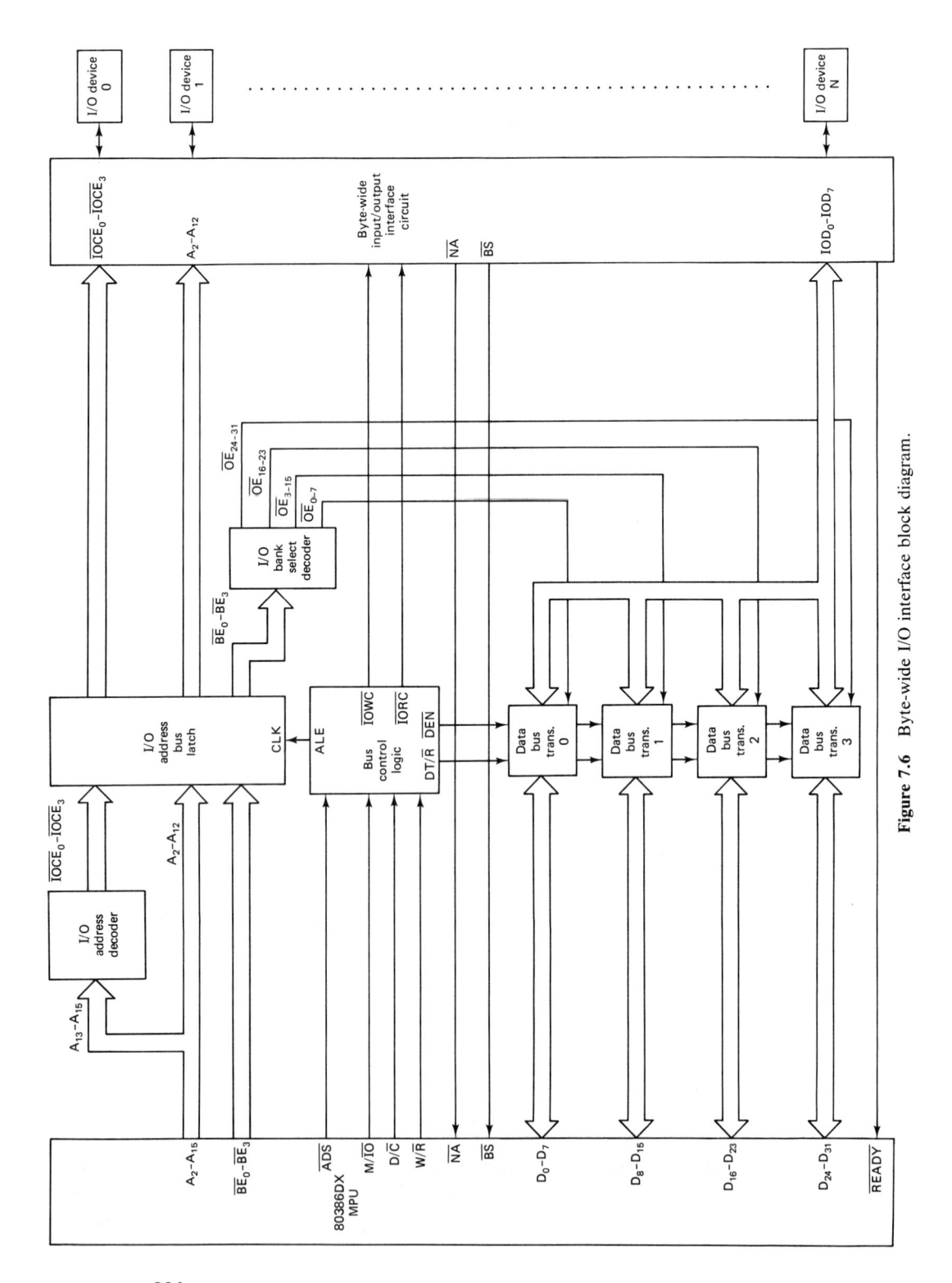

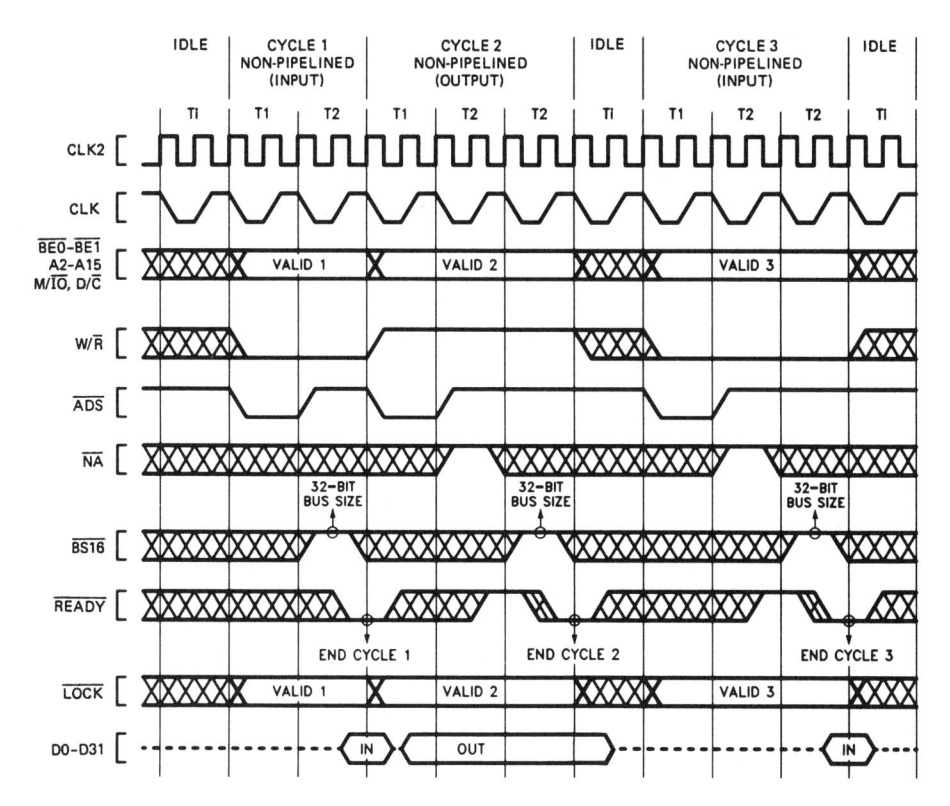

Figure 7.7 I/O read and I/O write bus cycles. (Reprinted by permission of Intel Corp. Copyright Intel Corp. 1987.)

Cycle 3 in Fig. 7.7 is also an input bus cycle. However, looking at the $\overline{\text{READY}}$ waveform, we find that this time it is not logic 0 at the end of the first T_2 state. Therefore, the input cycle is extended with a second T_2 state. Since some of the peripheral devices used with the 80386DX are older, slower devices, it is common to have several wait states in I/O bus cycles.

EXAMPLE 7.2

If the 80386DX that is executing cycle 3 in Fig. 7.7 is running at 20 MHz, what is the duration of this input cycle?

SOLUTION An 80386DX that is running at 20 MHz has a T state equal to 50 ns. Since the input bus cycle takes 3 T states, its duration is 150 ns.

Looking at the output bus cycle, cycle 2 in the timing diagram of Fig. 7.7, we see that the 80386DX puts the data that are to be output onto the data bus at the beginning of ϕ_2 in the T_1 state. This time the bus control logic switches $\overline{\text{IODEN}}$ to logic 0 and maintains IODT/$\overline{\text{R}}$ at the 1 level for transmit mode. From Fig. 7.4 we find that since $\overline{\text{IODEN}}$ is logic 0 and IODT/$\overline{\text{R}}$ is 1, the transceivers are enabled

Sec. 7.4 Input and Output Bus Cycle Timing

and set up to pass data from the 80386DX to the I/O devices. Therefore, the data outputs on the bus are available on the data inputs of the enabled I/O device. Finally, the signal $\overline{\text{IOWC}}$ is switched to logic 0. It is gated with $\overline{\text{BE}}_0$ through $\overline{\text{BE}}_3$ in the I/O bank write control logic to produce the needed bank write enable signals. These signals tell the I/O device that valid output data are on the bus. Now the I/O device must read the data off the bus before the bus control logic terminates the bus cycle. If the device cannot read data at this rate, it can hold $\overline{\text{READY}}$ at the 1 logic level to extend the bus cycle.

> **EXAMPLE 7.3**
>
> If the output bus cycles performed to byte-wide ports by an 80386DX running at 20 MHz are to be completed in a minimum of 250 ns, how many wait states are needed?
>
> **SOLUTION** Since each T state is 50 ns in duration, the bus cycle must last at least
>
> $$\text{number of T states} = \frac{250 \text{ ns}}{50 \text{ ns}} = 5$$
>
> A zero wait state output cycle lasts just two T states; therefore, all output cycles must include three wait states.

The same bus cycle requirements exist for data transfers for I/O ports as were found for memory. That is, all word and double-word data transfers to aligned port addresses take place in just one bus cycle. However, two bus cycles are required to perform data transfers for unaligned 16- or 32-bit I/O ports.

7.5 INPUT/OUTPUT INSTRUCTIONS

Input/output operations are performed by the 80386DX using special input and output instructions together with the I/O port addressing modes. The input and output instructions are listed in Fig. 7.8. This table provides the mnemonic, name,

Mnemonic	Meaning	Format	Operation
IN	Input direct	IN Acc, Port	(Acc) ← (Port) Acc = AL, AX, or EAX
	Input indirect (variable)	IN Acc, DX	(Acc) ← ((DX))
OUT	Output direct	OUT Port, Acc	(Acc) → (Port)
	Output indirect (variable)	OUT DX, Acc	(Acc) → ((DX))
INS	Input string byte	INSB	(ES:DI) ← ((DX)) E(DI) ← E(DI) ± 1
	Input string word	INSW	(ES:DI) ← ((DX)) E(DI) ← E(DI) ± 2
	Input string double	INSD	(ES:DI) ← ((DX)) E(DI) ← E(DI) ± 4
OUTS	Output string byte	OUTSB	(ES:SI) → ((DX)) E(SI) ± 1 → E(SI)
	Output string word	OUTSW	(ES:SI) → ((DX)) E(SI) ± 2 → E(SI)
	Output string double	OUTSD	(ES:SI) → ((DX)) E(SI) ± 4 → E(SI)

Figure 7.8 Input/output instructions.

and a brief description of operation for each instruction. Let us begin by looking at the *input* (IN) and *output* (OUT) instructions in more detail.

Notice that there are two different forms of IN and OUT instructions: the *direct I/O instruction* and *variable I/O instruction*. Either of these two types of instructions can be used to transfer a byte, a word, or a double word of data. All data transfers take place between an I/O device and the 80386DX's accumulator register. For this reason, this method of performing I/O is known as *accumulator I/O*. Byte transfers involve the AL register; word transfers, the AX register; and double-word transfers, the EAX register. In fact, specifying AL as the source or destination register in an I/O instruction indicates that it corresponds to a byte transfer. That is, byte-wide, word-wide, or double-word-wide input/output is selected by specifying the accumulator (Acc) in the instruction as AL, AX, or EAX, respectively.

In a direct I/O instruction, the address of the I/O port is specified as part of the instruction. Eight bits are provided for this direct address. For this reason, its value is limited to the address range from 0_{10} equal 0000_{16} to 255_{10} equal $00FF_{16}$. This range corresponds to page 0 in the I/O address space of Fig. 7.2.

EXAMPLE 7.4

How many aligned double-word addresses exist in page 0?

SOLUTION The number of aligned double-word addresses in page zero is found as

number of aligned double-word addresses = number of bytes in page 0/4

$$= 256/4$$

$$= 64$$

An example is the instruction

```
IN   AL, FEH
```

As shown in Fig. 7.8, execution of this instruction causes the contents of the byte-wide I/O port at address FE_{16} of the I/O address space to be input to the AL register. This data transfer takes place in one input bus cycle.

EXAMPLE 7.5

Write a sequence of instructions that will output FF_{16} to a byte-wide output port at address AB_{16} of the 80386DX's I/O address space.

SOLUTION First the AL register is loaded with a move instruction that includes the value FF_{16} as an immediate operand. This gives

```
MOV   AL, FFH
```

Now the data in AL can be output to the byte-wide output port with the instruction

```
OUT   ABH, AL
```

The difference between the direct and variable I/O instructions lies in the way in which the address of the I/O port is specified. We just saw that for direct I/O instructions, an 8-bit address is directly specified as part of the instruction. On the other hand, variable I/O instructions use an indirect 16-bit address that resides in the DX register within the 80386DX. The value in DX is not an offset. It is the actual physical address that is to be output on the address bus during the I/O bus cycle. Since this address is a full 16 bits in length, variable I/O instructions can access ports located anywhere in the 64K-byte I/O address space.

When using either type of I/O instruction, the data must be loaded into or removed from the AL, AX, or EAX register before another input or output operation can be performed. Moreover, in the case of the variable I/O instructions, the DX register must be loaded with an address. This requires execution of additional instructions. For instance, the instruction sequence

```
MOV  DX, A000H
IN   AL, DX
MOV  BL, AL
```

inputs the contents of the byte-wide input port at $A000_{16}$ of the I/O address space into AL and then saves it in BL.

EXAMPLE 7.6

Write a series of instructions that will output FF_{16} to an output port located at address $B000_{16}$ of the I/O address space.

SOLUTION The DX register must first be loaded with the address of the output port. This is done with the instruction

```
MOV  DX, B000H
```

Next the data that are to be output must be loaded into AL:

```
MOV  AL, FFH
```

Finally, the data are output with the instruction

```
OUT  DX, AL
```

EXAMPLE 7.7

Data are to be read in from two byte-wide input ports at addresses AA_{16} and $A9_{16}$, respectively, and then output as a word to a word-wide output port at address $B000_{16}$. Write a sequence of instructions to perform this input/output operation.

SOLUTION We first read the byte from the port at address AA_{16} into AL and then move it to AH. This is done with the instructions

```
                    IN    AL, AAH
                    MOV   AH, AL
```

Now the other byte can be read into AL by the instruction

```
                    IN    AL, A9H
```

The word is now held in AX. To write out the word of data, we load DX with the address $B000_{16}$ and use a variable output instruction. This leads to the following:

```
                    MOV   DX, B000H
                    OUT   DX, AX
```

In Fig. 7.8 we find that there are also input and output string instructions in the instruction set of the 80386DX. These two instruction forms are not supported in the 8086/8088 software architecture. Using these string instructions, a programmer can either input data from an input port to a storage location directly in memory or output data from a memory location to an output port.

The first instruction, which is called *input string*, can be denoted in three ways: INSB, INSW, or INSD. INSB stands for *input string byte*, INSW means *input string word*, and INSD represents *input string double word*. Let us now look at the operation performed by the INSB instruction. INSB assumes that the address of the input port that is to be accessed is in the DX register. This value must be loaded prior to executing the instruction. Moreover, the address of the memory storage location into which the byte of data is input is identified by the values in ES and DI; that is, when executed, the input operation performed is

$$(ES:DI) \leftarrow ((DX))$$

In the protected mode, DI is replaced by the extended destination index register EDI.

Just as for the other string instructions, the value in DI (or EDI) is either incremented or decremented by one after the data transfer takes place.

$$DI \leftarrow DI \pm 1$$

In this way it points to the next byte-wide storage location in memory. Whether the value in DI is incremented or decremented depends on the setting of the DF flag. Notice in Fig. 7.8 that INSW and INSD perform the same data transfer operation except that since the word or double-word contents of the I/O port are stored in memory, the value in DI (EDI) is incremented or decremented by two or four, respectively.

The INSB instruction performs the operation we just described on one data element, not an array of elements. This basic operation can be repeated to handle a block input operation. Block operations are done by inserting a repeat (REP) prefix in front of the string instruction. The operation performed by REP is described in Fig. 4.28. For example, the instruction

```
                    REP INSW
```

when executed in real mode will cause the contents of the word-wide port pointed to by the I/O address in DX to be input and saved in the memory location at address ES:DI. Then the value in DX is incremented by two (assuming that DF equals 0), the count in CX is decremented by one, and the value in CX is tested to determine if it is zero. As long as the value in CX is not zero, the input operation is repeated. When CX equals zero, all elements of the array have been input and the input string operation is complete. Remember that the count of the number of times the string operation is to be repeated must be loaded into the CX register prior to executing the repeat input string instruction.

In Fig. 7.8 we see that OUTSB, OUTSW, and OUTSD are the three forms of the *output string* instruction. These instructions operate in a similar way to the input string instructions; however, they perform an output operation. For instance, executing OUTSW causes the operation that follows:

$$(ES:SI) \rightarrow ((DX))$$
$$SI \pm 2 \rightarrow SI$$

That is, the word of data held at the memory location pointed to by address ES:SI (ES:ESI in protected mode) is output to the word-wide port pointed to by the I/O address in DX. After the output data transfer is complete, the value in SI (ESI) is either incremented or decremented by two.

An example of an output string instruction that can be used to output an array of data is

REPOUTSB

When executed in the real mode, this instruction causes the data elements of the array of data in memory pointed to by ES:SI to be output one after the other to the output port located at the I/O address in DX. Again, the count in CX defines the size of the array.

When the 80386DX is in the protected-address mode, the input/output instructions can be executed only if the current privilege level is greater than or equal to the I/O privilege level (IOPL). That is, the numerical value of CPL must be lower than or equal to the numerical value of IOPL. Remember that IOPL is defined by the code in bits 12 and 13 of the flags register. If the current privilege level is less than IOPL, the instruction is not executed; instead, a general protection fault occurs. The general protection fault is an example of an 80386DX exception and will be examined in more detail in Chapter 8.

In Chapter 5 we indicated that the task state segment (TSS) of a task includes a section known as the *I/O permission bit map*. This I/O permission bit map provides a second protection mechanism for the protected-mode I/O address space. Remember that the size of the TSS segment is variable. Its size is specified by the limit in the TSS descriptor. Figure 7.9 shows a typical task state segment. Here we see that the 16-bit *I/O map base* offset, which is held at word offset 66_{16} in the TSS, identifies the beginning of the I/O permission bit map. The upper end of the bit map is set by the limit field in the descriptor for the TSS. Let us now look at what the bits in the I/O permission bit map stand for.

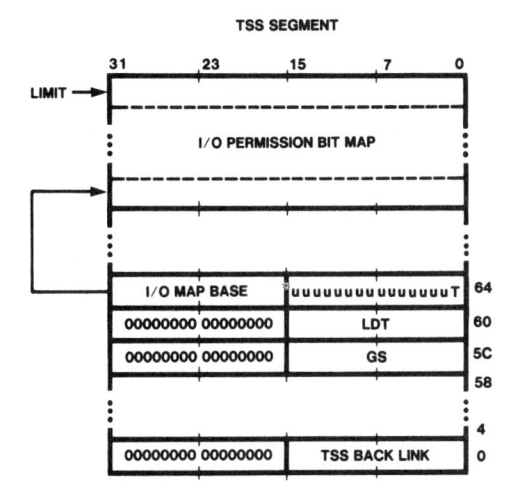

Figure 7.9 Location of the I/O permission bit map in the TSS. (Reprinted by permission of Intel Corp. Copyright Intel Corp. 1986.)

Figure 7.10 shows a more detailed representation of the I/O permission bit map. Notice that it contains one bit position for each of the 65,536 byte-wide I/O ports in the 80386DX's I/O address space. In the bit map we find that the bit position that corresponds to I/O port 0 (I/O address 0000_{16}) is the least significant bit at the address defined with the I/O bit map base offset. The rest of the bits in this first double word in the map represent I/O ports 1 through 31. Finally, the last bit in the table, which corresponds to port 65,535 and I/O address $FFFF_{16}$, is the most significant bit in the double word located at an offset of $1FFC_{16}$ from the I/O bit map base. In Fig. 7.10 we see that the byte address that follows the map must always contain FF_{16}. This is the least significant byte in the last double

Figure 7.10 Contents of the I/O permission bit map.

Sec. 7.5 Input/Output Instructions

word of the TSS. The value of the I/O map base offset must be less than $DFFF_{16}$; otherwise, the complete map may not fit within the TSS.

Using this bit map, restrictions can be put on input/output operations to each of the 80386DX's 65,536 I/O port addresses. In protected mode, the bit for the I/O port in the I/O permission map is checked only if the CPL when the I/O instruction is executed is less privileged than the IOPL. If logic 0 is found in a bit position, it means that an I/O operation can be performed to the port address. On the other hand, logic 1 inhibits the I/O operation. Any attempt to input or output data for an I/O address marked with a 1 in the I/O permission bit map by code with a CPL that is less privileged than the IOPL results in a general protection exception. In this way an operating system can detect attempts to access certain I/O devices and trap to special service routines for the devices through the general protection exception. In virtual 8086 mode, all I/O accesses reference the I/O permission bit map.

The I/O permission configuration defined by a bit map only applies to the task that uses the TSS. For this reason, many different I/O configurations can exist within a protected-mode software system. Actually, a different bit map could be defined for every task.

In practical applications, most tasks would use the same I/O permission bit map configuration. In fact, in some applications not all I/O addresses need to be protected with the I/O permission bit map. It turns out that any bit map position that is located beyond the limit of the TSS is interpreted as containing a 1. Therefore, all accesses to an I/O address that corresponds to a bit position beyond the limit of the TSS will produce a general protection exception. For instance, a protected-mode I/O address space may be set up with a small block of I/O address to which access is permitted at the low end of the I/O address space and with access to the rest of the I/O address space restricted. A smaller table can be set up to specify this configuration. By setting the values of the bit map base and TSS limit such that the bit positions for all the restricted addresses fall beyond the end of the TSS segment, they are caused to result in an exception. On the other hand, the bit positions that are located within the table are all made 0 to permit I/O accesses to their corresponding ports. Moreover, if the complete I/O address space is to be restricted for a task, the I/O permission map base address can simply be set to a value greater than the TSS limit.

7.6 82C55A PROGRAMMABLE PERIPHERAL INTERFACE (PPI)

The *82C55A* is an LSI peripheral designed to permit easy implementation of *parallel I/O* in the 80386DX microcomputer. It provides a flexible parallel interface which includes features such as: single-bit, 4-bit, and byte-wide input and output ports; level-sensitive inputs; latched outputs; strobed inputs or outputs; and strobed bidirectional input/outputs. These features are selected under software control.

A block diagram of the 82C55A is shown in Fig. 7.11(a) and its pin layout in Fig. 7.11(b). The left side of the block represents the *microprocessor interface*.

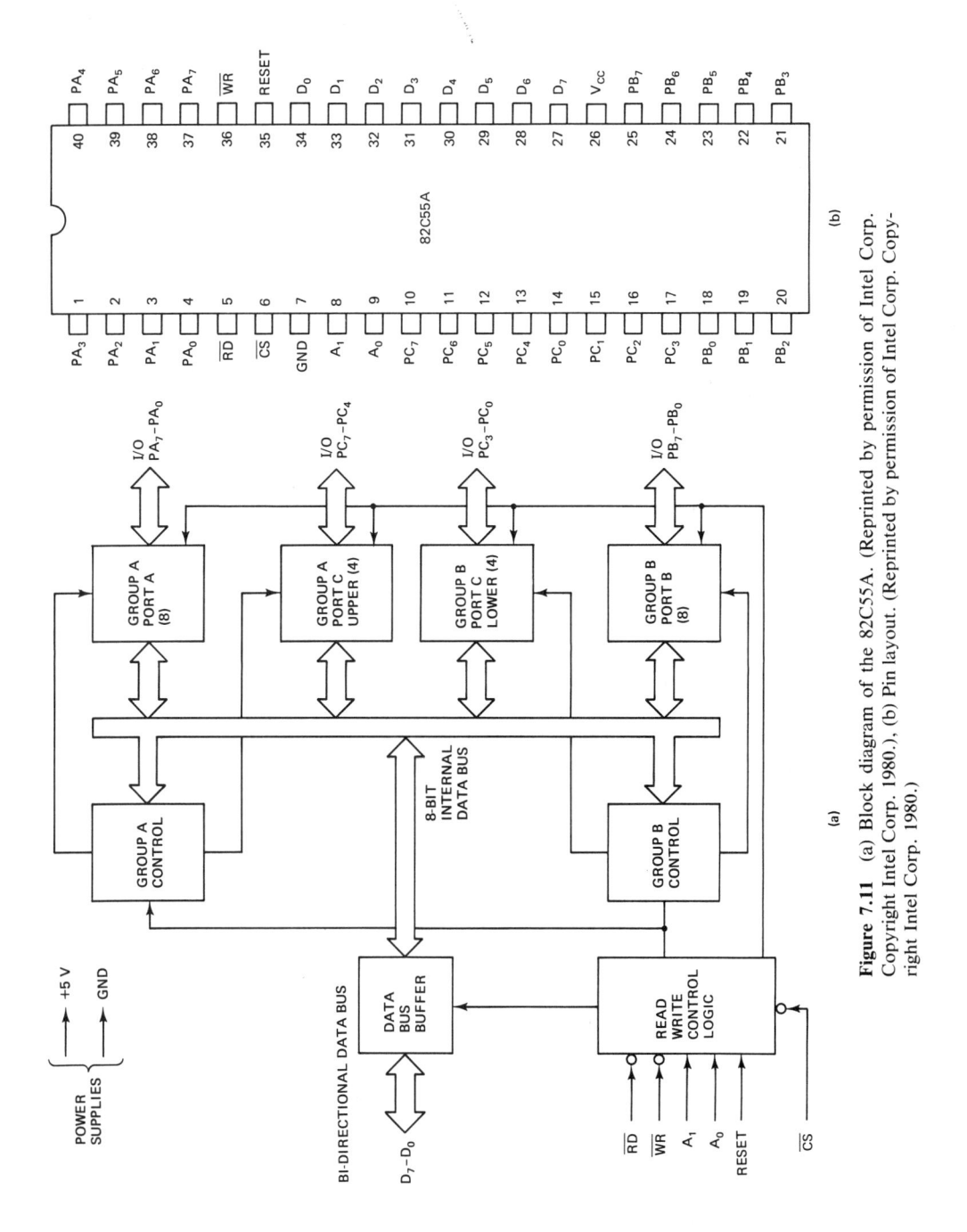

Figure 7.11 (a) Block diagram of the 82C55A. (Reprinted by permission of Intel Corp. Copyright Intel Corp. 1980.), (b) Pin layout. (Reprinted by permission of Intel Corp. Copyright Intel Corp. 1980.)

It includes an *8-bit bidirectional data bus* D_0 through D_7. Over these lines, commands, status information, and data are transferred between the 80386DX and 82C55A. These data are transferred whenever the 80386DX performs an input or output bus cycle to an address of a register within the device. Timing of the data transfers to the 82C55A is controlled by the *read/write* ($\overline{RD}$ and $\overline{WR}$) *control* signals.

The source or destination register within the 82C55A is selected by a 2-bit *register select code*. The 80386DX must apply this code to the *register select inputs*, A_0 and A_1 of the 82C55A. The *PORT A*, *PORT B*, and *PORT C* registers correspond to codes $A_1A_0 = 00$, $A_1A_0 = 01$, and $A_1A_0 = 10$, respectively.

Two other signals are shown on the microprocessor interface side of the block diagram. They are the *reset* (RESET) and *chip select* ($\overline{CS}$) inputs. $\overline{CS}$ must be logic 0 during all read or write operations to the 82C55A. It enables the microprocessor interface circuitry for an input or output operation.

On the other hand, RESET is used to initialize the device. Switching it to logic 0 at power-up causes the internal registers of the 82C55A to be cleared. *Initialization* configures all I/O ports for input mode of operation.

The other side of the block corresponds to three *byte-wide I/O ports*. They are called PORT A, PORT B, and PORT C and represent *I/O lines* PA_0 through PA_7, PB_0 through PB_7, and PC_0 through PC_7, respectively. These ports can be configured for input or output operation. This gives a total of 24 I/O lines.

We already mentioned that the operating characteristics of the 82C55A can be configured under software control. It contains an 8-bit internal control register for this purpose. This register is represented by the *group A* and *group B control blocks* in Fig. 7.11(a). Logic 0 or 1 can be written to the bit positions in this register to configure the individual ports for input or output operation and to enable one of its three modes of operation. The control register is write only and its contents are modified under software control by initiating a write bus cycle to the 82C55A with register select code $A_1A_0 = 11$.

The bits of the control register and their control functions are shown in Fig. 7.12. Here we see that bits D_0 through D_2 correspond to the group B control block in the diagram of Fig. 7.11(a). Bit D_0 configures the lower four lines of PORT C for input or output operation. Notice that logic 1 at D_0 selects input operation and logic 0 selects output operation. The next bit, D_1, configures PORT B as an 8-bit-wide input or output port. Again, logic 1 selects input operation and logic 0 selects output operation.

The D_2 bit is the mode select bit for PORT B and the lower 4 bits of PORT C. It permits selection of one of two different modes of operation called *MODE 0* and *MODE 1*. Logic 0 in bit D_2 selects MODE 0, while logic 1 selects MODE 1. These modes will be discussed in detail shortly.

The next four bits in the control register, D_3 through D_6, correspond to the group A control block in Fig. 7.11(a). Bits D_3 and D_4 of the control register are used to configure the operation of the upper half of PORT C and all of PORT A. These bits work in the same way as D_0 and D_1 configure the lower half of PORT C and PORT B. However, there are now two mode select bits D_5 and D_6 instead

CONTROL WORD

| D₇ | D₆ | D₅ | D₄ | D₃ | D₂ | D₁ | D₀ |

GROUP B

PORT C (LOWER)
1 = INPUT
0 = OUTPUT

PORT B
1 = INPUT
0 = OUTPUT

MODE SELECTION
0 = MODE 0
1 = MODE 1

GROUP A

PORT C (UPPER)
1 = INPUT
0 = OUTPUT

PORT A
1 = INPUT
0 = OUTPUT

MODE SELECTION
00 = MODE 0
01 = MODE 1
1X = MODE 2

MODE SET FLAG
1 = ACTIVE

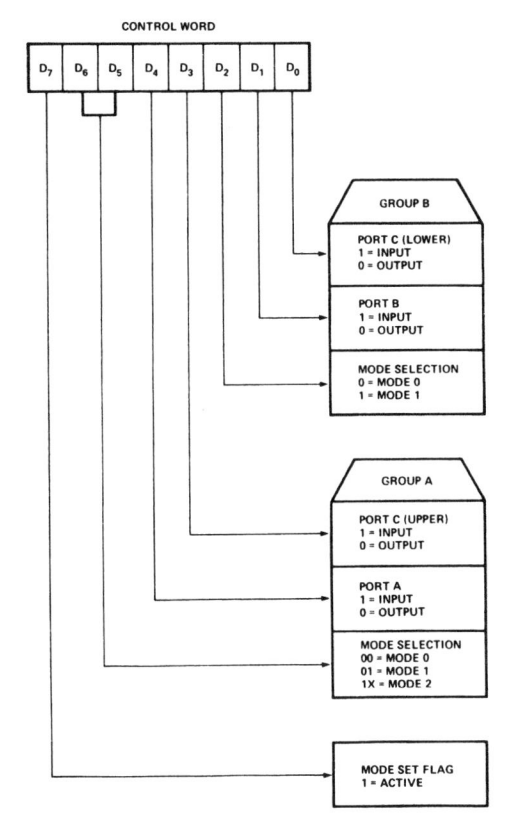

Figure 7.12 Control word bit functions. (Reprinted by permission of Intel Corp. Copyright Intel Corp. 1980.)

of just 1. They are used to select between three modes of operation known as *MODE 0*, *MODE 1*, and *MODE 2*.

The last control register bit, D_7, is the *mode set flag*. It must be at logic 1 (active) whenever the mode of operation is to be changed.

MODE 0 selects what is called *simple I/O operation*. By "simple I/O" we mean that the lines of the port can be configured as level-sensitive inputs or latched outputs. To set all ports for this mode of operation, load bit D_7 of the control register with logic 1, bits $D_6 D_5 = 00$, and $D_2 = 0$. Logic 1 at D_7 represents an active mode set flag. Now PORT A and PORT B can be configured as 8-bit input or output ports and PORT C can be configured for operation as two independent 4-bit input or output ports. This is done by setting or resetting bits D_4, D_3, D_1, and D_0.

For example, if $80_{16} = 10000000_2$ is written to the control register, the 1 in D_7 activates the mode set flag. MODE 0 operation is selected for all three ports because bits D_6, D_5, and D_2 are logic 0. At the same time, the zeros in D_4, D_3, D_1, and D_0 set up all port lines to work as outputs. This configuration is illustrated in Fig. 7.13(a).

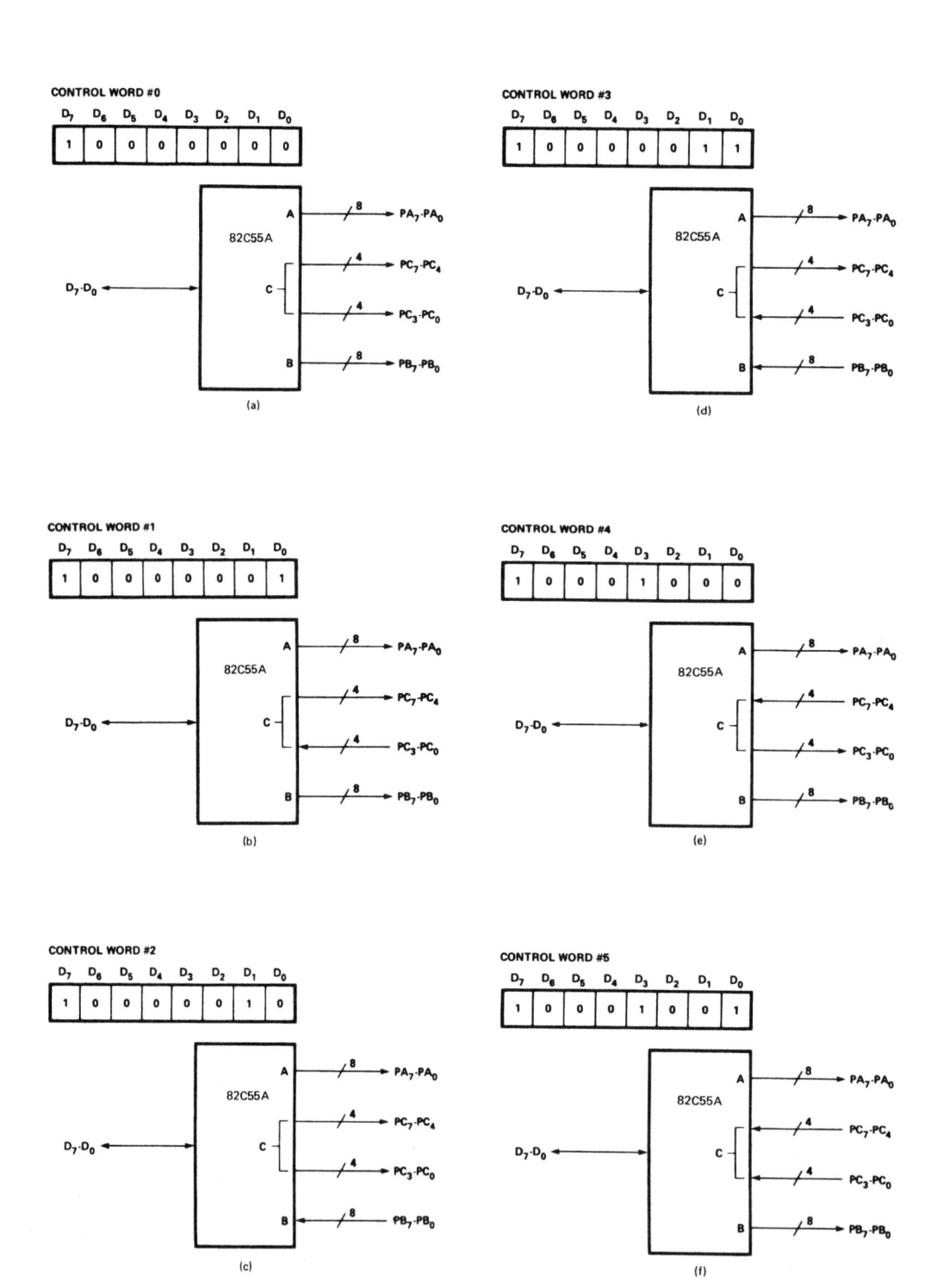

Figure 7.13 Mode 0 control words and corresponding input/output configuration. (Reprinted by permission of Intel Corp. Copyright Intel Corp. 1980.)

CONTROL WORD #6

D_7	D_6	D_5	D_4	D_3	D_2	D_1	D_0
1	0	0	0	1	0	1	0

(g)

CONTROL WORD #7

D_7	D_6	D_5	D_4	D_3	D_2	D_1	D_0
1	0	0	0	1	0	1	1

(h)

CONTROL WORD #8

D_7	D_6	D_5	D_4	D_3	D_2	D_1	D_0
1	0	0	1	0	0	0	0

(i)

CONTROL WORD #9

D_7	D_6	D_5	D_4	D_3	D_2	D_1	D_0
1	0	0	1	0	0	0	1

(j)

CONTROL WORD #10

D_7	D_6	D_5	D_4	D_3	D_2	D_1	D_0
1	0	0	1	0	0	1	0

(k)

CONTROL WORD #11

D_7	D_6	D_5	D_4	D_3	D_2	D_1	D_0
1	0	0	1	0	0	1	1

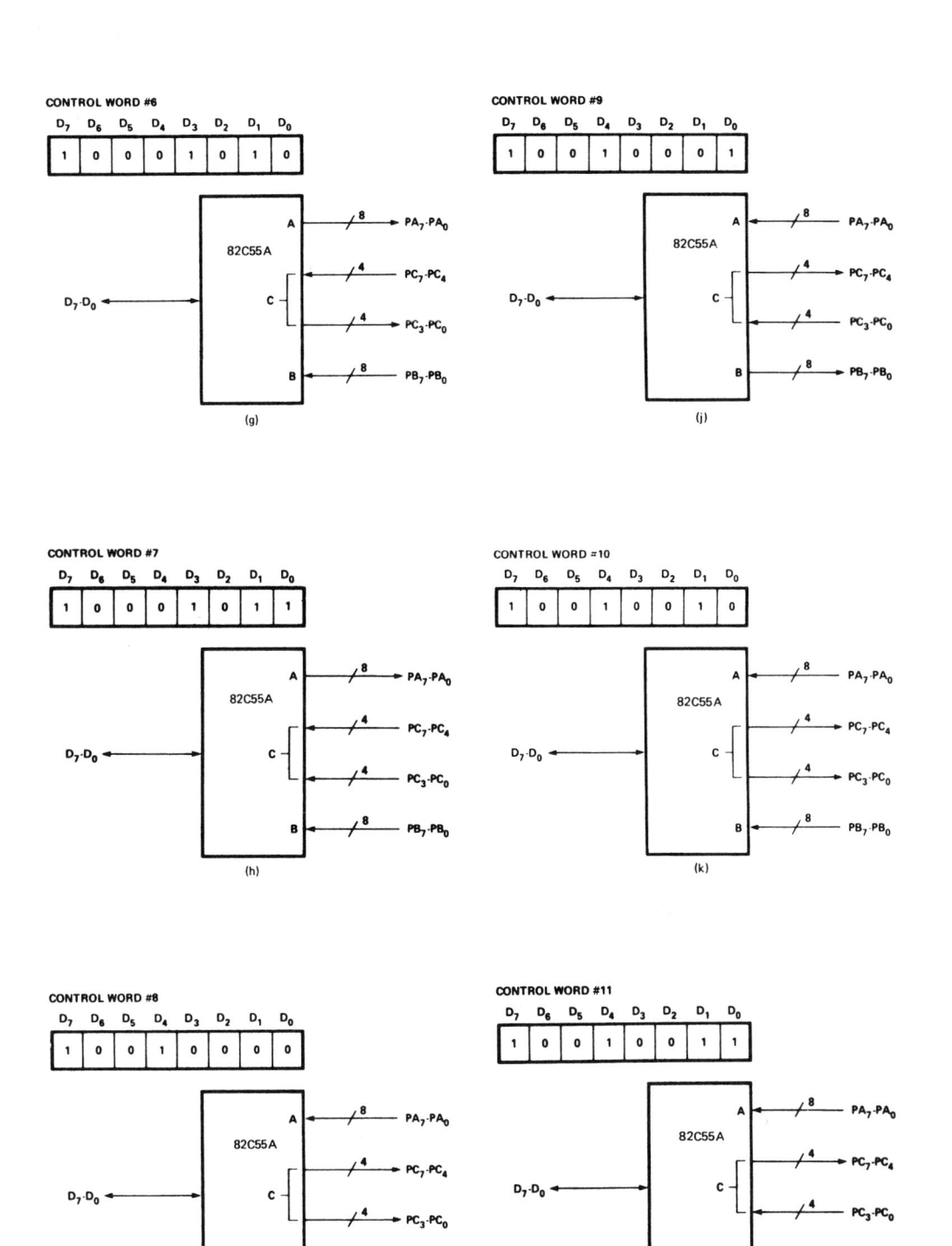

(l)

Figure 7.13 (*Continued*)

CONTROL WORD #12

D₇	D₆	D₅	D₄	D₃	D₂	D₁	D₀
1	0	0	1	1	0	0	0

(m)

CONTROL WORD #14

D₇	D₆	D₅	D₄	D₃	D₂	D₁	D₀
1	0	0	1	1	0	1	0

(o)

CONTROL WORD #13

D₇	D₆	D₅	D₄	D₃	D₂	D₁	D₀
1	0	0	1	1	0	0	1

(n)

CONTROL WORD #15

D₇	D₆	D₅	D₄	D₃	D₂	D₁	D₀
1	0	0	1	1	0	1	1

(p)

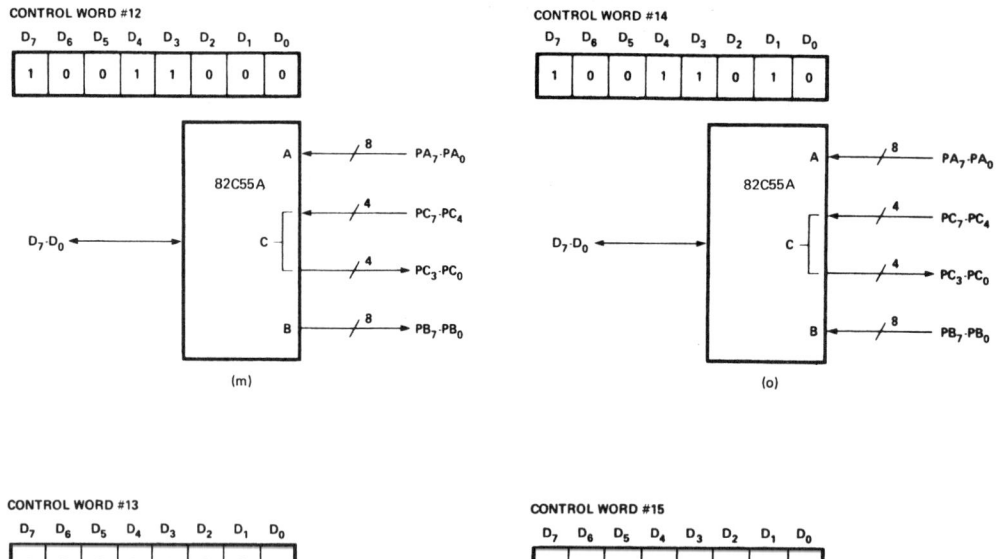

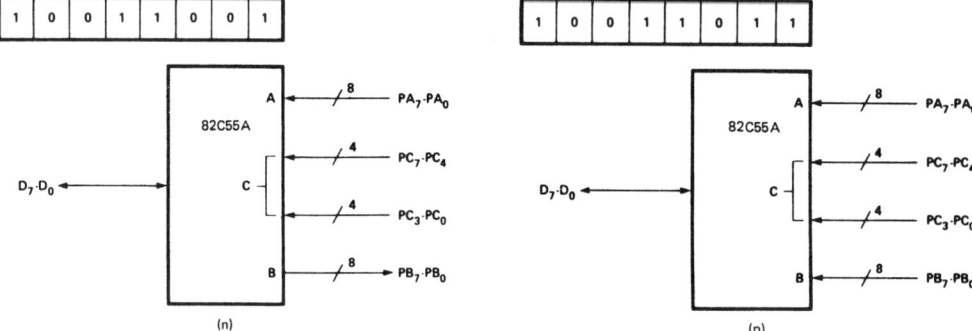

Figure 7.13 (*Continued*)

By writing different binary combinations into bit locations D_4, D_3, D_1, and D_0, any one of 16 different MODE 0 I/O configurations can be obtained. The control word and I/O setup for the rest of these combinations are shown in Fig. 7.13(b) through (p).

EXAMPLE 7.8

What is the mode and I/O configuration for ports A, B, and C of an 82C55A after its control register is loaded with 82_{16}?

SOLUTION Expressing the control register contents in binary form, we get

$$D_7D_6D_5D_4D_3D_2D_1D_0 = 10000010_2$$

Since D_7 is 1, the modes of operation of the ports are selected by the control word. The three least significant bits of the word configure PORT B and the lower four bits of PORT C. They give

$D_0 = 0$ Lower four bits of PORT C are outputs

$D_1 = 1$ PORT B are inputs

$D_1 = 0$ MODE 0 operation for both PORT B and the lower four bits of PORT C

The next four bits configure the upper part of PORT C and PORT A.

$D_3 = 0$ Upper four bits of PORT C are outputs

$D_4 = 0$ PORT A are outputs

$D_6 D_5 = 00$ MODE 0 operation for both PORT A and the upper part of PORT C

This MODE 0 I/O configuration is shown in Fig. 7.13(c).

MODE 1 operation represents what is known as *strobed I/O*. The ports of the 82C55A are put into this mode of operation by setting $D_7 = 1$ to activate the mode set flag and setting $D_6 D_5 = 01$ and $D_2 = 1$.

In this way the A and B ports are configured as two independent *byte-wide I/O ports* each of which has a *4-bit control/data port* associated with it. The control/data ports are formed from the lower and upper nibbles of PORT C, respectively.

When configured in this way, data applied to an input port must be strobed in with a signal produced in external hardware. Moreover, an output port is provided with handshake signals that indicate when new data are available at its outputs and when an external device has read these values.

As an example, let us assume for the moment that the control register of an 82C55A is loaded with $D_7 D_6 D_5 D_4 D_3 D_2 D_1 D_0 = 10111XXX$. This configures PORT A as a MODE 1 input port. Figure 7.14(a) shows the function of the signal lines for this example. Notice that PA_7 through PA_0 form an 8-bit input port. On the other hand, the function of the upper PORT C leads are reconfigured to provide the PORT A control/data lines. The PC_4 line becomes $\overline{STB}_A$ (*strobe input*), which is used to strobe data at PA_7 through PA_0 into the input latch. Moreover, PC_5 becomes IBF_A (*input buffer full*). Logic 1 at this output indicates to external circuitry that a word has already been strobed into the latch.

The third control signal is at PC_3 and is labeled $INTR_A$ (*interrupt request*). It switches to logic 1 as long as $\overline{STB}_A = 1$, $IBF_A = 1$, and an internal signal $INTE_A$ (*interrupt enable*) equals 1. $INTE_A$ is set to logic 0 or 1 under software control by using the bit set/reset feature of the 82C55A. Looking at Fig. 7.14(a) we see that logic 1 in $INTE_A$ enables the logic level of IBF_A to the $INTR_A$ output. This signal can be applied to an interrupt input of the 80386DX microcomputer to signal it that the new data are available at the input port. The corresponding interrupt service routine can read the data and clear the interrupt request.

As another example, let us assume that the contents of the control register are changed to $D_7 D_6 D_5 D_4 D_3 D_2 D_1 D_0 = 10100XXX$. This I/O configuration is shown in Fig. 7.14(b). Notice that PORT A is now configured for output operation

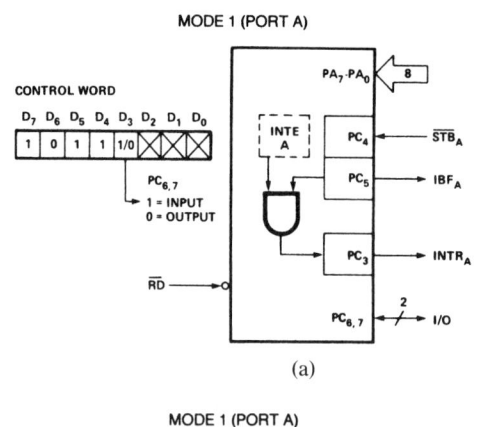

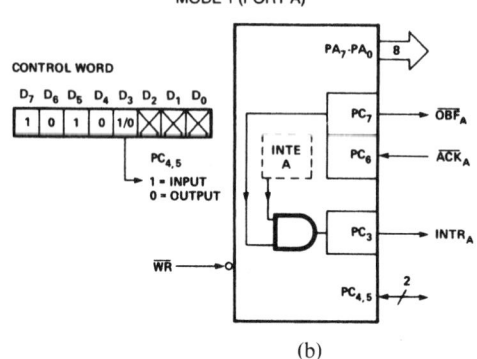

Figure 7.14 (a) Mode 1 port A input configuration. (Reprinted by permission of Intel Corp. Copyright Intel Corp. 1980.), (b) Mode 1 port A output configuration. (Reprinted by permission of Intel Corp. Copyright Intel Corp. 1980.)

instead of input operation. PA₇ through PA₀ are now an 8-bit output port. The control line at PC₇ is $\overline{OBF}_A$ (*output buffer full*). When data have been written into the output port, $\overline{OBF}_A$ switches to the 0 logic level. In this way it signals external circuitry that new data are available at the port outputs.

Signal line PC₆ becomes $\overline{ACK}_A$ (*acknowledge*), which is an input. An external device can signal the 82C55A that it has accepted the data provided at the output port by switching this input to logic 0. The last signal at the control port is output INTR_A (*interrupt request*), which is produced at the PC₃ lead. This output is switched to logic 1 when the $\overline{ACK}_A$ input is active. It is used to signal the 80386DX with an interrupt that indicates that an external device has accepted the data from the outputs. INTR_A switches to the 1 level when $\overline{OBF}_A = 1$, $\overline{ACK}_A = 0$, and INTE_A = 1. Again the interrupt enable (INTE_A) bit must be set to 1 under software control.

EXAMPLE 7.9

Figure 7.15(a) and (b) show how PORT B can be configured for MODE 1 operation. Describe what happens in Fig. 7.15(a) when the $\overline{STB}_B$ input is pulsed to logic 0. Assume that INTE_B is already set to 1.

SOLUTION As $\overline{STB}_B$ is pulsed, the byte of data at PB_7 through PB_0 are latched into the PORT B register. This causes the IBF_B output to switch to 1. Since $INTE_B$ is 1, $INTR_B$ also switches to logic 1.

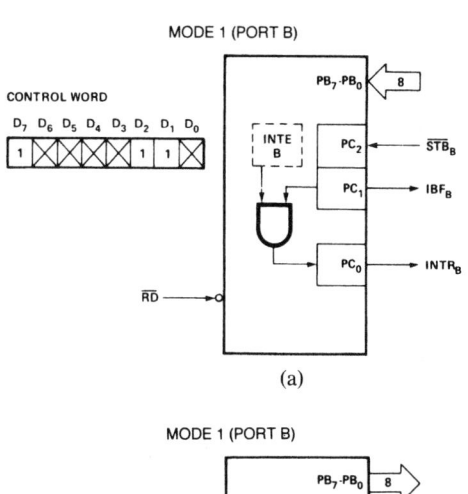

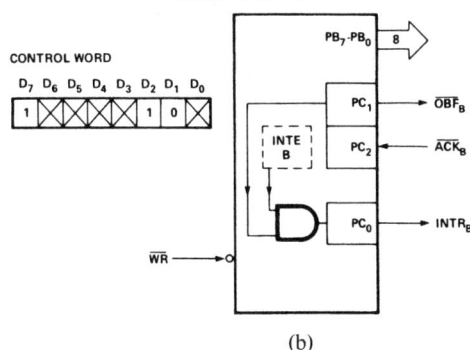

Figure 7.15 (a) Mode 1 port B input configuration. (Reprinted by permission of Intel Corp. Copyright Intel Corp. 1980.), (b) Mode 1 port B output configuration. (Reprinted by permission of Intel Corp. Copyright Intel Corp. 1980.)

The last mode of operation, MODE 2, represents what is known as *strobed bidirectional I/O*. The key difference is that now the port works as either inputs or outputs and control signals are provided for both functions. Only PORT A can be configured to work in this way.

To set up this mode, the control register is set to $D_7D_6D_5D_4D_3D_2D_1D_0$ = 11XXXXXX. The I/O configuration that results is shown in Fig. 7.16. Here we find that PA_7 through PA_0 operate as an *8-bit bidirectional port* instead of a unidirectional port. Its control signals are $\overline{OBF}_A$ at PC_7, $\overline{ACK}_A$ at PC_6, $\overline{STB}_A$ at PC_4, IBF_A at PC_5, and $INTR_A$ at PC_3. Their functions are similar to those already discussed for MODE 1. One difference is that $INTR_A$ is produced by either gating $\overline{OBF}_A$ with $INTE_1$ or IBF_A with $INTE_2$.

In our discussion of MODE 1, we mentioned that the *bit set/reset* feature could be used to set the INTE bit to logic 0 or 1. This feature also allows the individual bits of PORT C to be set or reset. To do this we write logic 0 to bit D_7 of the control register. This resets the bit set/reset flag. The logic level that is to

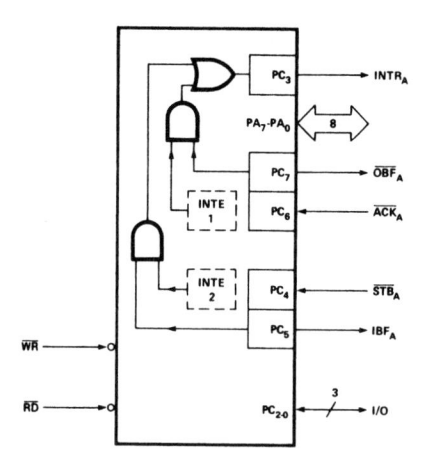

Figure 7.16 Mode 2 input/output configuration. (Reprinted by permission of Intel Corp. Copyright Intel Corp. 1980.)

be latched at a PORT C line is included as bit D_0 of the control word. This value is latched at the I/O line of PORT C, which corresponds to the three-bit code at $D_3D_2D_1$.

The relationship between the set/reset control word and input/output lines is illustrated in Fig. 7.17. For instance, writing $D_7D_6D_5D_4D_3D_2D_1D_0 = 00001111_2$ into the control register of the 82C55A selects bit 7 and sets it to 1. Therefore, output PC_7 at PORT C is switched to the 1 logic level.

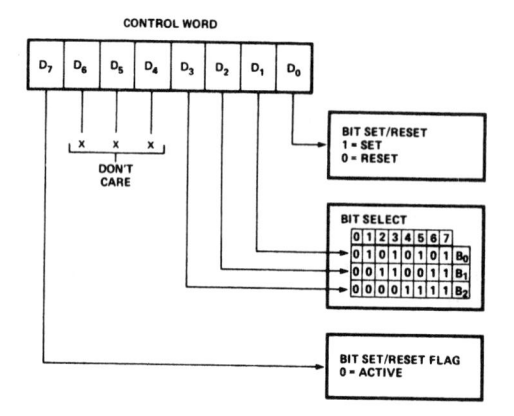

Figure 7.17 Bit set/reset format. (Reprinted by permission of Intel Corp. Copyright Intel Corp. 1980.)

EXAMPLE 7.10

The interrupt control flag $INTE_A$ is controlled by bit set/reset of PC_6. What command code must be written to the control register of the 82C55A to set its value to logic 1?

SOLUTION To use the set/reset feature, D_7 must be logic 0. Moreover, $INTE_A$ is to be set to logic 1; therefore D_0 must be logic 1. Finally, to select PC_6, the code at bits $D_3D_2D_1$ must be 110. The rest of the bits are don't-care states. This gives the control word

$$D_7D_6D_5D_4D_3D_2D_1D_0 = 0XXX1101_2$$

Replacing the don't-care states with the 0 logic level, we get

$$D_7D_6D_5D_4D_3D_2D_1D_0 = 00001101_2 = 0D_{16}$$

We have just described and given examples of each of the modes of operation that can be assigned to the ports of the 82C55A. In practice the A and B ports are frequently configured with different modes. For example, Fig. 7.18(a) shows the control word and port configuration of an 82C55A set up for bidirectional MODE 2 operation of PORT A and input MODE 0 operation of PORT B.

EXAMPLE 7.11

What control word must be written into the control register of the 82C55A such that PORT A is configured for bidirectional operation and PORT B is set up with MODE 1 outputs?

SOLUTION To configure the operating mode of the ports of the 82C55A, D_7 must be 1.

$$D_7 = 1$$

PORT A is set up for bidirectional operation by making D_6 logic 1. In this case, D_5 through D_3 are don't-care states.

$$D_6 = 1$$
$$D_5D_4D_3 = XXX$$

MODE 1 is selected for PORT B by logic 1 in bit D_2 and output operation by logic 0 in D_1. Since MODE 1 operation has been selected, D_0 is a don't-care state.

$$D_2 = 1$$
$$D_1 = 0$$
$$D_0 = X$$

This gives the control word

$$D_7D_6D_5D_4D_3D_2D_1D_0 = 11XXX10X_2$$

Assuming logic 0 for the don't-care states, we get

$$D_7D_6D_5D_4D_3D_2D_1D_0 = 11000100_2 = C4_{16}$$

This configuration is shown in Fig. 7.18(b).

EXAMPLE 7.12

Write the sequence of instructions needed to load the control register of an 82C55A with the control word formed in Example 7.11. Assume that the 82C55A resides at address $0F_{16}$ of the I/O address space and that the microcomputer uses the byte-wide I/O interface illustrated in Fig. 7.6.

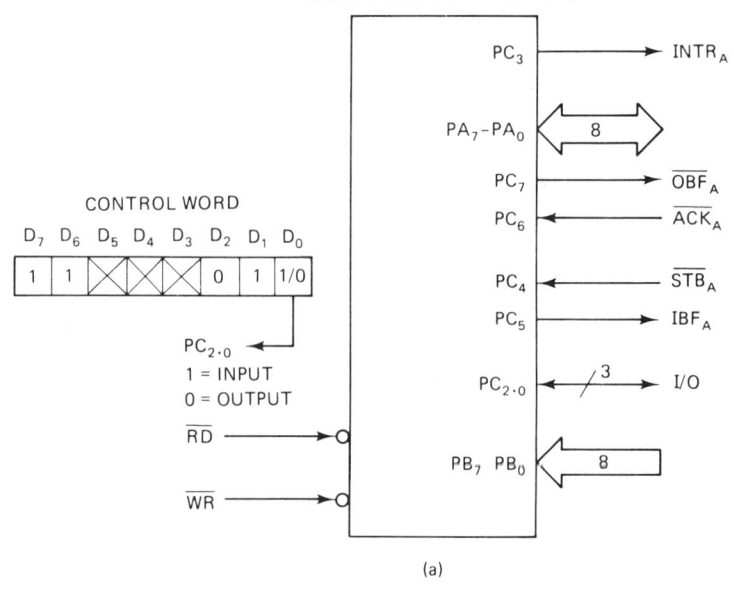

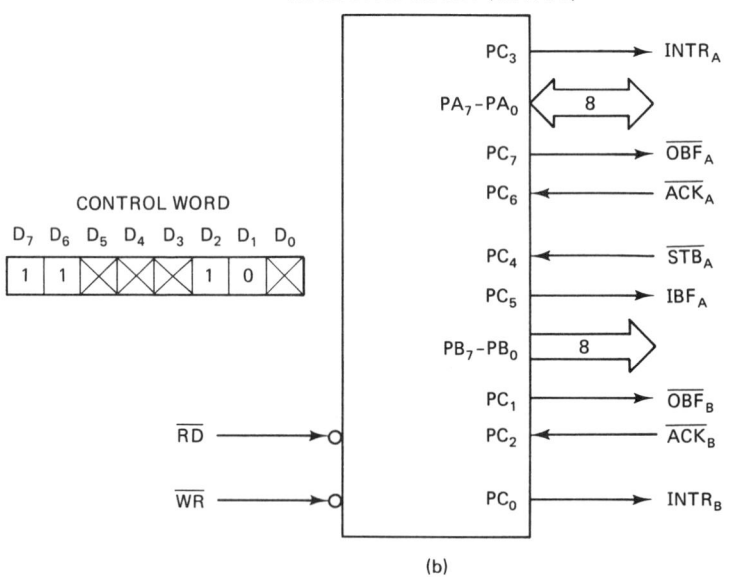

Figure 7.18 (a) Combined mode 2 and mode 0 (input) control word and I/O configuration. (Reprinted by permission of Intel Corp. Copyright Intel Corp. 1980.) (b) Combined mode 2 and mode 1 (output) control word and I/O configuration. (Reprinted by permission of Intel Corp. Copyright Intel Corp. 1980.).

324   Input/Output Interface of the 80386DX Microprocessor   Chap. 7

SOLUTION First we must load AL with $C4_{16}$. This is the value of the control word that is to be output to the control register at address $0F_{16}$. The move instruction used to load AL is

```
MOV  AL, C4H
```

This data are output to the control register with the OUT instruction

```
OUT  0FH, AL
```

7.7 IMPLEMENTING ISOLATED I/O PARALLEL INPUT/OUTPUT PORTS USING THE 82C55A

The circuit in Fig. 7.19 shows how 82C55A PPI devices can be connected to the bus of the 80386DX to implement isolated parallel input/output ports. Here we find that four groups, each with up to eight 82C55As, are connected to the data bus. Notice that each group has its own address decoder. The output of the decoder selects one of the devices in the group at a time for input or output of data. Each of these PPI devices provides up to three byte-wide ports. In the circuit they are labeled port A, port B, and port C. These ports can be individually configured as inputs or outputs through software. Therefore, each group is capable of implementing up to 192 individual I/O lines, for a total of 768 I/O lines.

Let us look more closely at the group 3 decoder circuit. Starting with the inputs of the 74F138 address decoder, we see that its enable inputs are $\overline{G}_{2B} = \overline{BE}_2$, $\overline{G}_{2A} = M/\overline{IO}$, and $G_1 = 1$. Whenever the 80386DX outputs an I/O address on the bus, $M/\overline{IO}$ is logic 0 and if $\overline{BE}_3$ also equals 0, the decoder is enabled for operation. Notice that the other three group decoders are connected in a similar way, except that their $\overline{G}_{2B}$ input is enabled by a different $\overline{BE}$ signal. For example, the $\overline{G}_{2B}$ input of the group 1 decoder is driven by $\overline{BE}_1$.

When the group 3 decoder is enabled for operation, the code at the CBA inputs causes one of the eight 82C55A PPIs attached to its outputs to be enabled for operation. Bits A_4 through A_6 of the I/O address are applied to these inputs of the decoder. When an I/O address is on the bus, the decoder responds by switching the output corresponding to this three-bit code to the 0 logic level. Decoder outputs O_0 through O_7 are applied to the chip select ($\overline{CS}$) inputs of the PPIs in group 3. For instance, $A_6A_5A_4 = 000$ switches output O_0 to logic 0. This enables the first 82C55A device, which is numbered 3 in Fig. 7.19.

At the same time that PPI 3 is selected, the two-bit code A_3A_2 at inputs A_1A_0 selects the port for which data are input or output. For example, A_3A_2 equal 00 indicates that port A is to be accessed. The input or output data transfer to this port takes place over data bus lines D_{24} through D_{31}. Timing of the read or write transfer is controlled by the signals $\overline{IORC}$ and $\overline{IOWC}$, respectively.

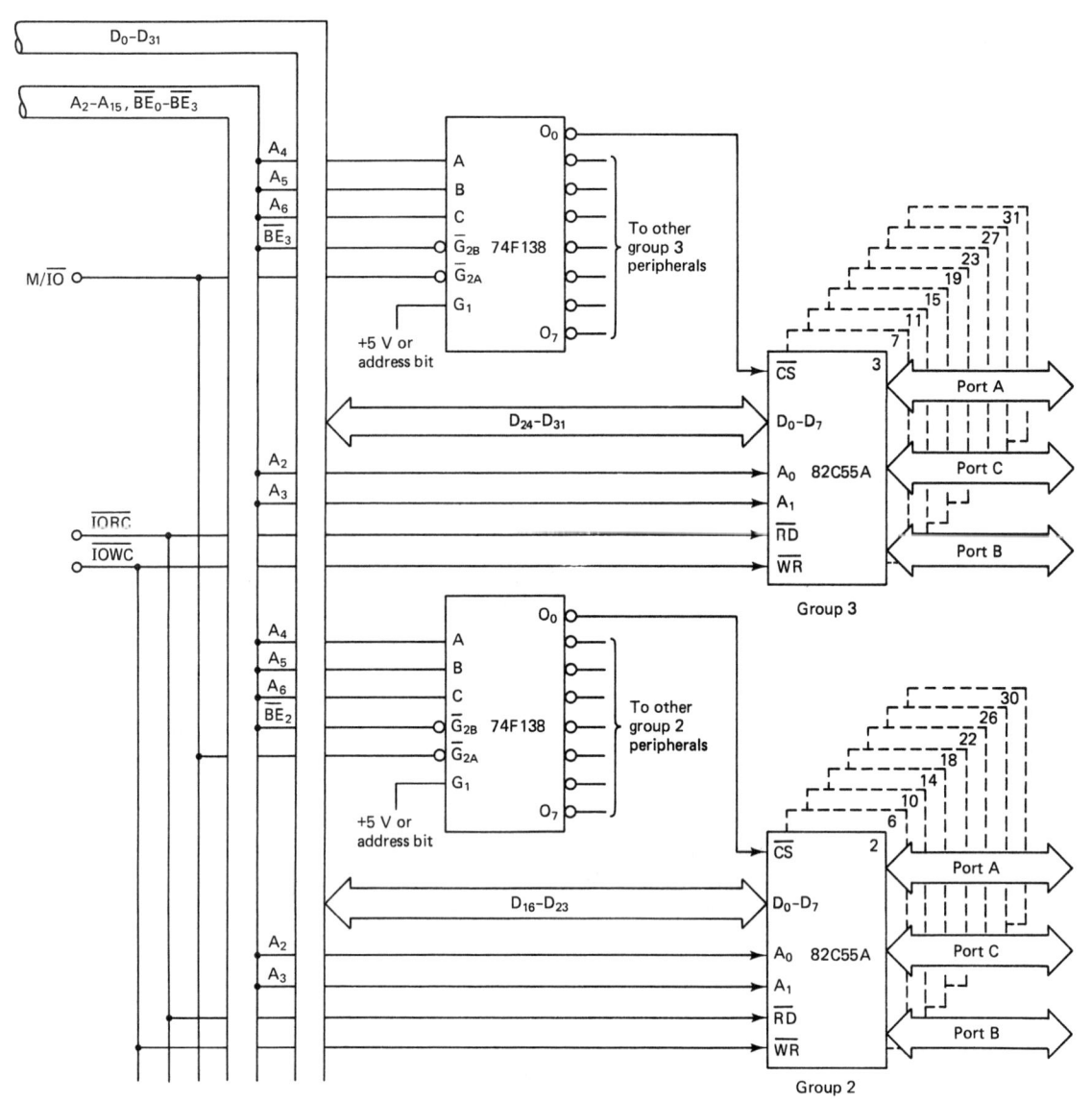

Figure 7.19 32-bit wide 82C55A isolated parallel I/O ports.

EXAMPLE 7.13

What must be the address inputs in Fig. 7.19 if port C of PPI 14 is to be accessed?

SOLUTION To enable PPI 14, the group 2 74F138 must be enabled for operation and its O_3 output switched to logic 0. This requires enable inputs $M/\overline{IO} = 0$ and $\overline{BE_2} = 0$ and chip select code $A_6 A_5 A_4 = 011$.

326 Input/Output Interface of the 80386DX Microprocessor Chap. 7

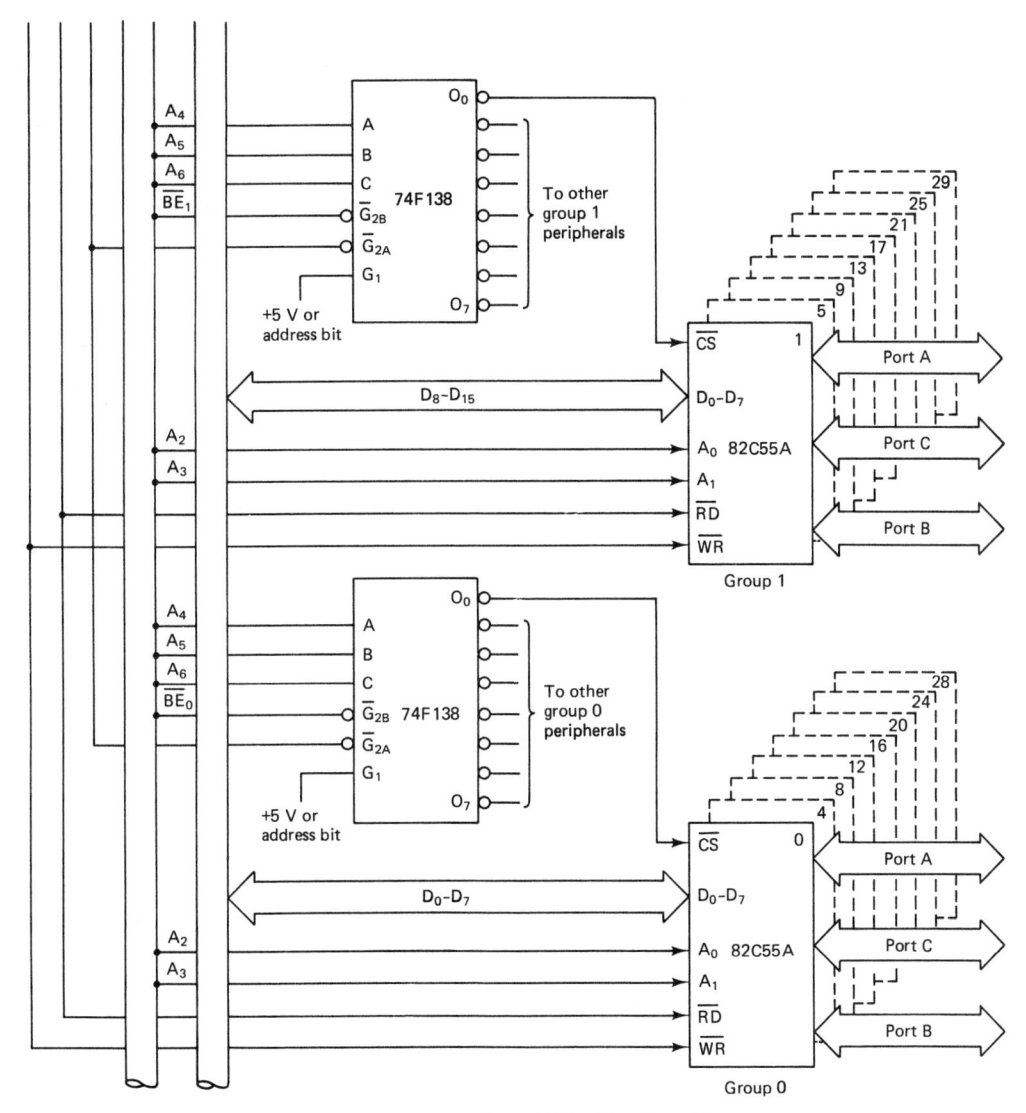

Figure 7.19 (*Continued*)

$$M/\overline{IO} = \overline{BE}_2 = 0 \quad \text{Enables 74F138 for group 2}$$

$$A_6A_5A_4 = 011 \quad \text{Selects PPI 14}$$

Port C of PPI 14 is selected with $A_3A_2 = 10$.

$$A_3A_2 = 10 \quad \text{Accesses port C}$$

The rest of the address bits are don't-care states.

Sec. 7.7　Implementing Isolated I/O Parallel Input/Output Ports

EXAMPLE 7.14

Assume that in the circuit of Fig. 7.19 PPI 14 is configured such that port A is an output port, ports B and C are both input ports, and that all three ports are set up for MODE 0 operation. Write a program that will input the data at ports B and C, find the difference C − B, and output this difference to port A.

SOLUTION From the circuit diagram in Fig. 7.19, we find that the addresses of the three I/O ports on PPI 14 are

$$\text{PORT A} = 0110010_2 = 32_{16}$$

$$\text{PORT B} = 0110110_2 = 36_{16}$$

$$\text{PORT C} = 0111010_2 = 3A_{16}$$

The data at ports B and C can be input with the instruction sequence

```
IN   AL, 36H    ;READ PORT B
MOV  BL, AL     ;SAVE DATA FROM PORT B
IN   AL, 3AH    ;READ PORT C
```

Now the data from port B are subtracted from the data at port C with the instruction

```
SUB  AL, BL     ;SUBTRACT B FROM C
```

Finally, the difference is output to port A with the instruction

```
OUT  32H, AL    ;WRITE TO PORT A
```

Notice in Fig. 7.19 that not all address bits are used in the I/O address decoding. Here only latched address bits A_4, A_5, and A_6 are decoded. Unused address bits are don't-care states. The address of the PPI is given in general by

$$XXXXXXXXXA_6A_5A_4A_3A_2$$

For this reason, many addresses decode to select each of the I/O ports. For instance, if all of the don't-care address bits are made 0, the address of the 82C55A labeled 0 is

$$A_{15} \cdots A_4 = 000000000000_2$$

$$\overline{BE_3}\,\overline{BE_2}\,\overline{BE_1}\,\overline{BE_0} = 1110_2$$

and the code at bits A_3A_2 selects port A, B, or C. However, if the don't-care address bits are all made equal to 1 instead of 0, the address is

$$A_{15} \cdots A_4 = 111111111000_2$$

$$\overline{BE_3}\,\overline{BE_2}\,\overline{BE_1}\,\overline{BE_0} = 1110_2$$

and still decodes to enable 82C55A number 0. In fact, every I/O address that has $A_6A_5A_4 = 000_2$ and $\overline{BE_0} = 0$ decodes to enable this 82C55A device.

7.8 IMPLEMENTING MEMORY-MAPPED I/O PARALLEL INPUT/OUTPUT PORTS USING THE 82C55A

The memory-mapped I/O interface of a real-mode 80386DX microcomputer is essentially the same as that employed in the isolated I/O circuit of Fig. 7.19. Figure 7.20 shows the equivalent memory-mapped circuit. Ports are still selected by an address on the address bus, and byte, word, or double words of data are trans-

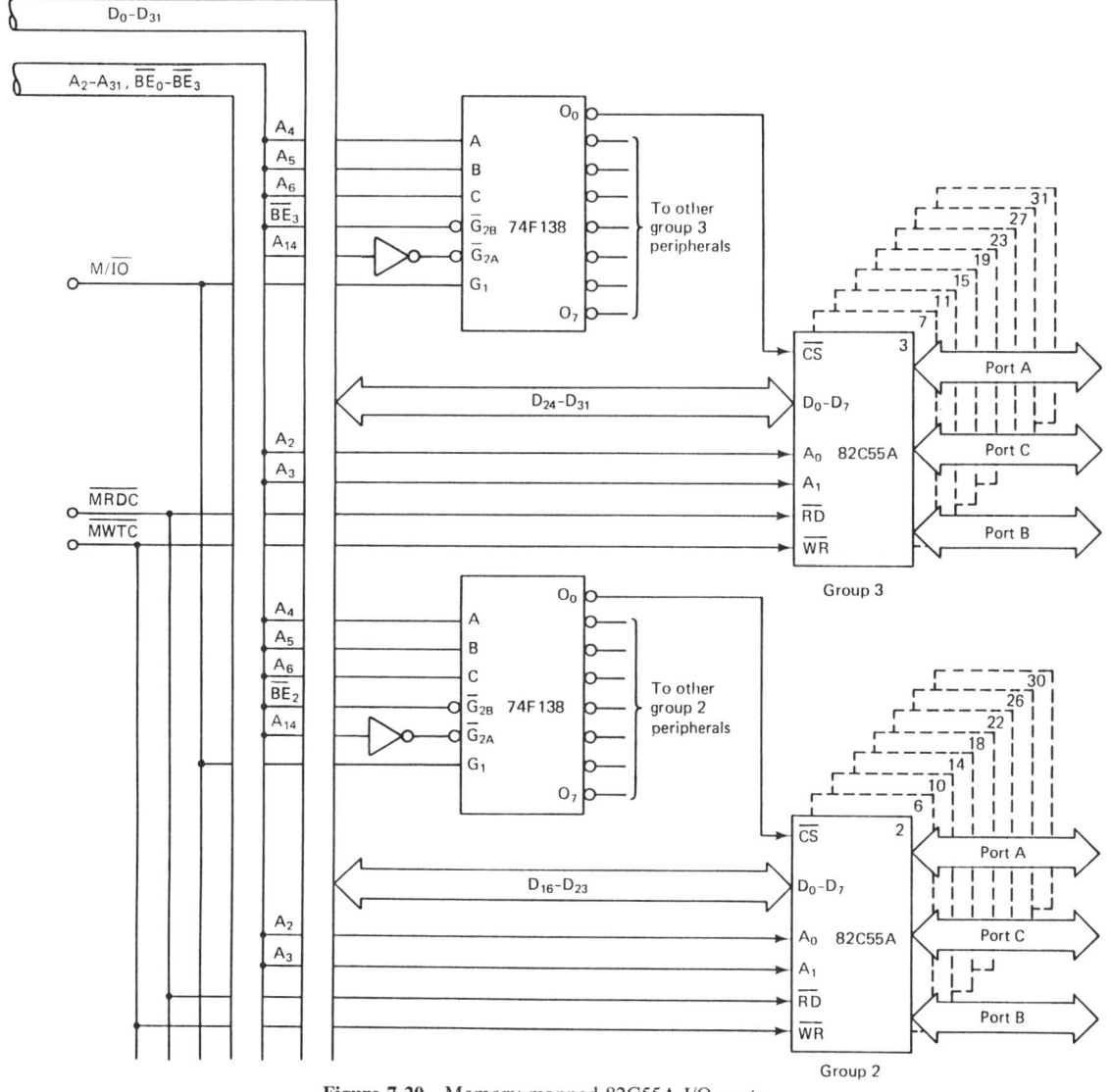

Figure 7.20 Memory-mapped 82C55A I/O ports.

Sec. 7.8 Implementing Memory-Mapped I/O Parallel Input/Output Ports 329

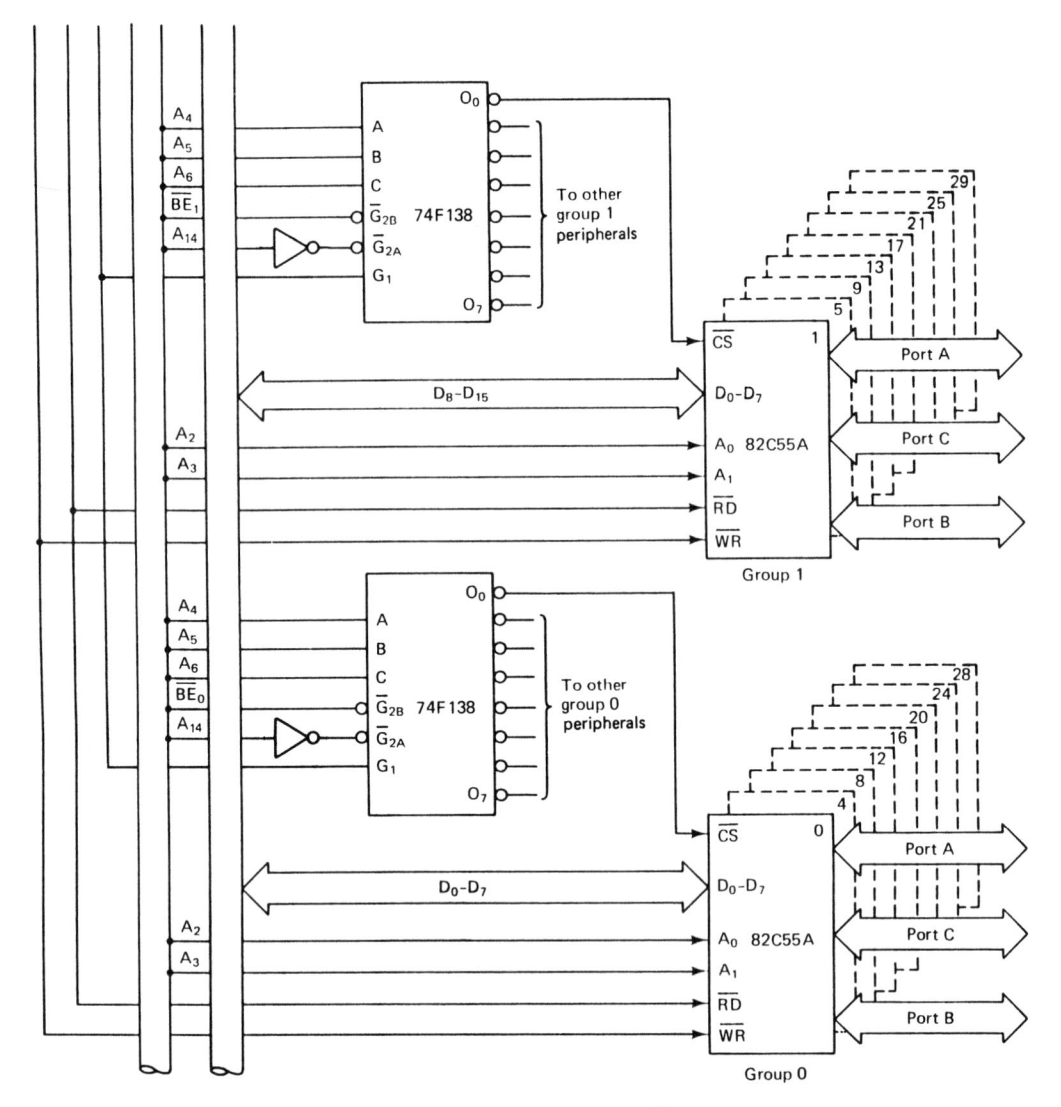

Figure 7.20 (*Continued*)

ferred between the MPU and I/O device over the data bus. One difference is that now the full 20-bit address is available for addressing I/O. Therefore, memory-mapped I/O devices can reside anywhere in the memory address space of the 80386DX.

Another difference is that during I/O operation memory read and write bus cycles are initiated instead of I/O bus cycles. This is because we are using memory instructions, not input/output instructions, to perform the data transfer. Furthermore, M/$\overline{\text{IO}}$ stays at the 1 logic level throughout the bus cycle. This indicates that a memory operation is in progress instead of an I/O operation.

Since memory-mapped I/O devices reside in the memory address space and are accessed with read and write cycles, additional I/O address latch, address buffer, data bus transceiver, and address decoder circuitry is not needed. The circuitry provided for the memory interface can be used. However, in some situations it may be practical to provide a separate I/O address decoder.

The key difference between the circuits in Fig. 7.19 and 7.20 is that $M/\overline{IO}$ is now used as the G_1 enable input and address bit A_{14} is inverted and applied to enable input $\overline{G}_{2A}$ of all decoders. Therefore, the I/O circuits are accessed whenever A_{14} is equal to logic 1 and $M/\overline{IO}$ is equal to logic 1.

EXAMPLE 7.15

Which I/O port in Fig. 7.20 is selected for operation when a byte access is performed to address 04002_{16}?

SOLUTION We begin by converting the address to binary form. This gives

$$A_{19} \cdots A_3 A_2 = 0000010000000000000_2$$

$$\overline{BE}_3 \overline{BE}_2 \overline{BE}_1 \overline{BE}_0 = 0100_2$$

In this address, bits $A_{14} = 1$ and $\overline{BE}_2 = 0$. Therefore, the group 74F138 address decoder is enabled.

$$M/\overline{IO} = 1 \quad \text{Enables the 74F138 decoder for group 2}$$

$$A_{14} = 1$$

$$\overline{BE}_2 = 0$$

A memory-mapped I/O operation takes place to the port selected by $A_6 A_5 A_4 = 000$. This input code switches decoder output O_0 to logic 0 and chip selects PPI 2 for operation.

$$A_6 A_5 A_4 = 000 \quad \text{Selects PPI 2}$$

$$O_0 = 0$$

The port select inputs of the PPI are $A_3 A_2 = 00$. These inputs cause PORT A to be accessed.

$$A_3 A_2 = 00 \quad \text{PORT A accessed}$$

Thus the address 04002_{16} selects PORT A on PPI 2.

EXAMPLE 7.16

Write the sequence of instructions needed to initialize the control register of PPI 0 in the circuit of Fig. 7.20 such that port A is an output port, ports B and C are input ports, and all three ports are configured for mode 0 operation.

SOLUTION Referring to Fig. 7.12, we find that the control byte required to provide this configuration is

Sec. 7.8 Implementing Memory-Mapped I/O Parallel Input/Output Ports

$10001011_2 = 8B_{16}$

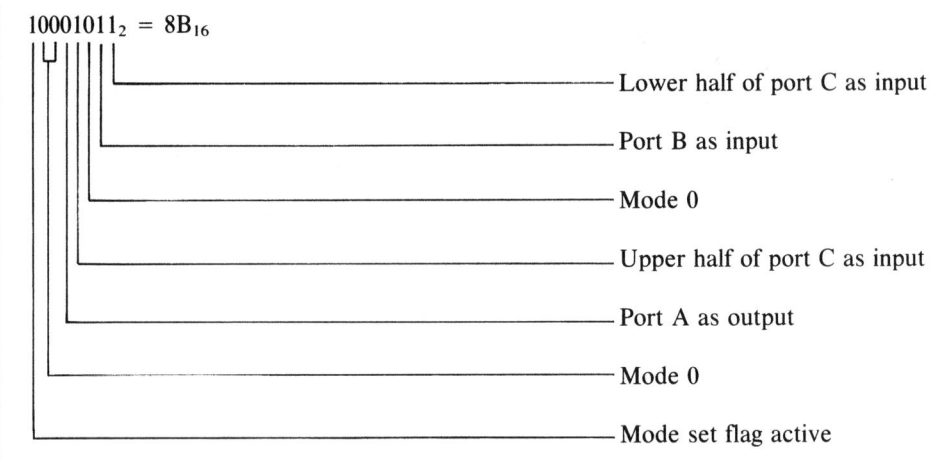

- Lower half of port C as input
- Port B as input
- Mode 0
- Upper half of port C as input
- Port A as output
- Mode 0
- Mode set flag active

From the circuit diagram, the memory address of the control register for PPI 0 is found to be

CONTROL REGISTER = 0000010000000001100_2 = $0400C_{16}$

Since PPI 0 is memory mapped, move instructions can be used to initialize the control register.

```
MOV   AX, 0H          ; CREATE DATA SEGMENT AT 00000₁₆
MOV   DS, AX
MOV   AL, 8BH         ; LOAD AL WITH CONTROL BYTE
MOV   [400CH], AL     ; WRITE CONTROL BYTE TO PPI 0
```

EXAMPLE 7.17

Assume that PPI 0 in Fig. 7.20 is configured as described in Example 7.16. Write a program that will input the contents of ports B and C, AND them together, and output the results to port A.

SOLUTION From the circuit diagram, we find that the addresses of the three I/O ports on PPI 0 are

$$PORT\ A = 04000_{16}$$

$$PORT\ B = 04004_{16}$$

$$PORT\ C = 04008_{16}$$

Now we set up a data segment at 00000_{16} and input the data from ports B and C.

```
MOV   AX, 0H          ; CREATE DATA SEGMENT AT 00000H
MOV   DS, AX
MOV   BL, [4004H]     ; READ PORT B
MOV   AL, [4008H]     ; READ PORT C
```

Next the contents of AL and BL must be ANDed and the result output to port A.

```
AND   AL, BL          ; AND DATA AT PORTS B AND C
MOV   [4000H], AL     ; WRITE TO PORT A
```

7.9 INPUT/OUTPUT POLLING AND HANDSHAKING

In some applications the microcomputer must synchronize the input or output of information to a peripheral device. Two examples of interfaces that normally require a synchronized data transfer are a serial communications interface and a parallel printer interface. For instance, an I/O service routine may repeatedly read the logic level of an input line and test it for a specific logic level. Normally, the I/O routine does not continue until the input under test switches to the appropriate logic level. Thus execution of the I/O routine has been synchronized to the occurrence of an event in external hardware. This mode of operation is known as *polling* an input.

Sometimes it is necessary as part of the I/O synchronization process first to poll an input from an I/O device and after receiving the appropriate level at the poll input, acknowledge this fact to the device with an output. This type of synchronization is achieved by implementing what is called *handshaking* as part of the input/output interface.

Polling an Input

Input polling is implemented through software. Let us now look at how a polling software routine is written. The first step in the polling operation is to read the contents of the input port. For example, the instructions needed to read the contents of PORT 0 of an 82C55A that is located at I/O address 8000_{16} are

```
POLL_I3:    MOV   DX, 8000H
            IN    AL, DX
```

We will assume that input I_3 at this port is the line that is polled. This is the reason the label POLL_I3 has been added to identify the beginning of the polling routine. After executing these instructions, the byte contents of port 0 are held in the AL register. Therefore, all other bits in AL are masked off with the instruction

```
            AND   AL, 08H
```

After this instruction is executed, the contents of AL will be either 00_{16} with the zero flag equal 1 or AL is 08_{16} with ZF equal 0. In this way we see that the zero flag reflects the complement of the logic level at input I_3. The state of the zero flag can be tested with the jump-on-zero instruction.

```
            JZ    POLL_I3
```

If zero flag is 1, a jump is initiated to POLL_I3 and the sequence repeats. On the other hand, if ZF is 0, the jump is not made; instead, the instruction following the jump instruction is executed. That is, the polling loop repeats until input I_3 is tested and found to be logic 1. Then the next instruction of the I/O routine is executed.

Input/Output Handshaking

A circuit diagram of a parallel printer interface is shown in Fig. 7.21(a). Here we find that the parallel interface consists of the eight data output lines D_0 through D_7, and the two handshake control signals strobe ($\overline{STB}$) and busy (BUSY). The MPU outputs data representing the character to be printed through the parallel printer interface. Character data are latched at the outputs of the parallel interface and are carried to the data inputs of the printer over data lines D_0 through D_7. The $\overline{STB}$ output of the parallel printer interface is used to signal the printer that new character data are available. Whenever the printer is already busy printing a character, it signals this fact to the MPU with the BUSY input of the parallel printer interface. The timing of these signals is illustrated in Fig. 7.21(b).

Let us now look at the sequence of events that take place at the parallel printer interface when data are output to the printer. Figure 7.21(c) is a flowchart of a subroutine that performs a parallel printer interface character transfer operation. First the BUSY input of the parallel printer interface is tested. Notice that this is done with a polling operation. That is, the MPU tests the logic level of BUSY repeatedly until it is found to be at the *not busy* logic level. *Busy* means that the printer is currently printing a character. On the other hand, *not busy* signals that the printer is ready to receive another character for printing.

After finding a not busy condition, the MPU outputs a character to the printer. The flowchart in Fig. 7.21(c) shows that a count of the number of characters in the printer buffer (microprocessor memory) is read, a byte of character data is read from the printer buffer, the character is output to the parallel interface, and then a pulse is produced at $\overline{STB}$. This pulse tells the printer to read the character off the data bus lines. The printer is again printing a character and signals this fact at BUSY. The handshake sequence is now complete.

Now the MPU prepares to output the next character. Figure 7.21(c) shows that it does this by updating the character address so that it points to the next character in the buffer, decrementing the count of the number of characters still to be printed, and checking this count to see whether the buffer contains more characters or is empty. If empty, the print operation is complete. Otherwise, the character transfer handshake sequence is repeated.

EXAMPLE 7.18

What are the addresses of the data lines, strobe output, and busy input in the circuit of Fig. 7.21(a)? Assume that all unused address bits are 0.

SOLUTION The I/O addresses that enable ports 0, 1, and 2 are found as

$$\text{Port } 0 = 1000000000000000_2 = 8000_{16}$$
$$\text{Port } 1 = 1001000000000000_2 = 9000_{16}$$
$$\text{Port } 2 = 1010000000000000_2 = A000_{16}$$

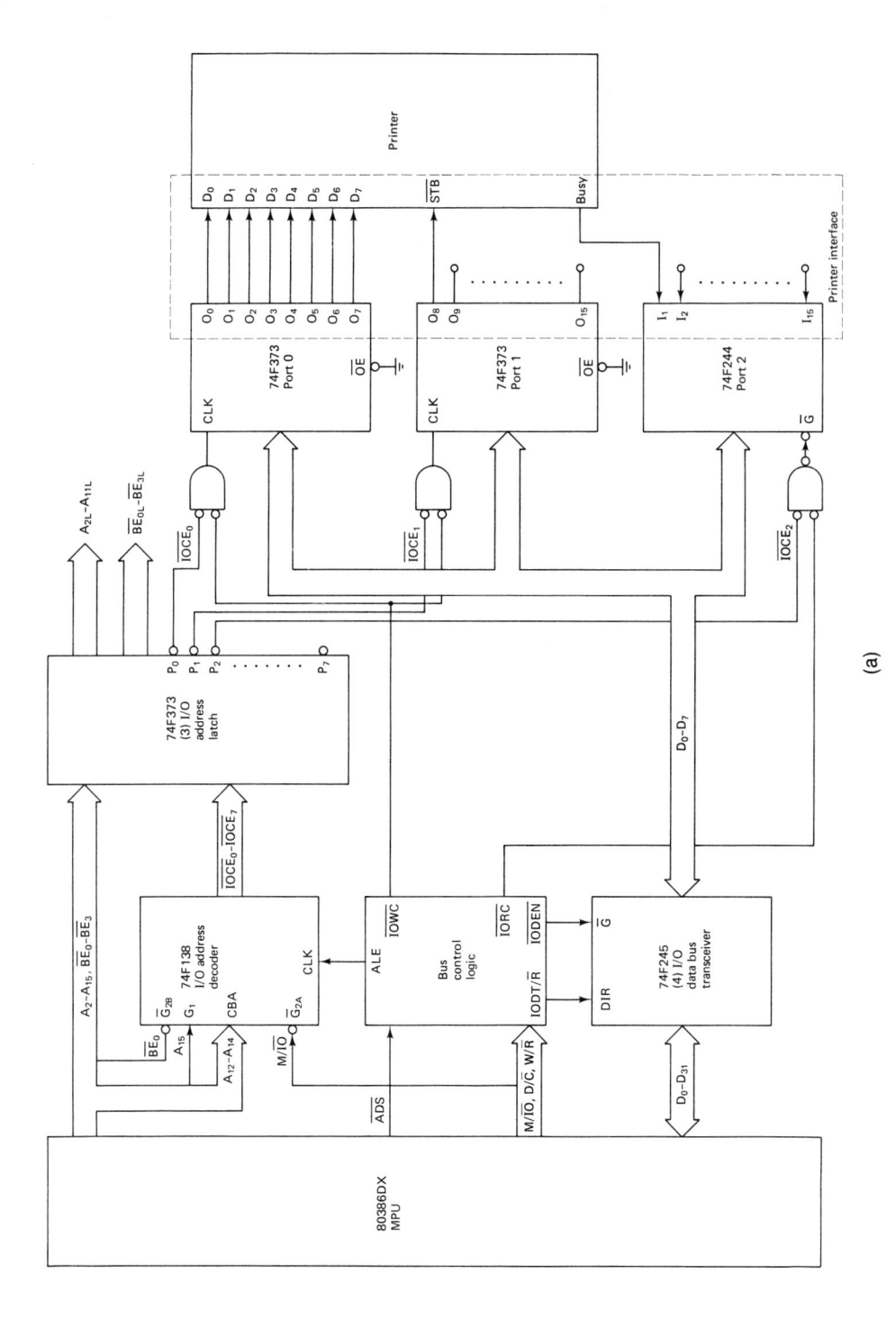

Figure 7.21 (a) Parallel printer interface circuit; (b) printer interface handshake signals; (c) handshake sequence flowchart.

(b)

(c)

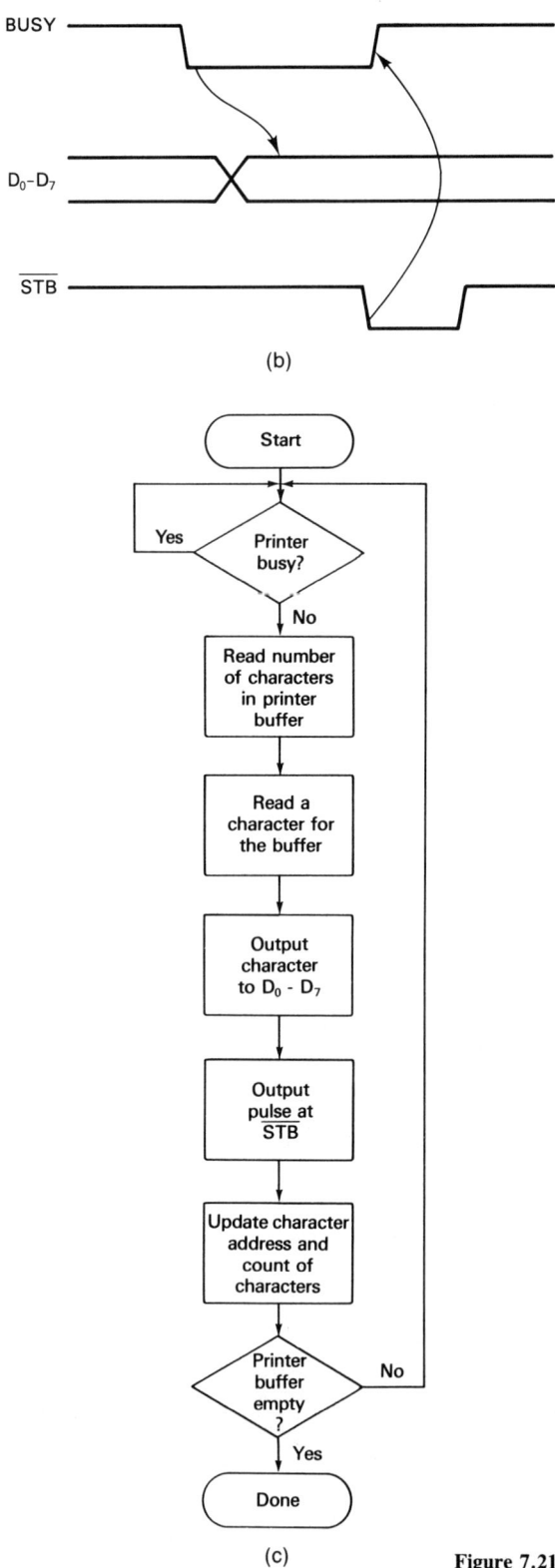

Figure 7.21 (*Continued*)

EXAMPLE 7.19

Write a program that will implement the sequence in Fig. 7.21(c) for the circuit in Fig. 7.21(a). Character data are held in memory starting at address PRNT_BUFF and the number of characters held in the buffer is identified by the count at address CHAR_COUNT. Use the port addresses from Example 7.18.

SOLUTION First the BUSY input is checked with the instructions

```
POLL_BUSY:   MOV   DX, A000H
             IN    AL, DX
             AND   AL, 01H
             JNZ   POLL_BUSY
```

Next the character count is loaded into the count register, the character is copied into AL, and then the character is output to port 0.

```
             MOV   CL, [CHAR_COUNT]
             MOV   SI, PRNT_BUFF
             MOV   AL, [SI]
             MOV   DX, 8000H
             OUT   DX, AL
```

Now a strobe pulse is generated at port 1 with the instructions

```
             MOV   AL, 00H
             MOV   DX, 9000H
             OUT   DX, AL
             MOV   BX, 0FH
STROBE:      DEC   BX
             JNZ   STROBE
             MOV   AL, 01H
             OUT   DX, AL
```

At this point, the value of PRNT_BUFF in SI must be incremented and the value of CHAR_COUNT in CL must be decremented. This is done with

```
             INC   SI
             DEC   CL
```

Finally, a check is made to see if the printer buffer is empty. To do this, we execute the instructions

```
             JNZ   POLL_BUSY
DONE:        ---
```

7.10 THE 82C54 PROGRAMMABLE INTERVAL TIMER

The 82C54 is an LSI peripheral designed to permit easy implementation of timer and counter functions in a microcomputer system. It contains three independent 16-bit counters that can be programmed to operate in a variety of ways to implement timing functions. For instance, they can be set up to work as a one-shot pulse generator, square wave generator, or rate generator.

Block Diagram of the 82C54

Let us begin our study of the 82C54 by looking at the signal interfaces shown in its block diagram of Fig. 7.22(a). The actual pin location for each of these signals is given in Fig. 7.22(b). In an 80386DX-based microcomputer system, the 82C54 is treated as a peripheral device. Moreover, it can be memory-mapped into the 80386DX's memory address space or I/O-mapped into the I/O address space. The microprocessor interface of the 82C54 allows the 80386DX to read from or write to its internal registers. In this way, it can configure the mode of operation for the timers, load initial values into the counters, or read the current value from a counter.

Now we will look at the signals of the microprocessor interface. The microprocessor interface includes an 8-bit bidirectional data bus, D_0 through D_7. It is over these lines that data are transferred between the 80386DX and 82C54. Register address inputs A_0 and A_1 are used to select the register to be accessed and control signals read ($\overline{RD}$) and write ($\overline{WR}$) indicate whether it is to be read from or written into, respectively. A chip select ($\overline{CS}$) input is also provided to enable the 82C54's microprocessor interface. This input allows the designer to locate the device at a specific memory or I/O address.

At the other side of the block in Fig. 7.22(a), we find three signals for each counter. For instance, counter 0 has two inputs that are labeled CLK_0 and $GATE_0$. Pulses applied to the clock input are used to decrement counter 0. The gate input is used to enable or disable the counter. $GATE_0$ must be switched to logic 1 to enable counter 0 for operation. For example, in the square-wave mode of operation the counter is to run continuously; therefore, $GATE_0$ is fixed at the 1 logic level and a continuous clock signal is applied to CLK_0. The 82C54 is rated for a maximum clock frequency of 10 MHz. Counter 0 also has an output line that is labeled OUT_0. The counter produces either a clock or a pulse at OUT_0 depending on the mode of operation selected. For instance, when configured for the square wave mode of operation, this output is a clock.

Architecture of the 82C54

The internal architecture of the 82C54 is shown in Fig. 7.23. Here we find the *data bus buffer*, *read/write logic*, *control word register*, and three *counters*. The data bus buffer and read/write control logic represent the microprocessor interface we just described.

The *control word register* section actually contains three 8-bit registers that

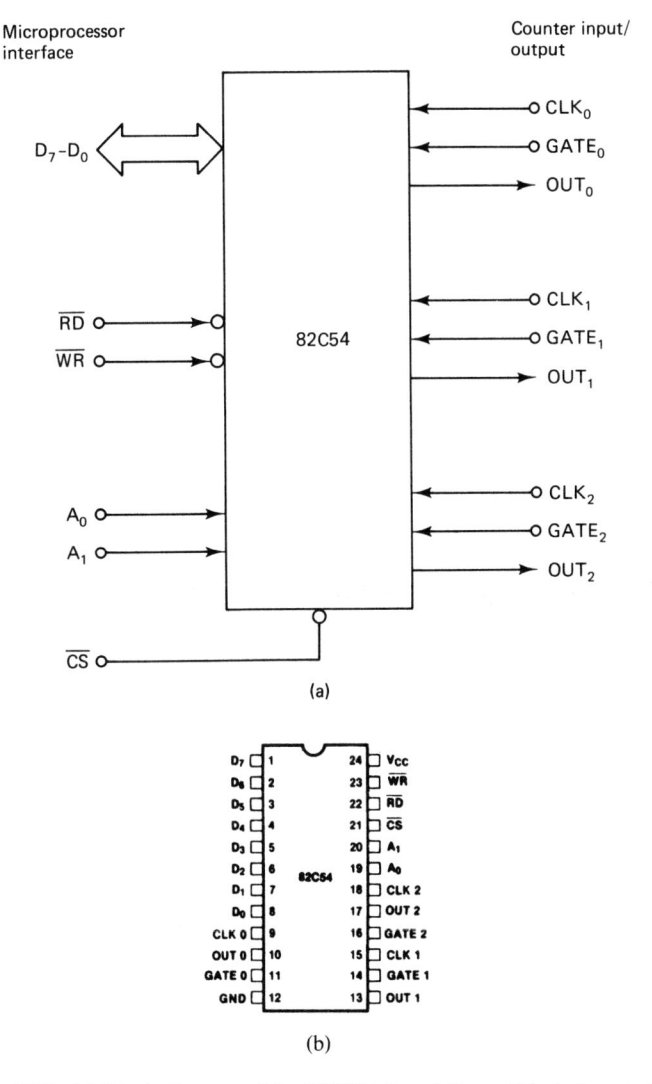

Figure 7.22 (a) Block diagram of the 82C54 interval timer; (b) pin layout. (Reprinted by permission of Intel Corp. Copyright Intel Corp. 1987.)

are used to configure the operation of counters 0, 1, and 2. The format of a *control word* is shown in Fig. 7.24. Here we find that the two most significant bits are a code that assigns the control word to a counter. For instance, making these bits 01 selects counter 1. Bits D_1 through D_3 are a three-bit mode select code, $M_2M_1M_0$, that selects one of six modes of counter operation. The least significant bit D_0 is labeled BCD and selects either binary or BCD mode of counting. For instance, if this bit is set to logic 1, the counter acts as a 16-bit binary counter. Finally, the two-bit code RW_1RW_0 is used to set the sequence in which bytes are read from or loaded into the 16-bit count registers.

Sec. 7.10   The 82C54 Programmable Interval Timer

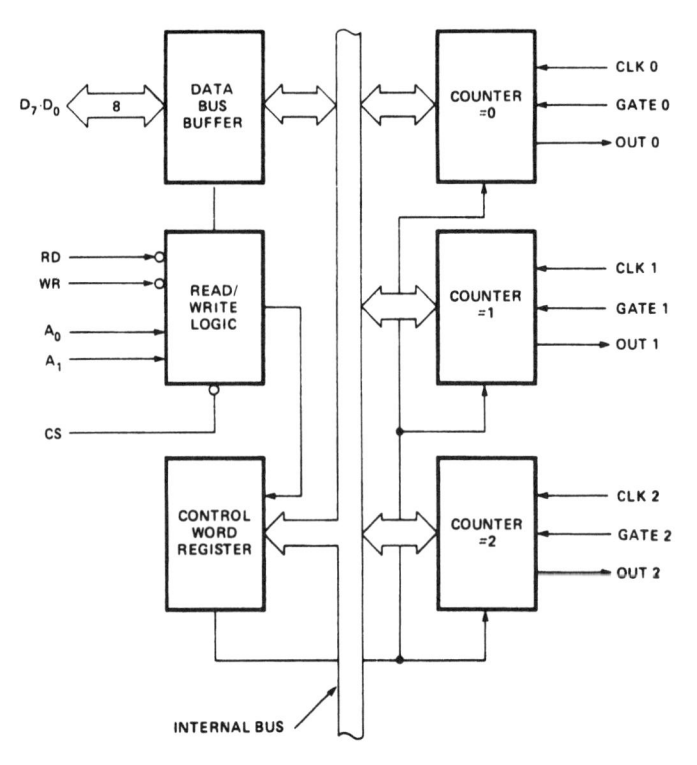

Figure 7.23 Internal architecture of the 82C54. (Reprinted by permission of Intel Corp. Copyright Intel Corp. 1987.)

EXAMPLE 7.20

An 82C54 receives the control word 10010000_2 over the bus. What configuration is set up for the counter?

SOLUTION Since the SC bits are 10, the rest of the bits are for setting up the configuration of counter 2. Following the format in Fig. 7.24, we find that 01 in the RW bits sets counter 2 for the read/write sequence identified as the least significant byte only. This means that the next write operation performed to counter 2 will load the data into the least significant byte of its count register. Next the mode code is 000 and this selects mode 0 operation for this counter. The last bit, BCD, is also set to 0 and selects binary counting.

The three counters shown in Fig. 7.23 are each 16 bits in length and operate as *down counters*. That is, when enabled by an active gate input, the clock decrements the count downward. Each counter contains a 16-bit *count register* that must be loaded as part of the initialization cycle. The value held in the count register can be read at any time through software.

To read from or write to the counters of the 82C54 or load its control word register, the microprocessor needs to execute instructions. Figure 7.25 shows the bus control information needed to access each register. For example, to write to

Control word format

D$_7$	D$_6$	D$_5$	D$_4$	D$_3$	D$_2$	D$_1$	D$_0$
SC$_1$	SC$_0$	RW/W$_1$	RW/W$_0$	M$_2$	M$_1$	M$_0$	BCD

Definition of control
SC-select counter:

SC$_1$	SC$_0$	
0	0	Select counter 0
0	1	Select counter 1
1	0	Select counter 2
1	1	Read back command

RW-read/write:

RW/W$_1$	RW/W$_0$	
0	0	Counter latch command
1	0	Read/write most significant byte only
0	1	Read/write least significant byte only
1	1	Read/write least significant byte first, then most significant byte

M-mode:

M$_2$	M$_1$	M$_0$	
0	0	0	Mode 0
0	0	1	Mode 1
X	1	0	Mode 2
X	1	1	Mode 3
1	0	0	Mode 4
1	0	1	Mode 5

BCD:

0	Binary counter 16-bits
1	Binary coded decimal (BCD) counter (4 decades)

Figure 7.24 Control word format. (Reprinted by permission of Intel Corp. Copyright Intel Corp. 1987.)

$\overline{CS}$	$\overline{RD}$	$\overline{WR}$	A$_1$	A$_0$	
0	1	0	0	0	Write into Counter 0
0	1	0	0	1	Write into Counter 1
0	1	0	1	0	Write into Counter 2
0	1	0	1	1	Write Control Word
0	0	1	0	0	Read from Counter 0
0	0	1	0	1	Read from Counter 1
0	0	1	1	0	Read from Counter 2
0	0	1	1	1	No-Operation (3-State)
1	X	X	X	X	No-Operation (3-State)
0	1	1	X	X	No-Operation (3-State)

Figure 7.25 Accessing the registers of the 82C54. (Reprinted by permission of Intel Corp. Copyright Intel Corp. 1987.)

the control register, the register address lines must be $A_1 A_0 = 11$, and the control lines $\overline{WR} = 0$, $\overline{RD} = 1$, and $\overline{CS} = 0$.

EXAMPLE 7.21

Write an instruction sequence to set up the three counters of an 82C54 located at I/O address 40H as follows:

Counter 0: Binary counter operating in mode 0 with an initial value of 1234H.
Counter 1: BCD counter operating in mode 2 with an initial value of 100H.
Counter 2: Binary counter operating in mode 4 with initial value of 1FFFH.

Assume that the device is attached to the I/O bus of the circuit in Fig. 7.6 and that inputs A_0 and A_1 are attached to address bits A_2 and A_3, respectively.

SOLUTION Since the base address of the 82C54 is 40H, the mode register is at address 4CH. The three counters 0, 1, and 2 are at addresses 40H, 44H, and 48H, respectively. Let us first determine the mode words for the three counters. Following the bit definitions in Fig. 7.24, we get

$$\text{Mode word for counter } 0 = 00110000_2 = 30_{16}$$

$$\text{Mode word for counter } 1 = 01010101_2 = 55_{16}$$

$$\text{Mode word for counter } 2 = 10111000_2 = B8_{16}$$

The following instruction sequence can be used to set up the 82C54 with the modes and counts:

```
MOV  AL, 30H     ;Set up counter 0 mode
OUT  4CH, AL
MOV  AL, 55H     ;Set up counter 1 mode
OUT  4CH, AL
MOV  AL, B8H     ;Set up counter 2 mode
OUT  4CH, AL
MOV  AL, 34H     ;Load counter 0
OUT  40H, AL
MOV  AL, 12H
OUT  40H, AL
MOV  AL, 01H     ;Load counter 1
OUT  44H, AL
MOV  AL, 00H
OUT  44H, AL
MOV  AL, FFH     ;Load counter 2
OUT  48H, AL
MOV  AL, 1FH
OUT  48H, AL
```

Earlier we pointed out that the contents of a count register can be read at any time. Let us now look at how this is done in software. One approach is to simply read the contents of the corresponding register with an input instruction.

In Fig. 7.25 we see that to read the contents of count register 0, the control inputs must be $\overline{CS} = 0$, $\overline{RD} = 0$, and $\overline{WR} = 1$, and the register address code must be $A_1A_0 = 00$. To assure that a valid count is read out of count register 0, the counter must be inhibited before the read operation takes place. The easiest way to do this is to switch the $GATE_0$ input to logic 0 before performing the input operation. The count is read as 2 separate bytes, low byte first followed by the high byte.

The contents of the count registers can also be read without first inhibiting the counter. That is, the count can be read on the fly. To do this in software, a command must first be issued to the mode register to capture the current value of the counter into a temporary storage register. In Fig. 7.24 we find that setting bits D_5 and D_4 of the mode byte to 00 specifies the latch mode of operation. Once this mode byte has been written to the 82C54, the contents of the temporary storage register for the counter can be read just as before.

EXAMPLE 7.22

Write an instruction sequence to read the contents of counter 2 on the fly. The count is to be loaded into the AX register. Assume that the 82C54 is located at I/O address 40H. Assume that the device is attached as described in Example 7.21.

SOLUTION First we will latch the contents of counter 2 and then this value is read from the temporary storage register. This is done with the following sequence of instructions:

```
MOV   AL, 10000000B    ;Latch counter 2
OUT   4CH, AL
IN    AL, 48H          ;Read the low byte
MOV   BL, AL
IN    AH, 48H          ;Read the high byte
MOV   AH, AL
MOV   AL, BL
```

Another mode of operation, called the *read-back mode*, permits a programmer to capture the current count values and status information of all three counters with a single command. In Fig. 7.24 we see that a read-back command has bits D_6 and D_7 both set to 1. The read-back command format is shown in more detail in Fig. 7.26. Notice that bits D_1 (CNT 0), D_2 (CNT 1), and D_3 (CNT 2) are made

A0, A1 = 11 $\overline{CS} = 0$ $\overline{RD} = 1$ $\overline{WR} = 0$

D_7	D_6	D_5	D_4	D_3	D_2	D_1	D_0
1	1	COUNT	STATUS	CNT 2	CNT 1	CNT 0	0

D_5: 0 = Latch count of selected counter(s)
D_4: 0 = Latch status of selected counter(s)
D_3: 1 = Select counter 2
D_2: 1 = Select counter 1
D_1: 1 = Select counter 0
D_0: Reserved for future expansion; must be 0

Figure 7.26 Read-back command format. (Reprinted by permission of Intel Corp. Copyright Intel Corp. 1990.)

logic 1 to select the counters, logic 0 in bit D_4 means that status information will be latched, and logic 0 in D_5 means that the counts will be latched. For instance, to capture the values in all three counters, the read-back command is 11011110_2 = DE_{16}. This command must be written into the control word register of the 82C54. Figure 7.27 shows some other examples of read-back commands. Notice that both count and status information can be latched with a single command.

\multicolumn{8}{c|}{Command}	Description	Results
D_7	D_6	D_5	D_4	D_3	D_2	D_1	D_0		
1	1	0	0	0	0	1	0	Read back count and status of Counter 0	Count and status latched for Counter 0
1	1	1	0	0	1	0	0	Read back status of Counter 1	Status latched for Counter 1
1	1	1	0	1	1	0	0	Read back status of Counters 2, 1	Status latched for Counters 1 and 2
1	1	0	1	1	0	0	0	Read back count of Counter 2	Count latched for Counter 2
1	1	0	0	0	1	0	0	Read back count and status of Counter 1	Count and status latched for Counter 1
1	1	1	0	0	0	1	0	Read back status of Counter 0	Status latched for Counter 0

Figure 7.27 Read-back command examples. (Reprinted by permission of Intel Corp. Copyright Intel Corp. 1990.)

Our read-back command example, DE_{16}, only latches the values of the three counters. These values must next be read by the programmer by issuing read commands for the individual counters. Once the value of a counter or status is latched, it must be read before a new value can be captured.

An example of a command that latches just the status for counters 1 and 2 is given in Fig. 7.27. This command is coded as

$$11101100_2 = EC_{16}$$

The format of the status information latched with this command is shown in Fig. 7.28. Here we find that bits D_0 through D_5 contain the mode control information that was written into the counter. These bits are identical to the six least significant bits of the control word in Fig. 7.24. In addition to this information, the status byte contains the logic state of the counters output pin in bit position D_7 and the value of the null count flip-flop in bit position D_6. Latched status information is also read by the programmer by issuing a read counter command to the 82C54.

The first command in Fig. 7.27, which is $11000010_2 = C2_{16}$, captures both the count and status information for counter 0. When both count and status information are captured with a read-back command, two read-counter commands

D_7	D_6	D_5	D_4	D_3	D_2	D_1	D_0
OUTPUT	NULL COUNT	RW1	RW0	M2	M1	M0	BCD

D_7 1 = Out Pin is 1
    0 = Out Pin is 0
D_6 1 = Null count
    0 = Count available for reading
D_5-D_0 Counter Programmed Mode

Figure 7.28 Status byte format. (Reprinted by permission of Intel Corp. Copyright Intel Corp. 1990.)

are required to return the information to the MPU. During the first read operation, the value of the count is read and the status information is transferred during the second read operation.

Operating Modes of 82C54 Counters

As indicated earlier, each of the 82C54's counters can be configured to operate in one of six modes. Figure 7.29 shows waveforms that summarize operation for each mode. Notice that mode 0 operation is known as interrupt on terminal count and mode 1 is called programmable one-shot. The GATE input of a counter takes on different functions depending on which mode of operation is selected. The effect of the gate input is summarized in Fig. 7.30. For instance in mode 0, GATE disables counting when set to logic 0 and enables counting when set to 1. Let us now discuss each of these modes of operation in more detail.

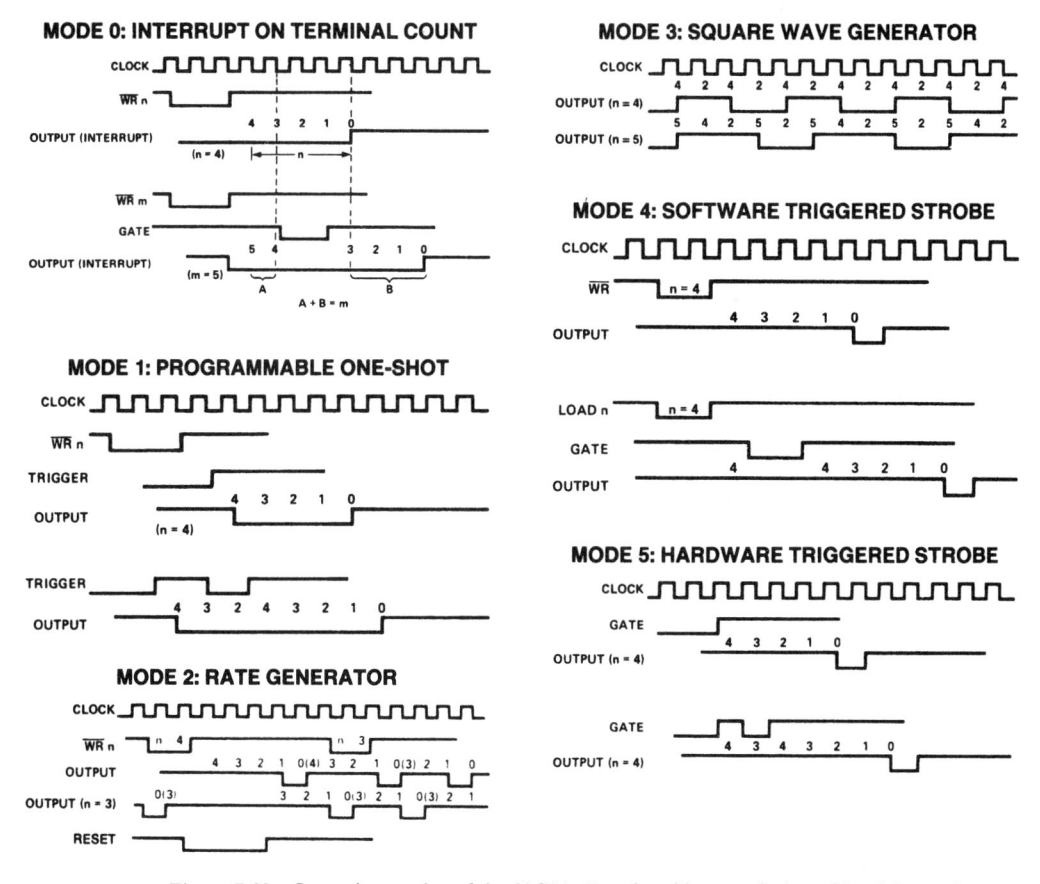

Figure 7.29 Operating modes of the 82C54. (Reprinted by permission of Intel Corp. Copyright Intel Corp. 1987.)

Sec. 7.10   The 82C54 Programmable Interval Timer

The *interrupt on terminal count* mode of operation is used to generate an interrupt to the microprocessor after a certain interval of time has elapsed. As shown in the waveforms for mode 0 operation in Fig. 7.29, a count of $n = 4$ is written into the count-register synchronously with the pulse at $\overline{WR}$. After the write operation is complete, the count is loaded into the counter on the next clock pulse and then the count is decremented by 1 for each clock pulse that follows. When the count reaches 0, the terminal count, a 0-to-1 transition occurs at OUTPUT. This occurs after $N + 1$ (five) clock pulses. This signal is used as the interrupt input to the microprocessor.

Earlier we found in Fig. 7.30 that GATE must be at logic 1 to enable the counter for interrupt on terminal count mode of operation. Figure 7.29 also shows waveforms for the case in which GATE is switched to logic 0. Here we see that the counter does not decrement below the value 4 until GATE returns to 1.

Signal Status Modes	Low Or Going Low	Rising	High
0	Disables counting	—	Enables counting
1	—	1) Initiates counting 2) Resets output after next clock	—
2	1) Disables counting 2) Sets output immediately high	Initiates counting	Enables counting
3	1) Disables counting 2) Sets output immediately high	Initiates counting	Enables counting
4	Disables counting	—	Enables counting
5	—	Initiates counting	—

Figure 7.30 Effect of the GATE input for each mode. (Reprinted by permission of Intel Corp. Copyright Intel Corp. 1987.)

EXAMPLE 7.23

The counter of Fig. 7.31 is programmed to operate in mode 0. Assuming that the value decimal 100 is written into the counter, compute the time delay (T_D) that occurs until the positive transition takes place at the counter 0 output. The counter is configured for BCD counting.

SOLUTION Once loaded, counter 0 needs to decrement down for 100 pulses at the clock input. During this period the counter is disabled by logic 0 at the GATE input for two clock periods. Therefore, the total time delay is calculated as

$$T_D = (n + 1 + d)(T_{CLK0})$$
$$= (100 + 1 + 2)\left(\frac{1\ \mu s}{1.19318}\right)$$
$$= 86.3\ \mu s$$

Mode 1 operation implements what is known as a *programmable one-shot*. As shown in Fig. 7.29, when set for this mode of operation, the counter produces a single pulse at its output. The waveforms show that an initial count, which in this example is the number 4, is written into the counter synchronous with a pulse at $\overline{WR}$. When GATE, called TRIGGER in the waveshapes, switches from logic 0-to-1, OUTPUT switches to logic 0 on the next pulse at CLOCK and the count begins to decrement with each successive clock pulse. The pulse is completed as OUTPUT returns to logic 1 when the terminal count, which is zero, is reached. In this way we see that the duration of the pulse is determined by the value loaded into the counter and the period of CLOCK.

The pulse generator produced with an 82C54 counter is what is called a *retriggerable one-shot*. By retriggerable we mean that if after an output pulse has been started another rising edge is experienced at TRIGGER, the count is reloaded and the pulse width is extended by the full pulse duration. The lower one-shot waveform in Fig. 7.29 shows this type of operation. Notice that when the count is 2 a second rising edge occurs at TRIGGER. On the next clock pulse the value 4 is reloaded into the counter to extend the pulse width to seven clock cycles.

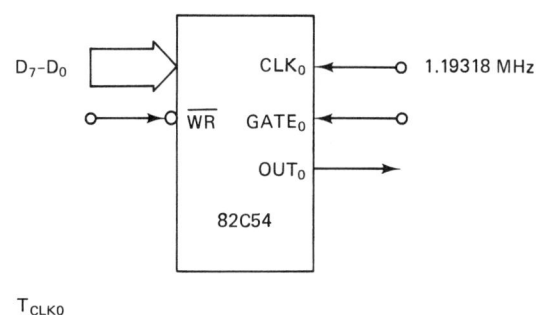

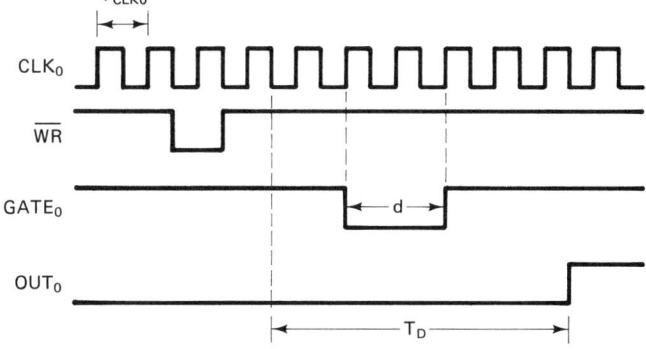

Figure 7.31 Mode 0 configuration.

Sec. 7.10 The 82C54 Programmable Interval Timer

EXAMPLE 7.24

Counter 1 of an 82C54 is programmed to operate in mode 1 and is loaded with the value decimal 10. The gate and clock inputs are as shown in Fig. 7.32. How long is the output pulse? Assume that the counter is configured for BCD counting.

SOLUTION The GATE input in Fig. 7.32 shows that the counter is operated as a nonretriggerable one-shot. Therefore, the pulse width is given by

$$T = (\text{counter contents}) \times (\text{clock period})$$

$$= 10 \times \frac{1}{1.19318 \text{ MHz}}$$

$$= 8.38 \text{ μs}$$

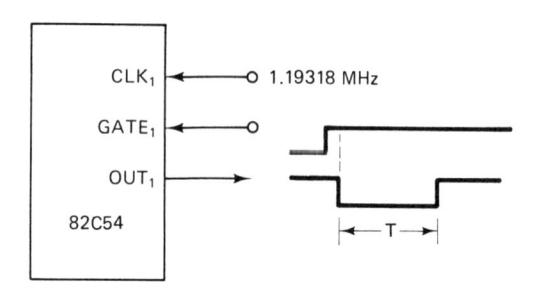

Figure 7.32 Mode 1 configuration.

When set for mode 2, *rate generator* operation, the counter within the 82C54 is set to operate as a divide-by-N counter. Here N stands for the value of the count loaded into the counter. Figure 7.33 shows counter 1 of an 82C54 set up in this way. Notice that the gate input is fixed at the 1 logic level. As shown in the table of Fig. 7.30, this enables counting operation. Looking at the waveforms for mode 2 operation in Fig. 7.29, we see that OUTPUT is at logic 1 until the count decrements to one. Then the output switches to the active 0 logic level for just one clock pulse width. When OUTPUT returns to logic 1, the count reloads and the counting sequence repeats. In this way we see that there is one clock pulse at the output for every N clock pulses at the input. This is why it is called a divide-by-N counter.

EXAMPLE 7.25

Counter 1 of the 82C54, as shown in Fig. 7.33, is programmed to operate in mode 2 and is loaded with decimal number 18. Describe the signal produced at OUT_1. Assume that the counter is configured for BCD counting.

SOLUTION In mode 2 the output goes low for one period of the input clock after the counter contents decrement to one. Therefore,

$$T_2 = \frac{1}{1.19318 \text{ MHz}} = 838 \text{ ns}$$

and

$$T = 18 \times T_2 = 15.094 \text{ μs}$$

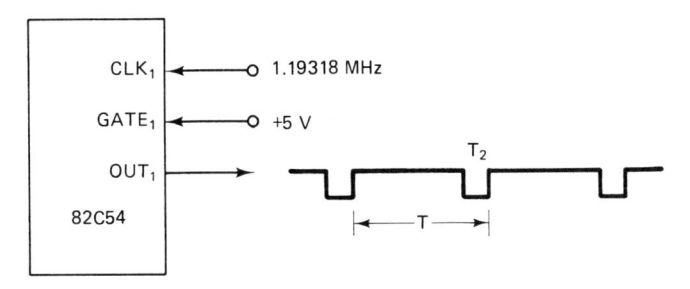

Figure 7.33 Mode 2 configuration.

Mode 3 sets the counter of the 82C54 to operate as a *square-wave generator*. In this mode, the output of the counter is a square wave with 50% duty cycle whenever the counter is loaded with an even number. That is, the output is at the 1 logic level for exactly the same amount of time that it is at the 0 logic level. As shown in Fig. 7.29, the count decrements by two with each pulse of the clock input. When the count reaches zero, the output switches logic levels, the original count (n = 4) is reloaded, and the count sequence repeats. Transitions of the output take place with respect to the negative edge of the input clock. The period of the symmetrical square wave at the output equals the number loaded into the counter multiplied by the period of the input clock.

If an odd number (N) is loaded into the counter instead of an even number, the time for which the output is high is given by (N + 1)/2 and the time for which the output is low is given by (N − 1)/2.

EXAMPLE 7.26

The counter in Fig. 7.34 is programmed to operate in mode 3 and is loaded with the decimal value 15. Determine the characteristics of the square wave at OUT_1. Assume that the counter is configured for BCD counting.

SOLUTION

$$T_{CLK1} = \frac{1}{1.19318 \text{ MHz}} = 838 \text{ ns}$$

$$T_1 = T_{CLK1} \frac{N + 1}{2} = 838 \text{ ns} \times \frac{15 + 1}{2}$$

$$= 6.704 \text{ } \mu s$$

$$T_2 = T_{CLK1} \frac{N - 1}{2} = 838 \text{ ns} \times \frac{15 - 1}{2}$$

$$= 5.866 \text{ } \mu s$$

$$T = T_1 + T_2 = 6.704 \text{ } \mu s + 5.866 \text{ } \mu s$$

$$= 12.57 \text{ } \mu s$$

Selecting mode 4 operation for a counter configures the counter to work as a *software-triggered strobed counter*. When in this mode, the counter automatically begins to decrement one clock pulse after it is loaded with the initial count through software. Again, it decrements at a rate set by the clock input signal. At

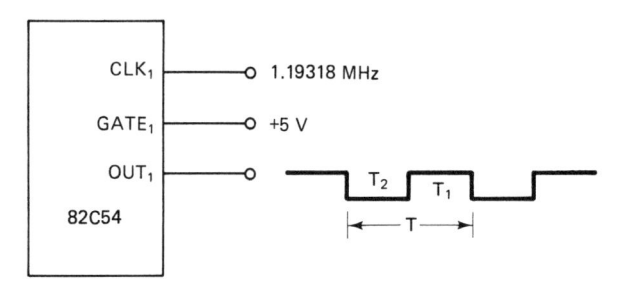

Figure 7.34 Mode 3 configuration.

the moment the terminal count is reached, the counter generates a single strobe pulse with duration equal to one clock pulse at its output. That is, a strobe pulse is produced at the output after N + 1 clock pulses. Here N again stands for the count loaded into the counter. This output pulse can be used to perform a timed operation. Figure 7.29 shows waveforms illustrating this mode of operation initiated by writing the value 4 into a counter. For instance, if CLOCK is 1.19318 MHz, the strobe occurs 4.19 μs after the count 4 is written into the counter. In the table of Fig. 7.30 we find that the gate input needs to be at logic 1 for the counter to operate in this mode.

This mode of operation can be used to implement a long duration interval timer or a free running timer. In either application, the strobe at the output can be used as an interrupt input to a microprocessor. In response to this pulse, an interrupt service routine can be used to reload the timer and restart the timing cycle. Frequently, the service routine also counts the strobes as they come in by decrementing the contents of a register. Software can test the value in this register to determine if the timer has timed out a certain number of times; for instance, to determine if the contents of the register have decremented to zero. When it reaches zero, a specific operation, such as a jump or call, can be initiated. In this way we see that software has been used to extend the interval of time at which a function occurs beyond the maximum duration of the 16-bit counter within the 82C54.

EXAMPLE 7.27

Counter 1 of Fig. 7.35 is programmed to operate in mode 4. What value must be loaded into the counter to produce a strobe signal 10 μs after the counter is loaded?

SOLUTION The strobe pulse occurs after counting down the counter to zero. The number of input clock periods required for a period of 10 μs is given by

$$N = \frac{T}{T_{CLK}}$$

$$= \frac{10 \text{ μs}}{1/1.19318 \text{ MHz}}$$

$$= 12_{16}$$

Thus the counter should be loaded with number $11_{16} = 0B_{16} = 00001011_2$ to produce a strobe pulse 10 μs after loading.

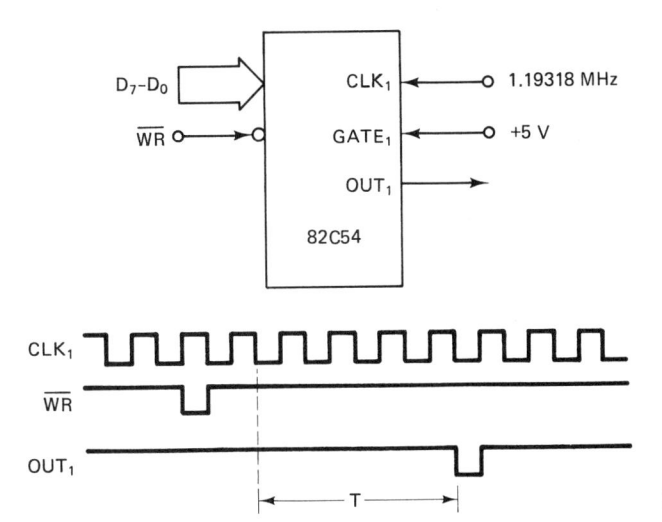

Figure 7.35 Mode 4 configuration.

The last mode of 82C54 counter operation, mode 5, is called *hardware-triggered strobe*. This mode is similar to mode 4 except that now counting is initiated by a signal at the gate input. That is, it is hardware triggered instead of software triggered. As shown in the waveforms of Fig. 7.29 and table of Fig. 7.30, a rising edge at GATE starts the countdown process. Just as for software triggered strobed mode of operation, the strobe pulse is output after the count is decremented to zero. But in this case, OUTPUT switches to logic 0 N clock pulses after GATE becomes active.

7.11 THE 82C37A PROGRAMMABLE DIRECT MEMORY ACCESS CONTROLLER

The 82C37A is the LSI controller IC that can be used to implement the *direct memory access* (DMA) function in 80386DX-based microcomputer systems. DMA capability permits devices, such as peripherals, to perform high-speed data transfers between either two sections of memory or between memory and an I/O device. In a microcomputer system, the memory or I/O bus cycles initiated as part of a DMA transfer are not performed by the MPU; instead, they are performed by a device known as a *DMA controller*, such as the 82C37A. DMA mode of operation is most frequently used when blocks or packets of data are to be transferred. For instance, disk controllers, local area network controllers, and communication controllers are devices that normally process data as blocks or packets. A single 82C37A supports up to four peripheral devices for DMA operation.

Microprocessor Interface of the 82C37A

A block diagram that shows the interface signals of the 82C37A DMA controller is given in Fig. 7.36(a). The pin layout in Fig. 7.36(b) identifies the pins at which these signals are available. Let us now look briefly at the operation of the microprocessor interface of the 82C37A.

In a microcomputer system, the 82C37A acts as a peripheral device and its operation must be initialized and controlled through software. This is done by reading from or writing to the bits of its internal registers. These data transfers take place through its microprocessor interface. Figure 7.37 shows how the 80386DX connects to the 82C37A's microprocessor interface.

Whenever the 82C37A is not in use by a peripheral device for DMA operation, it is in a state known as the *idle state*. When in this state, the microprocessor can issue commands to the DMA controller and read from or write to its internal registers. Data bus lines D_0 through D_7 are the paths over which these data transfers take place. Which register is accessed is determined by a 4-bit register address that is applied to address inputs A_0 through A_3. As shown in Fig. 7.37, these inputs are supplied by address lines A_2 through A_5 of the microprocessor.

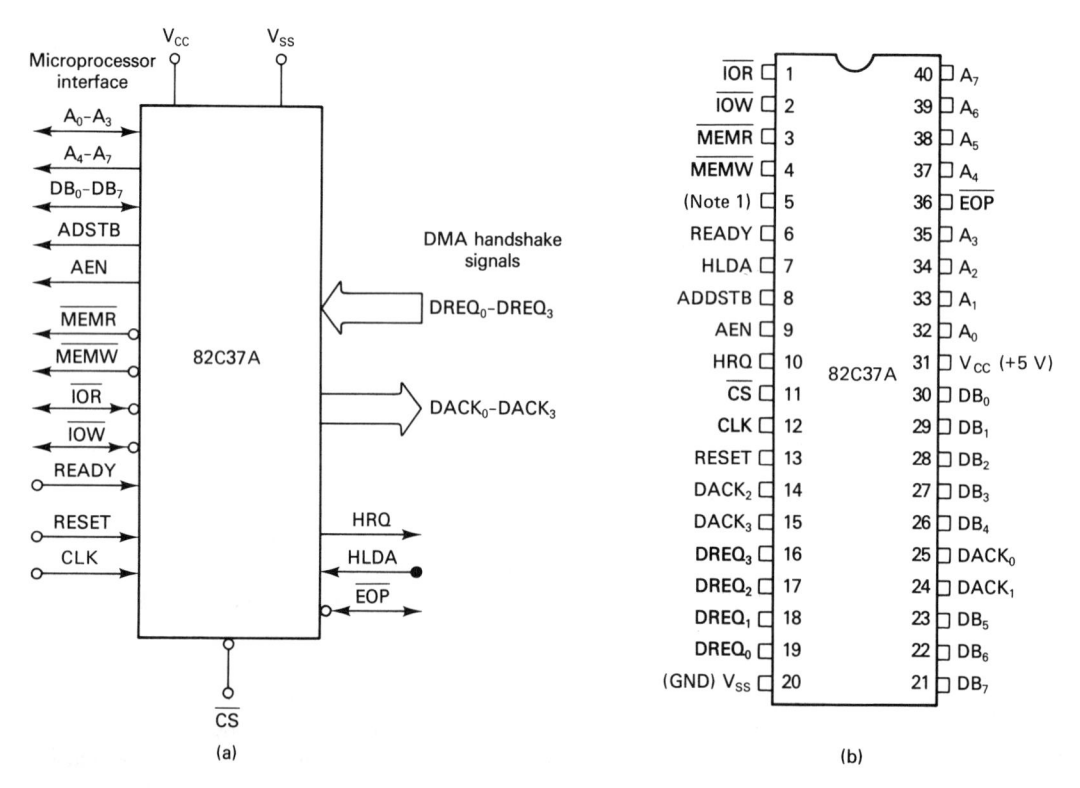

Figure 7.36 (a) Block diagram of the 82C37A DMA controller; (b) pin layout. (Reprinted by permission of Intel Corp. Copyright Intel Corp. 1987.)

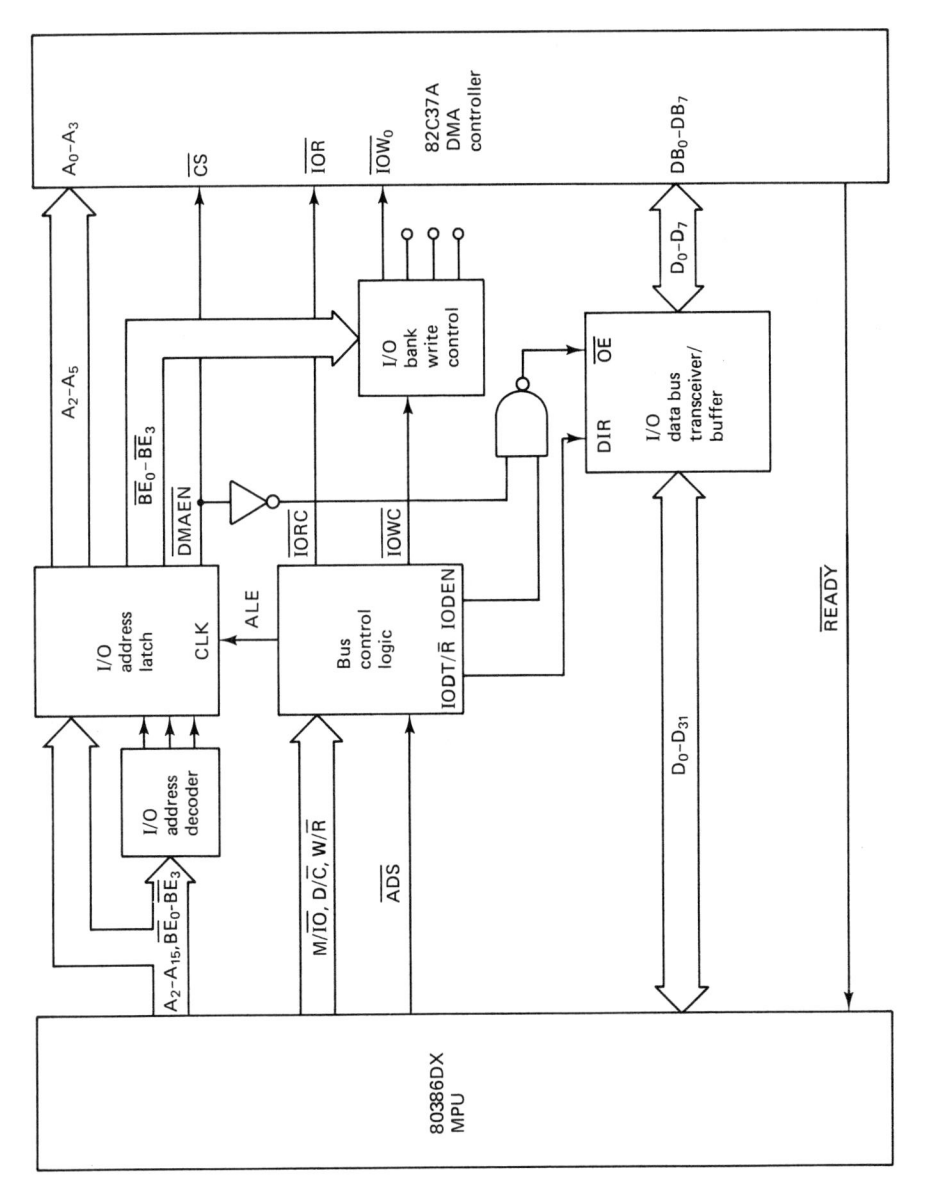

Figure 7.37 Interfacing the 82C37A to the 80386DX.

During the data transfer bus cycle, other bits of the address are decoded in external circuitry to produce a chip select ($\overline{CS}$) input for the 82C37A. When in the idle state, the 82C37A continuously samples this input, waiting for it to become active. Logic 0 at $\overline{CS}$ enables the microprocessor interface. The microprocessor tells the 82C37A whether an input or output bus cycle is in progress with the signal at $\overline{IOR}$ or $\overline{IOW}_0$, respectively. In this way, we see that the 82C37A is intended to be mapped into the I/O address space of the 80386DX microcomputer.

DMA Interface of the 82C37A

Now that we have described how a microprocessor talks to the registers of the 82C37A let us continue by looking at how peripheral devices initiate DMA service. The 82C37A contains four independent DMA channels called channel 0 through channel 3. Typically each of these channels is dedicated to a specific device, such as a peripheral. In Fig. 7.38 we see that the device has four DMA request inputs, denoted as $DREQ_0$ through $DREQ_3$. These DREQ inputs correspond to channels 0 through 3, respectively. In the idle state, the 82C37A continuously tests these inputs to see if one is active. When a peripheral device wants to perform DMA operations, it makes a request for service at its DREQ input by switching the input to logic 1.

In response to an active DMA request, the DMA controller switches the hold request (HRQ) output to logic 1. Normally, this output is supplied to the HOLD input of the 80386DX and signals the microprocessor that the DMA controller needs to take control of the system bus. When the 80386DX is ready to give up control of the bus, it puts its bus signals into the high-impedance state and signals this fact to the 82C37A by switching the HLDA (hold acknowledge) output to logic 1. HLDA of the 80386DX is applied to the HLDA input of the 82C37A and signals that the system bus is now available for use by the DMA controller.

When the 82C37A has control of the system bus, it tells the requesting peripheral device that it is ready by outputting a DMA acknowledge (DACK) signal. Notice in Fig. 7.38 that each of the four DMA request inputs, $DREQ_0$ through $DREQ_3$, has a corresponding DMA acknowledge output, $DACK_0$ through $DACK_3$. Once this DMA request/acknowledge handshake sequence is complete, the peripheral device gets direct access to the system bus and memory under control of the 82C37A.

During DMA bus cycles, the system bus is driven by the DMA controller, not the MPU. The 82C37A generates the address and all control signals needed to perform the memory or I/O data transfers. At the beginning of all DMA bus cycles, a 16-bit address is output on lines A_0 through A_7 and DB_0 through DB_7. The upper 8 bits of the address, which are available on the data bus lines, appear at the same time that address strobe (ADSTB) becomes active. Thus ADSTB is intended to be used to strobe the most significant byte of the address into an external address latch. This 16-bit address gives the 82C37A the ability to directly address up to 64K bytes of storage locations. The address enable (AEN) output

Figure 7.38 DMA interface to I/O devices.

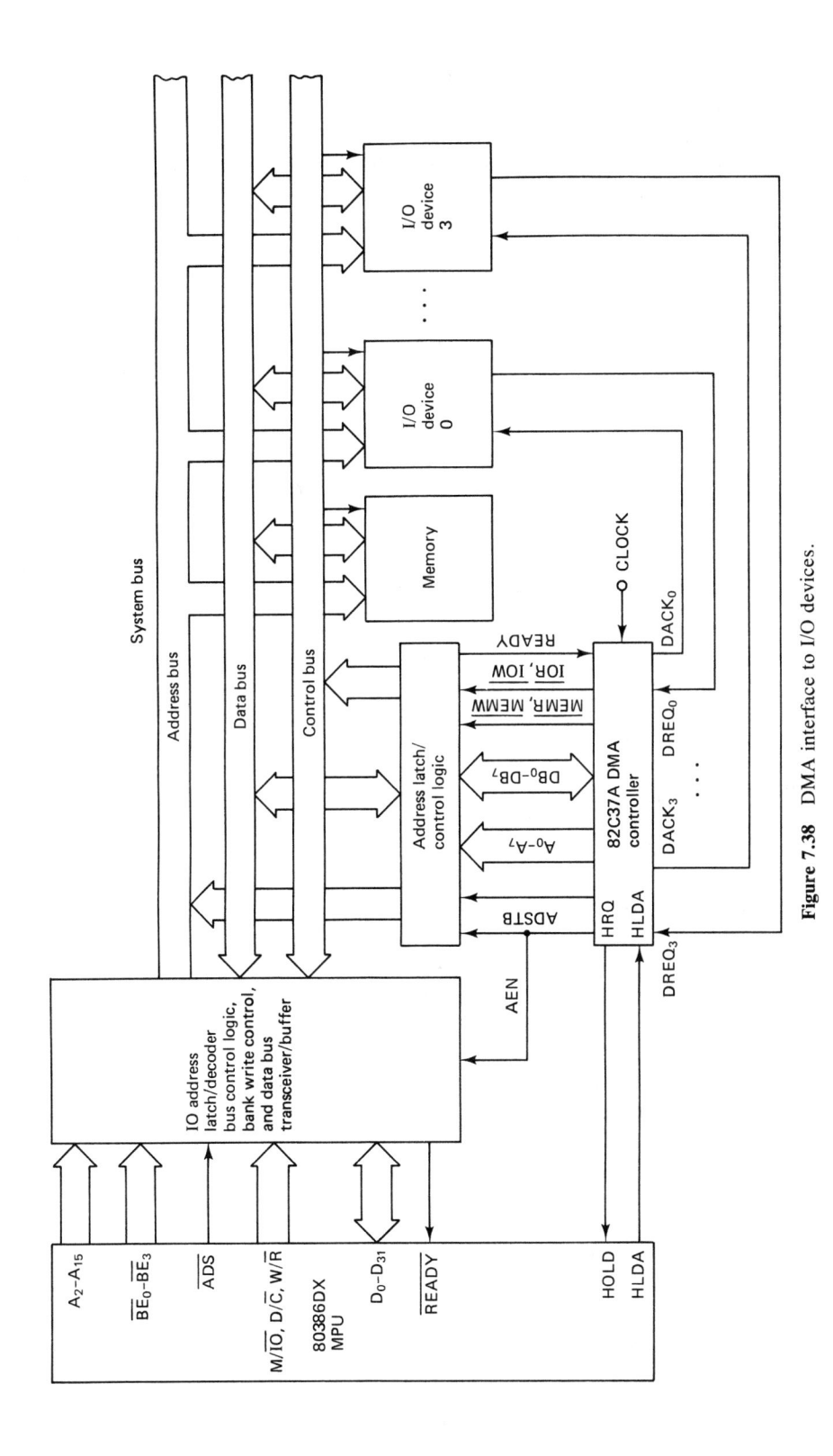

signal is active during the complete DMA bus cycle and can be used to both enable the address latch and disable other devices connected to the bus.

Let us assume for now that an I/O peripheral device is to transfer data to memory. That is, the I/O device wants to write data to memory. In this case, the 82C37A uses the $\overline{\text{IOR}}$ output to signal the I/O device to put the data onto data bus lines DB_0 through DB_7. At the same time, it asserts $\overline{\text{MEMW}}$ to signal that the data available on the bus are to be written into memory. In this case, the data are transferred directly from the I/O device to memory and do not go through the 82C37A.

In a similar way, DMA transfers of data can take place from memory to an I/O device. Now the I/O device reads data from memory. For this data transfer, the 82C37A activates the $\overline{\text{MEMR}}$ and $\overline{\text{IOW}}$ control signals.

The 82C37A performs both the memory-to-I/O and I/O-to-memory DMA bus cycles in just four clock periods. The duration of these clock periods is determined by the frequency of the clock signal applied to the CLOCK input. For instance, at 5 MHz the clock period is 200 ns and the bus cycle takes 800 ns.

The 82C37A is also capable of performing memory-to-memory DMA transfers. In such a data transfer both the $\overline{\text{MEMR}}$ and $\overline{\text{MEMW}}$ signals are utilized. Unlike the I/O-to-memory operation, this memory-to-memory data transfer takes eight clock cycles. This is because it is actually performed as a separate four-clock read bus cycle from the source memory location to a temporary register within the 82C37A and then another four-clock write bus cycle from the temporary register to the destination memory location. At 5 MHz a memory-to-memory DMA cycle takes 1.6 μs.

The READY input is used to accommodate for slow memory or I/O devices. READY must go active, logic 1, before the 82C37A will complete a memory or I/O bus cycle. As long as READY is at logic 0, wait states are inserted to extend the duration of the current bus cycle.

Internal Architecture of the 82C37A

Figure 7.39 is a block diagram of the internal architecture of the 82C37A DMA controller. Here we find the following functional blocks: the timing and control, the priority encoder and rotating priority logic, the command control, and 12 different types of registers. Let us now look briefly at the functions performed by each of these sections of circuitry and registers.

The timing and control part of the 82C37A generates the timing and control signals needed by the external bus interface. For instance, it accepts as inputs the READY and $\overline{\text{CS}}$ signals and produces as outputs signals such as ADDSTB and AEN. These signals are synchronized to the clock signal that is input to the controller. The highest speed version of the 82C37A available today operates at a maximum clock rate of 5 MHz.

If multiple requests for DMA service are received by the 82C37A, they are accepted on a priority basis. One of two priority schemes can be selected for the 82C37A under software control. They are called *fixed priority* and *rotating priority*. The fixed priority mode assigns priority to the channels in descending numeric

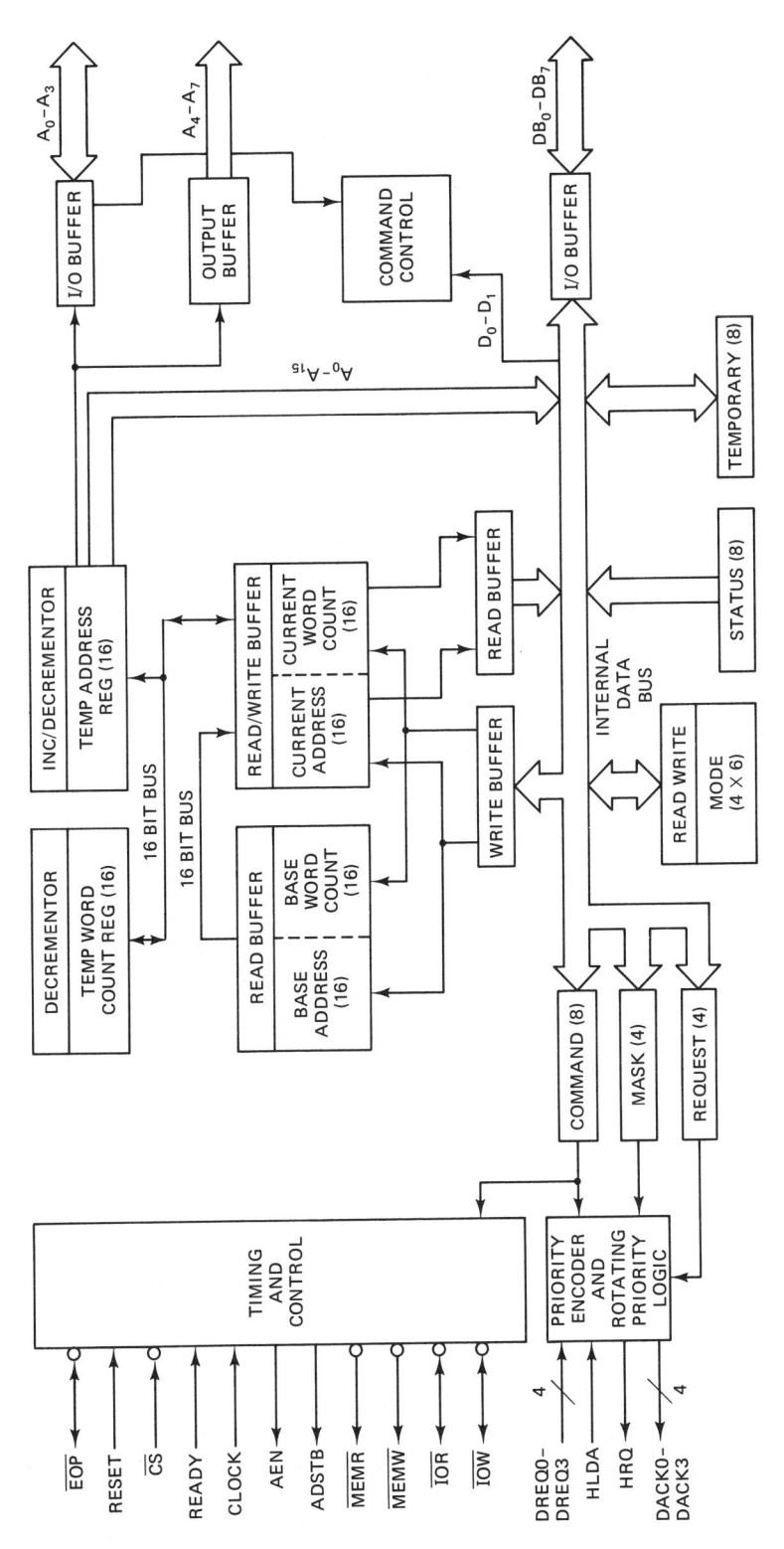

Figure 7.39 Internal architecture of the 82C37A. (Reprinted by permission of Intel Corp. Copyright Intel Corp. 1987.)

order. That is, channel 0 has the highest priority and channel 3 the lowest priority. Rotating priority starts with the priority levels initially the same way as in fixed priority. However, after a DMA request for a specific level gets serviced, priority is rotated such that the previously active channel is reassigned to the lowest priority level. For instance, assuming that channel 1, which was initially at priority level 1, was just serviced, then $DREQ_2$ is now at the highest priority level and $DREQ_1$ rotates to the lowest level. The priority logic circuitry shown in Fig. 7.39 resolves priority for simultaneous DMA requests from peripheral devices based on the enabled priority scheme.

The command control circuit decodes the register commands applied to the 82C37A through the microprocessor interface. In this way it determines which register is to be accessed and what type of operation is to be performed. Moreover, it is used to decode the programmed operating modes of the device during DMA operation.

Looking at the block diagram in Fig. 7.39, we find that the 82C37A has 12 different types of internal registers. Some examples are the current address register, current count register, command register, mask register, and status register. The names for all of the internal registers are listed in Fig. 7.40 along with their size and how many are provided in the 82C37A. Notice that there are actually four current address registers and they are all 16 bits long. That is, there is one current address register for each of the four DMA channels. We will now describe the function served by each of these registers in terms of overall operation of the 82C37A DMA controller. Addressing information for the internal registers is summarized in Fig. 7.41.

Name	Size	Number
Base Address Registers	16 bits	4
Base Word Count Registers	16 bits	4
Current Address Registers	16 bits	4
Current Word Count Registers	16 bits	4
Temporary Address Register	16 bits	1
Temporary Word Count Register	16 bits	1
Status Register	8 bits	1
Command Register	8 bits	1
Temporary Register	8 bits	1
Mode Registers	6 bits	4
Mask Register	4 bits	1
Request Register	4 bits	1

Figure 7.40 Internal registers of the 82C37A. (Reprinted by permission of Intel Corp. Copyright Intel Corp. 1987.)

Each DMA channel has two address registers. They are called the *base address register* and the *current address register*. The base address register holds the starting address for the DMA operation and the current address register contains the address of the next storage location to be accessed. Writing a value to the base address register automatically loads the same value into the current address register. In this way, we see that initially the current address register points to the starting I/O or memory address.

These registers must be loaded with appropriate values prior to initiating a DMA cycle. To load a new 16-bit address into the base register, we must write 2 separate bytes, one after the other, to the address of the register. The 82C37A has an internal flip-flop called the *first/last flip-flop*. This flip-flop identifies which

Channel(s)	Register	Operation	Register address	Internal FF	Data bus
0	Base and current address	Write	0_{16}	0 1	Low High
	Current address	Read	0_{16}	0 1	Low High
	Base and current count	Write	1_{16}	0 1	Low High
	Current count	Read	1_{16}	0 1	Low High
1	Base and current address	Write	2_{16}	0 1	Low High
	Current address	Read	2_{16}	0 1	Low High
	Base and current count	Write	3_{16}	0 1	Low High
	Current count	Read	3_{16}	0 1	Low High
2	Base and current address	Write	4_{16}	0 1	Low High
	Current address	Read	4_{16}	0 1	Low High
	Base and current count	Write	5_{16}	0 1	Low High
	Current count	Read	5_{16}	0 1	Low High
3	Base and current address	Write	6_{16}	0 1	Low High
	Current address	Read	6_{16}	0 1	Low High
	Base and current count	Write	7_{16}	0 1	Low High
	Current count	Read	7_{16}	0 1	Low High
All	Command register	Write	8_{16}	X	Low
All	Status register	Read	8_{16}	X	Low
All	Request register	Write	9_{16}	X	Low
All	Mask register	Write	A_{16}	X	Low
All	Mode register	Write	B_{16}	X	Low
All	Temporary register	Read	B_{16}	X	Low
All	Clear internal FF	Write	C_{16}	X	Low
All	Master clear	Write	D_{16}	X	Low
All	Clear mask register	Write	E_{16}	X	Low
All	Mask register	Write	F_{16}	X	Low

Figure 7.41 Accessing the registers of the 82C37A.

byte of the address is being written into the register. As shown in the table of Fig. 7.41, if the beginning state of the internal flip-flop (FF) is logic 0, then software must write the low byte of the address word to the register. On the other hand, if it is logic 1, the high byte must be written to the register. For example, to write the address 1234_{16} into the base address register and the current address register for channel 0 of a DMA controller located at I/O address 'DMA' (where DMA $\leq$ F0H), the following instructions may be executed:

```
MOV   AL, 34H              ;write low byte
OUT   DMA + 0H, AL
MOV   AL, 12H              ;write high byte
OUT   DMA + 0H, AL
```

This routine assumes that the internal flip-flop was initially set to 0. Looking at Fig. 7.41 we find that a command can be issued to the 82C37A to clear the internal flip-flop. This is done by initiating an output bus cycle to register address C_{16}. Assuming that the microcomputer uses the byte-wide I/O interface illustrated in Fig. 7.6, the address input A_0 must be driven by address line A_2. Therefore, the hardware address of the clear flip-flop would be DMA + 30_{16}.

If we read the contents of register address 0_{16}, the value obtained is the contents of the current address register. Once loaded, the value in the base address register cannot be read out of the device.

The 82C37A also has two word count registers for each of its DMA channels. They are called the *base count register* and the *current count register*. In Fig. 7.40 we find that these registers are also 16 bits in length and Fig. 7.41 identifies their register address as 1_{16}. The number of bytes of data that are to be transferred during a DMA operation is specified by the value in the base count register. Actually, the number of bytes transferred is always one more than the value programmed into this register. This is because the end of a DMA cycle is detected by the rollover of the current count from 0000_{16} to $FFFF_{16}$. At any time during the DMA cycle, the value in the current word count register tells how many bytes remain to be transferred.

The count registers are programmed in the same way as just described for the address registers. For instance, to program a count of $0FFF_{16}$ into the base and current count registers for channel 1 of a DMA controller located at address 'DMA,' the instructions that follow can be executed:

```
MOV   AL, FFH              ;write low byte
OUT   DMA + 8H, AL
MOV   AL, 0FH              ;write high byte
OUT   DMA + 8H, AL
```

Again we have assumed that the internal flip-flop was initially cleared and that the microcomputer employs the byte-wide I/O interface of Fig. 7.6.

In Fig. 7.40 we find that the 82C37A has a single eight-bit command register. The bits in this register are used to control operating modes that apply to all channels of the DMA controller. Figure 7.42 identifies the function of each of its control bits. Notice that the settings of the bits are used to select or deselect operating features such as memory-to-memory DMA transfer and the priority scheme. For instance, when bit 0 is set to logic 1, the memory-to-memory mode of DMA transfer is enabled and when it is logic 0, DMA transfers take place between I/O and memory. Moreover, setting bit 4 to logic 0 selects the fixed priority scheme for all four channels or logic 1 in this location selects rotating priority. Looking at Fig. 7.41 we see that the command register is loaded by outputting the command code to register address 8_{16}.

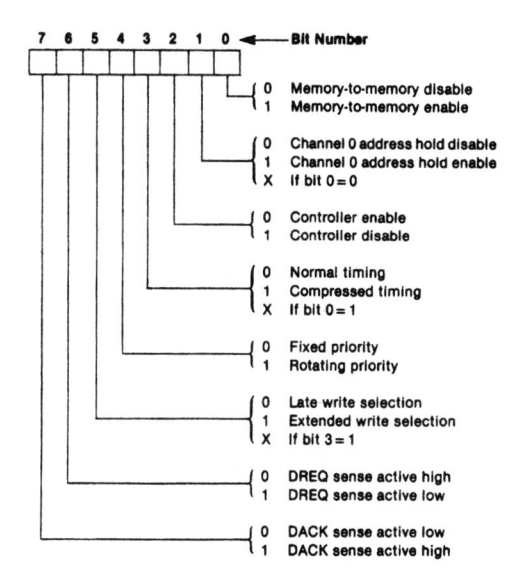

Figure 7.42 Command register format. (Reprinted by permission of Intel Corp. Copyright Intel Corp. 1987.)

EXAMPLE 7.28

If the command register of an 82C37A is loaded with 01_{16}, how does the controller operate?

SOLUTION Representing the command word as a binary number, we get

$$01_{16} = 00000001_2$$

Referring to Fig. 7.42 we find that the DMA operation can be described as follows:

Bit 0 = 1 = memory-to-memory transfers are disabled

Bit 1 = 0 = channel 0 address increments/decrements normally

Bit 2 = 0 = 82C37A is enabled

Bit 3 = 0 = 82C37A operates with normal timing

Bit 4 = 0 = channels have fixed priority, channel 0 having the highest priority and channel 3 the lowest priority

Bit 5 = 0 = write operation occurs late in the DMA bus cycle

$\Bigl\lbrack$ Bit 6 = 0 = DREQ is an active-high (logic 1) signal
Bit 7 = 0 = DACK is an active-low (logic 0) signal

The *mode registers* are also used to configure operational features of the 82C37A. In Fig. 7.40 we find that there is a separate mode register for each of the four DMA channels and that they are each six bits in length. Their bits are used to select various operational features for the individual DMA channels. A typical mode register command is shown in Fig. 7.43. As shown in the diagram, the two least significant bits are a two-bit code that identifies the channel to which the mode command byte applies. For instance, in a mode register command written for channel 1, these bits must be made 01. Bits 2 and 3 specify whether the channel is to perform data write, data read, or verify bus cycles. For example, if these bits are set to 01, the channel will only perform write data transfers (DMA data transfers from an I/O device to memory).

The next two bits of the mode register affect how the values in the current address and current count registers are updated at the end of a DMA cycle and DMA data transfer, respectively. Bit 4 enables or disables the autoinitialization function. When autoinitialization is enabled, the current address and current count registers are automatically reloaded from the base address and base count registers, respectively, at the end of a DMA cycle. In this way, the channel is prepared for the next cycle to begin. The setting of bit 5 determines whether the value in the current address register is automatically incremented or decremented at completion of each DMA data transfer.

The two most significant bits of the mode register select one of four possible modes of DMA operation for the channel. The four modes are called *demand mode*, *single mode*, *block mode*, and *cascade mode*. These modes allow for either one byte of data or a block of bytes to be transferred at a time. For example, when in the demand transfer mode, once the DMA cycle is initiated, bytes are continuously transferred as long as the DREQ signal remains active and the ter-

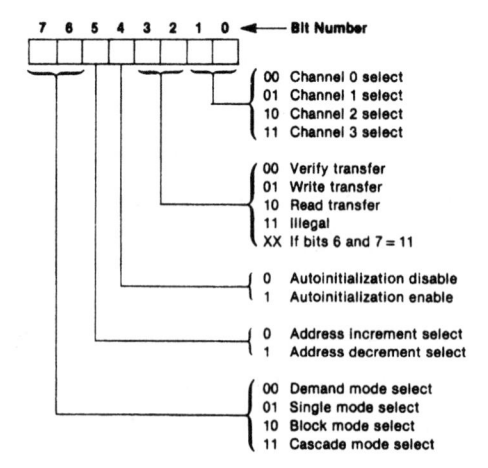

```
7 6 5 4 3 2 1 0 ◄── Bit Number

                  00  Channel 0 select
                  01  Channel 1 select
                  10  Channel 2 select
                  11  Channel 3 select

                  00  Verify transfer
                  01  Write transfer
                  10  Read transfer
                  11  Illegal
                  XX  If bits 6 and 7 = 11

                  0   Autoinitialization disable
                  1   Autoinitialization enable

                  0   Address increment select
                  1   Address decrement select

                  00  Demand mode select
                  01  Single mode select
                  10  Block mode select
                  11  Cascade mode select
```

Figure 7.43 Mode register format. (Reprinted by permission of Intel Corp. Copyright Intel Corp. 1987.)

minal count (TC) is not reached. By reaching the terminal count, we mean that the value in the current word count register, which automatically decrements after each data transfer, rolls over from 0000_{16} to $FFFF_{16}$.

Block transfer mode is similar to demand transfer mode in that once the DMA cycle is initiated data are continuously transferred until the terminal count is reached. However, they differ in that when in the demand mode, the return of DREQ to its inactive state halts the data transfer sequence. But, when in block transfer mode, DREQ can be released at any time after the DMA cycle begins and the block transfer will still run to completion.

In the single transfer mode, the channel is set up such that it performs just one data transfer at a time. At the completion of the transfer, the current word count is decremented and the current address either incremented or decremented (based on an option setting). Moreover, an autoinitialize, if enabled, will not occur unless the terminal count has been reached at the completion of the current data transfer. If the DREQ input becomes inactive before the completion of the current data transfer, another data transfer will not take place until DREQ once more becomes active. On the other hand, if DREQ remains active during the complete data transfer cycle, the HRQ output of the 82C37A is switched to its inactive 0 logic level to allow the microprocessor to gain control of the system bus for one bus cycle before another single transfer takes place. This mode of operation is typically used when it is necessary to not lock the microprocessor off the bus for the complete duration of the DMA cycle.

EXAMPLE 7.29

Specify the mode byte for DMA channel 2 if it is to transfer data from an input peripheral device to a memory buffer starting at address $A000_{16}$ and ending at $AFFF_{16}$. Ensure that the microprocessor is not completely locked off the bus during the DMA cycle. Moreover, at the end of each DMA cycle the channel is to be reinitialized so that the same buffer is to be filled when the next DMA operation is initiated.

SOLUTION For DMA channel 2, bit 1 and bit 0 must be loaded with 10.

$$B_1B_0 = 10$$

Transfer of data from an I/O device to memory represents a write bus cycle. Therefore, bit 3 and bit 2 must be set to 01.

$$B_3B_2 = 01$$

Selecting autoinitialization will set up the channel to automatically reset so that it points to the beginning of the memory buffer at completion of the current DMA cycle. This feature is enabled by making bit 4 equal to 1.

$$B_4 = 1$$

The address that points to the memory buffer must increment after each data transfer. Therefore, bit 5 must be set to 0.

$$B_5 = 0$$

Finally, to assure that the 80386DX is not locked off the bus during the complete DMA cycle, we will select the single transfer mode of operation. This is done by making bits B_7 and B_6 equal to 01.

$$B_7B_6 = 01$$

Thus the mode register byte is $01010110_2 = 56_{16}$.

Up to now we have just discussed how DMA cycles can be initiated with a hardware request at a DREQ input. However, the 82C37A is also able to respond to software initiated requests for DMA service. The *request register* has been provided for this purpose. Figure 7.40 shows that the request register has just four bits, one for each of the DMA channels. When the request bit for a channel is set, DMA operation is started and when reset the DMA cycle is stopped. Any channel used for software-initiated DMA must be programmed for block transfer mode of operation.

The bits in the request register can be set or reset by issuing software commands to the 82C37A. The format of a request register command is shown in Fig. 7.44. For instance, if a command is issued to the address of the request register with bits 0 and 1 equal 01 and with bit 2 at logic 1, a block mode DMA cycle is initiated for channel 1. In Fig. 7.41 we find that the request register is located at register address 9_{16}.

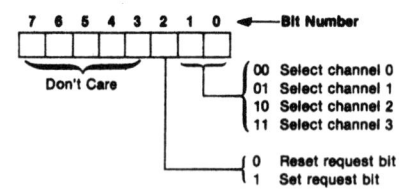

Figure 7.44 Request register format. (Reprinted by permission of Intel Corp. Copyright Intel Corp. 1987.)

A 4-bit *mask register* is also provided within the 82C37A. One bit is provided in this register for each of the DMA channels. When a mask bit is set, the DREQ input for the corresponding channel is disabled. That is, hardware requests to the channel are ignored and the channel is masked out. On the other hand, if the mask bit is cleared, the DREQ input is enabled and its channel can be activated by an external device.

The format of a software command that can be used to set or reset a single bit in the mask register is shown in Fig. 7.45(a). For example, to enable the DREQ input for channel 2, the command is issued with bits 0 and 1 set to 10 to select channel 2 and with bit 2 equal 0 to clear the mask bit. For this example, the software command byte could be 06_{16}. The table in Fig. 7.41 shows that this command byte must be issued to the 82C37A with register address A_{16}.

A second mask register command is shown in Fig. 7.45(b). This command can be used to load all 4 bits of the register at once. In Fig. 7.41 we find that this

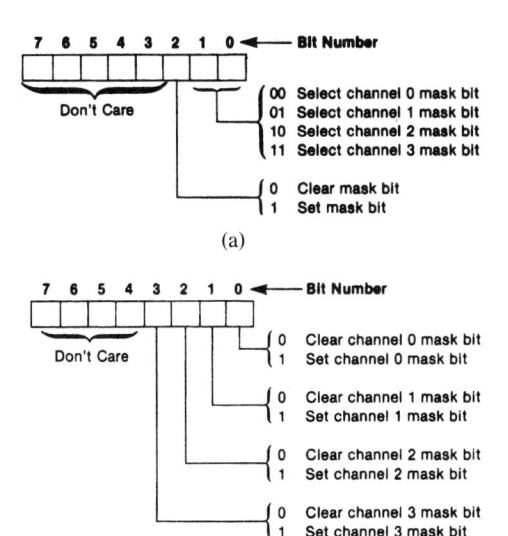

```
7 6 5 4 3 2 1 0  ◄─── Bit Number
```
```
     Don't Care    ┌ 00 Select channel 0 mask bit
                   │ 01 Select channel 1 mask bit
                   │ 10 Select channel 2 mask bit
                   └ 11 Select channel 3 mask bit

                   ┌ 0 Clear mask bit
                   └ 1 Set mask bit
```
(a)

```
7 6 5 4 3 2 1 0  ◄─── Bit Number
```
```
     Don't Care    ┌ 0 Clear channel 0 mask bit
                   └ 1 Set channel 0 mask bit

                   ┌ 0 Clear channel 1 mask bit
                   └ 1 Set channel 1 mask bit

                   ┌ 0 Clear channel 2 mask bit
                   └ 1 Set channel 2 mask bit

                   ┌ 0 Clear channel 3 mask bit
                   └ 1 Set channel 3 mask bit
```
(b)

Figure 7.45 (a) Single-channel mask register command format. (Reprinted by permission of Intel Corp. Copyright Intel Corp. 1987.), (b) Four-channel mask register command format. (Reprinted by permission of Intel Corp. Copyright Intel Corp. 1987.)

command is issued to register address F_{16} instead of A_{16}. For instance, to mask out channel 2 while enabling channels 0, 1, and 3, the command code 04_{16} is output to F_{16}. Either of these two methods can be used to mask or enable the DREQ input for a channel.

At system initialization, it is a common practice to clear the mask register. Looking at Fig. 7.41, we see that a special command is provided to perform this operation. Notice that the mask register can be cleared by executing an output cycle to register address E_{16}.

The 82C37A has a *status register* that contains information about the operating state of its four DMA channels. Figure 7.46 shows the bits of the status register and defines their functions. Here we find that the 4 least significant bits identify whether or not channels 0 through 3 have reached their terminal count. When the DMA cycle for a channel reaches the terminal count, this fact is recorded by setting the corresponding TC bit to the 1 logic level. The four most significant bits of the register tell if a request is pending for the corresponding channel. For

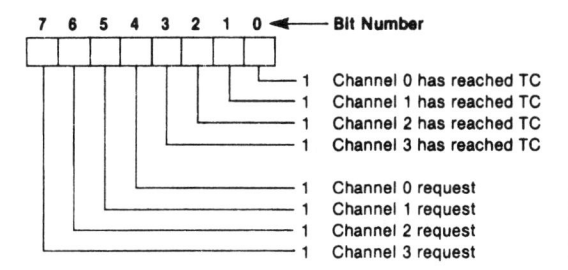

```
7 6 5 4 3 2 1 0  ◄─── Bit Number
```
```
             1 Channel 0 has reached TC
             1 Channel 1 has reached TC
             1 Channel 2 has reached TC
             1 Channel 3 has reached TC
             1 Channel 0 request
             1 Channel 1 request
             1 Channel 2 request
             1 Channel 3 request
```

Figure 7.46 Status register. (Reprinted by permission of Intel Corp. Copyright Intel Corp. 1987.)

instance, if a DMA request has been issued for channel 0 either through hardware or software, bit 4 is set to 1. The 80386DX can read the contents of the status register through software. This is done by initiating an input bus cycle for register address 8_{16}.

Earlier we pointed out that during memory-to-memory DMA transfers, the data read from the source address are held in a register known as the *temporary register* and then a write cycle is initiated to write the data to the destination address. At the completion of the DMA cycle, this register contains the last byte that was transferred. The value in this register can be read by the microprocessor.

EXAMPLE 7.30

Write an instruction sequence to issue a master clear to the 82C37A and then enable all its DMA channels. Assume that the device is located at I/O address 'DMA' in a byte-wide I/O interface such as the one shown in Fig. 7.6.

SOLUTION In Fig. 7.41 we find that a special software command is provided to perform a master reset of the 82C37A's registers. Since the contents of the data bus are a don't care state when executing the master clear command, it is performed by simply writing to register address D_{16} (hardware address 34H). For instance, the instruction

```
OUT    DMA + 34H, AL
```

can be used. To enable the DMA request inputs, all four bits of the mask register must be cleared. The clear mask register command is issued by performing a write to register address E_{16} (hardware address 38H). Again, the data put on the bus during the write cycle are a don't-care state. Therefore, the command can be performed with the instruction

```
OUT    DMA + 38H, AL
```

7.12 80386DX MICROCOMPUTER SYSTEM I/O CIRCUITRY

In Chapter 6 we examined the memory of the 80386DX-based microcomputer system shown in Fig. 7.47. This system includes an 82C54-2 timer/counter device in its I/O section. Moreover, the internal registers of the 8259A-2 interrupt controller are mapped into the I/O address space. Here we continue by studying the operation of this microcomputer's I/O interface.

Let us begin with the bus interface circuitry. The 74F139 memory address decoder device accepts inputs A_{31} and M/$\overline{\text{IO}}$ and decodes them to produce chip select outputs. When both M/$\overline{\text{IO}}$ and A_{31} equal 0, the $\overline{Y}_0$ output of the decoder is at its active 0 logic level. This signal indicates that an I/O address is on the bus. Notice that it is input to PAL 1 of the bus control logic as chip select I/O ($\overline{\text{CSIO}}$) and at the same time used to enable the I/O address decoder, which is the lower 74F139 device.

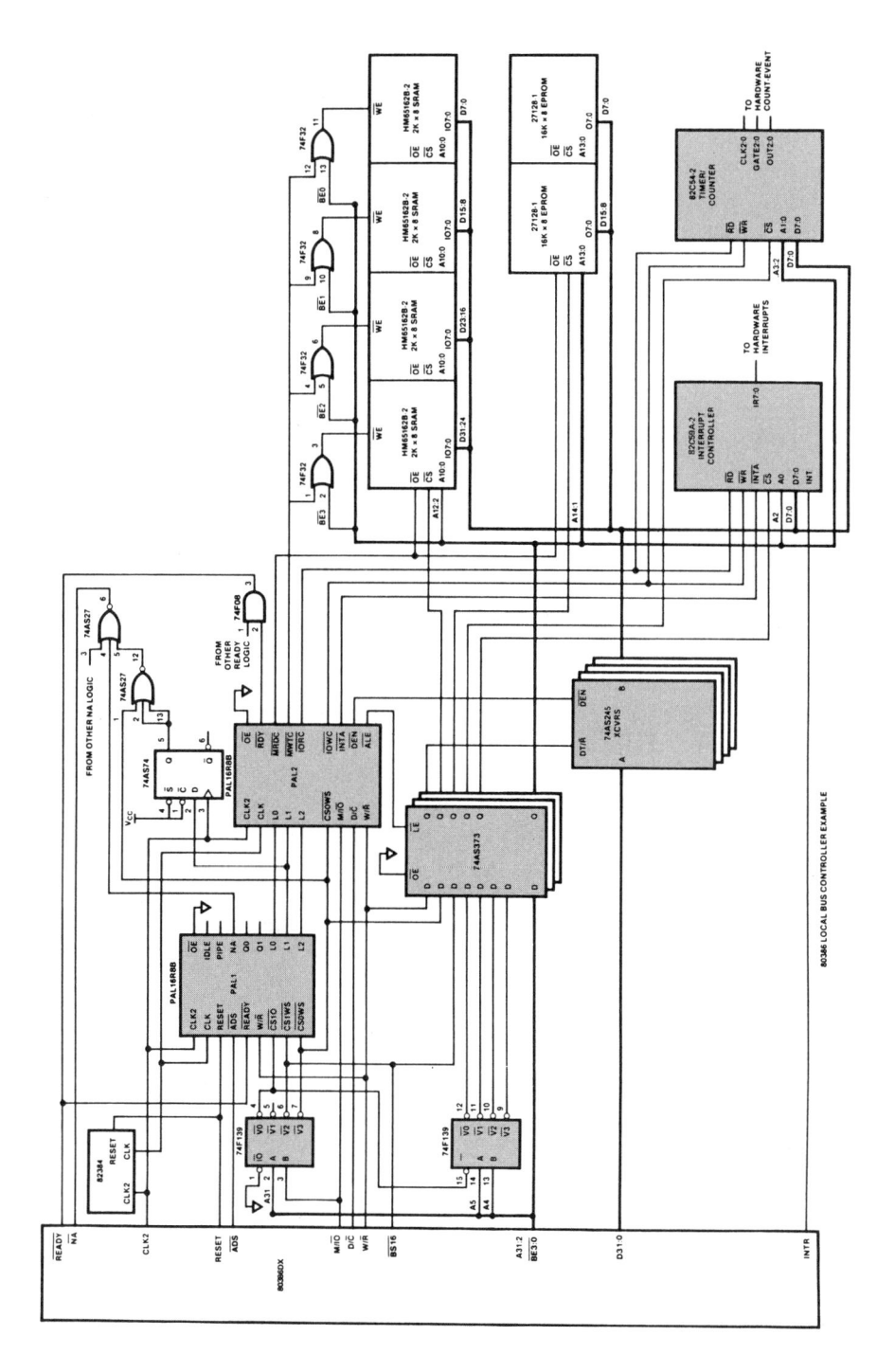

Figure 7.47 I/O interface of a typical 80386DX microcomputer. (Reprinted by permission of Intel Corp. Copyright Intel Corp. 1987.)

The I/O address decoder produces chip selects for the I/O peripherals. The control inputs of the decoder are address lines A_4 and A_5. When the decoder is enabled, these inputs are decoded to activate one of the four chip select signals, outputs $\overline{Y}_0$ through $\overline{Y}_3$. Tracing these signals in the circuit, we find that the $\overline{Y}_0$ output is supplied to the $\overline{CS}$ input of the 82C54-2 timer, and $\overline{Y}_1$ is sent to the $\overline{CS}$ input of the 8259A-2. Logic 0 must be applied to the $\overline{CS}$ input of these peripherals to enable their microprocessor interface for operation.

PAL 1 and PAL 2, which form the bus control logic, produce the control signals for the I/O bus interface. Notice that the inputs to this section of the circuit include the bus cycle definition signals $M/\overline{IO}$, $D/\overline{C}$, and $W/\overline{R}$, control signal $\overline{ADS}$, and chip selects $\overline{CSIO}$, $\overline{CS1WS}$, and $\overline{CS0WS}$. At the outputs they produce the signals needed to control the transfer of data between the 80386DX and I/O devices. These signals are address latch enable ($\overline{ALE}$), I/O read command ($\overline{IORC}$), I/O write command ($\overline{IOWC}$), data bus enable ($\overline{DEN}$), and ready ($\overline{RDY}$).

The I/O address signals, byte enable signals, chip select outputs of the I/O address decoder, and $W/\overline{R}$ are all latched with three 74AS373 latch devices. These latches are permanently enabled by the logic 0 applied at the $\overline{OE}$ input. Information is latched into the address latches synchronously with the $\overline{ALE}$ output from PAL 2 of the bus control logic.

The 82C54-2 timer supplies three independent 16-bit interval timers for use by the microcomputer. The inputs to the timers are clocks CLK_0 through CLK_2 and gates $GATE_0$ through $GATE_2$. The timer's outputs are OUT_0 through OUT_2. Earlier we pointed out that both the 8259A-2 and 82C54-2 are I/O mapped. For this reason, the read ($\overline{RD}$) and write ($\overline{WR}$) inputs on both devices are supplied by I/O read command ($\overline{IORC}$) and I/O write command ($\overline{IOWC}$), respectively, not $\overline{MRDC}$ and $\overline{MWTC}$. Moreover, the register to be accessed is selected by the code on A_2 and A_3 for the 82C54-2 and by just A_2 on the 8259A-2. Input and output data transfers between the 80386DX and the I/O peripherals are byte wide and take place over data bus lines D_0 through D_7. In the circuit diagram we see that the data bus lines are buffered with 74AS245 transceivers. The $\overline{DEN}$ output of the bus control logic enables the transceivers for operation whenever an I/O data transfer is in progress and the latched logic level of the $W/\overline{R}$ output of the 80386DX selects the direction in which data are transferred.

ASSIGNMENT

Section 7.2

1. Name the two types of input/output.
2. What type of I/O is in use when peripheral devices are mapped into the 80386DX's I/O address space?
3. Which type of I/O has the disadvantage that part of the memory address space must be given up to implement I/O ports?
4. Which type of I/O has the disadvantage that all I/O data transfers must take place through the AL, AX, or EAX register?

5. How many byte-wide I/O ports can exist in the 80386DX's I/O address space? Aligned word-wide ports? Aligned double-word-wide ports?
6. What is the address range of page 0?

Section 7.3

7. What are the functions of the 80386DX's address and data bus lines relative to input/output operation?
8. Which signal indicates to external circuitry that the current bus cycle is for the I/O interface and not for the memory interface?
9. If $\overline{BE_3}\overline{BE_2}\overline{BE_1}\overline{BE_0}$ = 1001 during an output bus cycle, is a byte, word, or double word of data being transferred? Over which data bus lines is it transferred?
10. What bus cycle indication code is output by the 80386DX during output bus cycles?
11. If the byte enable code $\overline{BE_3}\overline{BE_2}\overline{BE_1}\overline{BE_0}$ in the circuit of Fig. 7.4 is 0011, what are the $\overline{IOWR}$ signals for the current I/O write cycle?
12. If address lines $A_{15}A_{14}A_{13}$ = 100, which $\overline{IOCE}$ output is produced in the circuit of Fig. 7.4?
13. If in Fig. 7.4, just address lines A_2 through A_6 are used to select registers, what is the maximum number of registers that can be accessed?
14. Which signal in the circuit of Fig. 7.4 is used to enable the I/O data bus transceiver/buffer?
15. Describe briefly the function of each block in the I/O interface circuit in Fig. 7.6.
16. For the circuit in Fig. 7.6, if the address during an input bus cycle is A_{15} through A_2 = 0110000000011_2 and $\overline{BE_3}$ through $\overline{BE_0}$ = 1101, which chip enable output is active, which register is accessed in the enabled I/O device, and over which of the 80386DX's data bus lines is the data input?

Section 7.4

17. What is the minimum duration of I/O bus cycles for an 80386DX running at 25 MHz?
18. If an 80386DX-25 is running at full speed inserts two wait states into all I/O bus cycles, what is the duration of a nonpipelined bus cycle in which a byte of data are being output?
19. If the input cycles from byte-wide input ports of an 80386DX microcomputer running at 25 MHz are to be performed in a minimum of 250 ns, how many wait states are needed?
20. If the 80386DX in Problem 18 was outputting a word of data to a word-wide port at I/O address $1A3_{16}$, what would be the duration of the bus cycle?

Section 7.5

21. Describe the operation performed by the instruction IN AX, 1AH.
22. Write an instruction sequence to perform the same operation as that of the instruction in Problem 21, but this time use register DX to address the I/O port.
23. Describe the operation performed by the instruction OUT 2AH, AL.

24. Write an instruction sequence that will output the byte of data $0F_{16}$ to an output port at address 1000_{16}.
25. Write a sequence of instructions that will input the contents of the port at address $F004_{16}$, mask off all but the lower four bits, and output this value to the port at address $F000_{16}$.
26. Describe what happens when the instruction INSD is executed.
27. What operation is performed by the instruction sequence

```
MOV     DX, A000H
MOV     DI, 1001H
MOV     CX, 0FH
CLD
REPINSB
```

28. Write a sequence of instructions that will output the byte contents of memory addresses $ES:1001_{16}$ through $ES:100F_{16}$ to an output port at I/O address A010H.
29. What parameters identify the beginning of the I/O permission bit map in a TSS? At what address of the TSS is this parameter held?
30. At what double-word address in the I/O permission bit map is the bit for I/O port 64 held? Which bit of this double word corresponds to port 64?
31. To what logic level should the bit in Problem 30 be set if I/O operations are to be inhibited to the port in protected mode?

Section 7.6

32. What kind of input/output interface does a PPI implement?
33. How many I/O lines are available on the 82C55A?
34. What are the signal names of the I/O port lines of the 82C55A?
35. Describe the MODE 0, MODE 1, and MODE 2 I/O operations of the 82C55A PPI.
36. What is the function of the PORT B lines of the 82C55A when PORT A is configured for MODE 2 operation?
37. How is an 82C55A configured if its control register contains 9BH?
38. If the value $A4_{16}$ is written to the control register of an 82C55A, what is the mode and I/O configuration of PORT A? PORT B?
39. What should be the control word if ports A, B, and C of an 82C55A are all to be configured for MODE 0 operation? Moreover, ports A and B are to be used as inputs and C as an output.
40. What value must be written to the control register of the 82C55A to configure the device such that both PORT A and PORT B are to be configured for MODE 1 input operation?
41. If the control register of the 82C55A in Problem 39 is at I/O address 1000_{16}, write an instruction sequence that will load the control word.
42. Assume that the control register of an 82C55A resides at memory address 00100_{16}. Write an instruction sequence to load it with the control word formed in Problem 39.
43. What control word must be written to the control register of the 82C55A shown in Fig. 7.15(a) to enable the $INTR_B$ output? $INTE_B$ corresponds to bit PC_4 of PORT C.

44. If the value 03_{16} is written to the control register of an 82C55A set for MODE 2 operation, what bit at port C is affected by the bit set/reset operation? Is it set to 1 or cleared to 0?
45. Assume that the control register of the 82C55A for Problem 44 is at I/O address 0100_{16}. Write an instruction sequence that will load it with the bit set/reset value given in Problem 44.

Section 7.7

46. If I/O address $003D_{16}$ is applied to the circuit in Fig. 7.19 during a byte write cycle and the data output on the bus is 98_{16}, which 82C55A is being accessed? Are data being written into PORT A, PORT B, PORT C, or the control register of this device?
47. If the instruction

    IN   AL, 08H

    is executed to the I/O interface circuit in Fig. 7.19, what operation is performed?
48. What are the addresses of the A, B, and C ports of PPI 2 in the circuit of Fig. 7.19?
49. Assume that PPI 2 in Fig. 7.19 is configured as defined in Problem 39. Write a program that will input the data at ports A and B, add these values together, and output the sum to port C.

Section 7.8

50. Distinguish between memory-mapped I/O and isolated I/O.
51. What address inputs must be applied to the circuit in Fig. 7.20 to access PORT B of device 4? Assuming that all unused bits are 0, what would be the memory address?
52. Write an instruction that will load the control register of the port identified in Problem 51 with the value 98_{16}.
53. Repeat Problem 49 for the circuit in Fig. 7.20.

Section 7.9

54. Name a method that can be used to synchronize the input or output of information to a peripheral device.
55. List the control signals in the parallel printer interface circuit of Fig. 7.21(a). Identify whether it is an input or output of the printer and briefly describe its function.
56. Give an overview of what happens when a write bus cycle of byte-wide data is performed to I/O address 8000_{16}.
57. Show what push and pop instructions are needed in the program written in Example 7.19 to preserve the contents of registers used by the routine when it is used as a subroutine.

Section 7.10

58. What are the inputs and outputs of counter 2 of an 82C54?
59. Write a control word for counter 1 that selects the following options: load least significant byte only, mode 5 of operation, and binary counting.

60. What are the logic levels of inputs $\overline{CS}$, $\overline{RD}$, $\overline{WR}$, A_1, and A_0 when the byte in Problem 59 is written to an 82C54?
61. Write an instruction sequence that will load the control word in Problem 59 into an 82C54 that is located starting at address 01000_{16} of the memory address space. Assume that the device is attached to the I/O bus of the circuit in Fig. 7.6 and that address inputs A_0 and A_1 are supplied by address bits A_2 and A_3, respectively.
62. Write an instruction sequence that will write the value 12_{16} into the least significant byte of the count register for counter 2 of an 82C54 located starting at memory address 01000_{16}. Assume that the device is attached as in Problem 61.
63. Repeat Example 7.22 for the 82C54 located at memory address 01000_{16}, but this time just read the least significant byte of the counter.
64. What is the maximum time delay that can be generated with the timer in Fig. 7.31? What would be the maximum time delay if the clock frequency is increased to 2 MHz? Assume that it is configured for binary counting.
65. What is the resolution of pulses generated with the 82C54 in Fig. 7.31? What would the resolution be if the clock frequency is increased to 2 MHz?
66. Find the pulse width of the one-shot in Fig. 7.32 if the counter is loaded with the value 1000_{16}. Assume that the counter is configured for binary count operation.
67. What count must be loaded into the square-wave generator of Fig. 7.34 to produce a 25-kHz output?
68. If the counter in Fig. 7.35 is loaded with the value 120_{16}, how long a delay occurs before the strobe pulse is output?

Section 7.11

69. Are signal lines $\overline{MEMR}$ and $\overline{MEMW}$ of the 82C37A used in the microprocessor interface?
70. Overview the 82C37A's DMA request/acknowledge handshake sequence.
71. What is the total number of user-accessible registers in the 82C37A?
72. Write an instruction sequence that will read the value of the address from the current address register into the AX register.
73. Assuming that an 82C37A is located at I/O address 'DMA,' what is the address of the command register? Assume that the device is attached to the I/O bus in Fig. 7.6 and that address bit A_0 is attached to A_2.
74. Write an instruction sequence that will write the command word 00_{16} into the command register of an 82C37A that is located at address 'DMA' in the I/O address space.
75. Write an instruction sequence that will load the mode register for channel 2 with the mode byte obtained in Example 7.29. Assume that the 82C37A is located at I/O address 'DMA.'
76. What must be output to the mask register in order to disable all the DRQ inputs?
77. Write an instruction that will read the contents of the status register into the AL register. Assume that the 82C37A is located at address 'DMA.'

Section 7.12

78. Which Y output of the memory address decoder in the circuit of Fig. 7.47 is used as the signal $\overline{CSIO}$?

79. What code at M/$\overline{IO}$ A_{31} A_5 A_4 will make the Y output for the 82C54-2 become active?
80. What signal is applied to the $\overline{RD}$ inputs of the 8259-2 and 82C54-2?
81. Which data bus lines are used in the I/O interface?
82. Which address lines are used to select the registers within the 82C54-2 timer IC?

8

Interrupt and Exception Processing of the 80386DX Microprocessor

8.1 INTRODUCTION

In Chapter 7 we covered the input/output interface of the 80386DX-based microcomputer system. Here we will continue with a special input interface, the *interrupt interface*, and the *exception processing* capability of the 80386DX microprocessor. A list of the topics, in the order in which they are presented in this chapter, is as follows:

1. Types of interrupts and exceptions
2. Interrupt vector and descriptor tables
3. Interrupt instructions
4. Enabling/disabling of interrupts
5. External hardware interrupt interface
6. External hardware interrupt sequence
7. The 82C59A interrupt controller
8. Interrupt interface circuits using the 82C59A
9. Internal interrupts and exception functions

8.2 TYPES OF INTERRUPTS AND EXCEPTIONS

Interrupts provide a mechanism for quickly changing program environments. Transfer of program control is initiated by the occurrence of either an event internal to the 80386DX microprocessor or an event in its external hardware. For instance, when an interrupt signal occurs in external hardware indicating that an external device, such as a printer, requires service, the MPU must suspend what it is currently doing in the main part of the program and pass control to a special routine that performs the function required by the device. The section of program to which control is passed is called the *interrupt service routine*. In the case of our example of a printer, the routine is usually called the *printer driver*, which is the piece of software that when executed drives the printer output interface.

As shown in Fig. 8.1, interrupts supply a well-defined context switching mechanism for changing program environments. Here we see that interrupt 32

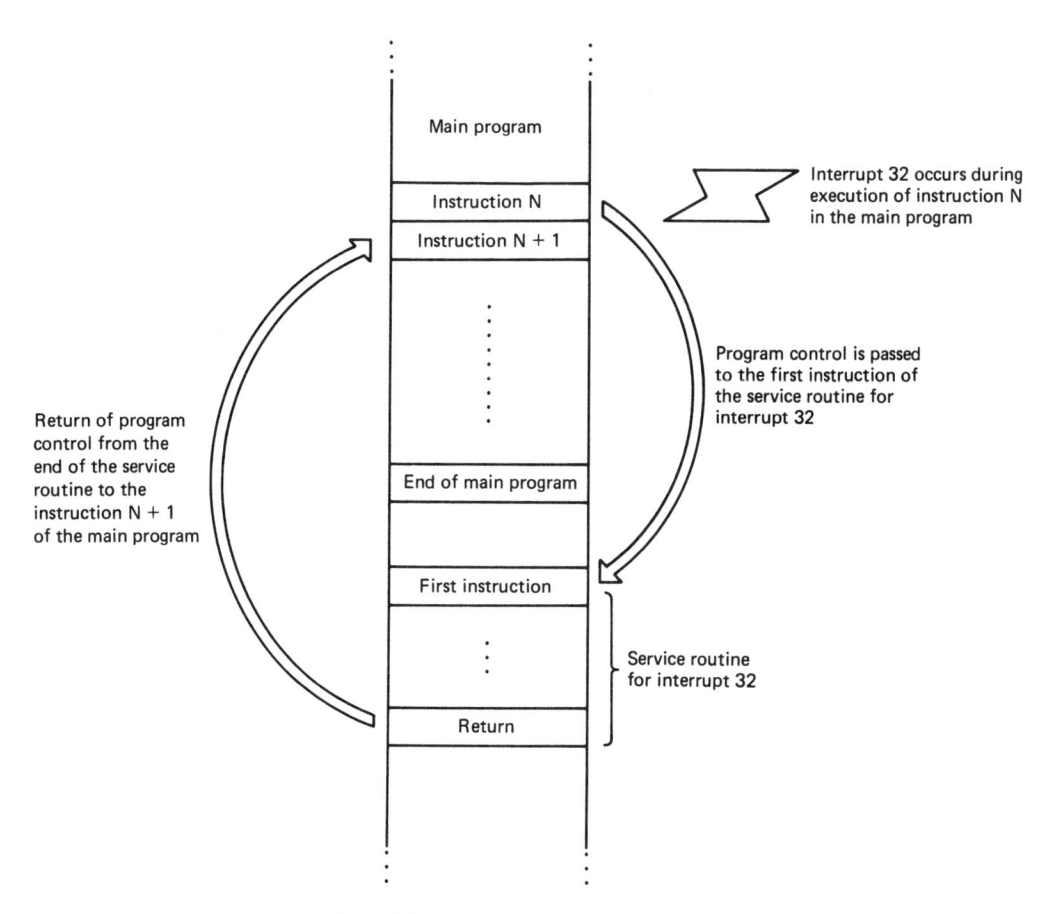

Figure 8.1 Interrupt program context switch mechanism.

occurs as instruction N of the program is being executed. When the 80386DX terminates execution of the main program in response to interrupt 32, it first saves information that identifies the instruction following the one where the interrupt occurred, which is instruction N + 1, and then picks up execution with the first instruction in the service routine. After this routine has run to completion, program control returns to the point where the MPU left the main program, instruction N + 1, and then execution resumes.

The 80386DX is capable of implementing any combination of up to 256 interrupts. As shown in Fig. 8.2, they are divided into five groups: *external hardware interrupts, nonmaskable interrupt, software interrupts, internal interrupts and exceptions*, and *reset*. The function of the external hardware, software, and nonmaskable interrupts are defined by the user. For instance, hardware interrupts are often assigned to devices such as the keyboard, printer, and timers. On the other hand, the functions of the internal interrupts and exceptions and reset are not user defined. They perform dedicated system functions.

Increasing priority

Reset
Internal interrupts and exceptions
Software interrupts
Nonmaskable interrupt
External hardware interrupts

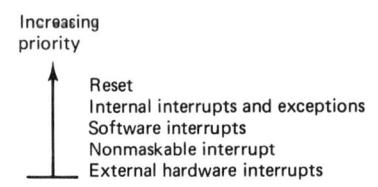

Figure 8.2 Types of interrupts and their priority.

Interrupts and exceptions are serviced on a *priority basis*. Priority is achieved in two ways. First the interrupt processing sequence implemented in the 80386DX tests for the occurrence of the various groups based on the hierarchy that follows: internal interrupts and exceptions, software interrupts, nonmaskable interrupt, and external hardware interrupts. Thus we see that internal interrupts and exceptions are the highest-priority group and the external hardware interrupts are the lowest-priority group.

Second, the various interrupts within a group are given different priority levels by assigning to each a *type number*. Type 0 identifies the highest-priority interrupt in the group and type 255 identifies the lowest-priority interrupt. Actually, a few of the type numbers are not available for use with software or hardware interrupts. This is because they are reserved for special interrupt functions of the 80386DX, such as the internal interrupts and exceptions. For instance, within the internal interrupt and exception group, the exception known as *divide error* is assigned to type number 0. Therefore, it has the highest priority of the exception functions. Another exception, called *general protection*, is assigned the type number 13.

The importance of priority lies in the fact that, if an interrupt service routine has been initiated to perform the function assigned to a specific priority level, only devices with higher priority are allowed to interrupt the active service routine. Lower-priority devices will have to wait until the current routine is completed before their request for service can be acknowledged. For hardware interrupts, this priority scheme is implemented in external hardware. For this reason, the user normally assigns tasks that must not be interrupted frequently to higher-priority levels, and those that can be interrupted, to the lower-priority levels.

An example of a high-priority service routine that should not be interrupted is that for a power failure. Once initiated this routine should be quickly run to completion to assure that the microcomputer goes through an orderly power down. A keyboard should also be assigned to a high-priority interrupt. This will assure that the keyboard buffer does not get full and lock out additional entries. On the other hand, devices such as the floppy disk or hard disk controller are typically assigned to a lower priority level.

We just pointed out that once an interrupt service routine is initiated, it can be interrupted only by a function that corresponds to a higher-priority level. For example, if a type 50 external hardware interrupt is in progress, it can be interrupted by the nonmaskable interrupt, all internal interrupts and exceptions, software interrupts, or any external hardware interrupt with type number less than 50. That is, external hardware interrupts with priority levels equal to 50 or greater are *masked out*.

8.3 INTERRUPT VECTOR AND INTERRUPT DESCRIPTOR TABLES

An address pointer table is used to link the interrupt type numbers to the location of their service routines in program storage memory. In a real-mode 80386DX-based microcomputer system, this table is called the *interrupt vector table*. In a protected-mode system, the table is referred to as the *interrupt descriptor table*. Figure 8.3 shows a map of the interrupt vector table in the memory of a real-mode 80386DX microcomputer. Looking at the table, we see that it contains 256 address pointers, which are identified as *Vector 0* through *Vector 255*. That is, one pointer corresponds to each interrupt type number, 0 through 255. These address pointers identify the starting locations of their service routines in program memory. The contents of these tables may be either held as firmware in EPROMs or loaded into RAM as part of the system initialization routine.

Notice that in Fig. 8.3 the interrupt vector table is located at the low-address end of the memory address space. It starts at address 00000_{16} and ends at $003FE_{16}$. This represents the first 1K bytes of memory. Actually, the interrupt vector table or interrupt descriptor table can be located anywhere in the memory address space. Its starting location and size are defined by the contents of a register within the 80386DX called the *interrupt descriptor table register* (IDTR). When the 80386DX is reset at power on, it comes up in the real mode with the bits of the base address in IDTR all equal to zero and the limit set to $03FF_{16}$. This positions the interrupt vector table as shown in Fig. 8.3. Moreover, when in the real mode, the value in IDTR is normally left at this initial value to maintain compatibility with 8086/8088-based microcomputer software.

Each of the 256 vectors requires two words (one double word) of memory. These words are always stored at a double-word aligned address boundary. The higher-addressed word of the two-word vector is called the *base address*. It iden-

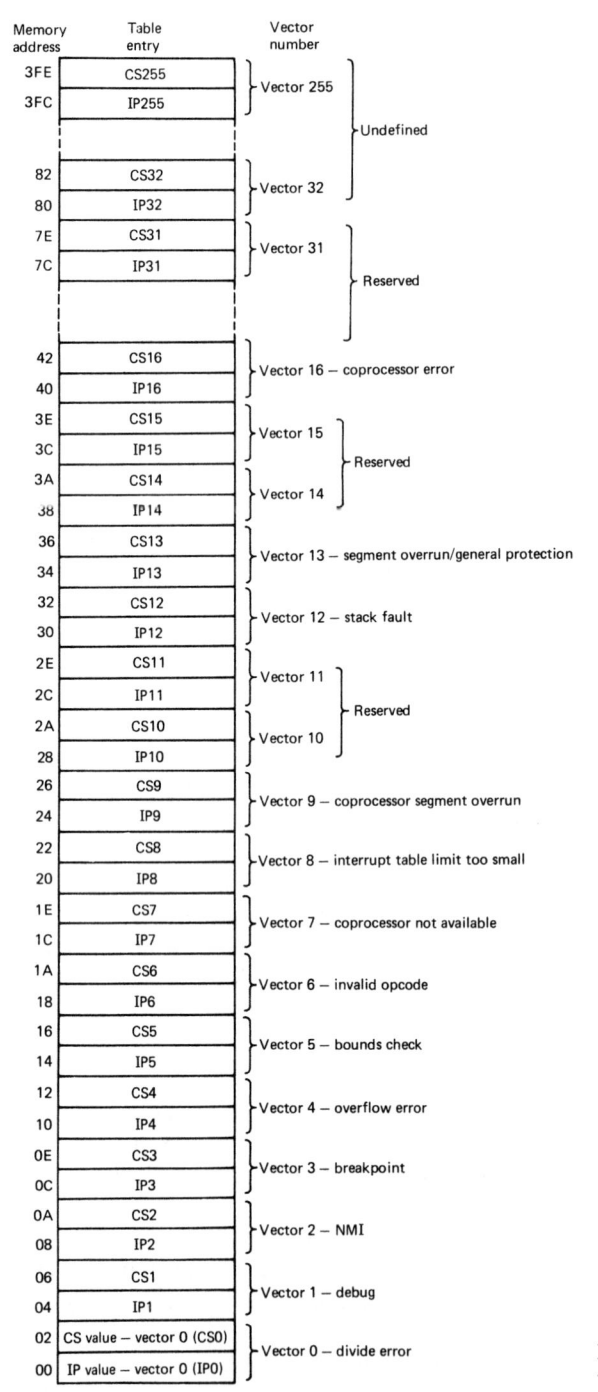

Memory address	Table entry	Vector number
3FE	CS255	Vector 255
3FC	IP255	
		Undefined
82	CS32	Vector 32
80	IP32	
7E	CS31	Vector 31
7C	IP31	
		Reserved
42	CS16	Vector 16 – coprocessor error
40	IP16	
3E	CS15	Vector 15
3C	IP15	Reserved
3A	CS14	Vector 14
38	IP14	
36	CS13	Vector 13 – segment overrun/general protection
34	IP13	
32	CS12	Vector 12 – stack fault
30	IP12	
2E	CS11	Vector 11
2C	IP11	Reserved
2A	CS10	Vector 10
28	IP10	
26	CS9	Vector 9 – coprocessor segment overrun
24	IP9	
22	CS8	Vector 8 – interrupt table limit too small
20	IP8	
1E	CS7	Vector 7 – coprocessor not available
1C	IP7	
1A	CS6	Vector 6 – invalid opcode
18	IP6	
16	CS5	Vector 5 – bounds check
14	IP5	
12	CS4	Vector 4 – overflow error
10	IP4	
0E	CS3	Vector 3 – breakpoint
0C	IP3	
0A	CS2	Vector 2 – NMI
08	IP2	
06	CS1	Vector 1 – debug
04	IP1	
02	CS value – vector 0 (CS0)	Vector 0 – divide error
00	IP value – vector 0 (IP0)	

Figure 8.3 Real-mode interrupt vector table.

tifies the program memory segment in which the service routine resides. For this reason it is loaded into the code segment (CS) register within the 80386DX. The lower-addressed word of the vector is the *offset* of the first instruction of the service routine from the beginning of the code segment defined by the base address loaded into CS. This offset is loaded into the instruction pointer (IP) register. For example, the vector for type number 255, IP_{255} and CS_{255}, is stored at word addresses $003FC_{16}$ and $003FE_{16}$. When loaded into the MPU, it points to the instruction at $CS_{255}:IP_{255}$.

Looking more closely at the table in Fig. 8.3, we find that the first 31 vectors either have dedicated functions or are reserved. For instance, pointers 0, 1, 3, and 4 are used by the 80386DX's internal interrupts or exceptions: *divide error, debug exception, breakpoint,* and *overflow error*, respectively. The remainder of the table, the 224 vectors in the address range 00080_{16} through $003FF_{16}$, are available to the user for storage of software or hardware interrupt vectors. These pointers correspond to type numbers 32 through 255. In the case of external hardware interrupts, each type number (priority level) is associated with an interrupt input in external hardware.

The protected-mode interrupt descriptor table can reside anywhere in the 80386DX's physical address space. The location and size of this table are again defined by the contents of the IDTR. Figure 8.4 shows that IDTR contains a 32-bit *base address* and a 16-bit *limit*. The base address identifies the starting point of the table in memory. On the other hand, the limit determines the number of bytes in the table.

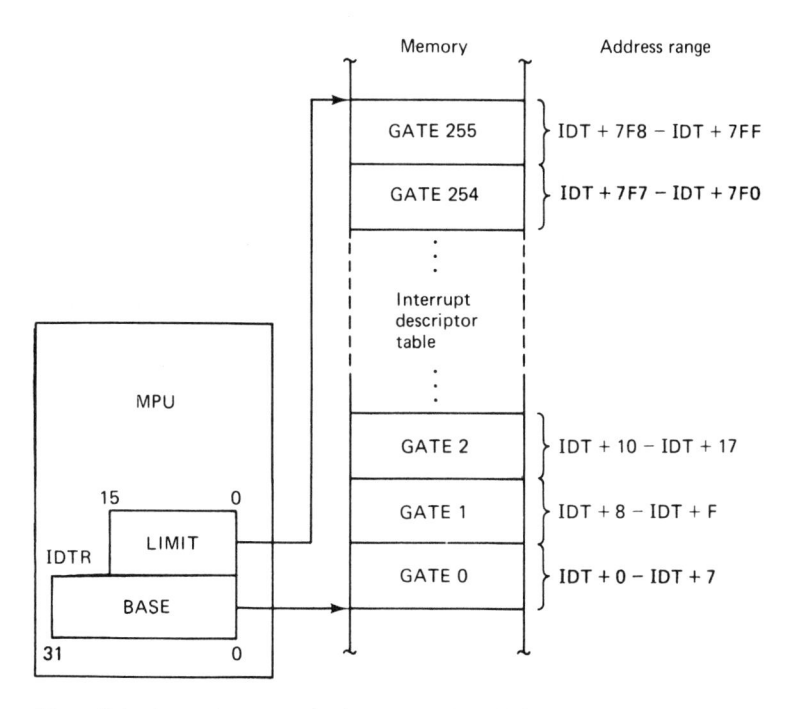

Figure 8.4 Accessing a gate in the protected-mode interrupt descriptor table.

Sec. 8.3   Interrupt Vector and Interrupt Descriptor Tables

The interrupt descriptor table contains gate descriptors, not vectors. In Fig. 8.4 we find that the table contains a maximum of 256 gate descriptors. These descriptors are identified as *Gate 0* through *Gate 255*. Each gate descriptor can be defined as a *trap gate, interrupt gate,* or *task gate.* Interrupt and trap gates permit control to be passed to a service routine that is located within the current task. On the other hand, the task gate permits program control to be passed to a different task.

Just as a real-mode vector, a protected-mode gate acts as a pointer that is used to redirect program execution to the starting point of a service routine. However, unlike an interrupt vector, a gate descriptor takes up eight bytes of memory. For instance, in Fig. 8.4 we see that Gate 0 is located at addresses IDT+0H through IDT+7H and Gate 255 at addresses IDT+7F8H through IDT+7FFH. If all 256 gates are not needed for an application, the limit can be set to a value lower than 07FF$_{16}$ to minimize the amount of memory reserved for the table.

Figure 8.5 illustrates the format of a typical interrupt or trap gate descriptor. Here we see that the two lower addressed words, 0 and 1, are the interrupt's *code offset 0–15* and *code segment selectors,* respectively. The highest-addressed word, word 3, is the interrupt's *code offset 16–31*. These three words identify the starting point of the service routine. The upper byte of word 2 of the descriptor is called the *access rights byte.* The settings of the bits in this byte identify whether or not this gate descriptor is valid, the privilege level of the service routine, and the type of gate. For example, the *present bit* (P) needs to be set to logic 1 if the gate descriptor is to be active. The next two bits, identified as DPL in Fig. 8.5, are used to assign a privilege level to the service routine. If these bits are made 00, level 0, which is the most privileged level, is assigned to the gate. Finally, the setting of the *type bit* (T) determines if the descriptor works as a trap gate or an interrupt gate. T equal to 0 selects the interrupt gate mode of operation. The only difference between the operation of these two types of gates is that when a trap gate context switch is performed, IF is not cleared to disable external hardware interrupts.

Normally, external hardware interrupts are configured with interrupt gate descriptors. Once an interrupt request has been acknowledged for service, the

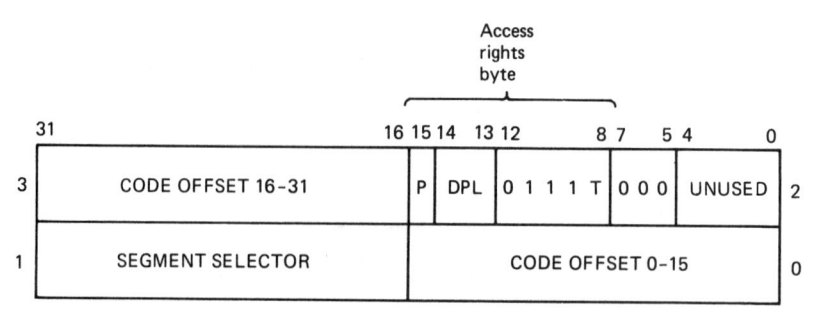

Figure 8.5 Format of a trap or interrupt gate descriptor.

external hardware interrupt interface is disabled with IF. In this way, additional external interrupts cannot be accepted unless the interface is reenabled under software control. On the other hand, internal interrupts, such as software interrupts, usually use trap gate descriptors. In this case the hardware interrupt interface is not affected when the service routine for the software interrupt is initiated. Sometimes low-priority hardware interrupts are assigned trap gates instead of an interrupt gate. This will permit higher-priority external events to interrupt their service routine easily.

8.4 INTERRUPT INSTRUCTIONS

A number of instructions are provided in the instruction set of the 80386DX for use with interrupt and exception processing. These instructions are listed with a brief description of their operation in Fig. 8.6. For instance, the first two instructions, which are CLI and STI, permit manipulation of the 80386DX's interrupt flag through software. STI stands for *set interrupt flag*. Execution of this instruction enables the external interrupt request (INTR) input for operation. That is, it

Mnemonic	Meaning	Format	Operation	Flags affected
CLI	Clear interrupt flag	CLI	$0 \rightarrow (IF)$	IF
STI	Set interrupt flag	STI	$1 \rightarrow (IF)$	IF
LIDT	Load interrupt descriptor table register	LIDT EA	$(EA) \rightarrow (LIMIT_{0-15})$ $(EA + 2) \rightarrow (BASE_{0-15})$ $(EA + 4) \rightarrow (BASE_{16-32})$	None
SIDT	Store interrupt descriptor table register	SIDT EA	$(LIMIT_{0-15}) \rightarrow (EA)$ $(BASE_{0-15}) \rightarrow (EA + 2)$ $(BASE_{16-32}) \rightarrow (EA + 4)$	None
INT n	Type n software interrupt	INT n	[Real Mode] $(Flags) \rightarrow ((SP) - 2)$ $0 \rightarrow TF, IF$ $(CS) \rightarrow ((SP) - 4)$ $(2 + 4 \cdot n) \rightarrow (CS)$ $(IP) \rightarrow ((SP) - 6)$ $(4 \cdot n) \rightarrow (IP)$	TF, IF
IRET	Interrupt return	IRET	[real mode] $((SP)) \rightarrow (IP)$ $((SP) + 2) \rightarrow (CS)$ $((SP) + 4) \rightarrow (Flags)$ $(SP) + 6 \rightarrow (SP)$	All
INTO	Interrupt on overflow	INTO	INT 4 steps	TF, IF
BOUND	Check array index against bounds	BOUND D, S	$(D) < (S) \rightarrow INT5$ or $(D) > (S + 2) \rightarrow INT5$	None
HLT	Halt	HLT	Wait for an external interrupt or reset to occur	None
WAIT	Wait	WAIT	Wait for $\overline{BUSY}$ to go inactive	None

Figure 8.6 Interrupt instructions.

sets interrupt flag (IF). On the other hand, execution of CLI (*clear interrupt flag*) disables the external interrupt input. It does this by resetting IF. When STI or CLI is executed in protected mode, the current privilege level is compared to the I/O privilege level (IOPL) in the flag register. If the current level is less than IOPL, the instruction is not executed; instead, a general protection exception results. General protection exceptions are explained later in the chapter.

Earlier we pointed out that the contents of IDTR determines the location and size of the interrupt descriptor table and that after reset the base address is initialized to 00000000_{16} and the limit to $000003FF_{16}$. The next two instructions in Fig. 8.6 let us modify or examine the contents of this register.

As its name implies, the *load interrupt descriptor table register* (LIDT) is the instruction used to modify the contents of IDTR. The general form of the instruction is given in Fig. 8.6 as

```
        LIDT   EA
```

Here EA stands for the effective address of the operand in memory. This operand is three words in length and contains the values of the base address and limit that are to be loaded into IDTR. Figure 8.7 shows the format of these data. Notice that the lowest addressed word is the 16-bit limit; the next two words are the 32-bit base address. For instance, executing the instruction

```
        LIDT   IDT_TABLE
```

causes IDTR to be loaded with the limit held at address IDT_TABLE and the base address at IDT_TABLE+2 and IDT_TABLE+4. These values must be stored in memory prior to execution of the LIDT instruction. In the protected mode, the LIDT instruction can only be executed when the 80386DX is operating at privilege level 0.

The instruction *store interrupt descriptor table register* (SIDT) can be used to examine the contents of IDTR. As shown in Fig. 8.6, its format is

```
        SIDT   EA
```

When executed it saves the current contents of the IDTR in the format shown in Fig. 8.6, starting in memory at the storage location pointed to by effective address EA. Once the value is stored in memory, it can be examined with additional software.

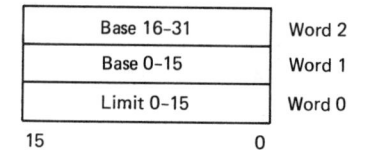

Base 16-31	Word 2
Base 0-15	Word 1
Limit 0-15	Word 0

15   0

Figure 8.7 LIDT instruction data type.

The next instruction listed in Fig. 8.6 is the *software interrupt* instruction INT n. It is used to initiate a vectored call to a service routine. Executing the instruction causes transfer of program control to the subroutine pointed to by the vector or gate for the number n specified in the instruction.

The operation outlined in Fig. 8.6 describes the effect of executing the INT instruction in the real mode. For example, execution of the instruction INT 50 initiates execution of a service routine whose starting point is identified by vector 50 in the pointer table in Fig. 8.3. First, the 80386DX saves the old flags on the stack, clears TF and IF, and saves the old program context, CS and IP, on the stack. Then it reads the values of IP_{50} and CS_{50} from addresses $000C8_{16}$ and $000CA_{16}$, respectively, in memory, loads them into the IP and CS registers, calculates the physical address $CS_{50}:IP_{50}$, and starts to fetch instructions from this new location in program memory.

An *interrupt return* (IRET) instruction must be included at the end of each interrupt service routine. It is required to pass control back to the point in the program where execution was terminated due to the occurrence of the interrupt. As shown in Fig. 8.6, when executed in the real mode, IRET causes the old values of IP, CS, and flags to be popped from the stack back into the internal registers of the 80386DX. This restores the original program environment.

INTO is the *interrupt on overflow* instruction. This instruction must be included after arithmetic instructions, such as divide, which can result in an overflow condition. It tests the overflow flag and if the flag is found to be set, a type 4 internal interrupt is initiated. In the real mode, this condition causes program control to be passed to an overflow service routine that is located at the starting address identified by the vector IP_4 at 00010_{16} and CS_4 at 00012_{16} of the pointer table in Fig. 8.3.

As its name implies, the *check array index against bounds* (BOUND) instruction can determine if the contents of a register, called the *array index*, lies within a set of minimum/maximum values, called the *upper bound* and *lower bound*. This type of operation is important when accessing elements of an array of data in memory. The format of the BOUND instruction is given in Fig. 8.6.

An example is the instruction

```
                BOUND    SI, LIMITS
```

Notice that the instruction contains two operands. The first operand represents the register whose word contents are to be tested to verify whether or not it lies within the boundaries. In our example this is the source index register (SI). The second operand is the effective relative address of the first of two word storage locations in memory that contain the values of the lower and upper boundaries. In the example the word of data starting at address LIMITS is the value of the lower boundary and that at address LIMITS + 2 is the value of the upper boundary.

When this BOUND instruction is executed, the contents of SI are compared to both the value of the lower bound at LIMITS and upper bounds at LIMITS + 2. If it is found to be either less than the lower bound or more than the upper bound, an exception occurs and control is passed to a service routine through the vector or gate for type number 5. Otherwise, the next sequential instruction is performed. The operands can also be 32 bits in length.

> **EXAMPLE 8.1**
>
> For the instruction
>
> BOUND EDI, LIMITS
>
> where LIMITS equals 101000_{16}, what are the addresses of the values of the upper and lower bounds for the value in EDI?
>
> **SOLUTION** The lower boundary is the 32-bit word starting at address 101000_{16} and the upper boundary is the double word at address 101004_{16}.

The last two instructions associated with the interrupt interface are *halt* (HLT) and *wait* (WAIT). They produce similar responses by the 80386DX and permit the operation of the MPU to be synchronized to an event in external hardware. For instance, when HLT is executed, the 80386DX suspends operation and enters the idle state. It no longer executes instructions; instead, it remains idle waiting for the occurrence of an external hardware interrupt, nonmaskable interrupt, or reset. With the occurrence of any of these events, the 80386DX resumes execution with the corresponding service routine. HLT is a privileged instruction and can only be executed at privilege level 0 in a protected-mode system.

If the WAIT instruction is used instead of the HLT instruction, the 80386DX checks the logic level of the $\overline{\text{BUSY}}$ input prior to going into the idle state. Only if $\overline{\text{BUSY}}$ is active, logic 0, will the MPU go into the idle state. While in the idle state, the 80386DX continues to check the logic level at $\overline{\text{BUSY}}$ looking for a transition back to its inactive level, logic 1. As $\overline{\text{BUSY}}$ switches back to 1, execution resumes with the next sequential instruction in the program. For example, the $\overline{\text{BUSY}}$ input is normally used by the 80387DX numerics coprocessor and is switched to logic 0 whenever a numeric operation is in progress. Therefore, WAIT can be used to determine when a numerics operation is completed.

8.5 ENABLING/DISABLING OF INTERRUPTS

Earlier we found that an interrupt enable flag is provided within the 80386DX and that it is identified as IF. The ability to initiate an external hardware interrupt at the INTR input is enabled by setting IF or masked out by resetting it. Through software, this can be done by executing the STI instruction or the CLI instruction, respectively. IF affects only the external hardware interrupt interface, not software interrupts, the nonmaskable interrupt, or internal interrupts or exceptions.

During the initiation sequence of a service routine for an external hardware interrupt, the 80386DX automatically clears IF. This masks out the occurrence of any additional external hardware interrupts. In some applications it may be necessary to permit other higher-priority external hardware interrupts to interrupt the active service routine. If this is the case, the interrupt flag bit can be set with an STI instruction located at the beginning of the service routine to reenable the INTR input.

8.6 EXTERNAL HARDWARE INTERRUPT INTERFACE

Up to this point in the chapter we have introduced the types of interrupts supported by the 80386DX, its interrupt vector and descriptor tables, interrupt instructions, and masking of interrupts. Earlier we pointed out that type numbers 32 through 255 can be used by external hardware interrupts. Let us now look at the external hardware interrupt interface of the 80386DX microcomputer in more detail.

A general interrupt interface for an 80386DX-based microcomputer system is illustrated in Fig. 8.8. Here we see that it includes the address and data buses, byte enable signals, bus cycle definition signals, lock output, and the ready and interrupt request inputs. Moreover, external circuitry is required to interface interrupt inputs, INT_{32} through INT_{255}, to the 80386DX's interrupt interface. This interface circuit must identify which of the pending active interrupts has the high-

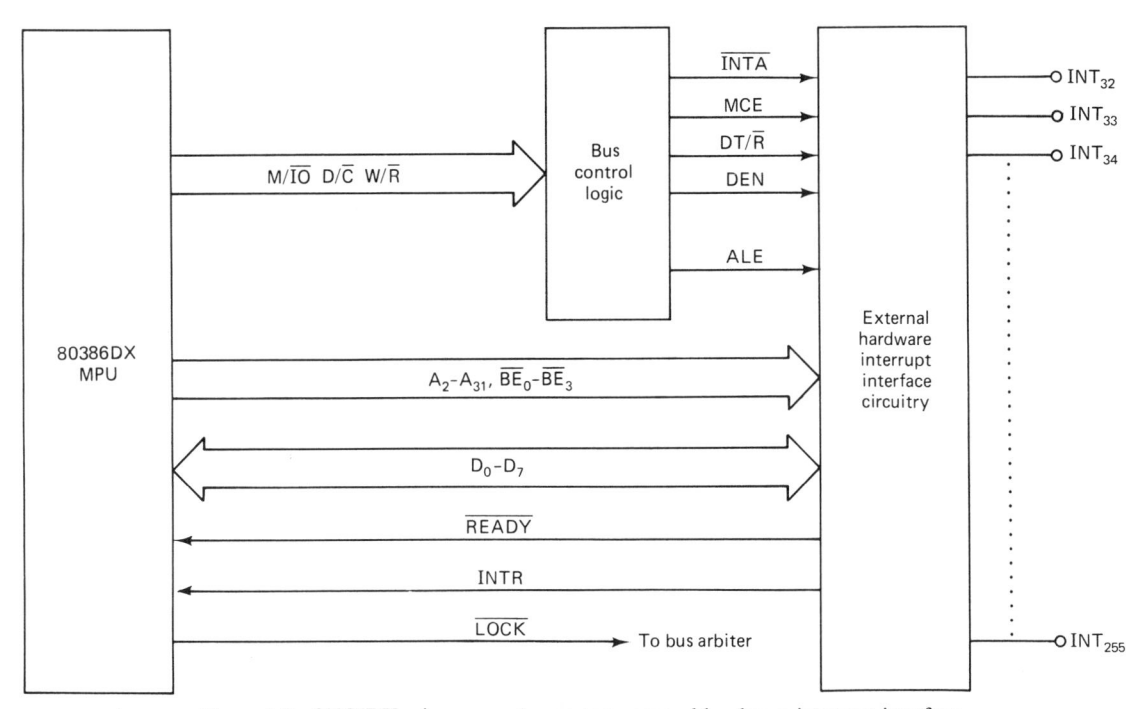

Figure 8.8 80386DX microcomputer system external hardware interrupt interface.

est priority, perform an interrupt request/acknowledge handshake, and then set up the bus to pass a type number to the 80386DX.

In this circuit we see that the key interrupt interface signals are *interrupt request* (INTR) and *interrupt acknowledge* ($\overline{INTA}$). The logic level input at the INTR line signals the 80386DX that an external device is requesting service. The 80386DX samples this input at the beginning of each instruction execution cycle, that is, at instruction boundaries. Logic 1 at INTR represents an active interrupt request. INTR is *level triggered*; therefore, its active level must be maintained by the external hardware until tested by the 80386DX. If it is not maintained, the request for service may not be recognized. For this reason, inputs INT_{32} through INT_{255} are normally latched. Moreover, the 1 at INTR must be removed before the service routine runs to completion; otherwise, the same interrupt may be acknowledged a second time.

When an interrupt request has been recognized by the 80386DX, it signals this fact to external circuitry by outputting the interrupt acknowledge bus cycle status code on $M/\overline{IOC}/\overline{DW}/\overline{R}$. This code, which equals 000_2, is highlighted in Fig. 8.9. Notice in Fig. 8.8 that this code is input to the bus control logic, where it is decoded to produce a pulse to logic 0 at the $\overline{INTA}$ output. Actually, two pulses are produced at $\overline{INTA}$ during the *interrupt acknowledge bus cycle sequence*. The first pulse, which is output during cycle 1, signals external circuitry that the interrupt request has been acknowledged and to prepare to send its type number to the 80386DX. The second pulse, which occurs during cycle 2, tells the external circuitry to put the type number on the data bus.

Notice that the lower eight lines of the data bus, D_0 through D_7, are also part of the interrupt interface. During the second cycle in the interrupt acknowledge bus cycle sequence, external circuitry puts the 8-bit type number of the highest-priority active interrupt request input onto this part of the data bus. The ready ($\overline{READY}$) input can be used to insert wait states into the bus cycle. The 80386DX reads the type number off the bus to identify which external device is requesting service. Then it uses the type number to generate the address of the interrupt's vector or gate in the interrupt vector or descriptor table, respectively.

Address lines A_2 through A_{31} and byte enable lines $\overline{BE_0}$ through $\overline{BE_3}$ are also shown in the interrupt interface circuit of Fig. 8.8. This is because LSI interrupt controller devices are typically used to implement most of the external circuitry. When a read or write bus cycle is performed to the controller, for

$M/\overline{IO}$	$D/\overline{C}$	$W/\overline{R}$	Type of Bus Cycle
0	0	0	Interrupt acknowledge
0	0	1	Idle
0	1	0	I/O data read
0	1	1	I/O data write
1	0	0	Memory code read
1	0	1	Halt/shutdown
1	1	0	Memory data read
1	1	1	Memory data write

Figure 8.9 Interrupt acknowledge bus cycle definition code.

example, to initialize its internal registers after system reset, some of the address bits are decoded to produce a chip select to enable the controller device, and other address bits are used to select the internal register that is to be accessed. The interrupt controller could be I/O mapped, instead of memory mapped; in this case only address lines A_0 through A_{15} are used in the interface.

Another signal shown in the interrupt interface of Fig. 8.8 is the *bus lock indication* ($\overline{LOCK}$) output of the 80386DX. $\overline{LOCK}$ is used as an input to the bus arbiter circuit in multiprocessor systems. The 80386DX switches this output to its active 0 logic level and maintains it at this level throughout the complete interrupt acknowledge bus cycle. In response to this signal, the arbitration logic assures that no other device can take over control of the system bus until the interrupt acknowledge bus cycle sequence is completed.

8.7 EXTERNAL HARDWARE INTERRUPT SEQUENCE

In the preceding section we showed the interrupt interface for external hardware interrupts in an 80386DX-based microcomputer system. Now we will continue by describing in detail the events that take place during the interrupt request, interrupt acknowledge bus cycle, and device service routine.

The interrupt sequence begins when an external device requests service by activating one of the interrupt inputs, INT_{32} through INT_{255}, of the external interrupt interface circuit in Fig. 8.8. For example, the INT_{50} input could be switched to the 1 logic level. This signals that the device associated with priority level 50 wants to be serviced.

The external circuitry evaluates the priority of this input. If there is no interrupt already in progress or if the new interrupt is of higher priority than that which is presently active, the external circuitry must issue a request for service to the MPU.

Let us assume that INT_{50} is the only active interrupt input. In this case the external circuitry switches INTR to logic 1. This tells the 80386DX that an interrupt is pending for service. To assure that it is recognized, the external circuitry must maintain INTR active until an interrupt acknowledge bus cycle definition code is output by the 80386DX.

Figure 8.10 is a flow diagram that outlines the events that take place when an 80386DX that is configured for real-mode operation processes an interrupt. The 80386DX tests for an active interrupt request at the end of the current instruction. Notice that it tests first for the occurrence of an internal interrupt or exception, then the occurrence of the nonmaskable interrupt, and finally checks the logic level of INTR to determine if an external hardware interrupt has occurred.

If INTR is at logic 1, a request for service is recognized. Before the 80386DX initiates the interrupt acknowledge sequence, it checks the setting of IF (interrupt flag). Notice that if IF is logic 0, external interrupts are masked out and the request is ignored. In this case, the next sequential instruction is executed. On the other

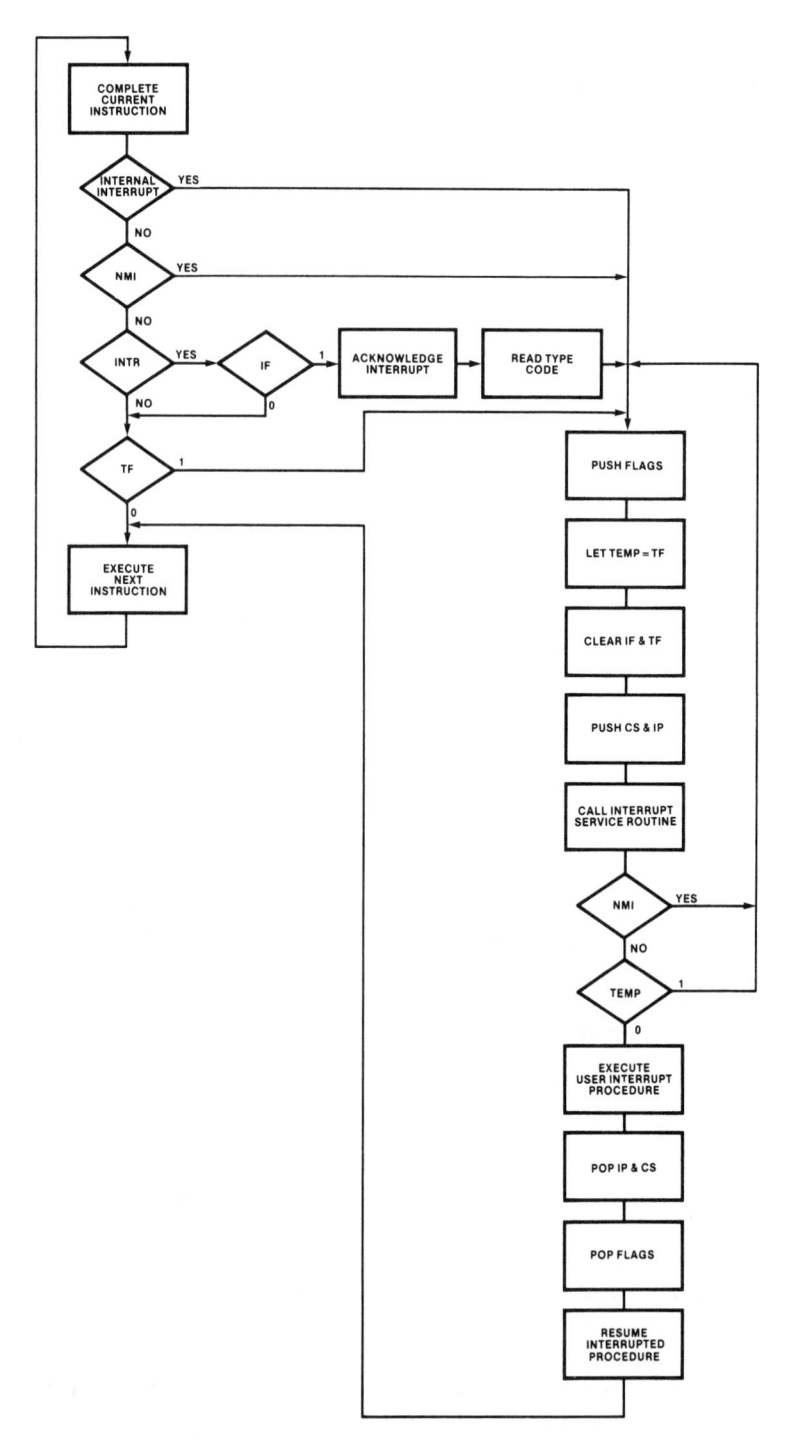

Figure 8.10 Real-mode interrupt processing sequence. (Reprinted by permission of Intel Corp. Copyright Intel Corp. 1979.)

hand, if IF is at logic 1, external hardware interrupts are enabled and the service routine is initiated.

Let us assume that IF is set to permit interrupts to occur when INTR is tested as 1. The 80386DX responds by initiating the interrupt acknowledge bus cycle sequence. This bus sequence is illustrated in Fig. 8.11. Here we see that at the beginning of T_1 of the first bus cycle (interrupt acknowledge cycle 1 in Fig. 8.11) the 80386DX switches $\overline{\text{LOCK}}$ to its active 0 logic level and holds it at this value for the complete bus cycle sequence. This locks the bus for uninterrupted use by the 80386DX. At the same time, address lines A_3 through A_{31} and $\overline{BE}_0$ are set to logic 0, while A_2 and $\overline{BE}_1$ through $\overline{BE}_3$ are set to 1. Moreover, the bus cycle definition code $M/\overline{IO}D/\overline{C}W/\overline{R} = 000$ is output to the bus control logic. These signals are latched into external circuitry with the pulse at $\overline{\text{ADS}}$. The code 000 is decoded by the bus control logic to produce a pulse at $\overline{\text{INTA}}$.

In the waveforms of Fig. 8.11, $\overline{\text{READY}}$ is shown to be logic 1 at the end of the first T_2 state of cycle 1. This signals not ready to the MPU. The response to this condition is that another T_2 state occurs. This T_2 state acts as a wait state to extend the current bus cycle. Notice that in the second T_2 state $\overline{\text{READY}}$ is logic 0 and interrupt acknowledge cycle 1 is complete. The data bus lines of the 80386DX are in the high-Z state during this cycle; therefore, any data on the bus are ignored.

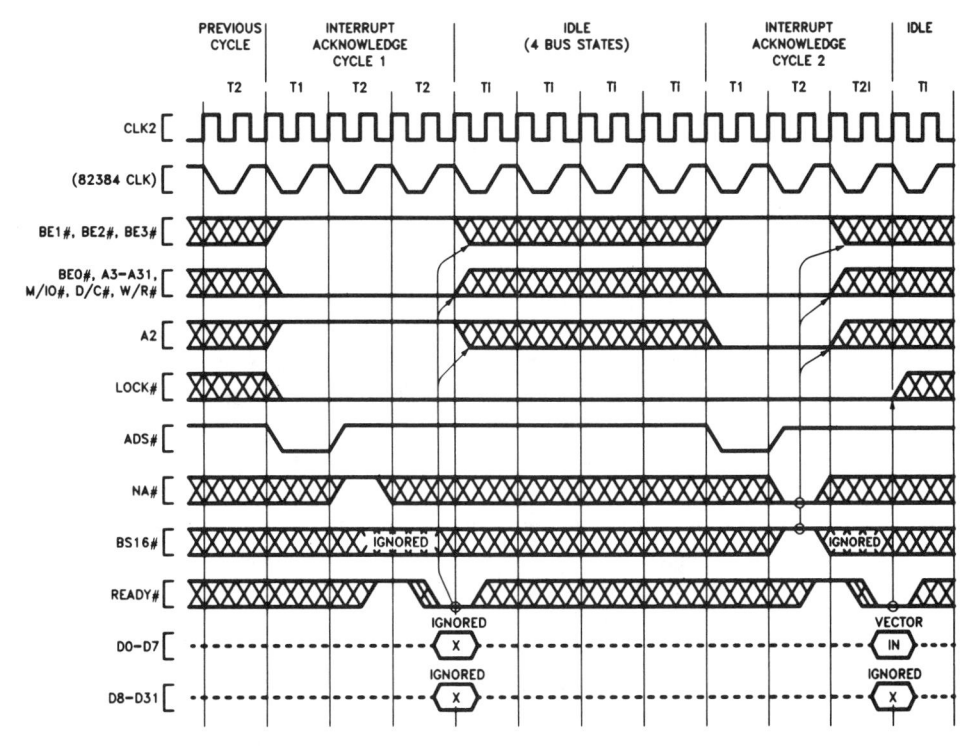

Figure 8.11 Interrupt acknowledge bus cycle. (Reprinted by permission of Intel Corp. Copyright Intel Corp. 1979.)

If a single interrupt controller is used in the interrupt interface circuit of an 80386DX-based microcomputer system, the MPU uses the first interrupt acknowledge bus cycle to acknowledge to external circuitry that an interrupt request has been accepted. On the other hand, systems that employ a master–slave interrupt controller configuration use this first interrupt acknowledge cycle to both signal external circuitry that an interrupt request has been acknowledged and to tell the master controller to communicate to the slave controllers which interrupt request has been accepted. The master interrupt controller outputs what is called the *cascade address* to all slave controllers in parallel. This address identifies the interrupt controller that is to supply the type number to the MPU.

At the completion of the first interrupt acknowledge bus cycle, the bus controller automatically inserts four idle states (T_I) before initiating the second interrupt acknowledge cycle. Looking at Fig. 8.11, we find that during interrupt acknowledge cycle 2 the levels and timing of all signals except A_2 are essentially the same as those for the first cycle. It is during this second bus cycle that the interrupt controller passes one of the interrupt type numbers, $32 = 20_{16}$ through $255 = FF_{16}$, from the interrupt interface circuit to the 80386DX. The type number, identified as vector in Fig. 8.11, is supplied to the MPU over data bus lines D_0 through D_7. For the case of INT_{50}, the code would be $00110010_2 = 32_{16}$. This completes the interrupt request/acknowledge handshake.

Looking at Fig. 8.10, we see that the 80386DX next saves the contents of the flags register by pushing it to the stack. This requires one write cycle. Then the TF and IF flags are cleared. This disables the single-step mode of operation if it happens to be active and masks out additional external hardware interrupts. Now the 80386DX automatically pushes the contents of CS and IP onto the stack. This requires two more write cycles to take place over the system bus. The current value of the stack pointer is decremented by 4 as each of these values is put onto the top of the stack.

Now the 80386DX knows the type number associated with the corresponding external device that is requesting service. It must next call the service routine by fetching the vector that defines its starting point from memory. The type number is internally multiplied by 4, and this result is used as the address of the first word of the interrupt vector in the vector table. A read operation is performed to read the two-word vector from memory. The lower-addressed word is loaded into IP and the higher-addressed word is loaded into CS. For instance, the words of the vector for INT_{50} would be read as a double word from address $000C8_{16}$.

The service routine is now initiated. That is, execution resumes with the first instruction of the service routine. It is located at the address generated from the new values in CS and IP. Figure 8.12 shows the structure of a typical interrupt service routine. Notice that the service routine includes PUSH instructions to save the contents of those internal registers that it will use. In this way their original contents are saved in the stack during execution of the routine. The contents of all registers can be saved simply by including a single PUSHA instruction instead of a separate PUSH instruction for each register.

At the end of the service routine, the original program environment must be restored. This is done by first popping the contents of the appropriate registers

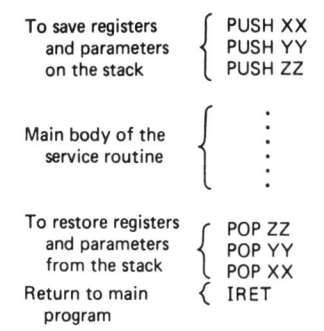

```
To save registers      ┌ PUSH XX
   and parameters      ┤ PUSH YY
     on the stack      └ PUSH ZZ

 Main body of the      ┌   .
  service routine      ┤   .
                       └   .

To restore registers   ┌ POP ZZ
   and parameters      ┤ POP YY
     from the stack    └ POP XX
    Return to main     { IRET
         program
```

Figure 8.12 Structure of an interrupt service routine.

from the stack by executing POP instructions (or POPA). Then an IRET instruction must be executed as the last instruction of the service routine. IRET reenables the interrupt interface and causes the old contents of the flags, CS, and IP to be popped from the stack back into the internal registers of the 80386DX. The original program environment has now been completely restored and execution resumes at the point in the program where it was interrupted.

If an 80386DX-based microcomputer is configured for the protected mode of operation, the interrupt processing sequence is different from that we just described. Actually, the interrupt request/acknowledge handshake sequence appears to take place exactly the same way in the external hardware; however, a number of changes do occur in the internal interrupt processing sequence of the 80386DX. Let us now look at how the protected mode 80386DX reacts to an interrupt request.

When processing interrupts in protected mode, the general protection mechanism of the 80386DX comes into play. The general protection rules dictate that program control can only be directly passed to a service routine that is in a segment with equal or higher privilege, that is, a segment with an equal or lower-numbered descriptor privilege level. Any attempt to transfer program control to a routine in a segment with lower privilege (higher-numbered descriptor privilege level) results in an exception unless the transition is made through a gate.

Typically, interrupt drivers are in code segments at a high privilege level, possibly level 0. Moreover, interrupts occur randomly; therefore, there is a good chance that the microprocessor will be executing application code that is at a low privilege level. In the case of interrupts, the current privilege level (CPL) is the privilege level assigned by the descriptor of the software that was executing when the interrupt occurred. This could be any of the 80386DX's valid privilege levels. The privilege level of the service routine is that defined in the interrupt or trap gate descriptor for the type number. That is, it is the descriptor privilege level (DPL).

When a service routine is initiated, the current privilege level may change. This depends on whether the software that was interrupted was in a code segment that was configured as *conforming* or *nonconforming*. If the interrupted code is in a conforming code segment, CPL does not change when the service routine is initiated. In this case the contents of the stack after the context switch are as

Sec. 8.7   External Hardware Interrupt Sequence

illustrated in Fig. 8.13(a). Since the privilege level does not change, the current stack (OLD SS:ESP) is used. Notice that as part of the interrupt initiation sequence the OLD EFLAGS, OLD CS, and OLD EIP are automatically saved on the stack. Actually, the *requested privilege level* (RPL) code is also saved on the stack. This is because it is part of OLD CS. RPL identifies the protection level of the interrupted routine.

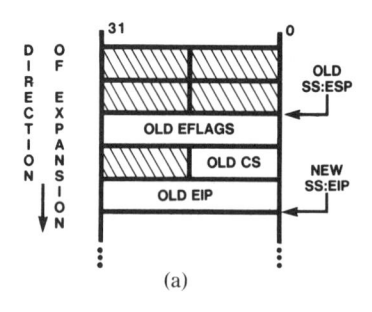

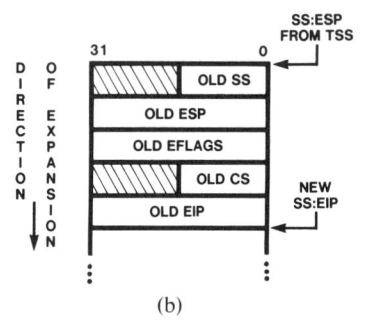

Figure 8.13 (a) Stack after context switch with no privilege level transition. (Reprinted by permission of Intel Corp. Copyright Intel Corp. 1986.) (b) Stack after context switch with a privilege level transition. (Reprinted by permission of Intel Corp. Copyright Intel Corp. 1986.)

However, if the segment is nonconforming, the value of DPL is assigned to CPL as long as the service routine is active. As shown in Fig. 8.13(b), this time the stack is changed to that for the new privilege level. The MPU is loaded with a new SS and new ESP from TSS and then the old stack pointer, OLD SS, and OLD ESP are saved on the stack followed by the OLD EFLAGS, OLD CS, and OLD EIP. Remember that for an interrupt gate, IF is cleared as part of the context switch, but for a trap gate, IF remains unchanged. In both cases, the TF flag is reset after the contents of the flag register are pushed to the stack.

Figure 8.14 shows the stack as it exists after an attempt to initiate an interrupt

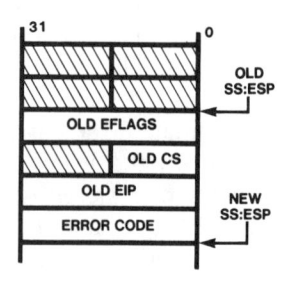

Figure 8.14 Stack contents after interrupt with an error. (Reprinted by permission of Intel Corp. Copyright Intel Corp. 1986.)

service routine that did not involve a privilege-level transition has failed. Notice that the context switch to the exception service routine caused an *error code* to be pushed onto the stack following the values of OLD EFLAGS, OLD CS, and OLD EIP.

One format of the error code is given in Fig. 8.15. This type of error code is known as an *IDT error code*. Here we see that the least significant bit, which is labeled EXT, indicates whether the error was for an externally or an internally initiated interrupt. For external interrupts, such as the hardware interrupts, the EXT bit is always set to logic 1. The next bit, which is labeled IDT, is set to 1 if the error is produced as the result of an interrupt. That is, it is the result of a reference to a descriptor in the IDT. If IDT is not set, the third bit indicates whether the descriptor is in the GDT (TI = 0) or the LDT (TI = 1). The next 14 bits contain the segment selector that produced the error condition. With this information available on the stack, the exception service routine can determine which interrupt attempt had failed and whether it was initiated internally or externally. A second format is used for errors that result from a protected-mode page fault. Figure 8.16 illustrates this error code and the function of its bits.

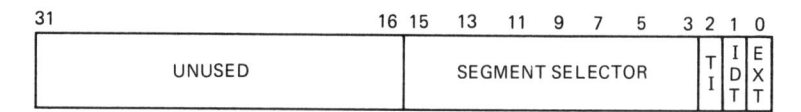

31	16 15 13 11 9 7 5 3 2 1 0
UNUSED	SEGMENT SELECTOR TI | IDT | EXT

Figure 8.15 IDT error code format.

Field	Value	Description
U/S	0	The access causing the fault originated when the processor was executing in supervisor mode.
	1	The access causing the fault originated when the processor was executing in user mode.
W/R	0	The access causing the fault was a read.
	1	The access causing the fault was a write.
P	0	The fault was caused by a not-present page.
	1	The fault was caused by a page-level protection violation.

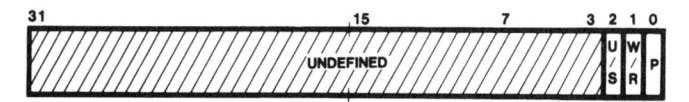

31	15	7	3 2 1 0
	UNDEFINED		U/S | W/R | P

Figure 8.16 Page fault error code format and bit functions. (Reprinted by permission of Intel Corp. Copyright Intel Corp. 1986.)

Sec. 8.7   External Hardware Interrupt Sequence

Just as in real mode, the IRET instruction is used to return from a protected-mode interrupt service routine. For service routines using an interrupt gate or trap gate, IRET is restricted to the return from a higher privilege level to a lower privilege level, for instance, from level 1 to level 3. Once the flags, OLD CS, and OLD EIP are returned to the 80386DX, the RPL bits of OLD CS are tested to see if they equal CPL. If RPL = CPL, an intralevel return is in progress. In this case the return is complete and program execution resumes at the point in the program where execution had stopped.

If RPL is greater than CPL, an interlevel return is taking place, not an intralevel return. During an interlevel return, checks are performed to determine if a protection violation will occur due to the protection-level transition. Assuming that no violations occur, the OLD SS and OLD ESP are popped from the stack into the MPU and then program execution resumes.

8.8 THE 82C59A PROGRAMMABLE INTERRUPT CONTROLLER

The 82C59A is an LSI peripheral IC that is designed to simplify the implementation of the interrupt interface in an 80386DX system. This device is known as a *programmable interrupt controller* or *PIC*. It is manufactured using the CMOS technology.

The operation of the PIC is programmable under software control and it can be configured for a wide variety of applications. Some of its programmable features are the ability to accept level-sensitive or edge-triggered inputs, the ability to be easily cascaded to expand from 8 to 64 interrupt inputs, and its ability to be configured to implement a wide variety of priority schemes.

Block Diagram of the 82C59A

Let us begin our study of the PIC with its block diagram in Fig. 8.17(a). We just mentioned that the 82C59A is treated as a peripheral in the 80386DX microcomputer. Therefore, its operation must be initialized by the MPU. The *host processor interface* is provided for this purpose. This interface consists of eight *data bus* lines D_0 through D_7 and control signals *read* ($\overline{RD}$), *write* ($\overline{WR}$), and *chip select* ($\overline{CS}$). The data bus is the path over which data are transferred between the 80386DX and 82C59A. These data can be command words, status information, or interrupt type numbers. Control input $\overline{CS}$ must be at logic 0 to enable the host processor interface. Moreover, $\overline{WR}$ and $\overline{RD}$ signal the 82C59A whether data are to be written into or read from its internal registers. They also control the timing of these data transfers.

Two other signals are identified as part of the host processor interface. They are INT and $\overline{INTA}$. Together, these two signals provide the handshake mechanism by which the 82C59A can signal the 80386DX of a request for service and receive an acknowledgment that the request has been accepted. INT is the interrupt request output of the 82C59A. It is applied directly to the INTR input of the

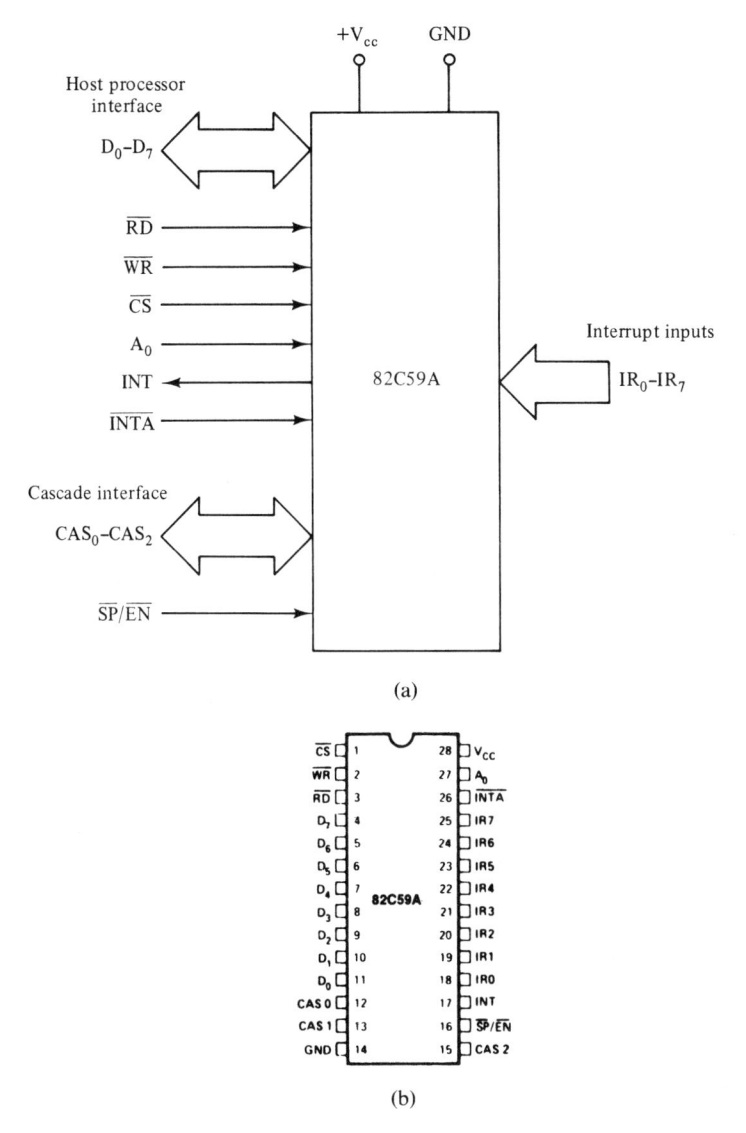

Figure 8.17 (a) Block diagram of the 82C59A; (b) pin layout. (Reprinted by permission of Intel Corp. Copyright Intel Corp. 1979.)

80386DX. Logic 1 is produced at this output whenever the interrupt controller receives a valid interrupt request.

On the other hand, $\overline{\text{INTA}}$ is an input of the 82C59A. It is connected to the $\overline{\text{INTA}}$ output of the 80386DX's bus control logic. This input of the 82C59A is pulsed to logic 0 twice during the interrupt acknowledge bus cycle, thereby signaling the 82C59A that the interrupt request has been acknowledged and that it should output the type number of the highest-priority active interrupt on data bus lines D_0 through D_7 such that it can be read by the MPU. The last signal line involved in the host processor interface is the A_0 input. This input is normally supplied by an address line of the 80386DX such as A_2. The logic level at this input is involved in the selection of the internal register that is accessed during read and write operations.

At the other side of the block in Fig. 8.17(a), we find the eight *interrupt inputs* of the PIC. They are labeled IR_0 through IR_7. It is through these inputs that external devices issue a request for service. One of the software options of the 82C59A permits these inputs to be configured for *level-sensitive* or *edge-triggered operation*. When configured for level-sensitive operation, logic 1 is the active level of the IR inputs. In this case, the request for service must be removed before the service routine runs to completion. Otherwise, the interrupt will be requested a second time and the service routine initiated again. Moreover, if the input returns to logic 0 before it is acknowledged by the 80386DX, the request for service will be missed.

Some external devices produce a short-duration pulse instead of a fixed logic level for use as an interrupt request signal. If the 80386DX is busy servicing a higher-priority interrupt when the pulse is produced, the request for service could be completely missed. To overcome this problem, the edge-triggered mode of operation is used.

Inputs of the 82C59A that are set up for edge-triggered operation become active on the transition from the inactive 0 logic level to the active 1 logic level. This represents what is known as a *positive edge-triggered input*. The fact that this transition has occurred at an IR line is latched internal to the 82C59A. If the IR input remains at the 1 logic level even after the service routine is completed, the interrupt is not reinitiated. Instead, it is locked out. To be recognized a second time, the input must first return to the 0 logic level and then be switched back to 1. The advantage of edge-triggered operation is that if the request at the IR input is removed before the 80386DX acknowledges service of the interrupt, its request is maintained latched internal to the 82C59A until it can be serviced.

The last group of signals on the PIC implement what is known as the *cascade interface*. As shown in Fig. 8.17(a), it includes bidirectional *cascading bus lines* CAS_0 through CAS_2 and a multifunction control line labeled $\overline{SP/EN}$. The primary use of these signals is in cascaded systems where a number of 82C59A ICs are interconnected in a master/slave configuration to expand the number of IR inputs from 8 to as high as 64. One of these 82C59A devices is configured as the *master* PIC and all others are set up as *slaves*.

In a cascaded system, the CAS lines of all 82C59As are connected together to provide a private bus between the master and slave devices. In response to the first $\overline{INTA}$ pulse during the interrupt acknowledge bus cycle, the master PIC outputs a 3-bit code on the CAS lines. This code identifies the highest-priority slave that is to be serviced. It is this device that is to be acknowledged for service. All slaves read this code off the *private cascading bus* and compare it to their internal ID code. A match condition at one slave tells that PIC that it has the highest-priority input. In response, it must put the type number of its highest-priority active input on the data bus during the second interrupt acknowledge bus cycle.

When the PIC is configured through software for the cascaded mode, the $\overline{\text{SP/EN}}$ line is used as an input. This corresponds to its $\overline{\text{SP}}$ (*slave program*) function. The logic level applied at $\overline{\text{SP}}$ tells the device whether it is to operate as a master or slave. Logic 1 at this input designates master mode and logic 0 designates slave mode.

If the PIC is configured for single mode instead of cascade mode, $\overline{\text{SP/EN}}$ takes on another function. In this case, it becomes an enable output which can be used to control the direction of data transfer through the bus transceiver that buffers the data bus. A pin layout of the 82C59A is given in Fig. 8.17(b).

Internal Architecture of the 82C59A

Now that we have introduced the input/output signals of the 82C59A, let us look at its internal architecture. Figure 8.18 is a block diagram of the PIC's internal circuitry. Here we find eight functional parts: the *data bus buffer, read/write logic, control logic, inservice register, interrupt request register, priority resolver, interrupt mask register,* and *cascade buffer/comparator*.

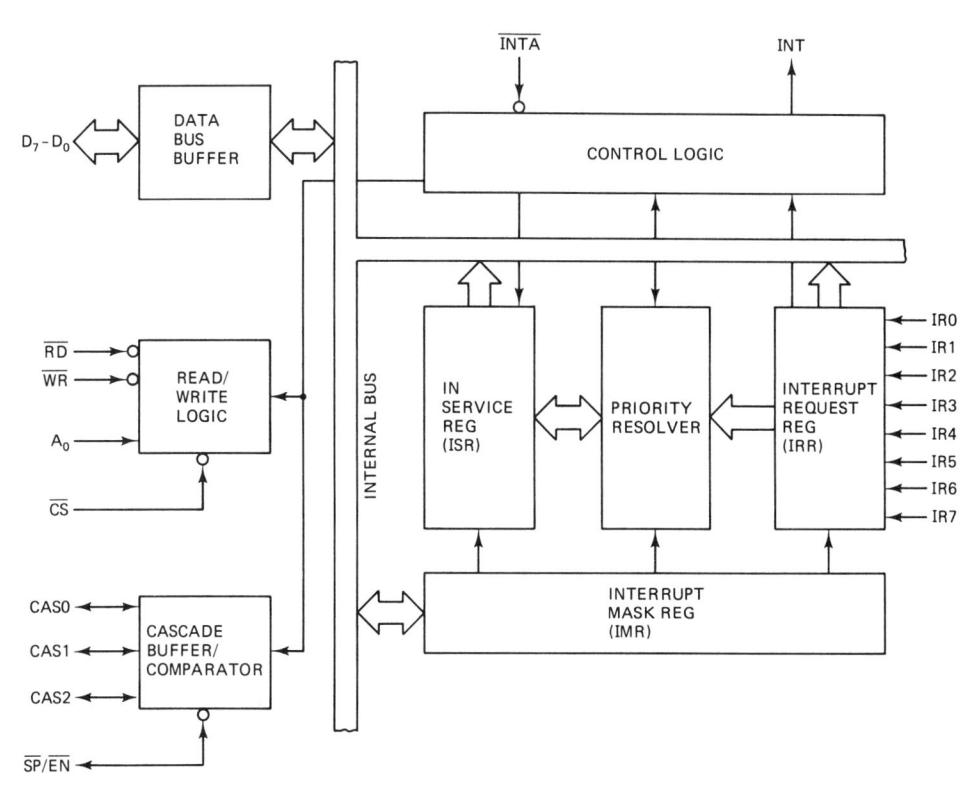

Figure 8.18 Internal architecture of the 82C59A. (Reprinted by permission of Intel Corp. Copyright Intel Corp. 1979.)

We will begin with the function of the data bus buffer and read/write logic sections. It is these parts of the 82C59A that let the 80386DX have access to the internal registers. Moreover, it provides the path over which interrupt type numbers are passed to the 80386DX. The data bus buffer is an 8-bit bidirectional three-state buffer that interfaces the internal circuitry of the 82C59A to the data bus of the 80386DX. The direction, timing, and source of destination for data transfers through the buffer are under control of the outputs of the read/write logic block. These outputs are generated in response to control inputs $\overline{RD}$, $\overline{WR}$, A_0, and $\overline{CS}$.

The interrupt request register, in-service register, priority resolver, and interrupt mask register are the key internal blocks of the 82C59A. The interrupt mask register (IMR) can be used to enable or mask out individually the interrupt request inputs. It contains eight bits identified by M_0 through M_7. These bits correspond to interrupt inputs IR_0 through IR_7, respectively. Logic 0 in a mask register bit position enables the corresponding interrupt input and logic 1 masks it out. This register can be read from or written into under software control.

On the other hand, the interrupt request register (IRR) stores the current status of the interrupt request inputs. It also contains one bit position for each of the IR inputs. The values in these bit positions reflect whether the interrupt inputs are active or inactive.

Which of the active interrupt inputs is identified as having the highest priority is determined by the priority resolver. This section can be configured to work using a number of different priority schemes through software. Following the selected scheme, it identifies which of the active interrupt inputs has the highest priority and signals the control logic that an interrupt is active. In response, the control logic causes the INT signal to be issued to the 80386DX.

The in-service register differs in that it stores the interrupt level that is presently being serviced. During the first $\overline{INTA}$ pulse in an interrupt acknowledge bus cycle, the level of the highest active interrupt is strobed into ISR. Loading of ISR occurs in response to output signals of the control logic section. This register cannot be written into by the microprocessor; however, its contents may be read as status.

The cascade buffer/comparator section provides the interface between master and slave 82C59As. As we mentioned earlier, it is this interface that permits easy expansion of the interrupt interface using a master/slave configuration. Each slave has an *ID code* that is stored in this section.

Programming the 82C59A

The way in which the 82C59A operates is determined by how the device is programmed. Two types of command words are provided for this purpose. They are the *initialization command words* (ICW) and the *operational command words* (OCW). ICW commands are used to load the internal control registers of the 82C59A. There are four such command words and they are identified as ICW_1, ICW_2, ICW_3, and ICW_4. On the other hand, the three OCW commands permit the 80386DX to initiate variations in the basic operating modes defined by the ICW commands. These three commands are called OCW_1, OCW_2, and OCW_3.

Depending upon whether the 82C59A is I/O-mapped or memory-mapped, the 80386DX issues commands to the 82C59A by initiating output or write cycles. This can be done by executing either the OUT or MOV instruction, respectively. The address put on the system bus during the output bus cycle must be decoded with external circuitry to chip select the peripheral. When an address assigned to the 82C59A is on the bus, the output of the decoder must produce logic 0 at the $\overline{CS}$ input. This signal enables the read/write logic within the PIC and data applied at D_0 through D_7 are written into the command register within the control logic section synchronously with a write strobe at $\overline{WR}$.

The interrupt request input (INTR) of the 80386DX must be disabled whenever commands are being issued to the 82C59A. This can be done by clearing the interrupt enable flag by executing the CLI (clear interrupt enable flag) instruction. After completion of the command sequence, the interrupt input must be reenabled. To do this, the 80386DX must execute the STI (set interrupt enable flag) instruction.

The flow diagram in Fig. 8.19 shows the sequence of events that the 80386DX must perform to initialize the 82C59A with ICW commands. The cycle begins

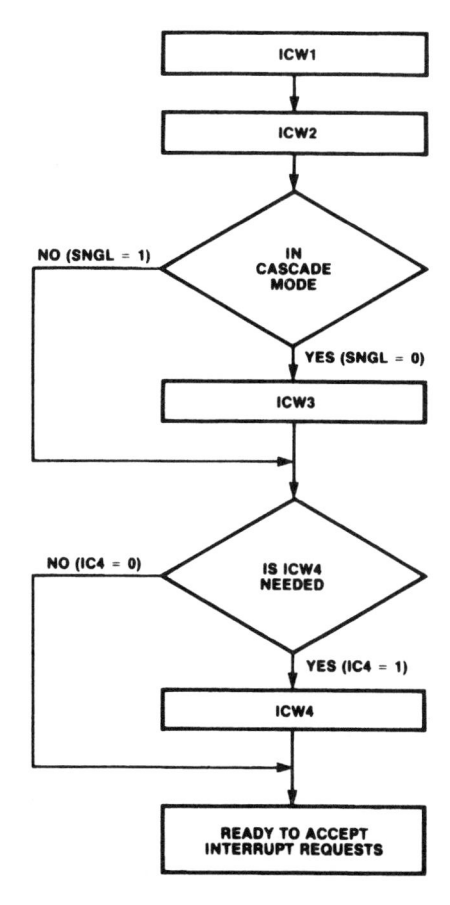

Figure 8.19 Initialization sequence of the 82C59A. (Reprinted by permission of Intel Corp. Copyright Intel Corp. 1979.)

Sec. 8.8   The 82C59A Programmable Interrupt Controller

with the MPU outputting initialization command word ICW₁ to the address of the 82C59A.

The moment that ICW₁ is written into the control logic section of the 82C59A certain internal setup conditions automatically occur. First the internal sequence logic is set up such that the 82C59A will accept the remaining ICWs as designated by ICW₁. It turns out that if the least significant bit of ICW₁ is logic 1, command word ICW₄ is required in the initialization sequence. Moreover, if the next least significant bit of ICW₁ is logic 0, the command word ICW₃ is also required.

In addition to this, writing ICW₁ to the 82C59A clears ISR and IMR. Also three operation command word bits, *special mask mode* (SMM) in OCW₃, *interrupt request register* (IRR) in OCW₃, and *end of interrupt* (EOI) in OCW₂, are cleared to logic 0. Furthermore, the *fully masked mode* of interrupt operation is entered with an initial priority assignment such that IR₀ is the highest-priority input and IR₇ the lowest-priority input. Finally, the edge-sensitive latches associated with the IR inputs are all cleared.

If the LSB of ICW₁ was initialized to logic 0, one additional event occurs. This is that all bits of the control register associated with ICW₄ are cleared.

In Fig. 8.19 we see that once the 80386DX starts initialization of the 82C59A by writing ICW₁ into the control register, it must continue the sequence by writing ICW₂ and then optionally ICW₃ and ICW₄ in that order. Notice that it is not possible to modify just one of the initialization command registers. Instead, all words that are required to define the device's operating mode must be output once again.

We found that all four words need not always be used to initialize the 82C59A. However, for its use in the 80386DX system, words ICW₁, ICW₂, and ICW₄ are always required. ICW₃ is optional and is needed only if the 82C59A is to function in the cascade mode.

Initialization Command Words

Now that we have introduced the initialization sequence of the 82C59A, let us look more closely at the functions controlled by each of the initialization command words. We will begin with ICW₁. Its format and bit functions are identified in Fig. 8.20(a). Notice that address bit A₀ is included as a ninth bit and it must be logic 0.

Here we find that the logic level of the LSB D₀ of the initialization word indicates to the 82C59A whether or not ICW₄ will be included in the programming sequence. As we mentioned earlier, logic 1 at D₀ (IC₄) specifies that it is needed. The next bit, D₁ (SNGL), selects between *single device* or *multidevice cascaded mode* of operation. When D₁ is set to logic 0, the internal circuitry of the 82C59A is configured for cascaded mode. Selecting this state also sets up the initialization sequence such that ICW₃ must be issued as part of the initialization cycle. Bit D₂ has functions specified for it in Fig. 8.20(a). However, it can be ignored when the 82C59A is being connected to the 80386DX and is a don't-care state. D₃, which is labeled LTIM, defines whether the eight IR inputs operate in the level-sensitive or edge-triggered mode. Logic 1 in D₃ selects level-triggered operation and logic

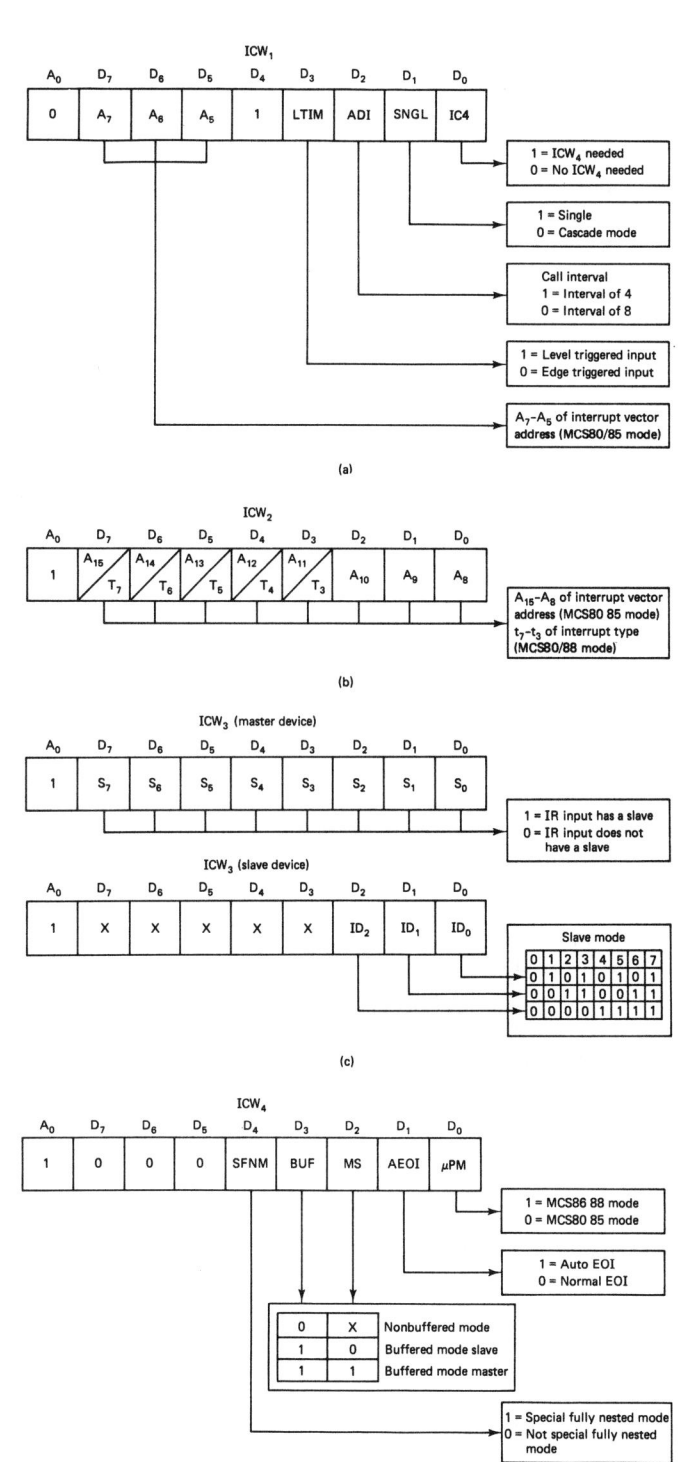

Figure 8.20 (a) ICW₁ format. (Reprinted by permission of Intel Corp. Copyright Intel Corp. 1979.) (b) ICW₂ format. (Reprinted by permission of Intel Corp. Copyright Intel Corp. 1979.) (c) ICW₃ format. (Reprinted by permission of Intel Corp. Copyright Intel Corp. 1979.) (d) ICW₄ format. (Reprinted by permission of Intel Corp. Copyright Intel Corp. 1979.)

Sec. 8.8 The 82C59A Programmable Interrupt Controller 401

0 selects edge-triggered operation. Finally, bit D_4 is fixed at the 1 logic level and the three MSBs D_5 through D_7 are not required in 80386DX-based systems.

EXAMPLE 8.2

What value should be written into ICW_1 in order to configure the 82C59A such that ICW_4 is needed in the initialization sequence, the system is going to use multiple 82C59As, and its inputs are to be level sensitive? Assume that all unused bits are to be logic 0. Give the result in both binary and hexadecimal form.

SOLUTION Since ICW_4 is to be initialized, D_0 must be logic 1.

$$D_0 = 1$$

For cascaded mode of operation, D_1 must be 0

$$D_1 = 0$$

and for level-sensitive inputs D_3 must be 1.

$$D_3 = 1$$

Bits D_2 and D_5 through D_7 are don't-care states and are all made logic 0.

$$D_2 = D_5 = D_6 = D_7 = 0$$

Moreover, D_4 must be fixed at the 1 logic level.

$$D_4 = 1$$

This gives the complete command word

$$D_7D_6D_5D_4D_3D_2D_1D_0 = 00011001_2 = 19_{16}$$

The second initialization word, ICW_2, has a single function in the 80386DX microcomputer. As shown in Fig. 8.20(b), its five most significant bits D_7 through D_3 define a fixed binary code T_7 through T_3 that is used as the most significant bits of its type number. Whenever the 82C59A puts the three-bit interrupt type number corresponding to its active input onto the bus, it is automatically combined with the value T_7 through T_3 to form an 8-bit type number. The three least significant bits of ICW_2 are not used. Notice that logic 1 must be output on A_0 when this command word is put on the bus.

EXAMPLE 8.3

What should be programmed into register ICW_2 if the type numbers output on the bus by the device are to range from $F0_{16}$ through $F7_{16}$?

SOLUTION To set the 82C59A up such that type numbers are in the range $F0_{16}$ through $F7_{16}$, its device code bits must be

$$D_7D_6D_5D_4D_3 = 11110_2$$

The lower three bits are don't-care states and all can be 0s. This gives the command word

$$D_7D_6D_5D_4D_3D_2D_1D_0 = 11110000_2 = F0_{16}$$

The information of initialization word ICW$_3$ is required by only those 82C59As that are configured for the cascaded mode of operation. Figure 8.20(c) shows its bits. Notice that ICW$_3$ is used for different functions depending on whether the device is a master or slave. In the case of a master, bits D$_0$ through D$_7$ of the word are labeled S$_0$ through S$_7$. These bits correspond to IR inputs IR$_0$ through IR$_7$, respectively. They identify whether or not the corresponding IR input is supplied by either the INT output of a slave or directly by an external device. Logic 1 loaded in an S position indicates that the corresponding IR input is supplied by a slave.

On the other hand, ICW$_3$ for a slave is used to load the device with a three-bit identification code ID$_2$ID$_1$ID$_0$. This number must correspond to the IR input of the master to which the slave's INT output is wired. The ID code is required within the slave so that it can be compared to the cascading code output by the master on CAS$_0$ through CAS$_2$.

EXAMPLE 8.4

Assume that a master PIC is to be configured such that its IR$_0$ through IR$_3$ inputs are to accept inputs directly from external devices but IR$_4$ through IR$_7$ are to be supplied by the INT outputs of slaves. What code should be used for the initialization command word ICW$_3$?

SOLUTION For IR$_0$ through IR$_3$ to be configured to allow direct inputs from external devices, bits D$_0$ through D$_3$ of ICW$_3$ must be logic 0.

$$D_3D_2D_1D_0 = 0000_2$$

The other IR inputs of the master are to be supplied by INT outputs of slaves. Therefore, their control bits must be all 1.

$$D_7D_6D_5D_4 = 1111_2$$

This gives the complete command word

$$D_7D_6D_5D_4D_3D_2D_1D_0 = 11110000_2 = F0_{16}$$

The fourth control word, ICW$_4$, which is shown in Fig. 8.20(d), is used to configure the device for use with the 80386DX and selects various features that are available in its operation. The LSB D$_0$, which is called microprocessor mode (μPM), must be set to logic 1 whenever the device is connected to the 80386DX. The next bit, D$_1$, is labeled AEOI for *automatic end of interrupt*. If this mode is enabled by writing logic 1 into the bit location, the EOI (*end of interrupt*) command does not have to be issued as part of the service routine.

Of the next two bits in ICW$_4$, BUF is used to specify whether or not the 82C59A is to be used in a system where the data bus is buffered with a bidirectional bus transceiver. When buffered mode is selected, the $\overline{SP/EN}$ line is configured as $\overline{EN}$. As indicated earlier, $\overline{EN}$ is a control output that can be used to control the direction of data transfer through the bus transceiver. It switches to logic 0 whenever data are transferred from the 82C59A to the 80386DX.

If buffered mode is not selected, the $\overline{SP/EN}$ line is configured to work as the master/slave mode select input. In this case, logic 1 at the $\overline{SP}$ input selects master mode operation and logic 0 selects slave mode.

Assume that the buffered mode was selected; then the $\overline{SP}$ input is no longer available to select between the master and slave modes of operation. Instead, the MS bit of ICW_4 defines whether the 82C59A is a master or slave device.

Bit D_4 is used to enable or disable another operational option of the 82C59A. This option is known as the *special fully nested mode*. This function is only used in conjunction with the cascaded mode. Moreover, it is enabled only for the master 82C59A, not for the slaves. This is done by setting the SFNM bit to logic 1.

The 82C59A is put into the fully nested mode of operation as command word ICW_1 is loaded. When an interrupt is initiated in a cascaded system that is configured in this way, the occurrence of another interrupt at the slave corresponding to the original interrupt is masked out even if it is of higher priority. This is because the bit in ISR of the master 82C59A that corresponds to the slave is already set; therefore, the master 82C59A ignores all interrupts of equal or lower priority.

This problem is overcome by enabling special fully nested mode of operation at the master. In this mode, the master will respond to those interrupts that are at lower or higher priority than the active level. The last three bits of ICW_4, D_5 through D_7, must always be logic 0.

Operational Command Words

Once the appropriate ICW commands have been issued to the 82C59A, it is ready to operate in the fully nested mode. Three operational command words are also provided for controlling the operation of the 82C59A. These commands permit further modifications to be made to the operation of the interrupt interface after it has been initialized. Unlike the initialization sequence, which requires that the ICWs be output in a special sequence after power-up, the OCWs can be issued under program control whenever needed and in any order.

The first operational command word, OCW_1, is used to access the contents of the interrupt mask register (IMR). A read operation can be performed to the register to determine its present status. Moreover, write operations can be performed to set or reset its bits. This permits selective masking of the interrupt inputs. Notice in Fig. 8.21(a) that bits D_0 through D_7 of command word OCW_1 are identified as mask bits M_0 through M_7, respectively. In hardware, these bits correspond to interrupt inputs IR_0 through IR_7, respectively. Setting a bit to logic 1 masks out the associated interrupt input. On the other hand, clearing it to logic 0 enables the interrupt input.

For instance, writing $F0_{16} = 11110000_2$ into the register causes inputs IR_0 through IR_3 to be enabled and IR_4 through IR_7 to be disabled. Input A_0 must be logic 1 whenever the OCW_1 command is issued.

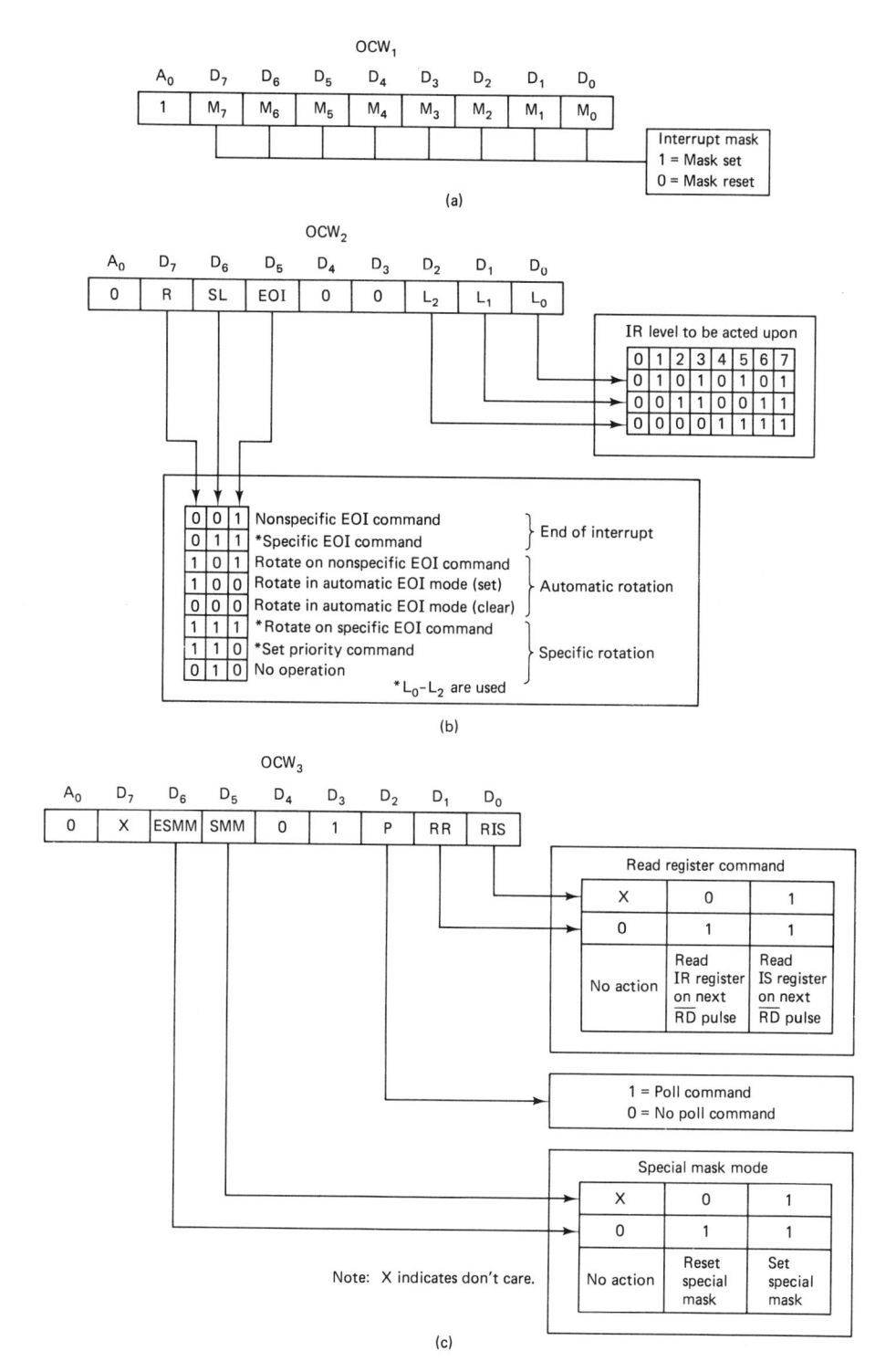

Figure 8.21 (a) OCW₁ format. (Reprinted by permission of Intel Corp. Copyright Intel Corp. 1979.) (b) OCW₂ format. (Reprinted by permission of Intel Corp. Copyright Intel Corp. 1979.) (c) OCW₃ format. (Reprinted by permission of Intel Corp. Copyright Intel Corp. 1979.)

EXAMPLE 8.5

What should be the OCW_1 code if interrupt inputs IR_0 through IR_3 are to be disabled and IR_4 through IR_7 enabled?

SOLUTION For IR_0 through IR_3 to be disabled, their corresponding bits in the mask register must be made logic 1.

$$D_3D_2D_1D_0 = 1111_2$$

On the other hand, in order for IR_4 through IR_7 to be enabled, D_4 through D_7 must be logic 0.

$$D_7D_6D_5D_4 = 0000_2$$

Therefore, the complete word for OCW_1 is

$$D_7D_6D_5D_4D_3D_2D_1D_0 = 00001111_2 = 0F_{16}$$

The second operational command word OCW_2 selects the appropriate priority scheme and assigns an IR level for those schemes that require a specific interrupt level. The format of OCW_2 is given in Fig. 8.21(b). Here we see that the three LSBs define the interrupt level. For example, using $L_2L_1L_0 = 000_2$ in these locations specifies interrupt level 0, which corresponds to input IR_0.

The other three active bits of the word D_7, D_6, and D_5 are called *rotation* (R), *specific level* (SL), and *end of interrupt* (EOI), respectively. They are used to select a priority scheme according to the table in Fig. 8.21(b). For instance, if these bits are all logic 1, the priority scheme known as *rotate on specific EOI command* is enabled. Since this scheme requires a specific interrupt, its value must be included in $L_2L_1L_0$. A_0 must be logic 0 whenever this command is issued to the 82C59A.

EXAMPLE 8.6

What OCW_2 must be issued to the 82C59A if the priority scheme rotate on nonspecific EOI command is to be selected?

SOLUTION To enable the rotate on nonspecific EOI command priority scheme, bits D_7 through D_5 must be set to 101. Since a specific level does not have to be specified, the rest of the bits in the command word can be 0. This gives OCW_2 as

$$D_7D_6D_5D_4D_3D_2D_1D_0 = 10100000_2 = A0_{16}$$

The last control word OCW_3, which is shown in Fig. 8.21(c), permits reading of the contents of the ISR or IRR registers through software, issue of the poll command, and enable/disable of the special mask mode. Bit D_1, which is called *read register* (RR), is set to 1 to initiate reading of either the in-service register (ISR) or interrupt request register (IRR). At the same time, bit D_0, which is labeled RIS, selects between ISR and IRR. Logic 0 in RIS selects IRR and logic 1 selects ISR. In response to this command, the 82C59A puts the contents of the selected register on the bus, where it can be read by the 80386DX.

If the next bit, D_2, in OCW_3 is logic 1, a *poll command* is issued to the

82C59A. The result of issuing a poll command is that the next $\overline{RD}$ pulse to the 82C59A is interpreted as an interrupt acknowledge. In turn, the 82C59A causes the ISR register to be loaded with the value of the highest-priority active interrupt. After this, a *poll word* is automatically put on the data bus. The 80386DX must read it off the bus.

Figure 8.22 illustrates the format of the poll word. Looking at this word, we

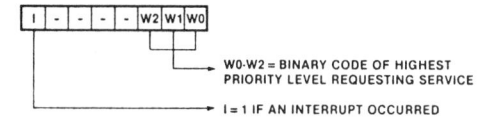

Figure 8.22 Poll word format. (Reprinted by permission of Intel Corp. Copyright Intel Corp. 1979.)

see that the MSB is labeled I for interrupt. The logic level of this bit indicates to the 80386DX whether or not an interrupt input was active. Logic 1 indicates that an interrupt is active. The three LSBs $W_2W_1W_0$ identify the priority level of the highest-priority active interrupt input. This poll word can be decoded through software and when an interrupt is found to be active a branch is initiated to the starting point of its service routine. The poll command represents a software method of identifying whether or not an interrupt has occurred; therefore, the INTR input of the 80386DX should be disabled.

D_5 and D_6 are the remaining bits of OCW_3 for which functions are defined. They are used to enable or disable the special mask mode. ESMM (*enable special mask mode*) must be logic 1 to permit changing of the status of the special mask mode with the SMM (*special mask mode*) bit. Logic 1 at SMM enables the special mask mode of operation. If the 82C59A is initially configured for the fully nested mode of operation, only interrupts of higher priority are allowed to interrupt an active service routine. However, by enabling the special mask mode, interrupts of higher or lower priority are enabled, but those of equal priority remain masked out.

EXAMPLE 8.7

Write a program that will initialize an 82C59A with the initialization command words ICW_1, ICW_2, and ICW_3 derived in Examples 8.2, 8.3, and 8.4, respectively. Moreover, ICW_4 is to be equal to $1F_{16}$. Assume that the 82C59A resides at address $A000_{16}$ in the memory address space. Assume that address bit A_2 of the 80386DX is applied to the A_0 input and that the device is attached to data bus lines D_0 through D_7.

SOLUTION Since the 82C59A resides in the memory address space, we can use a series of move instructions to write the initialization command words into its registers. However, before doing this, we must first disable interrupts. This is done with the instruction

```
CLI                    ;DISABLE INTERRUPTS
```

Next we will set up a data segment starting at address 00000_{16}.

```
        MOV   AX, 0H              ;CREATE A DATA SEGMENT AT 00000₁₆
        MOV   DS, AX
```

Now we are ready to write the command words to the 82C59A.

```
        MOV   AL, 19H             ;LOAD ICW 1
        MOV   [A000H], AL         ;WRITE ICW 1 TO 82C59A
        MOV   AL, F0H             ;LOAD ICW 2
        MOV   [A001H], AL         ;WRITE ICW 2 TO 82C59A
        MOV   AL, F0H             ;LOAD ICW 3
        MOV   [A001H], AL         ;WRITE ICW 3 TO 82C59A
        MOV   AL, 1FH             ;LOAD ICW 4
        MOV   [A001H], AL         ;WRITE ICW 4 TO 82C59A
```

Initialization is now complete and the interrupts can be enabled.

```
        STI                       ;ENABLE INTERRUPTS
```

8.9 INTERRUPT INTERFACE CIRCUITS USING THE 82C59A

Now that we have introduced the 82C59A programmable interrupt controller, let us look at how it is used to implement the interrupt interface in an 80386DX-based microcomputer system.

Figure 8.23(a) includes an interrupt interface circuit that uses a single 82C59A-2 programmable interrupt controller. Let us begin by looking at how the 82C59A's microprocessor interface is attached to the 80386DX. Notice that data bus lines, D_0 through D_7, of the 82C59A are connected through the 74AS245 transceiver to the 80386DX's data bus. It is over these lines that the MPU initializes the internal registers of the 82C59A, reads the contents of these registers, and reads the type number of the active interrupt input during the interrupt acknowledge bus cycle.

In this circuit the registers of the 82C59A are assigned to unique I/O addresses. During input or output bus cycles to one of these addresses, bits A_4 and A_5 of the address are decoded by the 74F139 I/O address decode logic to produce the $\overline{CS}$ input for the 82C59A. When this input is at its active, logic 0, level, the 82C59A's microprocessor interface is enabled for operation. At the same time, the A_2 output of the 74AS373 address latches is applied to the A_0 input of the interrupt controller. It is this signal that selects the register that is to be accessed.

The bus control logic produces the signals that identify whether an input or output data transfer is to take place. Notice that the I/O read ($\overline{IORC}$) and I/O write ($\overline{IOWC}$) control outputs from PAL2 of the control logic and are supplied to the $\overline{RD}$ and $\overline{WR}$ inputs of the 82C59A, respectively. Logic 0 at these outputs tells the 82C59A whether data are to be input or output over the bus, respectively.

Next we trace the sequence of events that take place as a device requests service through the interrupt interface circuit. The external interrupt request in-

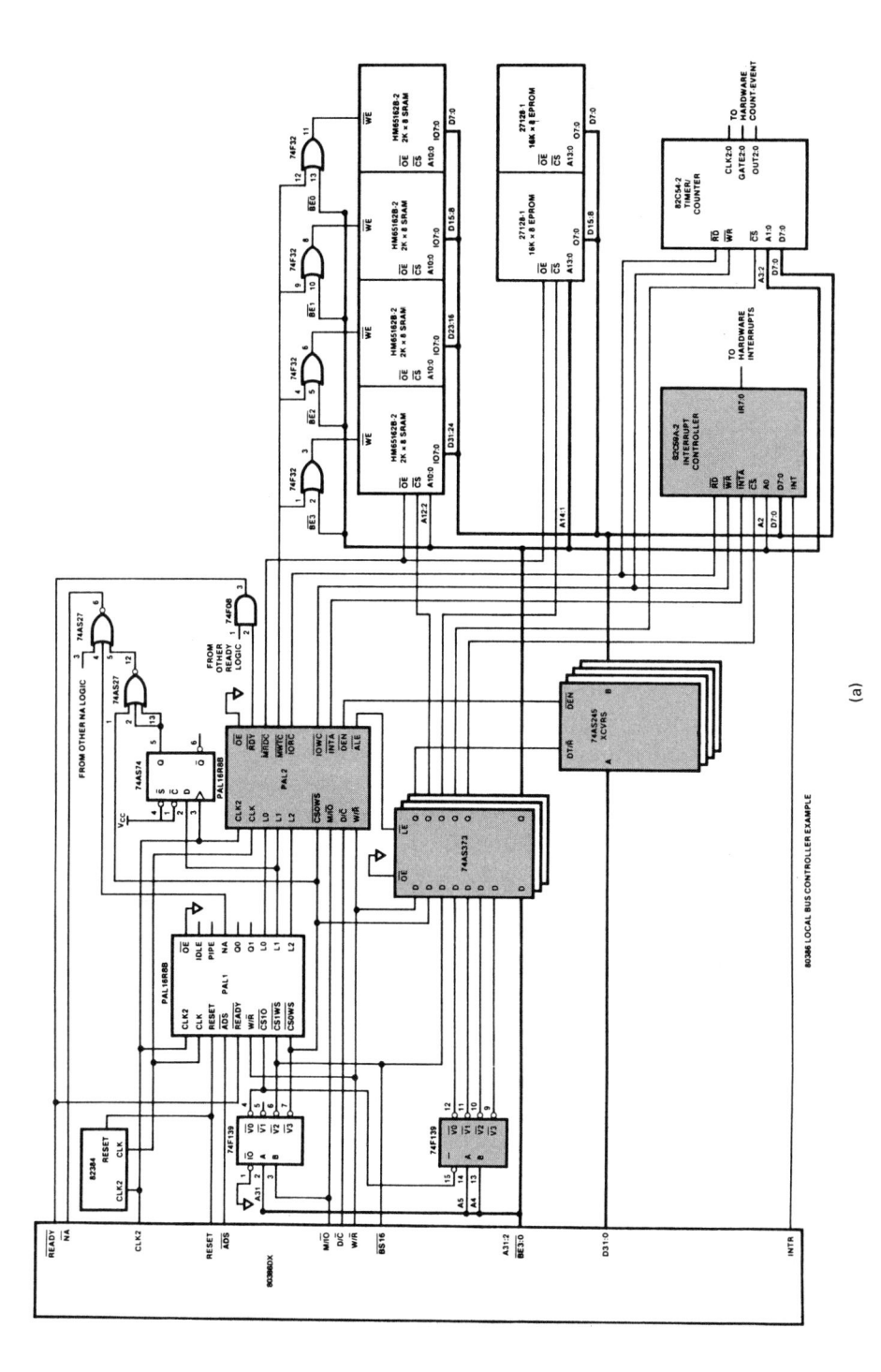

Figure 8.23 (a) Interrupt interface of a typical 80386DX based microcomputer system. (Reprinted by permission of Intel Corp. Copyright Intel Corp. 1990.) (b) Cascading 82C59As. (Reprinted by permission of Intel Corp. Copyright Intel Corp. 1979.)

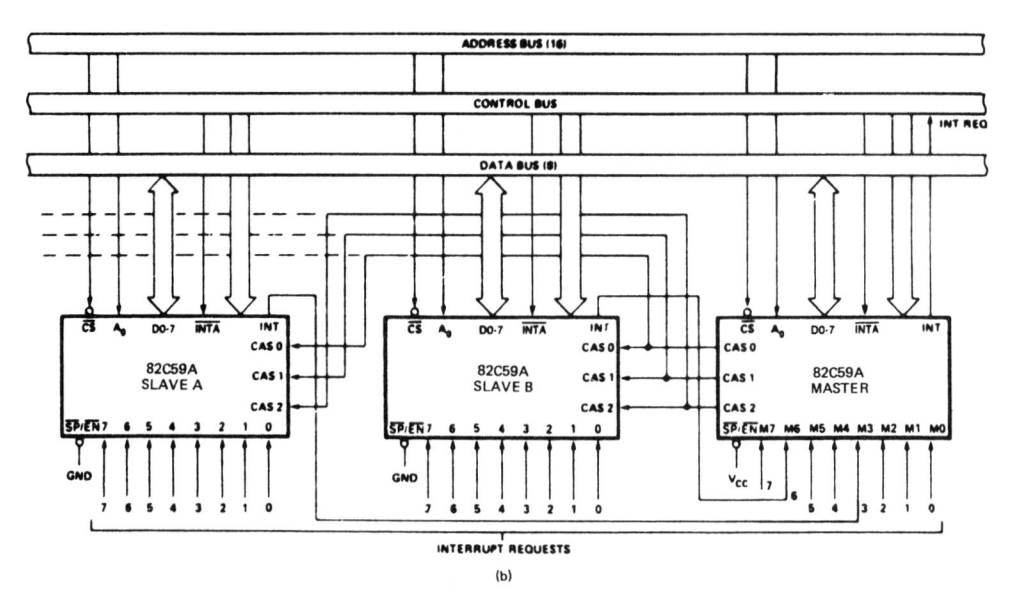

Figure 8.23 (*Continued*)

puts are identified as IR₀ through IR₇ in the circuit of Fig. 8.23(a). Whenever an interrupt input becomes active, and either no other interrupt is active or the priority level of the new interrupt is higher than that of the already active interrupt, the 82C59A switches its INT output to logic 1. This output is returned to the INTR input of the 80386DX. In this way it signals that an external device needs to be serviced.

As long as the interrupt flag within the 80386DX is set to 1, the interrupt interface is enabled. Assuming that IF is 1, the interrupt request is accepted and an interrupt acknowledge bus cycle sequence is initiated. During the first interrupt acknowledge bus cycle, the 80386DX outputs the status code $M/\overline{IO}D/\overline{C}W/\overline{R} = 000_2$ to the bus control logic. This input causes the $\overline{INTA}$ output of PAL2 to be pulsed to logic 0. $\overline{INTA}$ is applied to the $\overline{INTA}$ input of the 82C59A and when logic 0, it signals that the active interrupt request will be serviced.

As the second interrupt acknowledge bus cycle is executed, the status code at the inputs $M/\overline{IO}D/\overline{C}W/\overline{R}$ of PAL2 in the bus control logic is again 000_2 and another pulse is output at $\overline{INTA}$. This pulse signals the 82C59A to output the type number of its highest-priority active interrupt at D_0 through D_7. Then the signal $\overline{DEN}$ is switched to 0 to enable the 74AS245 bus transceivers for operation. At the same time, the $W/\overline{R}$ output of the 80386DX switches to 0. This signal is latched in the address latch and applied to the $DT/\overline{R}$ input of the transceivers. The transceivers are now set to pass the type number from data outputs D_0 through D_7 of the interrupt controller onto the data bus lines of the 80386DX. The 80386DX reads this number off the bus and initiates a vectored transfer of program control to the starting point of the corresponding service routine in program memory.

For applications that require more than eight interrupt request inputs, 82C59A devices are cascaded into a master–slave configuration. Figure 8.23(b) shows such a circuit. Here we find that the rightmost device is identified as the master and the devices to the left as slave A and slave B. Notice that the bus connections are similar to those explained for the circuit in Fig. 8.23(a).

At the interrupt request side of the devices, we find that slaves A and B are cascaded to the master 82C59A by attaching their INT outputs to the M_3 (IR_3) and M6 (IR_6) inputs, respectively. Moreover, the CAS lines of all three PICs are tied in parallel. It is over these CAS lines that the master signals the slaves whether or not their interrupt request has been acknowledged. As the first pulse is output at $\overline{INTA}$, the master PIC is signaled to output the three-bit cascade code of the device whose interrupt request is being acknowledged on the CAS bus. The slaves read this code and compare it to their internal code. In this way the slave corresponding to the code is signaled to output the type number of its highest-priority active interrupt onto the data bus during the second interrupt acknowledge bus cycle.

8.10 INTERNAL INTERRUPT AND EXCEPTION FUNCTIONS

Earlier we indicated that some of the 256 interrupt vectors of the 80386DX are dedicated to internal interrupt and exception functions. Internal interrupts and exceptions differ from external hardware interrupts in that they occur due to the result of executing an instruction, not an event that takes place in external hardware. That is, an internal interrupt or exception is initiated because an error condition was detected before, during, or after execution of an instruction. In this case a routine must be initiated to service the internal condition before resuming execution of the same or next instruction of the program.

Looking at Fig. 8.10, we find that internal interrupts and exceptions are not masked out with the interrupt enable flag. For this reason, occurrence of any one of these internal conditions is automatically detected by the 80386DX and causes an interrupt of program execution and a vectored transfer of control to a corresponding service routine. During the control transfer sequence, no interrupt acknowledge bus cycles are produced.

Figure 8.24 identifies the internal interrupts and exceptions that are active in real mode. Here we find internal interrupts such as breakpoint, and exception functions such as divide error, overflow error, and bounds check. Each of these functions is assigned a unique type number. Notice that the 32 highest-priority locations in the interrupt vector table, vectors 0 through 31, are reserved for internal functions.

Internal interrupts and exceptions are further categorized as a *fault*, *trap*, or *abort*, based on how the failing function is reported. In the case of an exception that causes a fault, the values of CS and IP saved on the stack point to the instruction that resulted in the fault. Therefore, after servicing the exception, the faulting instruction can be reexecuted. On the other hand, for those exceptions that result in a trap, the values of CS and IP that are pushed to the stack point

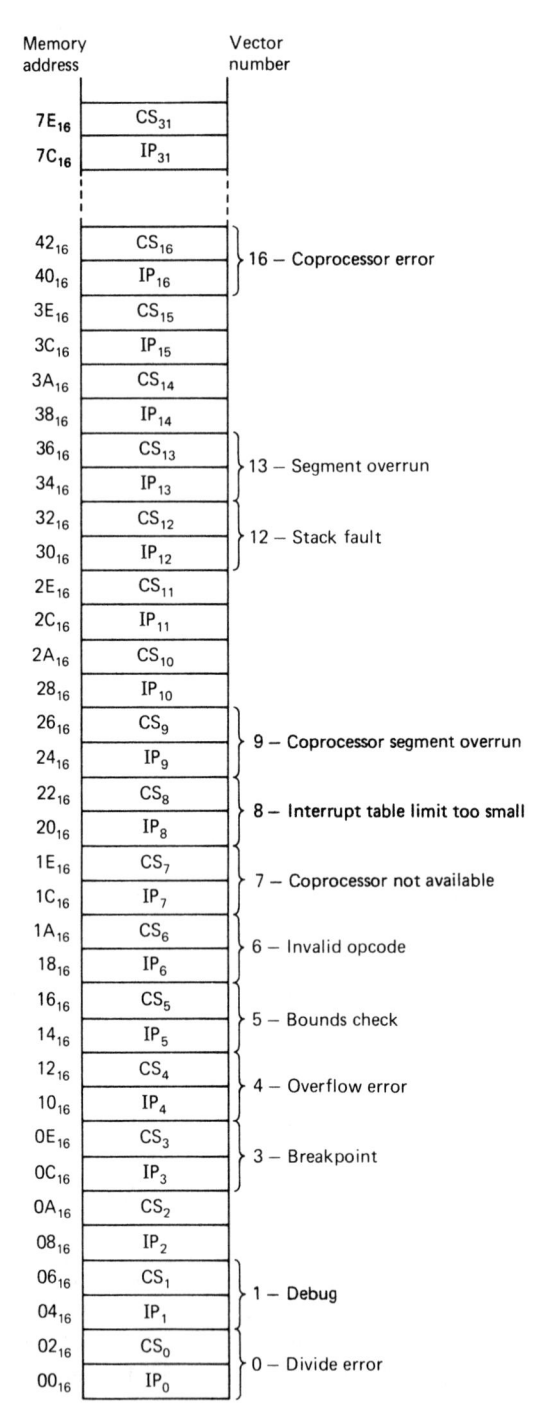

Memory address		Vector number
$7E_{16}$	CS_{31}	
$7C_{16}$	IP_{31}	
⋮	⋮	
42_{16}	CS_{16}	⎫ 16 – Coprocessor error
40_{16}	IP_{16}	⎭
$3E_{16}$	CS_{15}	
$3C_{16}$	IP_{15}	
$3A_{16}$	CS_{14}	
38_{16}	IP_{14}	
36_{16}	CS_{13}	⎫ 13 – Segment overrun
34_{16}	IP_{13}	⎭
32_{16}	CS_{12}	⎫ 12 – Stack fault
30_{16}	IP_{12}	⎭
$2E_{16}$	CS_{11}	
$2C_{16}$	IP_{11}	
$2A_{16}$	CS_{10}	
28_{16}	IP_{10}	
26_{16}	CS_9	⎫ 9 – Coprocessor segment overrun
24_{16}	IP_9	⎭
22_{16}	CS_8	⎫ 8 – Interrupt table limit too small
20_{16}	IP_8	⎭
$1E_{16}$	CS_7	⎫ 7 – Coprocessor not available
$1C_{16}$	IP_7	⎭
$1A_{16}$	CS_6	⎫ 6 – Invalid opcode
18_{16}	IP_6	⎭
16_{16}	CS_5	⎫ 5 – Bounds check
14_{16}	IP_5	⎭
12_{16}	CS_4	⎫ 4 – Overflow error
10_{16}	IP_4	⎭
$0E_{16}$	CS_3	⎫ 3 – Breakpoint
$0C_{16}$	IP_3	⎭
$0A_{16}$	CS_2	
08_{16}	IP_2	
06_{16}	CS_1	⎫ 1 – Debug
04_{16}	IP_1	⎭
02_{16}	CS_0	⎫ 0 – Divide error
00_{16}	IP_0	⎭

Figure 8.24 Real-mode internal interrupt and exception vector locations.

to the next instruction that is to be executed instead of the instruction that caused the trap. Therefore, upon completion of the service routine, program execution resumes with the instruction that follows the instruction that produced the trap. Finally, exceptions that produce an abort do not preserve any information that identifies the location that caused the error. In this case the system may need to be restarted. Let us now look at each of the real-mode internal interrupts and exceptions in more detail.

Divide Error Exception

The *divide error exception* represents an error condition that can occur as the result of execution of a division instruction. If the quotient that results from a DIV (divide) instruction or an IDIV (integer divide) instruction is larger than the specified destination, a divide error has occurred. This condition causes automatic initiation of a type 0 interrupt and passes control to a service routine whose starting point is defined by the values of IP_0 and CS_0, which are at addresses 00000_{16} and 00002_{16}, respectively, in the pointer table. Divide error produces a fault; therefore, the CS and IP pushed to the stack are those for the divide instruction.

Debug Exception

The *debug exception* relates to the debug mode of operation of the 80386DX. The 80386DX has a set of eight on-chip debug registers. Using these registers, the programmer can specify up to four breakpoint addresses and specify conditions under which they are to be active: for instance, the activating condition could be an instruction fetch from the address, data write to the address, or either a data read or write for the address, but not an instruction fetch. Moreover, for data accesses, the size of the data element can be specified as a byte, word, or double word. Finally, the individual addresses can be locally or globally enabled or disabled. If an access that matches any of these debug conditions is attempted, a debug exception occurs and control is passed to the service routine defined by IP_1 and CS_1 at word addresses 00004_{16} and 00006_{16}, respectively. The service routine could include a mechanism that allows the programmer to view the contents of the 80386DX's internal registers and its external memory.

If the trap flag (TF) bit in the flags register is set, the single-step mode of operation is enabled. This flag bit can be set or reset under software control. When TF is set, the 80386DX initiates a type 1 interrupt at the completion of execution of every instruction. This permits implementation of the single-step mode of operation that allows a program to be executed one instruction at a time.

Breakpoint Interrupt

The breakpoint function can also be used to implement a software diagnostic tool. A *breakpoint interrupt* is initiated by execution of the INT 3 breakpoint instruction. This instruction can be inserted at strategic points in a program that is being debugged to cause execution to be automatically stopped. Breakpoint interrupt can be used in a way similar to that of the debug exception when debugging programs. The breakpoint service routine can stop execution of the main program, permit the programmer to examine the contents of registers or memory, and allow for the resumption of execution of the program down to the next breakpoint. Breakpoint is an example of a trap. That is, upon execution of the INT 3 instruction, the CS and IP of the next instruction to be executed is saved on the stack.

Overflow Error Exception

The *overflow error exception* is an error condition similar to that of divide error. However, it can result from the execution of any arithmetic instruction. Whenever an overflow occurs, the overflow flag gets set. Unlike divide error, the transfer of program control to a service routine is not automatic at occurrence of the overflow condition. Instead, the INTO (interrupt on overflow) instruction must be executed to test the overflow flag and determine if an overflow has occurred. If OF is tested and found to be 1, program control traps to the overflow service routine. Its vector consists of IP_4 and CS_4, which are stored at 00010_{16} and 00012_{16}, respectively, in memory. The routine pointed to by this vector can be written to service the overflow condition. For instance, it could cause a message to be displayed to identify that an overflow has occurred.

Bounds Check Exception

Earlier we pointed out that the BOUND (check array index against bounds) instruction can be used to test an operand that is used as the index into an array to verify that it is within a predefined range. If the index is less than the lower bound (minimum value) or greater than the upper bound (maximum value), a *bounds check exception* has occurred and control is passed to the exception handler pointed to by $CS_5:IP_5$. The exception produced by the BOUND instruction is an example of a fault. Therefore, the values of CS and IP pushed to the stack represent the address of the instruction that produced the exception.

Invalid Opcode Exception

The exception-processing capability of the 80386DX permits detection of undefined opcodes. This feature of the 80386DX allows it to detect automatically whether or not the opcode to be executed as an instruction corresponds to one of the instructions in the instruction set. If it does not, execution is not attempted; instead, the opcode is identified as being undefined and the *invalid opcode exception* is initiated. In turn, control is passed to the exception handler identified by IP_6 and CS_6. This *undefined opcode detection mechanism* permits the 80386DX to detect errors in its instruction stream. Invalid opcode is an example of an exception that produces a fault.

Coprocessor Extension Not Available Exception

When the 80386DX comes up in the real mode, both the EM (emulate coprocessor) and MP (math present) bits of its machine status word are reset. This mode of operation corresponds to that of the 8088 or 8086 microprocessor. When set in this way, the *coprocessor extension not available exception* cannot occur. However, if the EM bit has been set to 1 under software control (do not monitor coprocessor) and the 80386DX executes an ESC (escape) instruction for the math coprocessor, a processor extension not present exception is initiated through the vector routine $CS_7:IP_7$. This service routine could pass control to a software emulation routine for the floating-point arithmetic operation. Moreover, if the MP and TS bits are set (meaning that a math coprocessor is available in the system and a task is in progress), when an ESC or WAIT instruction is executed an exception also takes place.

Interrupt Table Limit Too Small Exception

Earlier we pointed out that the LIDT instruction can be used to relocate or change the limit of the interrupt vector table in memory. If the real-mode table has been changed, for example, its limit is set lower than address $003FF_{16}$ and an interrupt is invoked that attempts to access a vector stored at an address higher than the new limit, the *interrupt table limit to small exception* occurs. In this case control is passed to the service routine by the vector $CS_8:IP_8$. This exception is a fault; therefore, the address of the instruction that exceeded the limit is saved on the stack.

Coprocessor Segment Overrun Exception

The *coprocessor segment overrun exception* signals that the 80387DX numerics coprocessor has overrun the limit of a segment while attempting to read or write its operand. This event is detected by the coprocessor data channel within the 80386DX and passes control to the service routine through interrupt vector 9. This exception handler can clear the exception, reset the 80387DX, determine the cause of the exception by examining the registers within the 80387DX, and then initiate a corrective action.

Stack Fault Exception

In the real mode, if the address of an operand access for the stack segment crosses the boundaries of the stack, a stack fault exception is produced. This causes control to be transferred to the service routine defined by CS_{12} and IP_{12}.

Segment Overrun Exception

This exception occurs in the real mode if an instruction attempts to access an operand that extends beyond the end of a segment. For instance, if a word access is made to the address CS:FFFFH, DS:FFFFH, or ES:FFFFH, a fault occurs to the *segment overrun exception* service routine.

Coprocessor Error Exception

As part of the handshake sequence between the 80386DX microprocessor and 80387DX math coprocessor, the 80386DX checks the status of its $\overline{\text{ERROR}}$ input. If the 80387DX encounters a problem performing a numerics operation, it signals this fact to the 80386DX by switching its $\overline{\text{ERROR}}$ output to logic 0. This signal is normally applied directly to the $\overline{\text{ERROR}}$ input of the 80386DX and signals that an error condition has occurred. Logic 0 at this input causes a *coprocessor error exception* through vector 16.

Protected-Mode Internal Interrupts and Exceptions

In protected mode, more internal conditions can initiate an internal interrupt or exception. Figure 8.25 identifies each of these functions and its corresponding type number.

ASSIGNMENT

Section 8.2

1. What are the five groups of interrupts supported on the 80386DX MPU?
2. What name is given to the special software routine to which control is passed when an interrupt occurs?
3. List the interrupt groups in order starting with the lowest priority and ending with the highest priority.
4. What is the range of type numbers assigned to the interrupts in a real-mode 80386DX microcomputer system?
5. Is the interrupt assigned to type number 21 at a higher or lower priority than the interrupt assigned to type number 35?

Section 8.3

6. What is the real-mode interrupt address pointer table called? Protected-mode address pointer table?
7. What is the size of a real-mode interrupt vector? Protected-mode gate?
8. The contents of which register determines the location of the interrupt address pointer table? To what value is this register initialized at reset?
9. What two elements make up a real-mode interrupt vector?

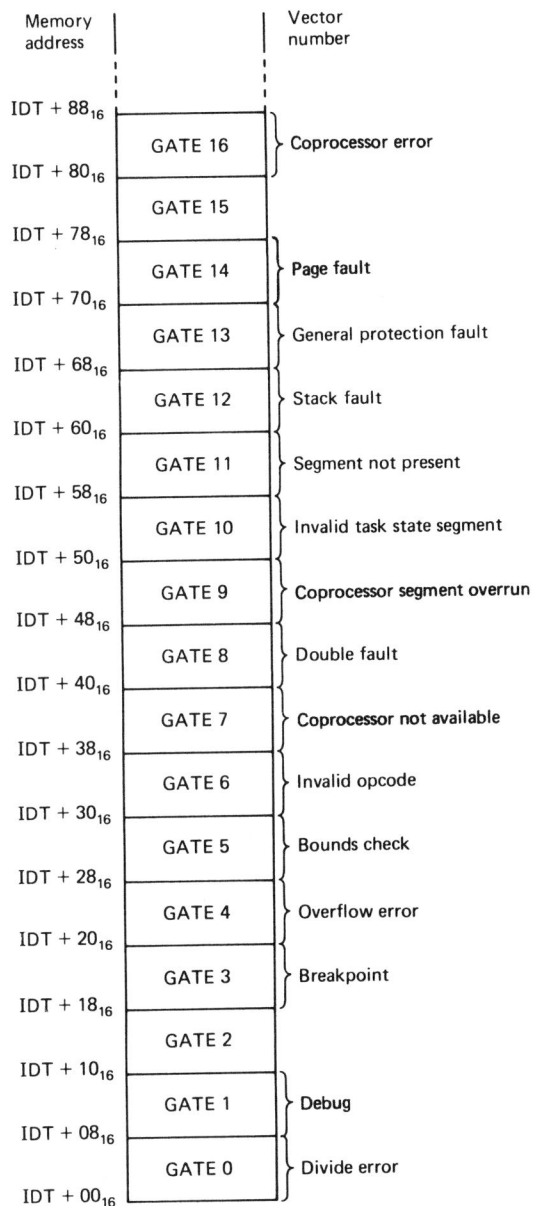

Memory address		Vector number	
IDT + 88₁₆	GATE 16	Coprocessor error	
IDT + 80₁₆	GATE 15		
IDT + 78₁₆	GATE 14	Page fault	
IDT + 70₁₆	GATE 13	General protection fault	
IDT + 68₁₆	GATE 12	Stack fault	
IDT + 60₁₆	GATE 11	Segment not present	
IDT + 58₁₆	GATE 10	Invalid task state segment	
IDT + 50₁₆	GATE 9	Coprocessor segment overrun	
IDT + 48₁₆	GATE 8	Double fault	
IDT + 40₁₆	GATE 7	Coprocessor not available	
IDT + 38₁₆	GATE 6	Invalid opcode	
IDT + 30₁₆	GATE 5	Bounds check	
IDT + 28₁₆	GATE 4	Overflow error	
IDT + 20₁₆	GATE 3	Breakpoint	
IDT + 18₁₆	GATE 2		
IDT + 10₁₆	GATE 1	Debug	
IDT + 08₁₆	GATE 0	Divide error	
IDT + 00₁₆			

Figure 8.25 Protected-mode internal exception gate locations.

10. The breakpoint routine in a real-mode 80386DX microcomputer system starts at address AA000₁₆ in the code segment located at address A0000₁₆. Specify how the breakpoint vector will be stored in the interrupt vector table.
11. At what addresses in the real-mode interrupt vector table are CS and IP for type number 40 stored?
12. At what addresses is the protected-mode gate for type number 20 stored in memory? Assume that the table starts at address 00000₁₆.

13. Assume that gate 3 consists of the following four words:

$$IDT + 8 = 1000H$$
$$IDT + A = B000H$$
$$IDT + C = AE00H$$
$$IDT + E = 0000H$$

  (a) Is the gate descriptor active?
  (b) What is the privilege level?
  (c) Is the gate a trap gate or an interrupt gate?
  (d) What is the starting address of the service routine?

Section 8.4

14. What does *STI* stand for?

15. If the values stored in memory at locations

$$IDT\_TABLE = 01FF_{16}$$
$$IDT\_TABLE + 2 = 0000_{16}$$
$$IDT\_TABLE + 4 = 0001_{16}$$

what address is loaded into the interrupt descriptor table register when the instruction LIDT IDT_TABLE is executed? What is the maximum size of the table? How many gates are provided for in this table?

16. Explain the operation performed by the following sequence of instructions:

```
INIT_IDTR   MOV    [IDT_TABLE], ILIMIT
            MOV    [IDT_TABLE + 2], IBASE_LOW
            MOV    [IDT_TABLE + 4], IBASE_HIGH
            LIDT   IDT_TABLE
            SIDT   IDT_COPY
            CMP    IDT_TABLE, IDT_COPY
            JNZ    INIT_IDTR
```

17. At what addresses is the gate for INT 50 stored in a protected-mode 80386DX microcomputer system?

18. Which type of instruction does INTO normally follow? Which flag does it test?

19. Explain the function of the bound instruction in the sequence

```
SCAN    DEC     DI
        BOUND   DI, LIMITS
          .
          .
          .
        JNZ     SCAN
```

if the contents of memory locations LIMITS and LIMITS + 2 are 0000_{16} and $00FF_{16}$.
20. What happens when the instruction HLT is executed?

Section 8.5

21. Explain how the CLI and STI instructions can be used to mask out external hardware interrupts during the execution of an uninterruptible subroutine.
22. How can the interrupt interface be reenabled during the execution of an interrupt service routine?

Section 8.6

23. What does $\overline{INTA}$ stand for?
24. Is the INTR input of the 80386DX edge triggered or level triggered?
25. Explain the function of the INTR and $\overline{INTA}$ signals in the circuit diagram of Fig. 8.8.
26. What bus cycle definition code must be decoded to produce the $\overline{INTA}$ signal?
27. During the interrupt acknowledge bus cycle, what does the first $\overline{INTA}$ pulse tell the external circuitry? The second pulse?
28. If the interface diagram in Fig. 8.8 represents an I/O-mapped interrupt controller interface, which address lines are part of the interface?
29. Over which lines does external circuitry send the type number of the active interrupt to the 80386DX?

Section 8.7

30. Give an overview of the events in the order in which they take place during the interrupt request, interrupt acknowledge, and interrupt vector fetch cycles of a real-mode 80386DX microcomputer system.
31. If a real-mode 80386DX microcomputer system is running at 20 MHz with no wait states in all I/O bus cycles, how long does it take to perform the interrupt-acknowledge bus cycle sequence?
32. How long does it take the 80386DX in Problem 31 to push the old flags, old CS, and old IP to the stack? How much stack space does this information take up?
33. How long does it take the 80386DX in Problem 31 to fetch the vector $CS_{NEW}:IP_{NEW}$ from memory?
34. What is the primary difference between the real-mode and protected-mode interrupt request/acknowledge handshake sequences?
35. If a program that is interrupted is in a conforming code segment, what is the new privilege level equal to? What stack is used? What information is saved on the stack?
36. If the program transition identified in Problem 35 fails, what additional information is pushed to the stack?
37. If an IDT error code has the value $0000F307_{16}$, what are the values of the EXT, IDT, and TI bits, and what do they stand for?
38. What is the relationship between CPL and RPL in an intralevel interrupt return? In an interlevel interrupt return?

Section 8.8

39. Specify the value of ICW$_1$ needed to configure an 82C59A as follows: ICW$_4$ not needed, single-device interface, and edge-triggered inputs.
40. Specify the value of ICW$_2$ if the type numbers produced by the 82C59A are to be in the range 70_{16} through 77_{16}.
41. Specify the value of ICW$_4$ such that the 82C59A is configured for use in an 80386DX system, with normal EOI, buffered-mode master, and special fully nested mode disabled.
42. Write a program that will initialize an 82C59A with the initialization command words derived in Problems 39, 40, and 41. Assume that the 82C59A resides at address $A000_{16}$ in the I/O address space. Assume that the contents of DS are 0000_{16}.
43. Write an instruction that when executed will read the contents of OCW$_1$ and place it in the AL register. Assume that the 82C59A has been configured by the software of Problem 42.
44. What priority scheme is enabled if OCW$_2$ equals 67_{16}?
45. Write an instruction sequence that when executed will toggle the state of the read register bit in OCW$_3$. Assume that the 82C59A is located at memory address $0A000_{16}$ and that the contents of DS are 0000_{16}.

Section 8.9

46. How many interrupt inputs can be directly accepted by the circuit in Fig. 8.23(a)?
47. How many interrupt inputs can be directly accepted by the circuit in Fig. 8.23(b)?
48. Overview the interrupt request/acknowledge handshake sequence for an interrupt initiated at an input to slave B in the circuit of Fig. 8.23(b).
49. What is the maximum number of interrupt inputs that can be achieved by expanding the number of slaves in the master–slave configuration of Fig. 8.23(b)?

Section 8.10

50. List the real-mode internal interrupts serviced by the 80386DX.
51. Internal interrupts and exceptions are categorized into groups based on how the failing function is reported. List the three groups.
52. Which real-mode vector numbers are reserved for internal interrupts and exceptions?
53. Into which reporting group is the invalid opcode exception classified?
54. What is the cause of a stack fault exception?
55. Which exceptions take on a new meaning or are only active in the protected mode?

9

80386DX PC/AT Microcomputer System Hardware

9.1 INTRODUCTION

Having learned about the 80386DX microprocessor, its memory, input/output, and interrupt interfaces, we now turn our attention to a microcomputer system designed using this hardware. The microcomputer we study in this chapter is one that implements an 80386DX-based PC/AT (personal computer/advanced technology). The material covered in this chapter is organized as follows:

1. Architecture of the system processor board in the original IBM PC/AT
2. High-integration PC/AT-compatible peripheral ICs
3. Core 80386DX microcomputer
4. 82345 data buffer
5. 82346 system controller
6. 82344 ISA controller
7. 82341 peripheral combo
8. 82077AA floppy disk controller

9.2 ARCHITECTURE OF THE SYSTEM PROCESSOR BOARD IN THE ORIGINAL IBM PC/AT

Before we begin our study of an 80386DX-based PC/AT compatible microcomputer system, we introduce the architecture implemented in IBM's original 80286-based PC/AT. The IBM PC/AT is a practical implementation of the 80286 microprocessor and its peripheral chip set as a general-purpose microcomputer. A block diagram of the system processor board (main circuit board) of the PC/AT is shown in Fig. 9.1(a). This diagram identifies the major functional elements of the PC/AT: MPU, PIC, DMA, PIT, parallel I/O, ROM, and RAM. Here we describe briefly the architecture of the PC/AT's microcomputer system.

The heart of the PC/AT's system processor board is the 80286 microprocessor unit (MPU). It is here that instructions of the program are fetched and executed. To interface to the peripherals and other circuitry such as memory, the 80286 microprocessor generates address, data, status, and control signals. Together, these signals form what is called the *local bus* in Fig. 9.1(a). Notice that the local address (A) lines are latched and buffered to form a 24-bit *system address bus* and the local data (D) lines are buffered to provide a 16-bit *system data bus* (SD).

Notice in Fig. 9.1(a) that the system address lines are further buffered by the *ROM/IO address buffer*. This *extended address bus* (XA) is distributed to ROM memory and the I/O peripherals. The system data bus is buffered by the *RAM/ROM data bus transceiver* to give the *memory data bus* (MD). Looking at Fig. 9.1(a), we see that the MD lines are the data path for both the ROM and RAM banks. In a similar way, the lower eight lines of the system data bus are buffered by the *I/O data bus transceiver* to give an 8-bit *extended data bus* (XD). The extended data bus is the data path for all of the I/O devices.

At the same time, the status and control lines of the local bus are decoded by the 82288 bus controller to generate the *system control bus*. This control bus consists of memory and I/O read and write control signals. The bus controller also produces the data bus control signals. For instance, it provides a signal that determines the direction of data transfer through the data bus transceivers: that is, the signal needed to make the data bus lines work as inputs to the microprocessor during memory and I/O read operations and as outputs during write operations.

The operation of the microprocessor and other devices in a microcomputer system must be synchronized. The circuitry of the clock generator block generates clock signals for this purpose. The clock generator section also produces a power-on reset signal that is needed to initiate initialization of the microprocessor and peripherals at power-up.

The clock generator section also works in conjunction with the wait-state logic to synchronize the MPU to slow peripheral devices. In Fig. 9.1(a) we see that the wait-state logic circuitry monitors the system control bus and the signals IO CH RDY and REFRESH. Depending on the state of these inputs, it generates a wait signal for input to the 82284 clock generator. In turn, the clock generator synchronizes this wait input with the system clock to produce a ready signal at its output. READY is input to the 80286 MPU and provides the ability to extend automatically bus cycles that are performed to slow devices by inserting wait states. For instance, all bus cycles to on-board memory take three clock periods. Therefore, they include one wait state. On the other hand, I/O bus cycles to the 8-bit LSI peripherals have four wait states and take a total of six clock periods.

The memory subsystem of the PC/AT system processor board has 256K

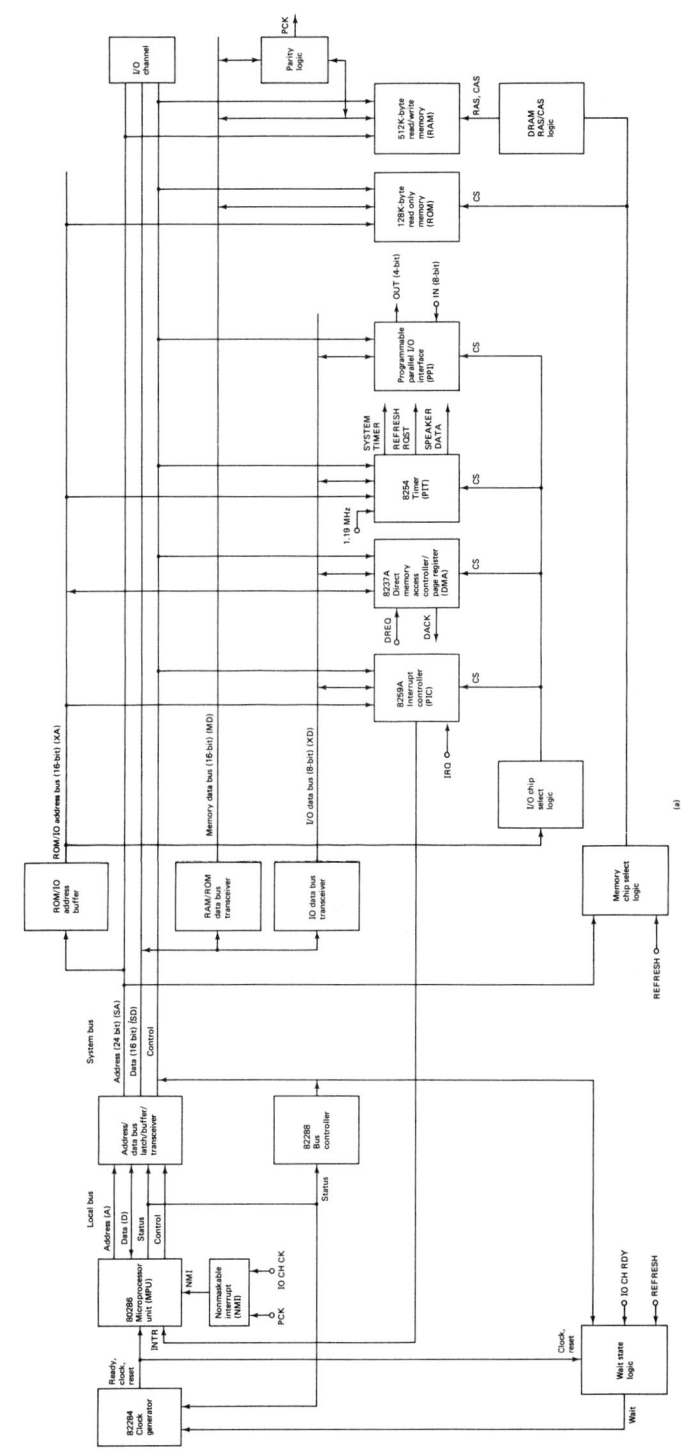

Figure 9.1 (a) IBM PC/AT microcomputer block diagram; (b) memory map; (c) input/output address map; (d) input/output functions; (e) interrupt levels and functions. (Parts b, c, and e courtesy of International Business Machines Corporation.)

Address	Name	Function
000000H to 07FFFFH	512KB system board	System board memory
080000H to 09FFFFH	128KB	I/O channel memory — IBM personal computer AT 128KB memory expansion option
0A0000H to 0BFFFFH	128KB video RAM	Reserved for graphics display buffer
0C0000H to 0DFFFFH	128KB I/O expansion ROM	Reserved for ROM on I/O adapters
0E0000H to 0EFFFFH	64KB reserved on system board	Duplicated code assignment at address FE0000H
0F0000H to 0FFFFFH	64KB ROM on the system board	Duplicated code assignment at address FF0000H
100000H to FDFFFFH	Maximum memory 15MB	I/O channel memory — IBM personal computer AT 512KB memory expansion option
FE0000H to FEFFFFH	64KB reserved on system board	Duplicated code assignment at address 0E0000H
FF0000H to FFFFFFH	64KB ROM on the system board	Duplicated code assignment at address 0F0000H

(b)

Address	Device
000H-01FH	DMA controller 1, 8237A-5
020H-03FH	Interrupt controller 1, 8259A, master
040H-05FH	Timer, 8254-2
060H-06FH	8042 (Keyboard)
070H-07FH	Real-time clock, NMI (nonmaskable interrupt) mask
080H-09FH	DMA page register, 74LS612
0A0H-0BFH	Interrupt controller 2, 8259A
0C0H-0DFH	DMA controller 2, 8237A-5
0F0H	Clear math coprocessor busy
0F1H	Reset math coprocessor
0F8H-0FFH	Math coprocessor
1F0H-1F8H	Fixed disk
200H-207H	Game I/O
278H-27FH	Parallel printer port 2
2F8H-2FFH	Serial port 2
300H-31FH	Prototype card
360H-36FH	Reserved
378H-37FH	Parallel printer port 1
380H-38FH	SDLC, bisynchronous 2
3A0H-3AFH	Bisynchronous 1
3B0H-3BFH	Monochrome display and printer adapter
3C0H-3CFH	Reserved
3D0H-3DFH	Color/graphics monitor adapter
3F0H-3F7H	Diskette controller
3F8H-3FFH	Serial port 1

(c)

Figure 9.1 (*Continued*)

Input/output	Signal	Function
OUT$_0$	TIM 2 GATE SPK	Speaker enable
OUT$_1$	SPKR DATA	Speaker data
OUT$_2$	ENB RAM PCK	RAM parity check enable
OUT$_3$	ENA IO CK	Enable I/O channel check
IN$_0$	TIM 2 GATE SPK	Speaker enable
IN$_1$	SPKR DATA	Speaker data
IN$_2$	ENB RAM PCK	RAM parity check enable
IN$_3$	ENA IO CK	Enable I/O channel check
IN$_4$	REF DET	Refresh detect
IN$_5$	OUT 2	Timer 2 output
IN$_6$	IO CH CK	I/O channel check
IN$_7$	PCK	Parity check

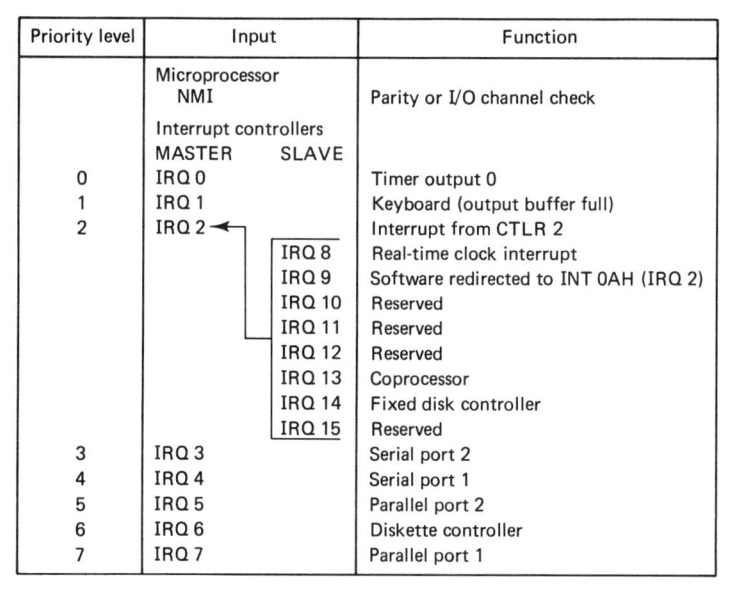

(d)

Priority level	Input	Function
	Microprocessor NMI	Parity or I/O channel check
	Interrupt controllers MASTER SLAVE	
0	IRQ 0	Timer output 0
1	IRQ 1	Keyboard (output buffer full)
2	IRQ 2	Interrupt from CTLR 2
	IRQ 8	Real-time clock interrupt
	IRQ 9	Software redirected to INT 0AH (IRQ 2)
	IRQ 10	Reserved
	IRQ 11	Reserved
	IRQ 12	Reserved
	IRQ 13	Coprocessor
	IRQ 14	Fixed disk controller
	IRQ 15	Reserved
3	IRQ 3	Serial port 2
4	IRQ 4	Serial port 1
5	IRQ 5	Parallel port 2
6	IRQ 6	Diskette controller
7	IRQ 7	Parallel port 1

(e)

Figure 9.1 (*Continued*)

bytes of dynamic R/W memory (RAM). This on-board RAM is implemented using 128K × 1-bit dynamic RAMs and can be expanded to 512K bytes by replacing the ICs with 256K-bit DRAMs. A memory map for the PC/AT's memory is shown in Fig. 9.1(b). From the map we find that the 512K bytes of system board RAM reside in the address range from 000000_{16} through $07FFFF_{16}$. In the microcomputer system, RAM is used to store operating system routines, application programs, and data that are to be processed. These programs and data are typically loaded into RAM from a mass storage device such as a diskette or hard disk.

Looking at Fig. 9.1(a), we see that the RAM subsystem includes parity logic. This circuit generates and adds a parity bit to all data written to RAM. Moreover, whenever data are read from RAM, parity is checked and if an error occurs, it signals this fact with the PCK signal.

The memory subsystem also contains 64K bytes of read-only memory (ROM). ROM is implemented with 32K × 8 EPROMs and is expandable to 128K bytes. The memory map shows that on-board ROM is located in two separate 64K-byte address ranges. The first 64K bytes are located from $0F0000_{16}$ to $0FFFFF_{16}$. Expansion is allowed by the second 64K-byte block of addresses from $FF0000_{16}$ to $FFFFFF_{16}$. This part of the memory subsystem contains the system ROM of the PC/AT. Included in these ROMs are fixed programs such as the *BASIC interpreter, power-on system procedures*, and *I/O device drivers* or *BIOS*, as they are better known.

The memory and I/O chip select logic sections, which are shown in the block diagram of Fig. 9.1(a), are used to select and enable the appropriate memory or peripheral device whenever a bus cycle takes place over the system bus. Here we see that they accept address information at their inputs and produce chip select signals at their outputs. To select a device in the I/O address space, such as the DMA controller, the I/O chip select logic decodes the extended address XA on the ROM/IO address bus to generate a chip select (CS) output for the corresponding I/O device. This chip select signal is applied to the I/O device and enables its microprocessor interface for operation.

The ROM and RAM chip selects are produced in a similar way. But this time they are produced by decoding the address on the system address (SA) bus. Notice that the RAM chip selects are converted to row address select (RAS) and column address select (CAS) signals before they are applied to the RAM array.

The LSI peripheral devices included on the PC/AT system processor board are the 8237A direct memory access (DMA) controller, 8254 programmable interval timer (PIT), and 8259A programmable interrupt controller (PIC). Notice that each of these devices is identified with a separate block in Fig. 9.1(a). These peripherals are all located in the 80286's I/O address space and their registers are accessed through software using the address ranges given in Fig. 9.1(c). For instance, the four registers within the PIT are located at addresses 0040_{16}, 0041_{16}, 0042_{16}, and 0043_{16}.

To support high-speed memory and I/O data transfers, two 8237A direct memory access controllers are provided on the PC/AT system board. DMA controller 1 contains four DMA channels, *DMA channel 0* through *DMA channel 3*. One channel, DMA 1, is dedicated to support an SDLC synchronous communications channel and the other three DMA channels are available at the I/O channel [industry standard architecture (ISA) bus] for use with peripheral devices. For instance, DMA 2 is used to support floppy disk drive data transfers over the ISA bus. DMA controller 2 provides *DMA channels 4 through 7*. Here channel 4 is used as a cascade input for DMA controller 1 and channels 5 through 7 are also available at the ISA bus.

The 8254-based timer circuitry is used to generate time-related functions and signals in the PC/AT. There are three 16-bit counters in the 8254 PIT and they are all driven by a 1.19-MHz clock input signal. Timer 0 is used to generate an interrupt to the microprocessor. This timing function is known as the *system timer*. On the other hand, timer 1 is used to produce a request to initiate refresh of the dynamic RAM every 15 μs. The last timer is used to generate programmable tones for driving the speaker.

The parallel I/O section of the PC/AT microcomputer in Fig. 9.1(a) provides both an 8-bit input port and a 4-bit output port. The signal and function for each of the I/O lines is identified in Fig. 9.1(d). Here we see that the I/O lines are used to output tones to the speaker (SPKR DATA), enable or disable I/O channel check (ENA IO CK) and RAM parity check (ENB RAM PCK), and input the state of signals such as PCK, IO CH CK, and OUT 2.

The circuitry in the nonmaskable interrupt (NMI) logic block allows a nonmaskable interrupt request derived from two sources to be applied to the microprocessor. As shown in Fig. 9.1(a), these interrupt sources are *R/W memory parity check* (PCK) and *I/O channel check* (IO CH CK). If either of these inputs is active, the NMI logic circuit outputs a request for service to the 80286 over the NMI signal line. The 80286 can determine the cause of the NMI request by reading the state of PCK and IO CH CK through the parallel I/O interface.

In addition to the nonmaskable interrupt, the PC/AT architecture provides for requests for service to the MPU by interrupts at another interrupt input called *interrupt request* (*INTR*). Notice in Fig. 9.1(a) that this signal is supplied to the 80286 by the output of the interrupt controller (PIC) block. Two 8259A LSI interrupt controllers are used in the PC/AT. Each provides for eight prioritized interrupt inputs. The inputs of the interrupt controller are supplied by peripherals such as the timer, keyboard, diskette drive, printer, and communication devices. Interrupt priority assignments for these devices are listed in Fig. 9.1(e). For example, the timer (actually just timer 0 of the 8254) is at the highest-priority level, which is IRQ_0.

I/O channel (*the ISA bus*), which is a collection of address, data, control, and power lines, is provided to support expansion of the PC/AT system. The chassis of the PC/AT has eight expansion slots. Six slots have both a 36-pin and a 62-pin card edge socket. The other two slots have just a 62-pin socket. In this way the system configuration can be expanded by adding special function adapter cards, such as boards to control a monochrome or color display, floppy disk drives, a hard disk drive, or expanded memory, or to attach a printer.

9.3 HIGH-INTEGRATION PC/AT-COMPATIBLE PERIPHERAL ICs

In the preceding section we provided an overview of the architecture of the main processor board of the original IBM PC/AT personal computer. The circuit implementation used on this board has become an architectural standard. PC/AT compatibles, whether they use the 80286, 80386SX, or 80386DX microprocessor, must provide equivalent circuitry to assure 100% hardware and 100% software compatibility. Here we begin our study of the circuit implementation techniques used in the design of modern 80386DX-based PC/AT-compatible microcomputer systems.

Since the introduction of the original IBM PC/AT, much work has been done to develop special-purpose ICs that simplify the design of PC/AT-compatible personal computers. The cornerstone of this effort has been the development of high-integration peripheral ICs. The objectives of these peripherals are to reduce the complexity and cost of the circuitry on the main processor board while increasing the functionality. In this section we begin a study of one of the more popular PC/AT-compatible peripheral IC chip sets and its use in the design of a modern PC/AT-compatible main processor board.

The *82340 PC/AT Chip Set* offers a four-chip high-integration solution for implementing 80386DX-based PC/AT-compatible personal computers. These devices, which are manufactured by Intel Corporation, implement most of the circuitry needed on the main processor board. That is, the 82340 chips, along with a few other ICs, produce a complete solution for the main processor board of a modern PC/AT-compatible microcomputer. In fact, they implement more functions than are provided on the original 80286-based PC/AT main processor board. Examples of additional functions included are two serial communication channels, a parallel printer port, and a hard disk drive interface.

Figure 9.2 is a block diagram of the architecture of a main processor board that uses the 82340 chip set. Here we find that at the core of the microcomputer are the 80386DX MPU, 80387DX numerics coprocessor, and 82385DX cache controller. Attached to the local bus of the MPU, we find three of the 82340 peripheral ICs: the 82345 data buffer, 82346 system controller, and 82344 ISA controller. At the other side, these peripherals connect to the ISA expansion bus. Notice that the 82341 peripheral combo does not connect directly to the 80386DX MPU; instead, it is tied to the ISA bus.

The 82340 chip set is very versatile. With its programmable features, the microcomputer design can be optimized for low cost, maximum performance, minimum physical size, or low power. The circuit diagram in Fig. 9.3 shows a typical design of a 33-MHz 80386DX PC/AT main processor board. Notice that some additional devices, such as PALs™, octal latches, octal transceivers, octal buffers, and logic gates, are needed along with the 82340 chips to complete the design.

EXAMPLE 9.1

What function is performed by the 82077AA device?

SOLUTION In Fig. 9.2 we find that the 82077AA is the floppy disk controller for the microcomputer system.

9.4 CORE 80386DX MICROCOMPUTER

Let us begin our study of the PC/AT main processor board in Fig. 9.3 with the circuitry of the core 80386DX microcomputer. This part of the microcomputer is shown on sheet 1 of the circuit diagram. Notice that it consists of the 80386DX

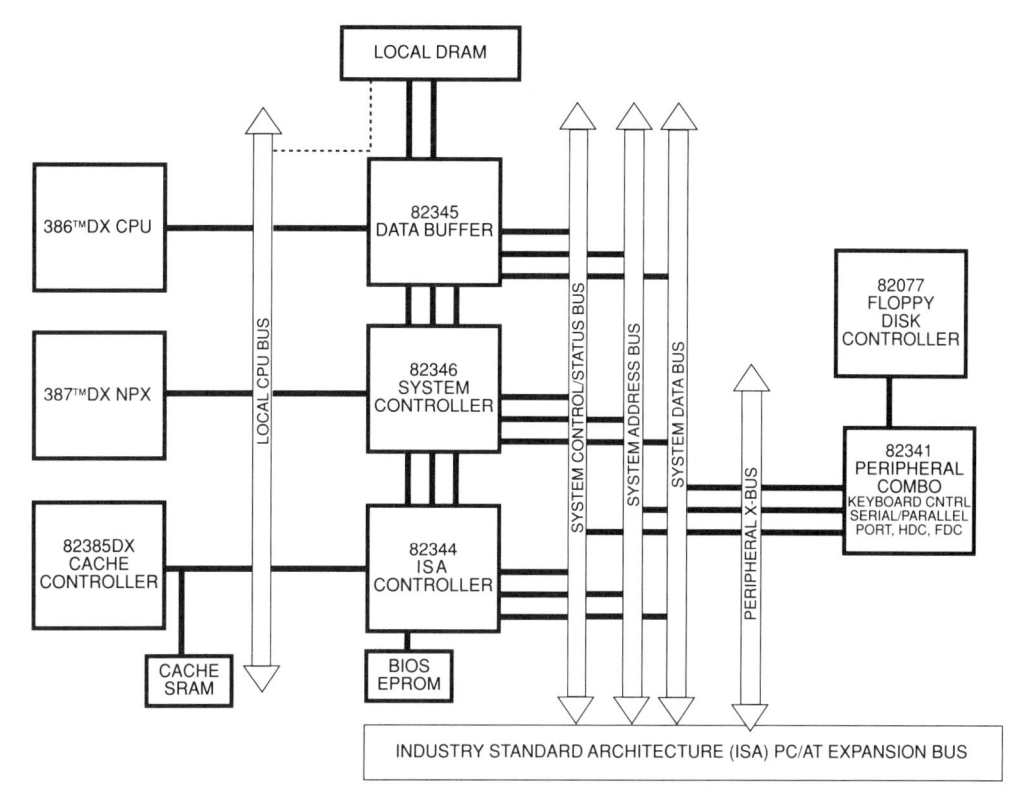

Figure 9.2 Block diagram of a PC/AT main processor board using the 82340 chip set. (Reprinted by permission of Intel Corp. Copyright Intel Corp. 1989.)

MPU (IC U_{12}), the 80387DX numerics coprocessor (U_{10}), the 82385DX cache controller (U_{15}), and the cache memory array (U_{16} and U_{17}). The 80386DX, 80387DX, and 82385DX are designed to attach to each other directly. Looking at their interconnection in Fig. 9.3, we see that address lines A_2 through A_{31} of the 80386DX attach directly to the corresponding address line of the 82385DX and data bus lines D_0 through D_{31} of the 80386DX connect directly to the corresponding data line on the 80387DX. Moreover, notice that the 80386DX's bus cycle definition outputs, D/$\overline{C}$, W/$\overline{R}$, and M/$\overline{IO}$, are applied directly to the D/$\overline{C}$, W/$\overline{R}$, and M/$\overline{IO}$ inputs of the 82385DX.

EXAMPLE 9.2

What signals supply the PEREQ, $\overline{ERROR}$, and $\overline{BUSY}$ inputs of the 80386DX? What are the sources of these signals?

SOLUTION From sheet 1 of Fig. 9.3, we find that the PEREQ, $\overline{ERROR}$, and $\overline{BUSY}$ inputs of the 80386DX are supplied by the signals PEREQ386, $\overline{ERROR386}$, and $\overline{BUSY386}$, respectively. These signals are produced at outputs of the 82346 system controller (sheet 3 of Fig. 9.3).

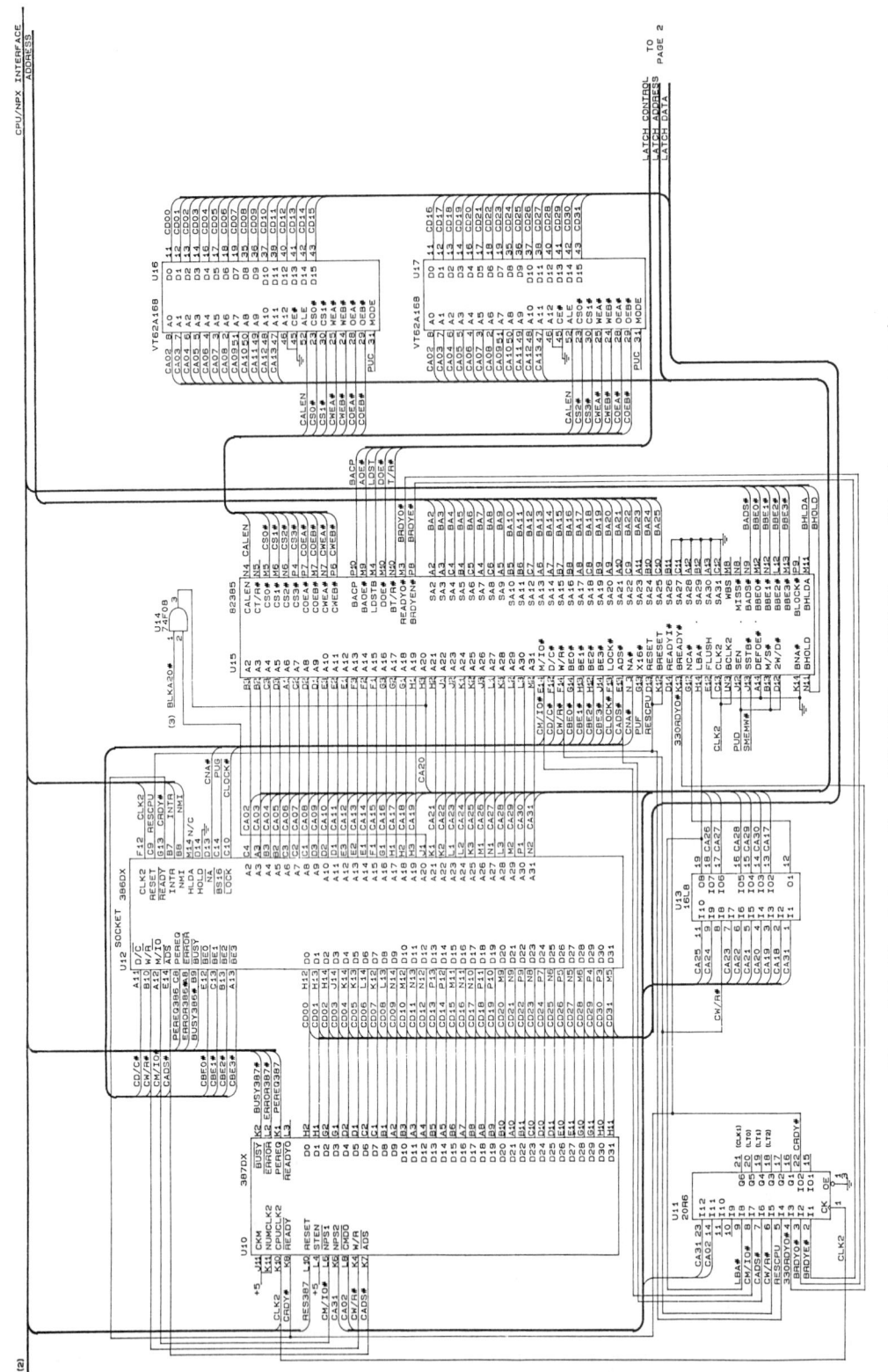

Figure 9.3 Circuit diagram for an 80386DX based PC/AT compatible microcomputer main processor board (Sheet 1 of 8). (Reprinted by permission of Intel Corp. Copyright Intel Corp. 1989.)

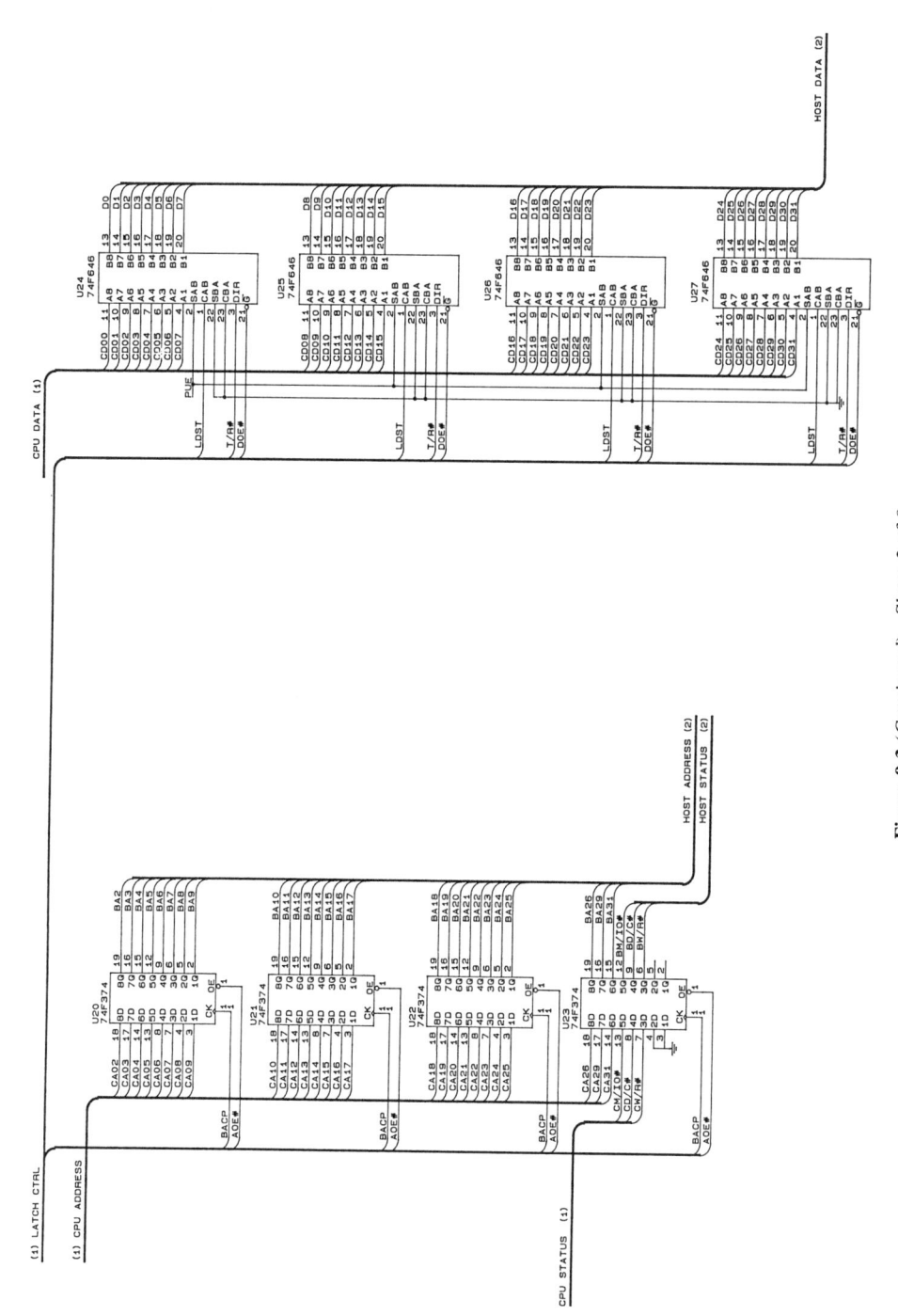

Figure 9.3 (*Continued*) Sheet 2 of 8.

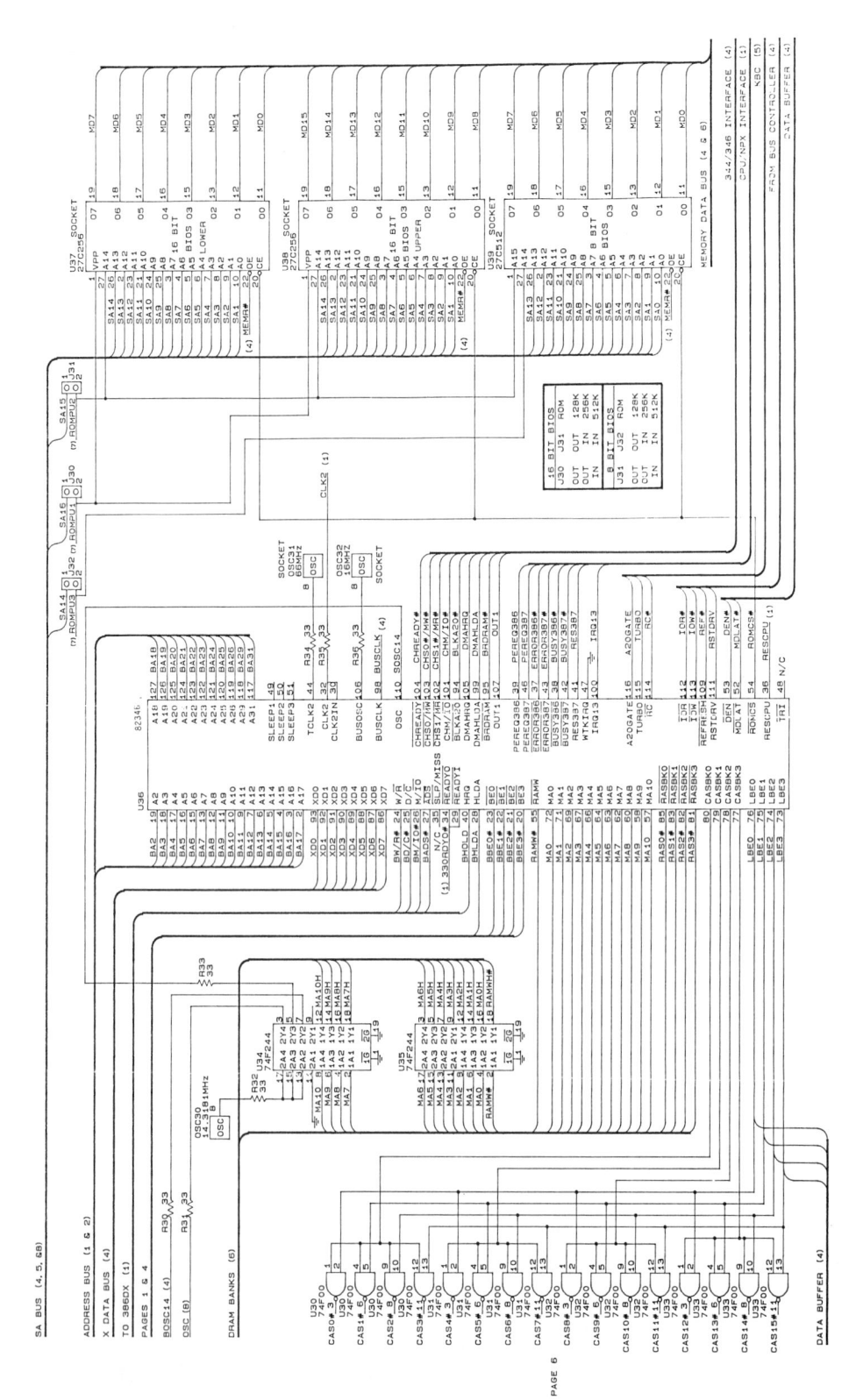

Figure 9.3 (*Continued*) Sheet 3 of 8.

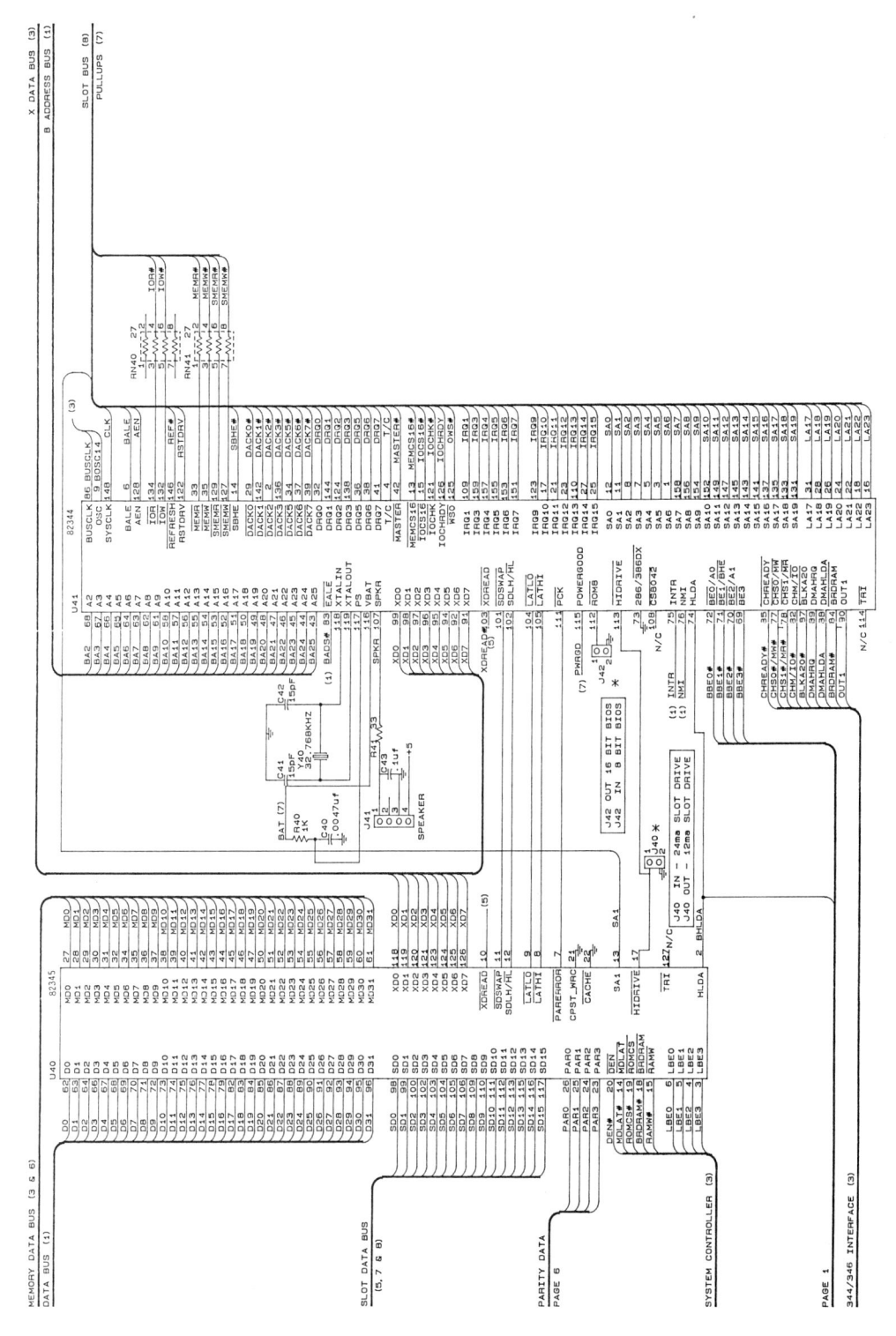

Figure 9.3 (*Continued*) Sheet 4 of 8.

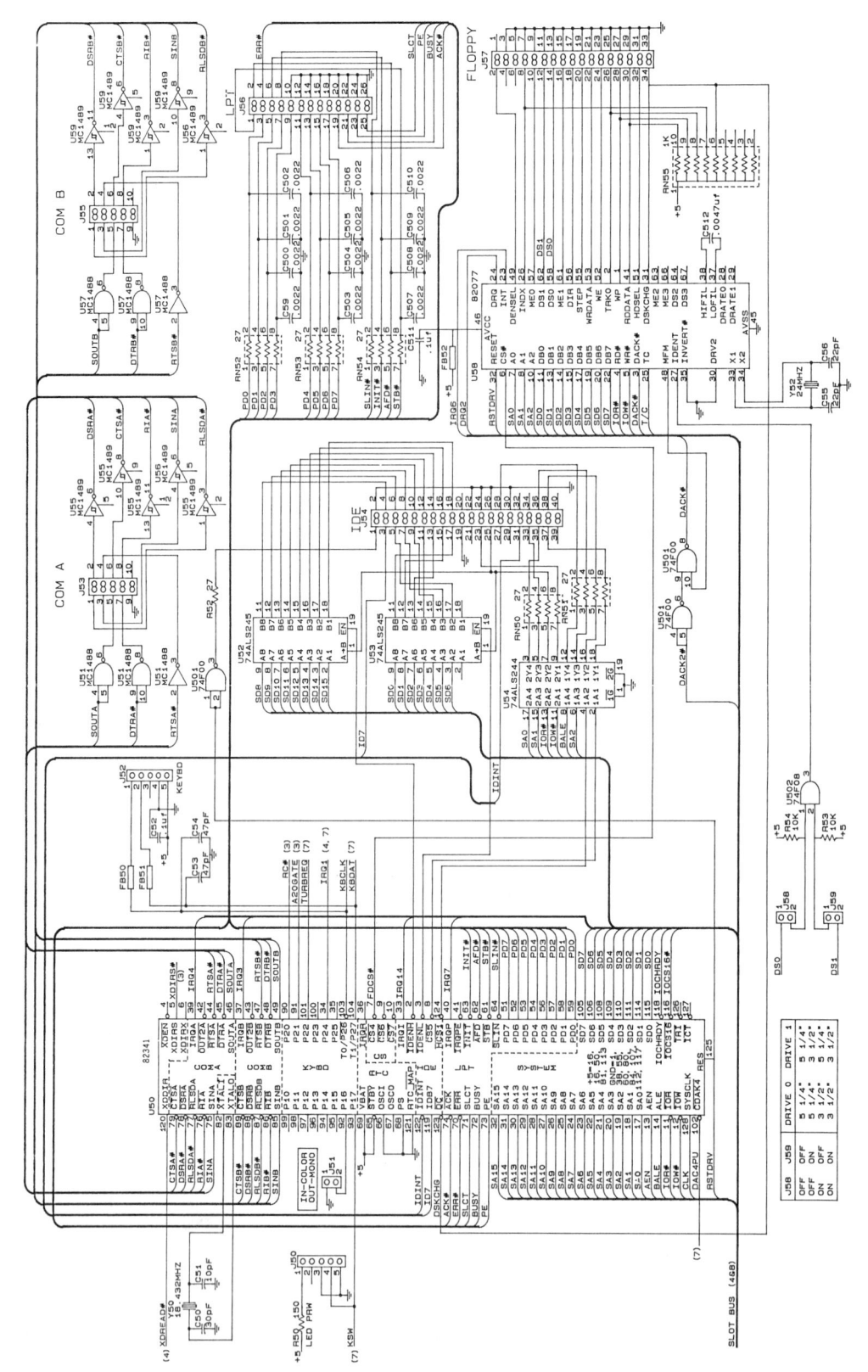

Figure 9.3 (*Continued*) Sheet 5 of 8.

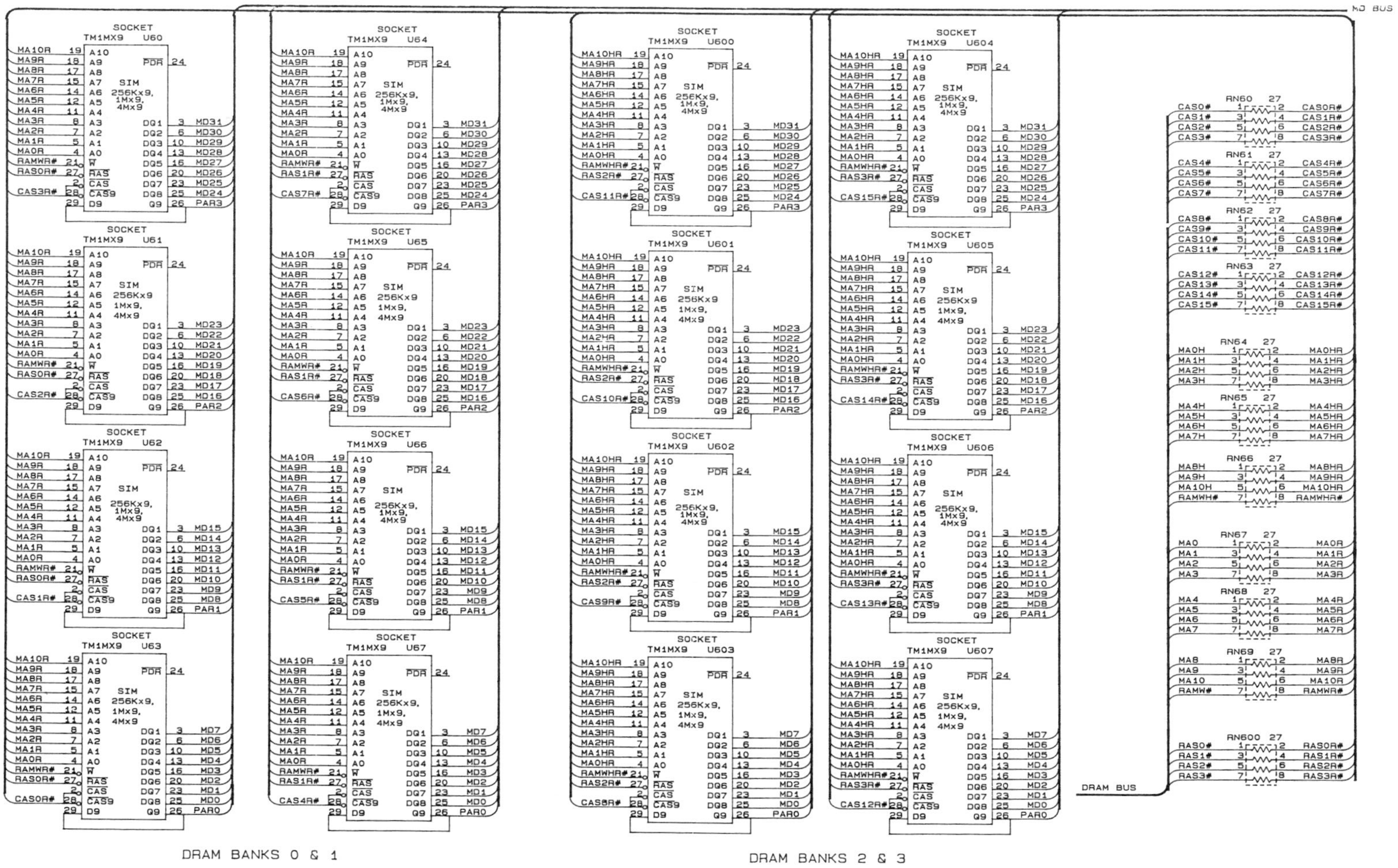

Figure 9.3 (*Continued*) Sheet 6 of 8.

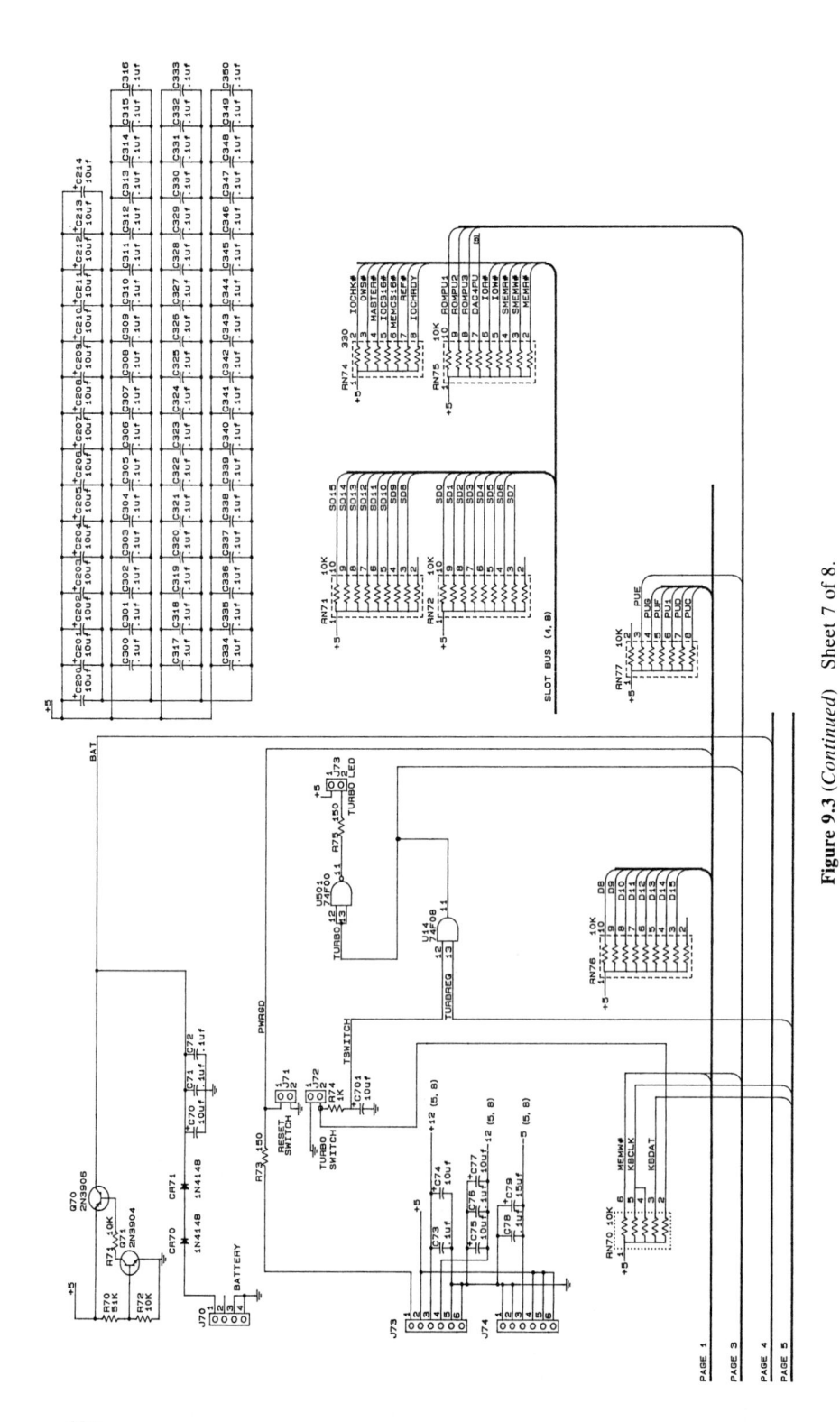

Figure 9.3 (*Continued*) Sheet 7 of 8.

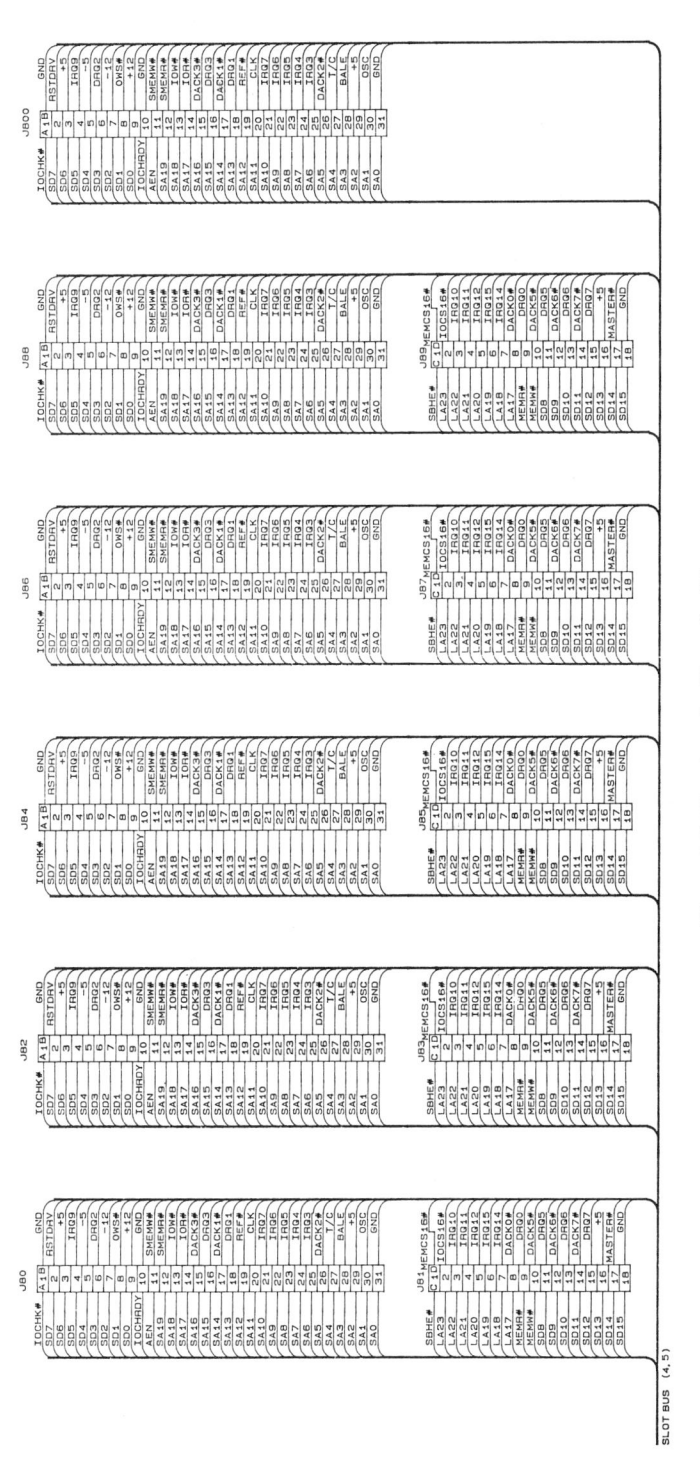

Figure 9.3 (*Continued*) Sheet 8 of 8.

The cache memory array is implemented with two VT62A16B 8K × 16-bit static RAMs. These SRAMs, which are labeled U_{16} and U_{17} on sheet 1 of Fig. 9.3, are specifically designed for use in an 82385DX-based cache memory subsystem. For this reason no external address latches or data bus transceivers are required in the cache memory interface. They are built within the SRAM ICs. Notice that cache memory control outputs CALEN, $\overline{\text{COEA}}$, $\overline{\text{COEB}}$, $\overline{\text{CWEA}}$, and $\overline{\text{CWEB}}$ are applied to equivalent inputs on both VT62A16Bs.

EXAMPLE 9.3

Is the 82385DX on sheet 1 in Fig. 9.3 configured to operate in direct-mapped or two-way set associative mode?

SOLUTION On sheet 1 of Fig. 9.3 we find that the $2W/\overline{D}$ input is supplied by the signal PUD. Tracing this signal back to sheet 7, we see that it supplies +5 V to the $2W/\overline{D}$ input through a 10-kΩ pull-up resistor. This is a logic 1 at the $2W/\overline{D}$ input and selects two-way set associative operation.

EXAMPLE 9.4

Which SRAM in the cache memory array is selected with chip select outputs $\overline{\text{CS}}_2$ and $\overline{\text{CS}}_3$ of the 82385DX cache controller?

SOLUTION Looking at sheet 1 in Fig. 9.3, we see that the $\overline{\text{CS}}_2$ and $\overline{\text{CS}}_3$ outputs of the 82385DX are applied to the $\overline{\text{CS}}_0$ and $\overline{\text{CS}}_1$ inputs, respectively, of SRAM U_{17}. Therefore, $\overline{\text{CS}}_2$ and $\overline{\text{CS}}_3$ enable the lower of the two SRAM ICs.

The ready/wait logic is implemented with a 20R6 PAL™. This is device U_{11} on sheet 1 of Fig. 9.3. Notice that it accepts signals such as CPU read/write ($\text{CW}/\overline{\text{R}}$), CPU address strobe ($\overline{\text{CADS}}$), CPU memory/IO ($\text{CM}/\overline{\text{IO}}$), and 82385DX bus ready ($\overline{\text{BREADY}}$) as inputs and decodes them to produce the appropriately timed CPU ready output ($\overline{\text{CRDY}}$). $\overline{\text{CRDY}}$ is applied in parallel to the $\overline{\text{READY}}$ input at pin G13 of the 80386DX, the $\overline{\text{READY}}$ input at pin K8 of the 80387DX, and $\overline{\text{READYI}}$ input at pin D14 of the 82385DX. Based on the input conditions, $\overline{\text{CRDY}}$ inserts the appropriate number of wait states into the bus cycle.

EXAMPLE 9.5

What is the source of the $\overline{\text{LBA}}$ input of the ready/wait state PAL™? To what other device is this signal also supplied?

SOLUTION Tracing the $\overline{\text{LBA}}$ line at pin 9 of U_{11}, we find that it is supplied by output O_8 at pin 19 of PAL™ U_{13}. O_8 is also connected to the $\overline{\text{LBA}}$ input at pin H14 of the 82385DX.

One other circuit is shown on sheet 1 of the circuit diagram in Fig. 9.3. This circuit, which is implemented with the 16L8 PAL™ U_{13}, is used to map the 80386DX's address space into cacheable and noncacheable regions. Notice that the noncacheable access ($\overline{\text{NCA}}$) input of the 82385DX is driven by output O_1 at pin 12 of U_{13}. Remember that if $\overline{\text{NCA}}$ is logic 0 during a bus cycle, the bus cycle is performed to main memory, not to cache memory; that is, a noncacheable bus cycle is in progress. The inputs to this PAL™ are address lines CA_{17} through CA_{30} and $CW/\overline{R}$.

The local address and data bus lines of the 80386DX are latched and buffered by the address latch and data bus transceiver circuits. On sheet 2 of Fig. 9.3 we find that four 74F374 octal latches are used to latch the address and bus cycle definition signals of the 80386DX. These devices are the ICs labeled U_{20} through U_{23}. For example, U_{20} latches *CPU address* inputs CA_{02} through CA_{09} to produce *buffered address* outputs BA_2 through BA_9. Address information is clocked into the latches with the bus address clock pulse (BACP) signal and enabled to the outputs with address output enable ($\overline{AOE}$). Both of these signals are produced by the 82385DX. The address output on BA_2 through BA_{31} is identified on sheet 2 of Fig. 9.3 as the *host address bus*.

EXAMPLE 9.6

Which 74F374 IC is used to latch the CPU's bus cycle definition signals? How are the buffered bus cycle definition signals labeled? What is the output status bus called?

SOLUTION Looking at sheet 2 in Fig. 9.3, we see that *CPU status* signals $CM/\overline{IO}$, $CD/\overline{C}$, and $CW/\overline{R}$ are inputs of the 74F374 U_{23} and are latched to produce the buffered bus cycle definition signals $BM/\overline{IO}$, $BD/\overline{C}$, and $BW/\overline{R}$. These signals produce what is called the *host status bus*.

The data bus buffer is formed from four 74F646 registered transceiver devices. These ICs are labeled U_{24} through U_{27} on sheet 2 of Fig. 9.3. The A bus side of the transceivers are attached to the *CPU data bus* lines and their B sides supply the *host data bus*.

Let us continue by looking at the connection of the control signals of the data bus transceivers. In Fig. 9.3 we find that control inputs S_{BA} and C_{BA} are both wired to ground (logic 0) and S_{AB} is supplied by the signal PUE. PUE is traced to sheet 7 of the circuitry in Fig. 9.3. Here it is found to supply +5 V through a 10-kΩ pull-up resistor. Therefore, S_{AB} is at logic 1. The other three control signals of the latches are supplied by the 82385DX. For example, the C_{AB} clock inputs are driven by the *load data strobe* (LDSTB) output of the 82385DX.

EXAMPLE 9.7

What signal supplies the direction input of the 74F646 transceivers? What is the source of this signal?

SOLUTION Sheet 2 of Fig. 9.3 shows that the DIR input of all four 74F646 transceivers is driven by the signal $T/\overline{R}$. This signal is traced back to pin N10 ($BT/\overline{R}$) of the 82385DX on sheet 1.

Next we examine how the transceivers operate during read bus cycles from main memory. Since S_{BA} is at logic 0, the 74F646s are set for real-time transfers of data at bus B to bus A. That is, when $\overline{DOE}$ and $T/\overline{R}$ are logic 0, data supplied to the host data bus by memory or an I/O device are immediately passed through the transceivers to the CPU data bus.

When data are written to memory or an I/O device, the data bus transceivers work differently. This is because S_{AB} equal 1 selects stored transfer mode of operation. In this case, $\overline{DOE}$ is 0, T/R is 1, and the write data are strobed from the A side of the transceivers to the B sides with a pulse on the LDST line.

9.5 82345 DATA BUFFER

The 82345 data buffer is a VLSI device that is used to perform data bus buffering and bus switching for the PC/AT microcomputer system. That is, it is responsible for automatic routing of data between the MPU's data bus and the microcomputer's BIOS ROM/DRAM data bus, I/O data bus, and system data bus. The 82345 also implements some other system functions. For instance, it contains both the parity generation and parity check logic. The 82345 is packaged in a high-density 128-pin *plastic quad flat package* (PQFP).

Block Diagram of the 82345

Figure 9.4 is a block diagram of the circuits contained within the 82345 data buffer. Notice that it includes several data bus transceivers, multiplexers, and control logic. For example, there is a separate transceiver and multiplexer in the write data path for the XD, SD, and MD data buses. These multiplexers are used to route data between the buses. For instance, the XD bus multiplexer accepts SD_0 through SD_{15} at its A port, D_0 through D_{31} at its B port, and MD_0 through MD_{31} at its C port. The control logic determines which of these inputs are passed to the IN inputs of the XD bus transceiver and onto the XD data bus.

> **EXAMPLE 9.8**
>
> Which multiplexers and latches are in the read data path?
>
> **SOLUTION** Read data are passed to the MPU's data bus (D_0 through D_{31}) from the IN inputs of the D bus transceiver. The multiplexers and latches in the data paths to this input are the D bus multiplexer, the 16/32 latch, the slot multiplexer, and the MD latch.

> **EXAMPLE 9.9**
>
> What are the sources of input data for the slot multiplexer?
>
> **SOLUTION** Looking at the slot multiplexer block in Fig. 9.4, we find that the source of read/input data for the A, B, and C inputs are the XD bus, SD bus, and MD bus, respectively.

The 82345 contains both a parity generator circuit and parity check circuit. Parity generation and checking applies only to memory accesses that take place

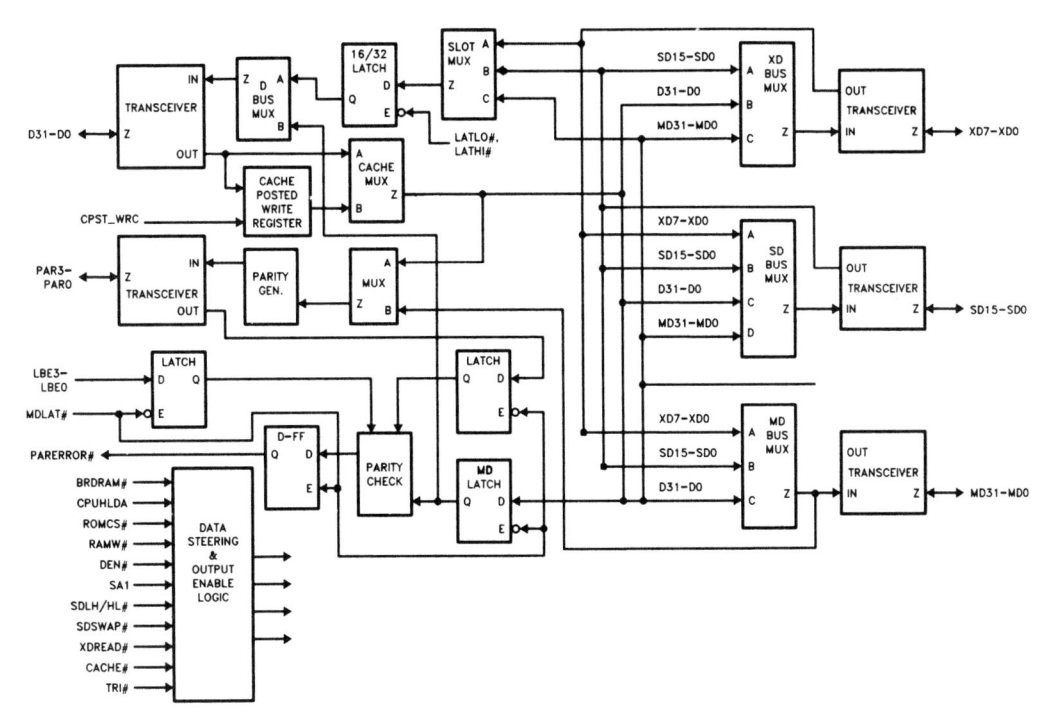

Figure 9.4 Block diagram of the 82345 data buffer. (Reprinted by permission of Intel Corp. Copyright Intel Corp. 1989.)

over the memory data (MD) bus. When a read bus cycle is in progress, the parity information held in the DRAM memory array is input to the 82345 over the *parity bit* (PAR$_0$ through PAR$_3$) lines. Notice that the outputs (OUT) of the parity transceiver are latched and applied as inputs to the parity checker circuit. At the same time, the *latched byte enable* (LBE$_0$ through LBE$_3$) signals are latched within the 82345 and passed to another set of inputs on the parity checker block. Finally, memory data from MD$_0$ through MD$_{31}$ are both multiplexed to the MPU's data bus and latched into the MD latch. The outputs of the MD latch are supplied to a third set of inputs on the parity checker. After these three sets of inputs are present, parity check operations take place on each byte of the 32-bit word that is selected by the byte enable code. If a parity error occurs in any of these bytes, the parity error D-type flip-flop is reset. In this way a parity error condition is signaled at the PARERROR output.

Parity generation takes place whenever data are written into memory attached to the MD bus of the 82345. If a write operation is taking place, write data from the MPU's data bus (D$_0$–D$_{31}$) are multiplexed to the input (IN) port of the MD bus transceiver. In Fig. 9.4 we find that these data are also applied to the B input port of the multiplexer for the parity generator circuit. When data are applied to the input of the parity generator, a 4-bit parity code (one parity bit for each byte of the 32-bit word) is generated and sent to the input (IN) port of the parity bit transceiver. Parity code $PAR_3 PAR_2 PAR_1 PAR_0$ is output along with the word of data on MD_0 through MD_{31}, and both are written into memory.

Inputs and Outputs of the 82345

The input and output signals of the 82345 are listed in Fig. 9.5. This table gives the name, pin number, type, and a brief description of function for each signal. Notice that the signals are grouped into six categories: CPU interface, cache interface, system controller interface, bus controller interface, buffer interface, and test mode pin. Let us next just briefly look at a few of these signals.

In the system controller interface signal group, we find a signal called *RAM write* ($\overline{RAMW}$). This input is supplied by the 82346 bus controller IC and tells the 82345 that a bus cycle is in progress in which data are being written into DRAM on the main processor board. Logic 0 at this input activates the parity generation logic and memory data bus circuitry. *ROM chip select* ($\overline{ROMCS}$) is another input of the 82345 that is supplied by the 82346. Logic 0 at this input means that data are being read from ROM. Finally, the signal *board memory selected* ($\overline{BRDRAM}$) is at its active 0 logic level whenever a data storage location in on-board DRAM is being accessed.

> **EXAMPLE 9.10**
>
> At which pins on the package of the 82345 are the signals $\overline{LATHI}$ and $\overline{LATHO}$ input?
>
> **SOLUTION** From Fig. 9.5 we find that $\overline{LATHI}$ is input at pin 8 and $\overline{LATHO}$ at pin 9 of the 82345's package.

Using the 82345 in the PC/AT Microcomputer

In the block diagram of Fig. 9.2 we find that the 82345 data buffer resides between the core microcomputer's local CPU bus and the ISA expansion bus. This data buffer performs the data bus buffer, multiplexer, and transceiver functions needed to implement the PC/AT's local DRAM data bus, peripheral X-data bus, and system (slot) data bus. The 82345 device is IC U_{40} on sheet 4 of the circuit diagram in Fig. 9.3. Notice that the interface between the 82345 and the core microcomputer is the 32 host data bus lines D_0 through D_{31}. The *memory data bus* to the local DRAM memory array and BIOS ROM is also 32 bits in length. These lines are labeled MD_0 through MD_{31}. At the ISA bus, the PC/AT microcomputer system provides just 16 data bus lines. These *system data bus* lines are identified as SD_0 through SD_{15}. They are called the *slot data bus*. Finally, the peripheral X-data bus, which is the I/O data bus part of the PC/AT's ISA expansion bus, is labeled XD_0 through XD_7.

> **EXAMPLE 9.11**
>
> At what pins of the 82345 are the eight peripheral X-bus lines located?
>
> **SOLUTION** On sheet 4 of Fig. 9.3, we find that XD_0 through XD_3 are at pins 118 through 121 and XD_4 through XD_7 are at pins 123 through 126.

SIGNAL DESCRIPTIONS

Signal Name	Pin Number	Signal Type	Signal Description
CPU INTERFACE			
HLDA	2	I-TTL	**CPU HOLD ACKNOWLEDGE, ACTIVE HIGH:** This is the hold acknowledge pin directly from the CPU. It indicates the CPU has given up the bus for either a DMA master or a slot bus master. It is used in the steering logic to determine data routing.
D31–D0	96–82, 79–66, 64–62	I/O-TTL	**CPU DATA BUS:** This is the data bus directly connected to the CPU. It is also referred to as the local data bus. This bus is output enabled by the DEN# signal.
CACHE INTERFACE			
CPST_WRC	21	I-TPU	**POSTED CACHE WRITE CLOCK:** This clock signal is driven by the cache controller and is needed to latch the write data during a posted cache write cycle. The data is latched on the rising edge of this signal. The latch inside of the Data Buffer is bypassed if the CACHE# input is high. Also, when CACHE# is high, the state of CPST_WRC determines on which bus (D or MD) system DRAM is accessed. When high, DRAM is accessed on the D bus. When low, DRAM is accessed on the MD bus. This pin is pulled up internally.
CACHE#	22	I-TPU	**CACHE ENABLE, ACTIVE LOW:** This signal is used to enable the cache posted write register. When there is not a cache in the system, data bypasses the register. When CACHE# is inactive (high) the state of the CPST_WRC pin determines whether the system DRAM is on the CPU's D bus or on the MD bus. This pin is pulled up internally.
SYSTEM CONTROLLER INTERFACE			
MDLAT#	14	I-TTL	**MEMORY DATA LATCH:** This latching signal serves two purposes simultaneously and is only activated during on-board memory read and write cycles. As a memory data latch, this transparent low signal allows read data to flow through to the CPU's local bus. It follows CAS# on early CAS# high read cycles and on the positive going edge, latches the memory data and holds it for the CPU to sample. As a parity clock, it clocks out PARERROR# on its falling edge an on the rising edge it latches the parity bits (PAR3–PAR0), the byte enables (LBE3–LBE0) and the memory data for parity error processing. Any parity errors will be reported on the next read cycle. It is the negative NOR of all CAS# signals gated by W/R#.
RAMW#	15	I-TTL	**RAM WRITE, ACTIVE LOW:** This signal is supplied by the System Controller to indicate to the Bus Controller that an on-board memory write cycle is occurring. It is used internally to direct the parity logic and to enable the MD bus outputs.
ROMCS#	19	I-TTL	**ROM CHIP SELECT:** This signal tells the Data Buffer when the ROM is to be accessed so that it can latch the data and convert it from 16 or 8 bits to 32 bits. This signal is driven by the System Controller.
BRDRAM#	18	I-TTL	**BOARD MEMORY SELECTED, ACTIVE LOW:** This signal is driven by the System Controller and indicates when on-board DRAM is being accessed.

Figure 9.5 82345 input/output signal descriptions. (Reprinted by permission of Intel Corp. Copyright Intel Corp. 1989.)

SIGNAL DESCRIPTIONS (Continued)

Signal Name	Pin Number	Signal Type	Signal Description
SYSTEM CONTROLLER INTERFACE (Continued)			
DEN#	20	I-TTL	**DATA ENABLE, ACTIVE LOW:** This is a control signal generated by the System Controller. It is used to enable data transfers on the local data bus and as an output enable for the D bus.
LBE3–LBE0	3–6	I-TTL	**LATCH BYTE ENABLES 3 THROUGH 0:** These signals are driven by the System Controller. They are used internally to enable the appropriate bytes (in a 4 byte wide memory configuration) for parity generation and checking.
PARERROR#	7	O	**PARITY ERROR, ACTIVE LOW:** This signal is the result of a parity check on the appropriate bytes being read from memory. It is generated on the falling edge of MDLAT#.
BUS CONTROLLER INTERFACE			
SA1	13	I-TTL	**SYSTEM ADDRESS BUS BIT 1:** This input will be driven by the Bus Controller or by the Controlling DMA or bus master. This signal is used for 16- to 32-bit conversion. When low, this signal indicates the low word is to be used.
SDLH/HL#	12	I-TTL	**SYSTEM DATA BUS LOW TO HIGH/HIGH TO LOW SWAP:** This signal is driven by the Bus Controller. It is used to establish the direction of byte swaps. (Similar to DIR245 in the existing PC/AT-type chip sets).
SDSWAP#	11	I-TTL	**SYSTEM DATA BUS BYTE SWAP ENABLE, ACTIVE LOW:** This signal is driven by the Bus Controller. It is the qualifying signal needed for SDLH/HL#. (It was formerly named GATE245 on the existing PC/AT-type chip sets).
XDREAD#	10	I-TTL	**PERIPHERAL DATA BUS (XD BUS) READ, ACTIVE LOW:** This signal is driven by the Bus Controller and it determines the direction of the XD bus data flow. (It is analogous to the XDATADIR control pin on the existing PC/AT-type chip sets). When this signal is high, the XD Bus is output enabled.
LATHI#	8	I-TTL	**SD BUS HIGH BYTE LATCH:** This signal is needed to latch the SD bus' high byte to the local data bus until the CPU is ready to sample the bus. When SA1 is low, the high byte is latched into both the one byte and the three byte of the 16/32 latch. When SA1 is high, the high byte is only latched into the three byte. This signal is driven by the Bus Controller.
LATLO#	9	I-TTL	**SD BUS LOW BYTE LATCH:** This signal is needed to latch the SD bus' low byte to the local data bus until the CPU is ready to sample the bus. When SA1 is low, the low byte is latched into both the zero byte and the two byte of the 16/32 latch. When SA1 is high, the high byte is only latched into the two byte. This signal is driven by the Bus Controller.
BUFFER INTERFACE			
MD31–MD0	61–50, 47–34, 32–27	I/O-TTL	**MEMORY DATA BUS:** This bus connects to the on-board DRAM and BIOS ROM. It is used to transfer data to/from memory during memory write/read bus cycles.
SD15–SD0	117–115, 103–109, 106–102, 100–98	I/O-TTL	**SYSTEM DATA BUS:** This bus connects directly to the slots. It is used to transfer data to/from local and system devices.

Figure 9.5 (*Continued*)

SIGNAL DESCRIPTIONS (Continued)

Signal Name	Pin Number	Signal Type	Signal Description
BUFFER INTERFACE (Continued)			
XD7–XD0	126–123 121–118	I/O-TTL	**PERIPHERAL DATA BUS:** This bus is connected to the Bus Controller and the System Controller. These I/O's are used to read and write to on-board 8-bit peripherals.
PAR3–PAR0	23–26	I/O-TTL	**PARITY BIT BYTES 3 THROUGH 0:** These bits are generated by the parity generation circuitry located on the Data Buffer chip. They are written to memory along with their corresponding bytes during memory write operations. During memory read operations, these bits become inputs and are used along with their respective data bytes to determine if a parity error has occurred. The generation and check of each bit is enabled only when their respective LBE3–LBE0 bits are active.
HIDRIVE#	17	I-TPU	**HIGH DRIVE ENABLE:** This pin is intended to be a wire option. When this pin is low, all bus drivers defined with an I_{OL} of 24 mA will sink the full 24 mA of current. When the input is high, all pins defined as 24 mA will have the output low drive capability cut in half to 12 mA. Note that al A.C. specifications are done with the outputs in the high drive mode and a 200 pF capacitive load. HIDRIVE# has an internal pull-up and can be left unconnected in 12 mA drive if desired. It should be tied low if 24 mA drive is desired.
TEST MODE PIN			
TRI#	127	I-TPU	**THREE-STATE:** This pin is used to drive all outputs to a high impedance state. When TRI# is low, all outputs and bidirectional pins are three-stated. This pin should be pulled up via a 10 kΩ pull-up resistor in a standard system configuration.

SIGNAL TYPE LEGEND

Signal Code	Signal Type
I-TTL	TTL Level Input
I-TPD	Input with 30 kΩ Pull-Down Resistor
I-TPU	Input with 30 kΩ Pull-Up Resistor
I-TSPU	Schmitt-Trigger Input with 30 kΩ Pull-Up Resistor
I-CMOS	CMOS Level Input
I/O-TTL	TTL Level Input/Output
IT-OD	TTL Level Input/Open Drain Output
I/O-OD	Input or Open Drain, Slow Turn On
O	CMOS and TTL Level Compatible Output
O-TTL	TTL Level Output
O-TS	Three-State Level Output
I1	Input Used for Testing Purposes
GND	Ground
PWR	Power

Figure 9.5 (*Continued*)

A number of signals are input to the 82345 from the 82346 system controller. They are identified as the system controller interface signals in Fig. 9.5. On sheet 4 of Fig. 9.3, the system controller interface is located at the lower left corner of the 82345 IC. Here we find signals such as *memory data latch* ($\overline{\text{MDLAT}}$) and *data enable* ($\overline{\text{DEN}}$). A brief description of each of these signals is supplied in Fig. 9.5. For example, the $\overline{\text{MDLAT}}$ input is at its active 0 logic level during all read and write bus cycles to memory on the main processor board. During read cycles, logic 0 at this input enables the read data to the MPU's data bus. This signal is also used to clock in the *parity bits* (PAR_0 through PAR_3), *latched byte enables* (LBE_0 through LBE_3), and clock out the *parity error* ($\overline{\text{PARERROR}}$) signal.

The cache controller interface of the 82345 consists of just two signals: *cache enable* ($\overline{\text{CACHE}}$) and *posted cache write clock* (CPST_WRC). In Figs. 9.4 and 9.5 we see that $\overline{\text{CACHE}}$ is a control input that is used to enable or disable operation of the cache posted write register within the 82345. From Fig. 9.3 we find that $\overline{\text{CACHE}}$ is fixed at the 0 logic level. According to the description of $\overline{\text{CACHE}}$ in the table of Fig. 9.5, this logic level enables the cache posted write register. Looking at the block diagram in Fig. 9.4, we see that write data available on D_0 through D_{31} are output at OUT of the CPU data bus transceiver. These outputs directly feed the A port of the cache multiplexer and the inputs of the cache posted write register. When posted write operation is enabled, write data are clocked into the cache posted write register with the signal CPST_WRC and the cache multiplexer is set up to select the input at the B port for output to memory. Notice that data at the output of the cache multiplexer are sent as D_0 through D_{31} to the XD, SD, and MD bus multiplexers and also to the A input port of the multiplexer to the parity generation circuit. If the data are written into the on-board DRAM, parity information is also saved in memory.

EXAMPLE 9.12

If $\overline{\text{CACHE}}$ is logic 1, what is the input path of write data from the D bus of the MPU?

SOLUTION Looking at the block diagram in Fig. 9.4, we see that data from the MPU's data bus are passed to OUT of the transceiver. When a posted write operation is disabled, the data applied to the A port of the cache multiplexer are selected. These data are not latched as in a posted write operation. Instead, they are passed directly to the inputs of the parity generation circuit and the XD, SD, and MD multiplexers in parallel.

The last group of interface signals on the 82345 we will consider are those called the bus controller interface signals. In the table of Fig. 9.5, we find that this interface consists of six input signals: SA_1, SDLH/$\overline{\text{HL}}$, $\overline{\text{SDSWAP}}$, $\overline{\text{XDREAD}}$, $\overline{\text{LATHI}}$, and $\overline{\text{LATLO}}$. The source for each of these six signals is a corresponding output on the 82344 ISA bus controller. For example, Fig. 9.3 shows that the $\overline{\text{SDSWAP}}$ input at pin 11 of the 82345 is attached to the $\overline{\text{SDSWAP}}$ output at pin 101 of the 82344 (U_{41}). Another example is $\overline{\text{XDREAD}}$, which is a control signal that determines the direction of data flow on the XD data bus. This input is supplied by the $\overline{\text{XDREAD}}$ output at pin 103 of the ISA controller.

9.6 82346 SYSTEM CONTROLLER

The second VLSI device that we consider from the 82340 PC/AT chip set is the 82346 system controller. In the block diagram of the PC/AT microcomputer system in Fig. 9.2, we see that the system controller is the heart of the 82340 chip set. It attaches to the local bus of the 80386DX, the ISA expansion bus, the 82345 data buffer, and the 82344 ISA controller. The 82346 is responsible for performing a number of functions for the microcomputer system. For instance, it generates the clock signals, produces addresses for the DRAM array, control signals for all on-board memory, creates the reset signal for the MPU, and interfaces the numeric coprocessor to the 80386DX. Just like the 82345, the 82346 is packaged in a 128-pin plastic PQFP.

Block Diagram of the 82346

A block diagram of the circuits within the 82346 system controller is given in Fig. 9.6. Here we find that it contains the *clock control logic, hold/hold acknowledge arbitration logic, bus controller, address decoder, DRAM controller, remap logic, ready control logic, numeric coprocessor interface, reset control logic,* and *configuration control registers*. Let us next look briefly at the function of some of these blocks.

We begin with the clock control logic block. This section of circuitry is used to produce the CPU clock and bus clock signals for the 80386DX-based microcomputer system. The clock signal that is applied to the $TCLK_2$ input must be generated in external circuitry. Within the 82346, this signal is divided by 2 and buffered to CMOS levels to produce system clock output CLK_2. For instance, in a 33-MHz 80386DX microcomputer, $TCLK_2$ must be driven by a 66-MHz clock. The 33-MHz CLK_2 output is used to drive the 80386DX MPU.

The TURBO input can be used to scale the operating frequency of the MPU. When TURBO is active (logic 1), the MPU runs at full speed. That is, CLK_2 equals $\frac{1}{2} TCLK_2$. However, if TURBO is switched to logic 0, CLK_2 is decreased to a lower frequency. This is done by dividing it by a scale factor that has been loaded into one of the internal configuration registers of the 82346. In this way the microcomputer can be set up in a low-power, low-performance mode waiting for an entry to be made at the keyboard. When an entry is detected, TURBO can be switched to 1 and the microcomputer runs at full speed (high-performance mode).

A second clock input, bus oscillator (BUSOSC), is used to produce the bus clock (BUSCLK) signal that is needed for the ISA bus. For this reason the BUSCLK output is supplied to the 82344 ISA bus controller device.

Next we examine the function of the bus controller, address decoder, DRAM controller, remap logic, and ready control blocks of the 82346. The circuitry within these blocks is used to produce control signals for on-board memory, RAS and CAS address and strobe signals for the on-board DRAM array, and a control bus interface for communicating with the 82344 ISA bus controller. Notice in Fig. 9.6 that the 80386DX's bus cycle definition signals $M/\overline{IO}$, $D/\overline{C}$, and $W/\overline{R}$ are inputs

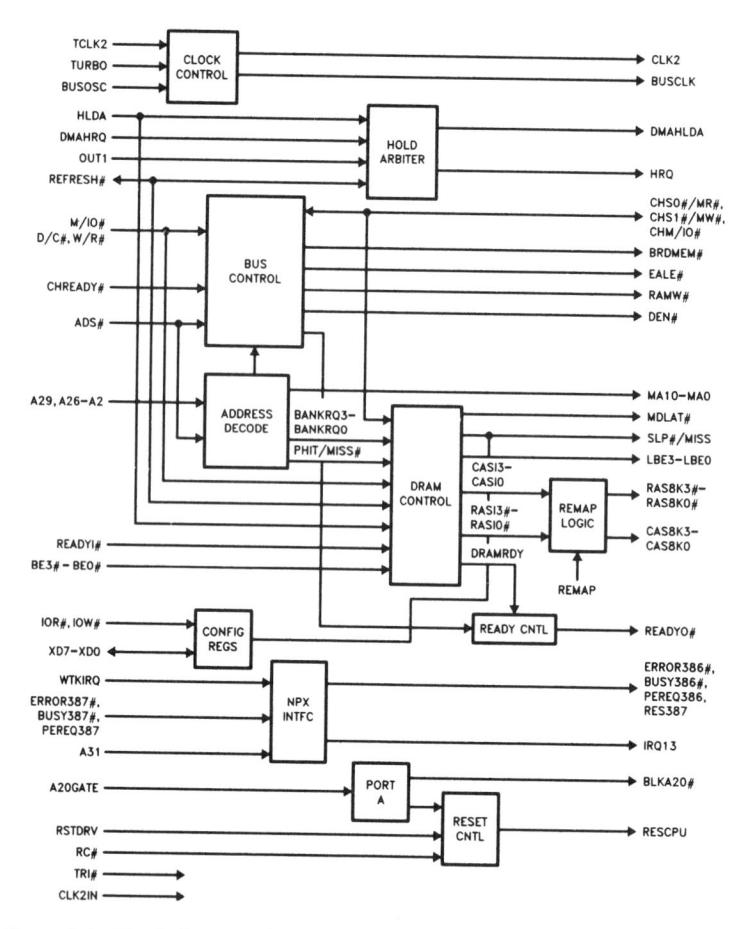

Figure 9.6 Block diagram of the 82346 system controller. (Reprinted by permission of Intel Corp. Copyright Intel Corp. 1989.)

to the bus controller. Another input is channel ready ($\overline{\text{CHREADY}}$) from the 82344 ISA bus controller. These signals are strobed into the bus control logic with the address strobe ($\overline{\text{ADS}}$) signal. The bus controller converts them into appropriately timed memory control outputs. For example, board memory selected ($\overline{\text{BRDMEM}}$) and RAM write ($\overline{\text{RAMW}}$) are outputs of the bus controller. These memory control signals are used to drive the 82345 data buffer.

At the same time, $\overline{\text{ADS}}$ strobes address information into the address decoder. Here the address is converted into appropriately timed row and column addresses. The RAS and CAS addresses are output on memory address lines MA_0 through MA_{10}. This address is supplied to the DRAM array.

Looking at Fig. 9.6, we see that the DRAM controller accepts the bank request outputs of the address decoder, byte enable signals from the 80386DX or 82385DX, and a number of control signals as inputs. At its output it generates row address strobes $\overline{\text{RASI}}_0$ through $\overline{\text{RASI}}_3$ and column address strobes CASI_0 through CASI_3. These RAS/CAS signals are mapped to the RAS and CAS bank outputs ($\overline{RASBK_0}$–$\overline{RASBK_3}$ and $CASBK_0$–$CASBK_3$) by the remap logic circuit. The RAS/CAS bank outputs are used to select DRAMs in the memory array.

Finally, the reset control logic produces the CPU reset signal for the microcomputer. It accepts the various signals that can be used to initiate a reset of the microcomputer at its inputs. In Fig. 9.6 we find that these signals are reset control ($\overline{RC}$), reset drive (RSTDRV), and A20GATE (which is supplied through the port A circuit). These inputs initiate a hardware reset of the 80386DX MPU by activating the reset CPU (RESCPU) output.

Inputs and Outputs of the 82346 and Their Use in the PC/AT Microcomputer

Now that we have introduced the function and block diagram of the 82346 let us turn our attention to the inputs and outputs of the device and their connection in an 80386DX microcomputer system.

Figure 9.7 lists the input and output signals of the 82345. In this table the signals are arranged into seven groups. They are called the *CPU interface signals, on-board memory system interface signals, coprocessor signals, bus control signals, peripheral interface signals, bus interface signals,* and *test mode pin.* Again, the table includes the mnemonic, pin number, type, name, and a brief description of each signal. We will begin by examining signals from the CPU interface.

Looking at the list of CPU interface signals in Fig. 9.7, we find that address lines A_2 through A_{26}, A_{29}, and A_{31}, byte enables $\overline{BE_0}$ through $\overline{BE_3}$, and control signals W/$\overline{R}$, D/$\overline{C}$, M/$\overline{IO}$, $\overline{ADS}$, $\overline{READYI}$, and $\overline{HLDA}$ are all inputs of the 82346. These signals are supplied by either the 80386DX MPU or 82385DX cache controller. For example, the address inputs of the 82346 on sheet 3 of the circuit diagram in Fig. 9.3 are supplied by the outputs of the 80386DX's address bus latch on sheet 2. On the other hand, the byte enable inputs are traced back to the 82385DX bus byte enable outputs of the cache controller on sheet 1.

EXAMPLE 9.13

What is the function of the $\overline{ADS}$ input of the 82346? Is this signal supplied by the 80386DX or by the 82385DX in the microcomputer of Fig. 9.3?

SOLUTION In the table of Fig. 9.7 we find that logic 0 at this input tells the 82346 that the information at the address and control inputs of the CPU interface is valid.

The signal applied to this input in Fig. 9.3 (sheet 3) is $\overline{BADS}$. On sheet 1 of Fig. 9.3, $\overline{BADS}$ is found to be produced at pin N9 of the 82385DX cache controller.

The CPU interface also includes the CLK_{2IN} and $TCLK_2$ clock inputs and the CLK_2 output. In the block diagram of Fig. 9.6 we see that $TCLK_2$ is one input of the clock control logic. Earlier we pointed out that this input is supplied by an external crystal oscillator and that the frequency of this oscillator must be twice that of the microcomputer system clock, CLK_2. From the circuits on sheet 3 of Fig. 9.3, we find that the $TCLK_2$ input of the 82346 is driven by a 66-MHz oscillator (OSC_{31}). This input produces a 33-MHz clock at the CLK_2 output.

SIGNAL DESCRIPTIONS

Signal Name	Pin Number	Signal Type	Signal Description
CPU INTERFACE SIGNALS			
A31, A29, A26	119–117	I-TPU	A26 is used to prevent aliasing above 64 MByte. A29 is used to separate upper BIOS accesses from Weitek 3167 accesses. A 31 is used to determine 387DX accesses. These address lines are driven only by the CPU. When HLDA is active, these signals are held low internally. Since externally these pins are three-stated, this is required in order to prevent errant Bus Master and DMA accesses to on-board memory.
A25–2	120–127, 2–7, 9–13, 15–19	I-TTL	Address bits driven by the CPU when it is Bus Master. They are driven by the 82344 Bus Controller whenever HLDA is active. These bits allow direct access of up to 64 Mbytes of memory.
BE3# – BE0#	20–23	I-TTL	**BYTE ENABLES 3 THROUGH 0, ACTIVE LOW:** These signals are driven by the CPU or the 82344.
W/R#	24	I-TPU	WRITE or active low READ enable driven by the CPU. W/R# is decoded with the remaining CPU control signals to indicate the type of bus cycle requested. The bus cycle types include: Interrupt Acknowledge, Halt, Shutdown, I/O Reads and Writes, Memory Data Reads and Writes, and Memory Code reads. WR# is internally pulled up.
D/C#	25	I-TPU	DATA or active low CODE enable driven by the CPU. D/C# is decoded with the remaining CPU control signals to indicate the type of bus cycle requested. See W/R# definition for bus cycle types. D/C# is internally pulled up.
M/IO#	26	I-TPU	MEMORY or active low I/O enable driven by the CPU. M/IO# is decoded with the remaining CPU control signals to indicate the type of bus cycle requested. See W/R# definition for bus cycle types. M/IO# is internally pulled up.
ADS#	27	I-TPU	**ADDRESS STROBE, ACTIVE LOW:** Driven by the CPU as an indicator that the address and control signals currently supplied by the CPU are valid. Ths signal is used internally to indicate that the data and command are valid and to determine the beginning of a memory cycle. ADS# is internally pulled up.
CLK2IN	30	I-CMOS	This is the main clock input to the System Controller and is connected to the CLK2 signal that is output by the System Controller. This signal is used internally to clock the System Controller's logic.
TCLK2	44	I-TTL	This input is connected to a crystal ocscillator whose frequency is equal to two times the system frequency. The TTL level oscillator output is converted internally to CMOS levels and sent to the CLK2 output.
CLK2	32	O	This output signal is a CMOS level converted TCLK2 signal. It is output to the CPU and other on-board logic for synchronization.
SLP#/MISS	35	IO-od	As a "power on reset" default, this bit is an output that reflects the inverse state of the SLEEP[7] configuration register bit. It is active low when sleep mode is active. Sleep mode is activated by setting SLEEP[7] = 1. When configuration register CTRL1(0) = 1, this pin becomes a MISS input for use with a future 82340 compatible product.

Figure 9.7 82346 input/output signal descriptions. (Reprinted by permission of Intel Corp. Copyright Intel Corp. 1989.)

SIGNAL DESCRIPTIONS (Continued)

Signal Name	Pin Number	Signal Type	Signal Description
CPU INTERFACE SIGNALS (Continued)			
READYO#	34	O	**READY OUT, ACTIVE LOW:** This signal is an indication that the current memory or I/O bus cycle is complete. It is generated from the internal DRAM controller or the synchronized version of CHREADY# for slot bus accesses. Outside the chip it is ORed with any other local bus I/O or master such as a coprocessor or cache controller. The culmination of these ORed READY signals is sent to the 386DX and is also connected to the System Controller's ReadyL# input.
READYI#	29	I-TTL	**READY INPUT, ACTIVE LOW:** This signal is the ORed READY signals from the coprocessor, cache controller, or other optional add-in device. See the READYO# description for more details on how the signal is used inside the System Controller.
HLDA	28	I-TTL	**HOLD ACKNOWLEDGE, ACTIVE HIGH:** This signal is issued by the CPU in response to the HRQ driven by the System Controller. It indicates that the CPU is floating its outputs to the high impedance state so that another master can take control of the bus. When HLDA is active, the memory control is generated from CHS1#/MR# and CHS0#/MW# rather than CPU status signals.
HRQ	40	O	**HOLD REQUEST, ACTIVE HIGH:** Driven by the System Controller to the CPU, this output indicates that a bus master, such as a DMA or AT channel master, is requesting control of the bus. HRQ is a result of the DMAHRQ input or a coupled refresh cycle. It is synchronized to CLK2.
RESCPU	36	O	**RESET CPU, ACTIVE HIGH:** This signal is sent to the CPU by the System Controller. It is issued in response to the control bit for software reset located in the Port A register or a dummy read to I/O port EFh. It is also issued in response to signals on the RSTDRV or RC inputs and in response to System Controller detection of a shutdown command. In all cases, it is synchronized to CLK2.
ERROR386#	37	O	**ERROR 386, ACTIVE LOW:** This signal is sent to the 386DX. On any CPU reset it is pulled low to set the 386DX to 32-bit coprocessor interface mode.
BUSY386#	38	O	**BUSY 386, ACTIVE LOW:** This signal is sent to the 386DX. The state of BUSY387# is always passed through to BUSY387# indicating that the 387DX is processing a command. On occurrence of an ERROR387# signal, it is latched and held active until an occurrence of a write to ports F0h, F1h, or RES387. The former case is the normal mechanism used to reset the active latched signal. The latter two are resets. Since ERROR387# generates IRQ13 for PC/AT-compatibility, BUSY386# is held active to prevent software access of the 387DX until the interrupt service routine writes F0h. The System Controller also activates BUSY386# for 16 CLK2 cycles when no 387DX is connected and I/O writes to the coprocessor space are detected.

Figure 9.7 (*Continued*)

SIGNAL DESCRIPTIONS (Continued)

Signal Name	Pin Number	Signal Type	Signal Description
CPU INTERFACE SIGNALS (Continued)			
PEREQ386	39	O	**PROCESSOR EXTENSION REQUEST 386, ACTIVE HIGH:** Sent to the CPU in response to a PEREQ387, which is issued by the coprocessor to the System Controller. It indicates to the CPU that the coprocessor is requesting a data operand to be sent to or from memory by the CPU. For PC/AT-compatiblity, PEREQ386 is returned active on occurrence of ERROR387# after BUSY387# has gone inactive. A write to F0h by the interrupt 13 handler returns control of the PEREQ386 signal to directly follow the PEREQ387 input.
ON-BOARD MEMORY SYSTEM INTERFACE SIGNALS			
RAMW#	55	O	**RAM ACTIVE LOW WRITE OR ACTIVE HIGH READ:** Output to the 82345 Data Buffer and DRAM memory to control the direction of data flow of the on-board memory. It is a result of the address and bus control decode. It is active during memory write cycles and is high at all other times.
MA10–MA0	57, 58, 60, 62–64, 66, 67, 69, 71, 72	O	**MEMORY ADDRESSES 10 THROUGH 0:** These address bits are the row and column addresses sent to on-board memory. They are buffered and multiplexed versions of the bus master addresses. Along with LBE3–LBE0 they allow addressing of up to 16 MBytes per bank.
RASBK3#– RASBK0#	81–83, 85	O	**ROW ADDRESS STROBE BANK 0 THROUGH 3, ACTIVE LOW:** These signals are sent to their respective RAM banks to strobe in the row address during on-board memory bus cycles. The active period for this signal is completely programmable.
CASBK3– CASBK0	77–80	O	**COLUMN ADDRESS STROBE BANK 0 THROUGH 3:** These signals are the respective column address strobes for each of the banks. These signals are externally gated (NAND) with the LBE signals to generate the CAS# strobes for each byte of a DRAM memory bank.
LBE3–LBE0	73–76	O	**LATCHED BYTE ENABLE 0 THROUGH 3, ACTIVE HIGH:** These signals select one of four banks to access memory data from when an on-board memory access is activated. They are the latched version of the CPU's BE3#–BE0# signals when the CPU is bus master or is the latched version of SA1, SA0, and BHE# when the master or DMA is in control.
REFRESH#	109	I-CMOS/ O-OD	**REFRESH SIGNAL, ACTIVE LOW:** This output is used by the System Controller to initiate an off-board DRAM refresh operation in coupled refresh mode. In decoupled mode, the Bus Controller drives refresh active to indicate to the System Controller that it has decoded a refresh request command and is initiating an off-board refresh cycle.

Figure 9.7 (*Continued*)

SIGNAL DESCRIPTIONS (Continued)

Signal Name	Pin Number	Signal Type	Signal Description
ON-BOARD MEMORY SYSTEM INTERFCE SIGNALS (Continued)			
ROMCS#	54	O	**ROM CHIP SELECT:** This output is active in CPU mode only (CPUHLDA is negated). It is active anytime the address on the A bus selects the address range between AFFE0000–AFFFFFFF, BFFE0000-BFFFFFFF, EFFE0000–EFFFFFFF, or FFFE0000h–FFFFFFFFh. It is also active during a memory read of 000E0000h–000FFFFFh when RAM-MAP[7] = 1. On reset, it also decodes the middle BIOS space between 00FE0000h–00FFFFFFh. However, this decode space can be changed via internal configuration register to System Board DRAM space after RESET if desired. NOTE: The lower ROM are from 00FE0000h–00FFFFFFh is impacted by shadow and/or EMS. Any 16k segments for which EMS is active (00EXXXXh only) or for which the shadow code had been changed from its 00b default are mapped out of the -ROMCS space.
COPROCESSOR SIGNALS			
PEREQ387	46	I-TPD	**COPROCESSOR EXTENSION REQUEST, ACTIVE HIGH:** this input signal is driven by the coprocessor and indicates that it needs transfer of data operands to or from memory. For PC/AT-compatibility, this signal is gated with the internal ERROR/BUSY control logic before being output to the CPU as PEREQ386.
ERROR387#	43	I-TPU	This is an active low numerics signal which is driven by the coprocessor to indicate that an error has occurred in the previous instruction. This signal is decoded internally with BUSY387# to produce IRQ13.
BUSY387#	42	I-TPU	This is an active low numerics input signal which is driven by the coprocessor to indicate that it is currently executing a previous instruction and is not ready to accept another. This signal is decoded internally to produce IRQ13 and to control PEREQ386 and BUSY386#.
RES387	41	O	**RESET 387, ACTIVE HIGH:** This output is connected to the 387DX reset input. It is triggered through an internally generated system reset or via a write to port F1h. In the case of a system reset, the CPURESET signal is also activated. A write to port F1h only resets the coprocessor. A software FNINT signal must occur after an F1h generated reset before the coprocessor is reset to the same internal state that a 287 is put into by a hardware reset alone. For, compatibility, the F1h reset may be disabled by setting bit 6 of MISCSET to 1.
WTKIRQ	47	I-TPD	**WEITEK 3167 INTERRUPT REQUEST, ACTIVE HIGH:** An input from the Weitek 3167 coprocessor.
IRQ13	100	O	**INTERRUPT REQUEST 13, ACTIVE HIGH:** This signal is driven to the Bus Controller to indicate than an error has occurred within the coprocessor. This signal is a decode of the BUSY387# and ERROR387# inputs ORed with the WTKIRQ input.
BUS CONTROL SIGNALS			
CHREADY#	104	I-CMOS	**CHANNEL READY, ACTIVE LOW:** This signal is issued by the Bus Controller as an indication that the current channel bus cycle is complete. This signal is synchronized internally then combined with ready signals from the coprocessor and DRAM controller to form the final version of READYO# which is sent to the CPU.

Figure 9.7 (*Continued*)

SIGNAL DESCRIPTIONS (Continued)

Signal Name	Pin Number	Signal Type	Signal Description
BUS CONTROL SIGNALS (Continued)			
CHS0#/MW#	103	IO-TTL	**CHANNEL SELECT 0/MEMORY WRITE, ACTIVE LOW:** This signal is a decode of the 386DX's bus control signals and is sent to the Bus Controller. When combined with CHS1# and CHM/IO# and decoded, the bus cycle type is defined for the Bus Controller. Activation of CPUHLDA reverses this signal to become an input from the Bus Controller. It is then a MEMW# signal for DMA or bus master access to system memory.
CHS1#/MR#	102	IO-TTL	**CHANNEL SELECT 1/MEMORY READ, ACTIVE LOW:** This signal is a decode of the 386DX's bus control signals and is sent to the Bus Controller. When combined with CHS0# and CHM/IO# and decoded, the bus cycle type is defined for the Bus Controller. Activation of CPUHLDA reverses this signal to become an input from the Bus Controller. It is the a MEMR# signal for DMA or bus master access to system memory.
CHM/-IO	101	O	**CHANNEL MEMORY/ACTIVE LOW IO:** A decode of the M/IO# signal sent by the CPU to the System Controller. It is an indicator that the current bus cycle is a channel access. When combined with CHS0#, and CHM/IO# and decoded, the bus cycle type is defined for the Bus Controller.
BLKA20#	94	O	**BLOCK A20, ACTIVE LOW:** Driven to the Bus Controller to deactivate address bit 20. It is a decode of the A20GATE signal and Port A bit 1 indicating the dividing line of the 1 MByte memory boundary. Port A bit 1 may be directly written or set by a dummy read of I/O port EEh. BLKA20# is forced high when HLDA is active. (Refer to the "Sleep Mode Control Subsystem" section.)
BUSOSC	106	I-TTL	**BUS OSCILLATOR:** This signal is supplied from an external oscillator. It is supplied to the Bus Controller when the System Controller's internal configuration registers are set for asynchronous slot bus mode. This signal is two times the AT bus clock speed (SYSCLK).
BUSCLK	98	O-TTL	**BUS CLOCK:** This is the source clock used by the Bus Controller to drive the slot bus. It is two times the AT bus clock (SYSCLK). It is a programmable division from CLK2 or BUSOSC when in a synchronous bus mode.
DMAHRQ	105	I-CMOS	**DMA HOLD REQUEST, ACTIVE HIGH:** This input is sent by the Bus Controller, it is internally synchronized by the System Controller before it is sent out to the CPU as the HRQ signal. It is the indicator of the DMA controller or an other bus masters' desire to control the bus.
DMAHLDA	99	O	**DMA HOLD ACKNOWLEDGE, ACTIVE HIGH:** This output to the Bus controller indicates that the current hold acknowledge state is for the DMA controller or an other bus master.
BRDRAM#	95	O	**BOARD DRAM, ACTIVE LOW:** An output to Bus Controller and Data Buffer to indicate that on-board DRAM is being addressed.
OUT1	107	I-CMOS	Indicate a refresh request from the Bus Controller.

Figure 9.7 (*Continued*)

SIGNAL DESCRIPTIONS (Continued)

Signal Name	Pin Number	Signal Type	Signal Description
PERIPHERAL INTERFACE SIGNALS			
A20GATE	116	I-TTL	**ADDRESS BIT 20 ENABLE:** This is an input from the keyboard controller and is used internally along with Port A bit 1 to determine if address bit 20 from the CPU is true or gated low. It also determines the state of BLKA20#.
TURBO	115	I-TTL	**TURBO, ACTIVE HIGH:** This input to the System Controller determines the speed at which the system board operates. It is normally the externally ANDed signal from the keyboard controller and a turbo switch. It is internally ANDed with a software settable latch. When high, operation is full speed. When low, CLK2 is divided by the value coded in configuration register MISCSET. A range is provided that allows slow operation at or below 8 MHz for any valid CPU speed. Slow speed takes precedence. When any one request for slow mode is present, slow mode is active. Turbo mode is active only when all TURBO requests are active.
RC#	114	I-TTL	**RESET CONTROL, ACTIVE LOW:** The falling edge of this signal causes a RESCPU signal. RC# is generated by the keyboard controller and its inverse is ORed with Port A bit 0 to form RESCPU.
SLEEP1	49	O-OD	**SLEEP SIGNAL 1, ACTIVE HIGH:** This pin is the logical OR of the enable and external control bits (bits 1 and 7) of the sleep indexed configuration register. It can be used with external interface logic to control external devices. The pin is always active while in sleep mode but can also be controlled via software when sleep mode is inactive. It is pulled low when inactive and three-states when active. An external pull-up is required. This allows an external interface to control logic operation at voltages different than VDD.
SLEEP2	50	O-OD	**SLEEP SIGNAL 2, ACTIVE HIGH:** This pin is the logical OR of the enable and external control bits (bits 2 and 7) of the sleep indexed configuration register. It can be used with external interface logic to control external devices. The pin is always active while in sleep mode but can also be controlled via software when sleep mode is inactive. It is pulled low when inactive and three-states when active. An external pull-up is required. This allows an external interface to control logic operation at voltages different than VDD.
SLEEP3	51	O-OD	SLEEP SIGNAL 3, ACTIVE HIGH: This pin is the logical OR of the enable and external control bits (bits 3 and 7) of the Sleep indexed configuration register. It can be used with external interface logic to control external devices. The pin is always active while in sleep mode but can also be controlled via software when sleep mode is inactive. It is pulled low when inactive and three-states when active. An external pull-up is required. This allows an external interface to control logic operation at voltages different than VDD.

Figure 9.7 (*Continued*)

SIGNAL DESCRIPTIONS (Continued)

Signal Name	Pin Number	Signal Type	Signal Description
BUS INTERFACE SIGNALS			
XD7–XD0	86–93	IO–TTL	**PERIPHERAL DATA BUS:** This bus is used to read and write the internal configuration registers.
DEN#	53	O	**DATA ENABLE, ACTIVE LOW:** This signal is an output to the 82345 Data Buffer to enable data transfers on the local bus. This signal is low during any CPU read cycles or INTA cycles.
IOR#	112	I-TTL	**I/O READ CYCLE, ACTIVE LOW:** Driven by the Bus Controller to indicate to the 82346 that an I/O read cycle is occurring on the bus. Whenever an I/O cycle occurs, the memory interface signals are inactive.
IOW#	113	I-TTL	**I/O WRITE CYCLE, ACTIVE LOW:** Driven by the Bus Controller to indicate to the 82346 that an I/O write cycle is occurring on the bus. Whenever an I/O cycle occurs, the memory interface signals are inactive.
RSTDRV	111	I-TTL	**RESET DRIVE, ACTIVE HIGH:** This reset signal is output by the Bus Controller. It indicates that a hardware reset signal has been activated. This is the same signal which is output to the channel. This signal is used to reset internal logic and to derive the RESCPU which is output by the System Controller.
MDLAT#	52	O	**MEMORY DATA BUS LATCH:** This is an output signal to the Data Buffer. On the rising edge, the Data Buffer latches the memory data bus. MDLAT# is low anytime one of the CASBK signals is high. When low, the Data Buffer latches are transparent.
OSC	110	I-TTL	**OSCILLATOR:** This is the buffered input of the external 14.318 MHz oscillator.
TEST MODE PIN			
TRI#	48	I1	**THREE-STATE:** This pin is used to drive all outputs to a high impedance state. When TRI# is low, all outputs and bidirectional pins are three-stated.

Signal Type Legend

Signal Code	Signal Type
I-TTL	TTL Level Input
I-TPD	Input with 30 kΩ Pull-Down Resistor
I-TPU	Input with 30 kΩ Pull-Up Resistor
I-TSPU	Schmitt-Trigger Input with 30 kΩ Pull-Up Resistor
I-CMOS	CMOS Level Input
IO-TTL	TTL Level Input/Output
IT-OD	TTL Level Input/Open

Signal Code	Signal Type
IO-OD	Input or Open Drain, Slow Turn On
O	CMOS and TTL Level Compatible Output
O-TTL	TTL Level Output
O-TS	Three-State Level Output
I1	Input used for Testing Purposes
GND	Ground
PWR	Power

Figure 9.7 (*Continued*)

EXAMPLE 9.14

Where is the CLK$_2$ output of the 82346 system controller on sheet 3 of Fig. 9.3 distributed to?

SOLUTION Clock output CLK$_2$ is supplied to the CLK$_2$ input at pin F12 of the 80386DX on sheet 1 of Fig. 9.3.

Most of the signals in the on-board memory system interface group are used to drive the DRAM array on the main processor board. Figure 9.7 indicates that memory address lines MA$_0$ through MA$_{10}$ of the 82346 carry the row and column addresses to the on-board DRAMs. Looking at sheet 3 of Fig. 9.3, we find that these signals and buffered versions MA$_{0H}$ through MA$_{10H}$ are output to the DRAM array on sheet 6. Notice on sheet 6 that MA$_0$ through MA$_{10}$ are applied through resistor networks RN$_{67}$ through RN$_{69}$ to address inputs A$_0$ through A$_{10}$ of all the DRAM SIM modules in banks 0 and 1.

EXAMPLE 9.15

What is the destination of the $\overline{RAS_0}$ row address strobe bank output of the 82346 on sheet 3 in Fig. 9.3?

SOLUTION In Fig. 9.3 we find that $\overline{RAS_0}$ is sent to the DRAM array. In the array on sheet 6, $\overline{RAS_0}$ is passed through resistor network RN$_{600}$ to the $\overline{RAS}$ input of each DRAM SIM in bank 0. They are the four DRAM SIMs, labeled U$_{60}$, U$_{61}$, U$_{62}$, and U$_{63}$.

The column address strobe bank (CASBK$_0$ through CASBK$_3$) outputs and latched byte enable (LBE$_0$ through LBE$_3$) outputs of the 82346 are not supplied directly to the DRAM array. Instead, they are decoded to produce a separate column address strobe signal for each of the DRAM SIMs in the on-board RAM array. This decoder circuit is formed with NAND gates U$_{30}$, U$_{31}$, U$_{32}$, and U$_{33}$ on sheet 3 of Fig. 9.3. For example, the LBE$_0$ output of the 82346 is applied to the input at pin 2 of each of these four NAND gates. On the other hand, the CASBK$_0$ output is applied to one input of each gate on NAND gate U$_{30}$. If LBE$_0$ and CASBK$_0$ are both logic 1 and all of the other byte enable and column address strobe signals are logic 0, the only column address strobe output of the decoder that is active is $\overline{CAS_0}$. Notice in the DRAM array on sheet 6 of Fig. 9.3 that $\overline{CAS_0}$ is applied as $\overline{CAS_{0R}}$ to the $\overline{CAS}$ and $\overline{CAS_9}$ inputs of DRAM SIM U$_{63}$. If this type of CAS strobe follows a $\overline{RAS_0}$ strobe, it represents a byte data access from bank 0 DRAM SIM U$_{63}$. This data transfer takes place over memory data bus lines MD$_0$ through MD$_7$. Remember that the memory data bus is interfaced to the 80386DX's local bus by the 82345 data buffer. The logic level of the $\overline{RAMW}$ output of the 82346 signals the DRAM array whether a read or write data transfer is in progress.

EXAMPLE 9.16

Assume that during a memory bus cycle to the on-board DRAM array, the CASBK$_1$ and all four latched byte enable outputs of the 82346 are at logic 1. Which outputs of the CAS decoder circuit are active? To which DRAMs are they applied? Over which data bus lines are data carried? What size data transfer is taking place?

SOLUTION Looking at the CAS decoder circuit on sheet 3 of Fig. 9.3, we find that $CASBK_1$ is applied to one input of all four NAND gates in IC U_{31}. Since LBE_0, LBE_1, LBE_2, and LBE_3 are all equal to 1, the four outputs of U_{31}, $\overline{CAS}_4$, $\overline{CAS}_5$, $\overline{CAS}_6$, and $\overline{CAS}_7$, are all at their active 0 logic level.

On sheet 6 of Fig. 9.3, these CAS lines are found to be applied to DRAM SIMs U_{67}, U_{66}, U_{65}, and U_{64}, respectively. Therefore, all four DRAM SIMs in bank 1 of the DRAM array are accessed and the data transfer takes place over the complete memory data bus, MD_0 through MD_{31}. This transfer represents a double word data access.

A few of the 80387DX's handshake signals are interfaced to the 80386DX MPU through the 82346 system controller. In Fig. 9.7 the coprocessor signals are found to include four input signals, $PEREQ_{387}$, $\overline{ERROR}_{387}$, $\overline{BUSY}_{387}$, and WTKIRQ. The 80387DX numeric coprocessor uses the first three signals to tell the 80386DX MPU that it needs to perform data operations, that an error has occurred in a numerical calculation, and that it is busy making a numerical calculation, respectively. These conditions are signaled to the 80386DX through the $PEREQ_{386}$, $\overline{ERROR}_{386}$, and $\overline{BUSY}_{386}$ outputs of the 82346's CPU interface.

EXAMPLE 9.17

What signals are listed as coprocessor outputs of the 82346 in the table of Fig. 9.7?

SOLUTION The coprocessor output signals are reset 387 (RES_{387}) and interrupt request 13 (IRQ_{13}).

9.7 82344 ISA CONTROLLER

The 82344 ISA controller is the last of the three 82340 PC/AT chip set devices that attach between the local bus and ISA expansion bus. The 82344 is another VLSI device and is manufactured in a 160-pin PQFP. In Fig. 9.3 (sheet 4) we find that the 82344 connects to the buffered address outputs of the 80386DX core microcomputer on the local bus side and the 82346 system bus controller over the 344/346 communication interface lines. The primary function of the 82344 is to supply most of the signals of the ISA PC/AT expansion bus. It is also used to select between an 8- or a 16-bit data bus for the BIOS EPROMs and to drive the speaker.

Block Diagram of the 82344

Figure 9.8 is a block diagram of the circuits within the 82344 IC. From the block diagram we find that it contains four main sections, called the *data conversion and wait state control, 284/288 ready and bus control, peripheral control,* and *address decoder*. Let us next look briefly at the functions of these blocks.

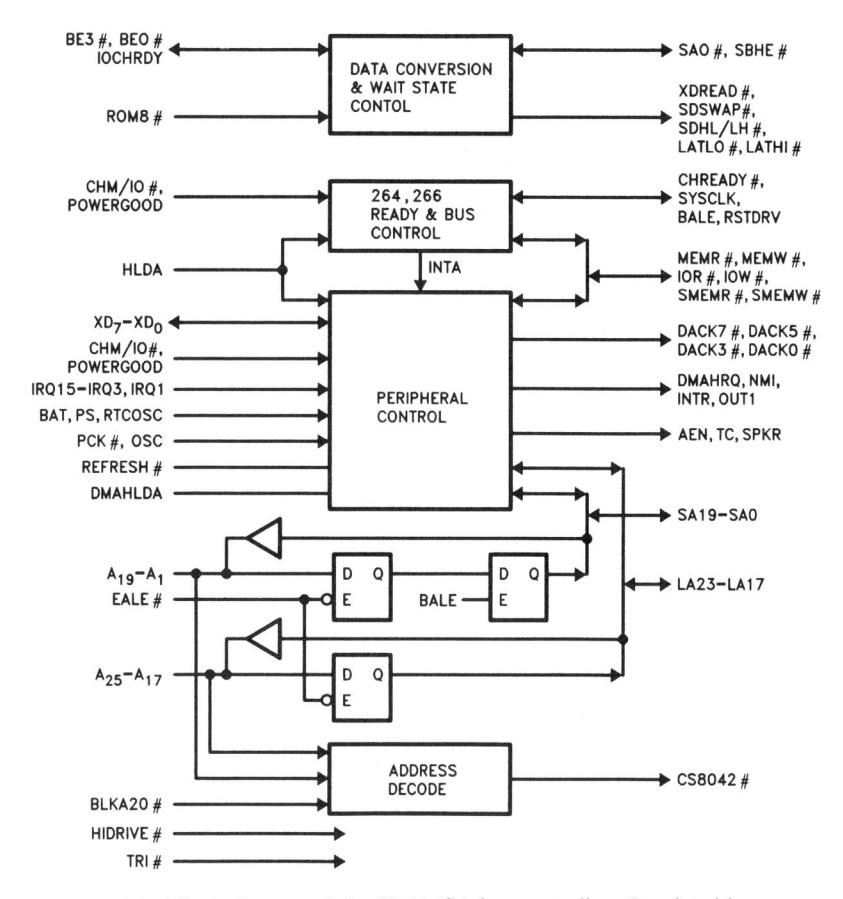

Figure 9.8 Block diagram of the 82344 ISA bus controller. (Reprinted by permission of Intel Corp. Copyright Intel Corp. 1989.)

One function that is performed by the data conversion and wait-state control block is to translate the byte enable signals of the 80386DX to the 80286-compatible odd- and even-byte select signals needed for the ISA expansion bus. In Fig. 9.8 we see that it accepts as inputs byte enable signals $\overline{BE}_0$ through $\overline{BE}_3$ from the MPU, the $\overline{ROM}_8$ strap input, and IOCHRDY from the ISA bus. In response to these signals, the bus controller produces ISA odd- and even-byte select signals $\overline{SA}_0$ and $\overline{SBHE}$. Notice that the byte enables and byte selects are shown as bidirectional lines. This is because they act as inputs when supplied by the MPU or as outputs when driven by the 82344's on-chip DMA controllers.

The 284/288 ready and bus control logic section also produces control signals for the ISA expansion bus. For example, its SYSCLK, BALE, and RSTDRV outputs are distributed to the expansion bus. These two blocks also produce data buffer interface signals, such as $\overline{XDREAD}$, $\overline{SDSWAP}$, and $\overline{CHREADY}$. They are used for communication between the 82344, 82345, and 82346 ICs.

Sec. 9.7 82344 ISA Controller

The peripheral control section represents much of the circuitry within the 82344. This block contains the LSI peripheral devices that are used to implement the microcomputer in the PC/AT. Figure 9.9 shows the circuitry within the peripheral control block in more detail. Here we find that it contains the *82C59A interrupt controllers, 82C37A DMA controllers, 74LS612 page address register, 82C54 programmable timer/counter, speaker driver, parallel I/O port B, real-time clock*, and *refresh counter*.

Finally, the address decoder block decodes address inputs to produce chip select signals $\overline{CS8042}$ and $\overline{ROMCS}$. The $\overline{CS8042}$ output is used to enable the 8042 keyboard controller device. This keyboard controller is optional in an 82340-based PC/AT system. On the other hand, $\overline{ROMCS}$ is used to enable the BIOS EPROMs, which are located on the main processor board.

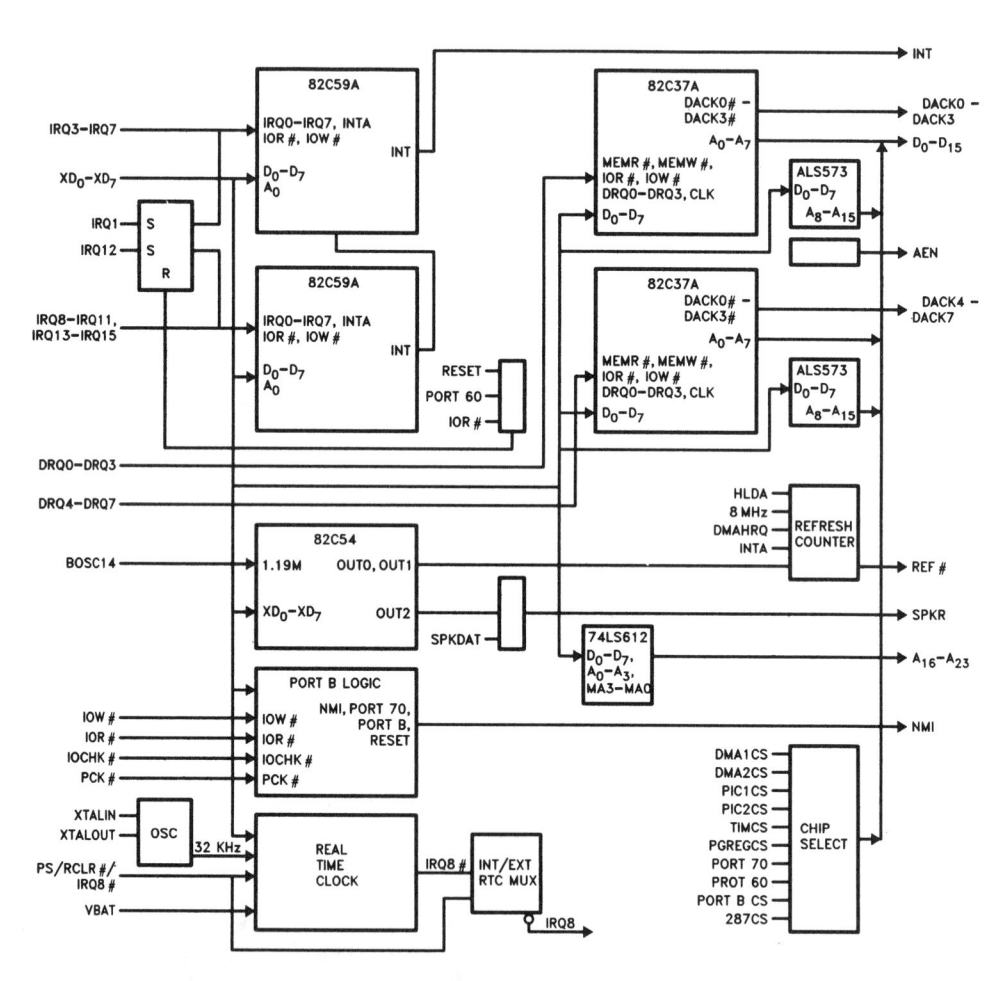

Figure 9.9 Block diagram of the peripheral control block of the 82344. (Reprinted by permission of Intel Corp. Copyright Intel Corp. 1989.)

460   80386DX PC/AT Microcomputer System Hardware   Chap. 9

Inputs and Outputs of the 82344 and Their Use in the PC/AT Microcomputer

Now that we have introduced the 82344 ISA bus controller and its block diagram, let us continue by examining its signal interfaces. Figure 9.10 lists the signals at each of the 82344's interfaces. Notice that the signals are divided into seven groups: the *CPU interface, system controller interface, ROM interface, bus interface, peripheral interface, data buffer interface*, and *test mode pin*. We continue by examining some of these interface signals and their connections in the PC/AT microcomputer system.

The 82344 is designed for use in either an 80286- or 80386DX-based PC/AT microcomputer system. For this reason, the device can be set up to operate with CPU interface signals that are either 80286 or 80386DX compatible. The mode of operation is selected with the *CPU 286/386* ($C286/\overline{386DX}$) input. Looking at Fig. 9.10, we find that applying logic 0 at this input enables 80386DX mode of operation.

On sheet 4 of Fig. 9.3, $C286/\overline{386DX}$, which is pin 73, is connected to ground to select 80386DX-compatible interface signals. This means that signal lines $\overline{BE_2}/A_1$, $\overline{BE_1}/\overline{BHE}$, and $\overline{BE_0}/A_0$ are configured to act like byte enable lines. In the circuit diagram of Fig. 9.3, the address lines of the 82344 are traced back to the outputs of the address buffers on sheet 2 and the byte enable lines are traced back to the bus byte enable lines of the 82385DX cache controller.

> **EXAMPLE 9.18**
>
> In the microcomputer of Fig. 9.3, what are the destinations of the INTR and NMI outputs of the 82344?
>
> **SOLUTION** In the circuit diagram of Fig. 9.3, the INTR and NMI CPU interface outputs of the 82344 are found to be applied directly to the corresponding input of the 80386DX MPU.

Let us next look at the ROM interface and BIOS EPROMs. $\overline{ROM_8}$ is an option select input of the 82344. Notice on sheet 4 of Fig. 9.3 that jumper J_{42} is used to fix the setting of this input. Figure 9.10 identifies that if this input is made logic 0, the microcomputer is configured for eight-bit-wide BIOS memory. This is done by connecting jumper J_{42}. In this case a single 27C512 EPROM is used to hold the BIOS software for the microcomputer. Sockets for both eight- and sixteen-bit BIOS are illustrated on sheet 3 of the circuit diagram in Fig. 9.3. U_{39} is populated with the 27C512 EPROM when eight-bit BIOS mode is selected.

Now we will look more closely at the connection of the 27C512 BIOS EPROM in the PC/AT microcomputer system. Storage locations in BIOS EPROM U_{39} are addressed by system address bus signals SA_0 through SA_{15}. These address lines are outputs of the 82344 ISA bus controller. The occurrence of a read bus cycle to the BIOS EPROM is signaled to the EPROM by another output of the 82344. Notice that the signal $\overline{MEMR}$ is applied to the $\overline{OE}$ input at pin 22 of the 27C512. Logic 0 at $\overline{OE}$ enables the outputs of the EPROM for operation. At the

SIGNAL DESCRIPTIONS

Name	Pin Number	Type	Description
CPU INTERFACE			
A25, A24	43–44	O-TS	Address Bus—These pins are outputs during DMA, master, or standard refresh modes. They are high impedance at all other times. A25 and A24 are driven from the alternate 612 registers during DMA and refresh cycles and are driven low during master cycles.
A23–A2	45–58, 61–68	IO-TTL	Address Bus—These pins are outputs during DMA, master, or standard refresh modes. They are inputs at all other times. As inputs, they are passed to the SA and LA buses and A15–A2 are used to address I/O registers internal to the bus control chip. As outputs, they are driven from different sources depending on which mode the Bus Controller is in. While in refresh mode, these pins are driven from the 612 and refresh address counter. While in DMA mode, they are driven from the 612 and DMA controller subsection. If the Bus Controller is in master mode, the pins A23–A17 are driven from the inputs LA23–LA17 and the pins A16–A2 are driven from the inputs SA16–SA2.
—BE3	69	IO-TTL	Byte Enable 3, active low—This pin is an output during DMA, master, or standard refresh modes. It is an input at all other times. As an input in 386DX mode, it is decoded along with the other byte enable signals to generate SA1, SA0 and —SBHE. As an output in 386DX mode SA1, SA0, and —SBHE are used to determine the value of —BE3. This pin should be left unconnected when using this part in 286 mode. The pin has an internal pull-up.
—BE2/A1	70	IO-TTL	Byte Enable 2, active low, or A1—This pin has a dual function depending on the state of the 286/–386DX input. If 286/–386DX is high (286 mode), then the pin is treated as address bit 1. If 286/–386DX is low (386DX mode), the pin is treated as —BE2. This pin is an output during DMA, master, or standard refresh modes. It is an input at all other times. As an input in 386DX mode, it is decoded along with the other byte enable signals to generate SA1, SA0, and —SBHE. SA1, SA0, and —SBHE are used to determine the value of —BE2. When in 286 mode, it is interpreted as address A1 and passed to SA1. As an output in 286 mode it is driven from the SA1 input.
—BE1/—BHE	71	IO-TTL	Byte Enable 1 or Byte High Enable, active low—This pin has a dual function depending on the state of the 286/—386DX input. If 286/—386DX is high (286 mode), then the pin is treated as —BHE. If 286/—386 is low (386 mode), the pin is treated as —BE1. This pin is an output during DMA, master, or standard refresh modes. It is an input at all other times. As an input in 386 mode, it is decoded along with the other byte enable signals to generate SA1, SA0, and —SBHE. As an output in 386 mode, SA1, SA0, and —SBHE are used to determine the value of —BE1. When in 286 mode, it is interpreted as —BHE and passed to —SBHE. As an output in 286 mode, it is driven from the —SBHE input.

Figure 9.10 82344 input/output signal descriptions. (Reprinted by permission of Intel Corp. Copyright Intel Corp. 1989.)

SIGNAL DESCRIPTIONS (Continued)

Name	Pin Number	Type	Description
CPU INTERFACE (Continued)			
—BE0/A0	72	IO-TTL	Byte Enable 0, active low, or A0—This pin has a dual function depending on the state of the 286/—386DX input. If 286/—386DX is high (286 mode), then the pin is treated as address bit 0. If 286/—386DX is low (386 mode), the pin is treated as —BE0. This pin is an output during DMA, master, or standard refresh modes. It is an input at all other times. As an input in 386 mode, it is decoded along with the other Byte Enable signals to generate SA1, SA0, and —SBHE. As an output in 386 mode, SA1, SA0, and —SBHE are used to determine the value of —BE0. When in 286 mode, it is interpreted as A0 and passed to SA0. As an output in 286 mode, it is driven from the SA0 input.
286/—386DX	73	I-TPU	CPU is 286 or 386DX—This pin defines the type of address bus to which the bus controller chip is interfaced. If the pin is tied high, the address bus is assumed to be emulating 286 signals. In this mode, A25, A24, and —BE3 would be left unconnected. The pins —BE2/A1, —BE1/—BHE and —BE0/A0 would take on the 286 functions. If the pin is tied low, A25, A24 can be used to generate up to 64 Mbyte addressing for DMA, and the byte enable pins will take on the normal 386DX addressing functions. This pin has an internal pull-up to cause the chip to default to 286 mode if left unconnected. This pin is a hard wiring option and must not be changed dynamically during operation. When strapped for 286 mode, the Bus Controller is assumed to be interfaced to the 82343 System Controller which in turn may be strapped for 286 or 386SX operation. The 82344 is strapped for 286 operation when used with the 82343 strapped for 386SX operation.
HLDA	74	I-TTL	Hold Acknowledge—This is the hold acknowledge pin directly from the CPU. It is used to control direction on address and command pins. When HLDA is low, the Bus Controller is defined as being in the CPU mode. In the CPU mode, the local address bus (A bus) pins are inputs. The system address bus (SA and LA) pins along with the command pins (—MEMR, —MEMW, —IOR and —IOW) are outputs. When HLDA is high, the Bus Controller can be in DMA, refresh, or master modes. In both DMA and refresh modes, the commands and all address buses (A, SA and LA) are outputs. In master mode, the commands and system address bus (SA and LA) pins are inputs and the local address bus (A bus) pins are outputs. The SA bus is passed directly to the A bus except bits 17, 18, and 19 are ignored. LA23–LA17 is passed directly to A23–A17.
INTR	75	O	Interrupt Request—INTR is used to interrupt the CPU and is generated by the 8259 megacells any time a valid interrupt request input is received.
NMI	76	O	Non-Maskable Interrupt—This output is used to drive the NMI input to the CPU. This signal is asserted by either a parity error (indicated by —PCK being asserted after the ENPARCK bit in Port B has been asserted), or an I/O channel error (indicated by —IOCHCK being asserted after the ENIOCK bit in Port B has been asserted). The NMI output is enabled by writing a 0 to bit D7 of I/O port 70h. NMI is disabled on reset.

Figure 9.10 (*Continued*)

SIGNAL DESCRIPTIONS (Continued)

Name	Pin Number	Type	Description
SYSTEM CONTROLLER INTERFACE			
—CHS0/—MW	77	IO-TTL	Channel Status 0 or active low Memory Write—This input is used along with —CHS1 and CHM/—IO to determine what type of bus cycle the Bus Controller is to perform. This input has the same meaning and timing requirements as the S0 signal for a 286 microprocessor. —CHS0 going active indicates a write cycle unless —CHS1 is also active. When both status inputs are active it indicates an interrupt acknowledge cycle. This input is synchronized to the BUSCLK input. Activation of CPUHLDA reverses this signal to become an output to the System Controller. It is then a —MEMW signal for DMA or bus master access to system memory.
—CHS1/—MR	78	IO-TTL	Channel Status 1 or active low Memory Read—This input is used along with —CHS0 and CHM/—IO to determine the bus cycle type. This input has the same meaning and timing requirements as the S1 signal for a 286 microprocessor —CHS1 going active indicates a read cycle unless —CHS0 is also active. When both status inputs are active it indicates an interrupt acknowledge cycle. This input is synchronized to the BUSCLK input. Activation of CPUHLDA reverses this signal to become an output to the System Controller. It is then a —MEMR signal for DMA or bus master access to system memory.
CHM/—IO	82	I-TTL	Channel Memory or active low I/O select—This input is used along with —CHS0 and —CHS1 to determine the bus cycle type. This input has the same meaning and timing requirements as the M/—IO signal for a 286 microprocessor. CHM/—IO is sampled anytime —CHS0 or —CHS1 is active. If sampled high, it indicates a memory read or write cycle. If sampled low, an I/O read or write cycle should be executed. This input is synchronized to the BUSCLK input.
—EALE	83	I-TTL	Early Address Latch Enable, active low—This input is used to latch the A25–A2 and Byte Enable signals. The latches are open when —EALE is low and hold their value when —EALE is high. The latched addresses are fed directly to the LA23–LA17 bus to provide more address setup time on the bus before a command goes active. The lower latched addresses are latched again with an internal ALE signal as soon as —CHS0 or —CHS1 is sampled active and fed to the SA19–SA0 and —SBHE outputs. In a 386DX system, this input is connected directly to the —ADS output from the CPU. In a 286 system, this input is connected to the —EALE output from the 82343 System Controller.
—BRDRAM	84	I-TTL	On-board DRAM, active low—An input from the System Controller indicating that the on-board DRAM is being addressed.

Figure 9.10 (*Continued*)

SIGNAL DESCRIPTIONS (Continued)

Name	Pin Number	Type	Description
SYSTEM CONTROLLER INTERFACE (Continued)			
—CHREADY	85	O	Channel Ready, active low—This output is maintained in the active state when no bus accesses are active. This indicates that the Bus Controller is ready to accept a new command. During normal bus accesses, —CHREADY is negated as soon as a valid bus requested is sampled on the —CHS0 and —CHS1 inputs. It is asserted again to indicate that the Bus Controller is ready to complete the current cycle. The bus command signals are then terminated on the next falling edge of the BUSCLK input.
BUSCLK	86	I-CMOS	Bus Clock—This is the main clock input for the Bus Controller. It runs at twice the frequency desired for the SYSCLK output. All inputs are synchronous with the falling edge of this input.
—BLKA20	87	I-TTL	Block A20, active low—This input is used while CPUHLDA is low to force the LA20 and SA20 outputs low anytime it is active. When —BLKA20 is negated LA20 and SA20 are generated from A20.
DMAHRQ	89	O	Hold Request—This output is generated by the DMA controller any time a valid DMA request is received. It is connected to the DMAHRQ pin on the System Controller.
DMAHLDA	88	I-TTL	DMA Hold Acknowledge—An input from the System Controller which indicates that the current hold acknowledge state is for the DMA controller or other bus master.
OUT1	90	O	Output 1—Indicates a refresh request to the System Controller. This is the 15 μs output of timer channel 1.
ROM INTERFACE			
—ROM8	112	I-TPU	8/16 bit ROM select—This input indicates the width of the ROM BIOS. If —ROM8 is low, the Bus Controller chip generates 8- to 16-bit conversions for ROM accesses. Data buffer controls are generated assuming the ROM is on the MD bus. If —ROM8 is high, data buffer controls are generated assuming 16-bit wide ROMs are on the MD bus.
BUS INTERFACE			
—IOR	134	IO-TTL	I/O Read, active low—This signal is an input when CPUHLDA is high and —MASTER is low. It is an output at all other times. When CPUHLDA is low, —IOR is driven from the 288 bus controller megacell. When CPUHLDA is high and —MASTER is high, it is driven by the 8237 DMA controller megacells. This pin requires an external 10 KΩ pull-up resistor.
—IOW	132	IO-TTL	I/O Write, active low—This signal is an input when CPUHLDA is high and —MASTER is low. It is an output at all other times. When CPUHLDA is low, —IOW is driven from the 288 bus controller megacell. When CPUHLDA is high and —MASTER is high, it is driven by the 8237 DMA controller megacells. This pin requires an external 10 KΩ pull-up resistor.

Figure 9.10 (*Continued*)

SIGNAL DESCRIPTIONS (Continued)

Name	Pin Number	Type	Description
BUS INTERFACE (Continued)			
—MEMR	33	IO-TTL	Memory Read, active low—This signal is an input when CPUHLDA is high and —MASTER is low. It is an output at all other times. When CPUHLDA is low, —MEMR is driven from the 288 bus controller megacell. When CPUHLDA is high and —MASTER is high, it is driven by the 8237 DMA controller megacells. This signal does not pulse low for DMA addresses above 16 Mbytes. DMA above 16 Mbytes is only performed to the system board, never to the slot bus. This pin requires an external 10 KΩ pull-up resistor.
—MEMW	35	IO-TTL	Memory Write, active low—This signal is an input when CPUHLDA is high and —MASTER is low. It is an output at all other times. When CPUHLDA is low, —MEMW is driven from the 288 bus controller megacell. When CPUHLDA is high and —MASTER is high, it is driven by the 8237 DMA controller megacells. This pin requires an external 10 KΩ pull-up resistor.
—SMEMR	129	IO-TTL	Memory Read, active low—This signal is an input when CPUHLDA is high and —MASTER is low. It is an output at all other times. When CPUHLDA is low, —MEMR is driven from the 288 bus controller megacell. When CPUHLDA is high and —MASTER is high, it is driven by the 8237 DMA controller megacells. —SMEMR is active on memory read cycles to addresses below 1 Mbyte. This pin requires an external 10 KΩ pull-up resistor.
—SMEMW	127	IO-TTL	Memory Write, active low—This signal is an input when CPUHLDA is high and —MASTER is low. It is an output at all other times. When CPUHLDA is low, —MEMW is driven from the 288 bus controller megacell. When CPUHLDA is high and —MASTER is high, it is driven by the 8237 DMA controller megacells. —SMEMW is active on memory write cycles to addresses below 1 Mbyte. This pin requires an external 10 KΩ pull-up resistor.
LA23–LA17	16, 18, 22, 24, 26, 28, 31	IO-TTL	Latchable Address bus—This bus in an input when CPUHLDA is high and —MASTER is low. It is an output bus at all other times. When CPUHLDA is low, the LA bus is driven by the latched values for the A bus. When CPUHLDA is high and —MASTER is high, the SA bus is driven by the 612 memory mapper for DMA cycles and normal refresh. The LA bus is latched internally with the —EALE input.
SA19–SA0	131, 133, 135, 137, 141, 143, 145, 147, 149, 152, 154, 156, 158, 1, 3, 5, 7, 8, 11, 12	IO-TTL	System Address bus—This bus is an input when CPUHLDA is high and —MASTER is low. It is an output bus at all other times. When CPUHLDA is low, the SA bus is driven by the latched values from the A bus. When CPUHLDA is high and —MASTER is high, the SA bus is driven by the 8237 DMA controller megacells or refresh address generator. The SA bus will become valid in the middle of the status cycle generated by the —CHS0 and —CHS1 inputs. They are latched with an internally generated ALE signal.

Figure 9.10 (*Continued*)

SIGNAL DESCRIPTIONS (Continued)

Name	Pin Number	Type	Description
BUS INTERFACE (Continued)			
—SBHE	14	IO-TTL	System Byte High Enable, active low—This pin is controlled the same way as the SA bus. It is generated from a decode of the —BE inputs in CPU mode. It is forced low for 16-bit DMA cycles and forced to the opposite value of SA0 for 8-bit DMA cycles.
—REFRESH	146	IT-OD	Refresh signal, active low—This I/O signal is pulled low whenever a decoupled refresh command is received from the System Controller. It is used as an input to sense refresh requests from external sources such as the System Controller for coupled refesh cycles or bus masters. It is used internally to clock the refresh address counter and select a location in the memory mapper which drives A23–A17. —REFRESH is an open drain output capable of sinking 24 mA and requires an external pull-up resistor.
SYSCLK	148	O	System Clock—This output is half the frequency of the BUSCLK input. The bus control outputs BALE and the —IOR, —IOW, —MEMR and —MEMW are synchronized to SYSCLK.
OSC	9	I-TTL	Oscillator—This is the buffered input of the external 14.318 MHz oscillator.
RSTDRV	122	O	Reset Drive, active high—This output is a system reset generated from the POWERGOOD input. RSTDRV is synchronized to the BUSCLK input.
BALE	6	O	Buffered Address Latch Enable, active high—A pulse which is generated at the beginning of any bus cycle initiated from the CPU. BALE is forced high anytime CPUHLDA is high.
AEN	128	O	Address Enable—This output goes high anytime the inputs CPUHLDA and —MASTER are both high.
T/C	4	O	Terminal Count—This output indicates that one of the DMA channels terminal count has been reached. This signal directly drives the system bus.
—DACK7-—DACK5, —DACK3-—DACK0	39, 37, 34, 136, 2, 142, 29	O	DMA Acknowledge, active low—These outputs are the acknowledge signals for the corresponding DMA requests. The active polarity of these lines is set active low on reset. Since the 8237 megacells are internally cascaded together, the polarity of the —DACK signals must not be changed. This signal directly drives the system bus.
DRQ7-DRQ5 DRQ3-DRQ0	41, 38, 36, 138, 124, 144, 32	I-TSPU	DMA Request—These asynchronous inputs are used by an external device to indicate when they need service from the internal DMA controllers. DRQ0-DRQ3 are used for transfers from 8-bit I/O adapters to/from system memory. DRQ5-DRQ7 are used for transfers from 16-bit I/O adapters to/from system memory. DRQ4 is not available externally as it is used to cascade the two DMA controllers together. All DRQ pins have internal pull-ups.

Figure 9.10 (*Continued*)

SIGNAL DESCRIPTIONS (Continued)

Name	Pin Number	Type	Description
BUS INTERFACE (Continued)			
IRQ15–IRQ9, IRQ7–IRQ3, IRQ1	25, 27, 110, 23, 21, 17, 123, 151, 153, 155, 157, 159, 109	I-TPSU	Internal Request—These are the asynchronous interrupt request inputs for the 8259 megacells. IRQ0, IRQ2, and IRQ8 are not available as external inputs to the chip, but are used internally. IRQ0 is connected to the output of the 8254 counter 0. IRQ2 is used to cascade the two 8259 megacells together. IRQ8 is output from the RTC megacell to the 8259 megacell. All IRQ input pins are active high and have internal pull-ups.
—MASTER	42	I-TTL	Master, active low—This input is used by an external device to disable the internal DMA controllers and get access to the system bus. When asserted it indicates that an external bus master has control of the bus.
—MEMCS16	13	I-TTL	Memory Chip Select 16-bit—This input is used to determine when a 16-bit to 8-bit conversion is needed for CPU accesses. A 16 to 8 conversion is done anytime the System Controller requests a 16-bit memory cycle and —MEMCS16 is sampled high.
—IOCS16	15	I-TTL	I/O Chip Select 16-bit—This input is used to determine when a 16-bit to 8-bit conversion is needed for CPU accesses. A 16 to 8 conversion is done anytime the System Controller requests a 16-bit I/O cycle and —IOCS16 is sampled high.
—IOCHK	121	I-TTL	I/O Channel Check, active low—This input is used to indicate that an error has taken place on the I/O bus. If I/O checking is enabled, an —IOCHK assertion by a peripheral device generates an NMI to the processor. The state of the —IOCHK signal is read as data bit D6 of the Port B register.
IOCHRDY	126	I-TTL	I/O Channel Ready—This input is pulled low in order to extend the read or write cycles of any bus access when required. The cycle can be initiated by the CPU, DMA controllers or refresh controller. The default number of wait states for cycles intiated by the CPU are four wait states for 8-bit peripherals, one wait state for 16-bit peripherals and three wait states for ROM cycles. One DMA wait state is inserted as the default for all DMA cycles. Any peripheral that cannot present read data, or strobe-in write data in this amount of time must use —IOCHRDY to extend these cycles.
—WS0	125	I-TTL	Wait State 0, active low—This input is pulled low by a peripheral on the S bus to terminate a CPU controlled bus cycle earlier than the default values defined internally on the chip.
POWERGOOD	115	I-TSPU	System power on reset—This input signals that power to the board is stable. A Schmitt-trigger input is used. This allows the input to be connected directly to an RC network.
PERIPHERAL INTERFACE			
—CS8042	108	O	Chip select for 8042. This output is active any time an SA address is decoded at 60h or 64h. It is intended to be connected to the chip select of the keyboard controller. If BUSCTL[6] = 1, this pin is also active for RTC accesses at 70h and 71h. This is for use when the internal RTC is disabled and an external RTC is used.

Figure 9.10 (*Continued*)

SIGNAL DESCRIPTIONS (Continued)

Name	Pin Number	Type	Description
PERIPHERAL INTERFACE (Continued)			
XTALIN	118	I-CMOS	Crystal Input—An internal oscillator input for the real time clock crystal. It requires a 32.768 KHz external crystal or stand-alone oscillator.
XTALOUT	119	O	Crystal Output—An internal oscillator output for the real time clock crystal. See XTALIN. This pin is a no connect when an external oscillator is used.
PS/—RCLR/ IRQ8	117	I-TSPU	The Power Sense input (active high) is used to reset the status of the Valid RAM and Time (VRT) bit. This bit is used to indicate that the power has failed, and that the contents of the RTC may not be valid. This pin is connected to an external RC network. When BUSCTL[6] = 1, this pin becomes —IRQ8 input for use with an external RTC.
VBAT	116	I	Voltage Battery—Connected to the RTC hold-up battery between 2.4 and 5V.
SPKR	107	O	Speaker—This output drives an externally buffered speaker. This signal is created by gating the output of timer 2. Bit 1 of Port B, 61H, is used to enable the speaker output, and bit 0 is used to gate the output timer.
DATA BUFFER INTERFACE			
XD7–XD0	91, 92, 94–99	IO-TTL	Peripheral data bus—The bidirectional X data bus outputs data on an INTA cycle or I/O read cycle to any valid address within the Bus Controller. It is configured as an input at all other times.
—SDSWAP	101	O	System Data Swap, active low during some 8-bit accesses—It indicates that the data on the SD bus must be swapped from low byte to high byte or vice versa depending on the state of the SDLH/—HL pin. —SDSWAP is active for 8-bit DMA cycles when an odd address access occurs for data more than one byte wide. For non-DMA accesses. —SDSWAP is active for any bus cycle to an 8-bit peripheral that is addressing the odd byte.
SDLH/—HL	102	O	System Data Low to High, or High to Low—This signal is used to determine which direction data bytes must be swapped when —SDSWAP is active. When SDLH/—HL is high, it indicates that data on the low byte must be transferred to the high byte. When SDLH/—HL is low, it indicates that data on the high byte must be transferred to the low byte. SDLH/—HL is low for 8-bit DMA memory read cycles. For non-DMA accesses, SDLH/—HL is low for any memory write or I/O write when —SBHE is low. SDLH/—HL is high at all other times.
—XDREAD	103	O	Peripheral Data Read—This output is active low any time an INTA cycle occurs or an I/O read occurs to the address space from 0000h to 00FFh, which is defined as being resident on the peripheral bus.

Figure 9.10 (*Continued*)

SIGNAL DESCRIPTIONS (Continued)

Name	Pin Number	Type	Description
DATA BUFFER INTERFACE (Continued)			
—LATLO	104	O	Latch Low byte—This output is generated for all I/O read and memory read bus accesses to the low byte. It is active with the same timing as the read command and returns high at the same time as the read command. This signal latches the data into the data buffer chip so that it can be presented to the CPU at a later time. This step is required due to the asynchronous interface between the System Controller and Bus Controller.
—LATHI	105	O	Latch High byte—This output is generated for all I/O read and memory read bus accesses to the high byte. It is active with the same timing as the read command and returns high at the same time as the read command. This signal latches the data into the data buffer chip so that it can be presented to the CPU at a later time. This step is required due to the asynchronous interface between the System Controller and Bus Controller.
—PCK	111	I-TPU	Party Check input, active low with pull-up—Indicates that a parity error has occurred in the on-board memory array. Assertion of this signal (if enabled) generates an NMI to the processor. The state of the —PCK signal is read as data bit D7 of the Port B register.
—HIDRIVE	113	I-TPU	High Drive Enable—This pin is a wire strap option. When this input is low, all bus drivers defined with an IOL spec of 24 mA will sink the full 24 mA of current. When this input is high, all pins defined as 24 mA have the output low drive capability cut in half to 12 mA. Note that all AC specifications are done with the outputs in the high drive mode and a 200 pF capacitive load —HIDRIVE has an internal pull-up and can be left unconnected if 12 mA drive is desired. It is tied low if 24 mA drive is desired.
TEST MODE PIN			
—TRI	114	I-TPU	Three-state—This pin is used to control the three-state drive of all outputs and bidirectional pins on the chip. If this pin is pulled low, all pins on the chip except XTALOUT are in a high impedance mode. This is useful during system test when test equipment or other chips drive the signals or for hardware fault tolerant applications. —TRI has an internal pull-up.

Figure 9.10 (*Continued*)

same time, the $\overline{\text{ROMCS}}$ output of the 82346 is supplied to the $\overline{\text{CE}}$ input of the EPROM and selects U_{39} for operation. Finally, information held at the addressed storage location in the EPROM are output on memory data bus lines MD_0 through MD_7. This byte of data is passed through the 82345 data buffer to data bus lines D_0 through D_7 of the MPU.

EXAMPLE 9.19

A chart for the settings of BIOS EPROM address jumpers J_{30}, J_{31}, and J_{32} is given on sheet 3 of the circuit diagram in Fig. 9.3. What should be the settings of the jumpers if eight-bit mode is selected and a single 27C512 EPROM is used?

Signal Type Legend

Signal Code	Signal Type
I-TTL	TTL Level Input
I-TPD	Input with 30 KΩ Pull-Down Resistor
I-TPU	Input with 30 KΩ Pull-Up Resistor
I-TSPU	Schmitt-Trigger Input with 30 KΩ Pull-Up Resistor
I-CMOS	CMOS Level Input
IO-TTL	TTL Level Input/Output
IT-OD	TTL Level Input/Open Drain Output
IO-OD	Input or Open Drain, Slow Turn On
O	CMOS and TTL Level Compatible Output
O-TTL	TTL Level Output
O-TS	Three-State Level Output
I1	Input used for Testing Purposes
GND	Ground
PWR	Power

Figure 9.10 (*Continued*)

SOLUTION The chart indicates that the jumper setting should be

$$J_{30} = \text{Don't care}$$
$$J_{31} = \text{IN}$$
$$J_{32} = \text{IN}$$

The ISA expansion bus of the PC/AT is a collection of address, data, control, and power lines that are provided to support expansion of the microcomputer system. The main processor board design of Fig. 9.3 includes connectors for six expansion slots. The pin layout of these sockets are illustrated on sheet 8 of the circuit diagram. Notice that five slots have both a 62-pin and a 36-pin card edge connector. They are full 16-bit data bus expansion slots. An example is the slot made by the connectors labeled J_{80} and J_{81}. The sixth slot, which has just a 62-pin connector (J_{90}), only supports an eight-bit data bus. Using these slots, the microcomputer system is expanded by plugging in special function adapter cards, such as boards to control a monochrome or color display, expanded memory, or communication ports.

The pins of these connectors are attached to inputs or outputs of the 82345 data buffer, 82344 ISA controller, or power supply. Let us now examine the source of some of these signals. Sheet 4 of Fig. 9.3 shows that system data bus lines SD_0 through SD_{15} at J_{80} and J_{81} are tied to the system data bus lines of the 82345 data buffer. On the other hand, the address and control signals of the ISA bus are all supplied by the 82344 ISA controller. For instance, address signal SA_{19} at pin A12 of connector J_{80} is attached to pin 131 of the 82344, which is also labeled

SA_{19}. Another example is control signal $\overline{MEMR}$ at pin C9 of J_{81}, which connects to $\overline{MEMR}$ at pin 33 of the 82344. The signal names for each of the signals at the ISA expansion bus are listed in Fig. 9.11.

> **EXAMPLE 9.20**
>
> What power supply voltage is available at pin B5 of J_{80}?
>
> **SOLUTION** From sheet 8 of the circuit diagram, we find that -5 V dc is supplied from pin B5.

> **EXAMPLE 9.21**
>
> What does the signal mnemonic RSTDRV stand for? Is it active when at the 0 or 1 logic level?
>
> **SOLUTION** In Fig. 9.10 we see that RSTDRV means reset drive and is active when at the 1 logic level.

The peripheral interface corresponds to a number of special purpose inputs and outputs of the 82344. Earlier we identified one of these signals, $\overline{CS8042}$, as

Mnemonic	Name	Function
AEN	Address enable	O
BALE	Buffered address latch enable	O
CLK	Clock	O
$\overline{DACK}_0$-$\overline{DACK}_3$ $\overline{DACK}_5$-$\overline{DACK}_7$	DMA acknowledge 0-3 and 5-7	O
DRQ_0-DRQ_3 DRQ_5-DRQ_7	DMA request 1-3 and 5-7	I
$\overline{I/O\ CH\ CK}$	I/O channel check	I
I/O CH RDY	I/O channel ready	I
I/O CS16	I/O 16-bit chip select	I
$\overline{IOR}$	I/O read command	I/O
$\overline{IOW}$	I/O write command	I/O
IRQ_3-IRQ_7, IRQ_9-IRQ_{12} and IRQ_{14}-IRQ_{15}	Interrupt request 3-7, 9-12, and 14-15	I
LA_{17}-LA_{23}	System address lines 17-23	I/O
$\overline{MASTER}$	Master	I
$\overline{MEM\ CS16}$	Memory 16 chip select	I
$\overline{MEMR}$	Memory read command	I/O
$\overline{MEMW}$	Memory write command	I/O
OSC	Oscillator	O
$\overline{REFRESH}$	Refresh	I/O
RESET DRV	Reset drive	O
SA_0-SA_{19}	System address lines 0-19	I/O
SBHE	System high byte enable	O
SD_0-SD_{15}	Data lines 0-15	I/O
$\overline{SMEMR}$	System memory read command	O
$\overline{SMEMW}$	System memory write command	O
T/C	Terminal count	O
OWS	Zero wait state	I

Figure 9.11 ISA bus signals.

an output that is used to enable the 8042 keyboard controller. Three other peripheral interface signals are PS/$\overline{\text{RCLR}}$/IRQ$_8$, VBAT, and SPKR. In the circuitry of sheet 4 of Fig. 9.3, we find that SPKR is output to pin 1 of connector J$_{41}$. J$_{41}$ is used to connect the speaker to the microcomputer. VBAT and PS/$\overline{\text{RCLR}}$/IRQ$_8$ are inputs of the 82344. For instance, in the circuit diagram, power sense (PS) is found to be the voltage across capacitor C$_{40}$.

Earlier we found that a special interface was provided on the 82345 data buffer for communication with the 82344 ISA controller. Figure 9.10 shows that this data buffer interface includes peripheral data bus lines XD$_0$ through XD$_7$. This part of the interface is used to input data from or output data to the registers of the peripherals within the 82344 and to transfer type numbers from the interrupt controllers to the MPU during interrupt acknowledge bus cycles.

This interface also includes a number of signals that tell the 82345 data buffer how to set its multiplexers to switch data that are being transferred over the system data bus. For instance, the $\overline{\text{SDSWAP}}$ and SDLH/$\overline{\text{HL}}$ signals are used together to indicate that byte data on the bus must be swapped and whether the data must be swapped from the low-byte data bus lines to the high-byte lines, or vice versa.

EXAMPLE 9.22

To what logic level is SDLH/$\overline{\text{HL}}$ set to signal the 82345 that a byte of data on the system data bus must be swapped from the low-byte lines to the high-byte lines?

SOLUTION Figure 9.10 indicates that SDLH/$\overline{\text{HL}}$ must be set to 1 when data are to be transferred from the low-byte lines to the high-byte lines.

A special interface is also provided on the 82344 for communication with the 82346 system controller. It includes signals that identify the type of bus cycle that is taking place, on-board DRAM is being accessed, a request and acknowledgment for DMA, and a request for refresh. On sheet 4 of Fig. 9.3 we find that these signals are called the 344/346 interface; however, in Fig. 9.10 they are identified as those for the system controller interface. The function of each of these signals is described briefly in Fig. 9.10. Let us trace the DMA handshake signal sequence that takes place between the 82344 and 82346. Notice that the signal DMAHRQ at pin 89 is the DMA hold request output of the 82344. Logic 1 at this output signals that the DMA controllers within the 82344 want access to the system bus. This signal is input to the 82346 system controller at pin 105 (DMAHRQ). When the bus is available to the DMA controller, the 82346 signals this fact to the 82344 by switching its DMA hold acknowledge (DMAHLDA) output at pin 99 to logic 1. DMAHLDA is input to the 82344 at pin 88. Now a DMA controller within the 82344 has control of the system bus and can initiate data transfer.

9.8 82341 HIGH-INTEGRATION PERIPHERAL COMBO

The 82341 IC implements a number of the peripheral functions that are needed in the microcomputer of the PC/AT. It contains circuitry for two *16C450-compatible asynchronous serial communication ports*, a *parallel printer port*, a *real-*

time clock, a *scratchpad RAM, keyboard and mouse controllers*, an *integrated drive electronics (IDE) hard disk interface*, and *programmable chip selects*. For this reason it is known as the *PC/AT peripheral combo chip*. This device is packaged in a 128-pin PQFP.

Some of the functions provided in the 82341 did not reside on the main processor board of IBM's original 80286-based PC/AT microcomputer. Instead, they were provided with add-in cards that connect to the ISA bus. This is the reason that the 82341 peripheral combo is shown attached to the ISA expansion bus in the 82340 PC/AT block diagram of Fig. 9.2.

Block Diagram of the 82341

The circuitry within the 82341 is illustrated with the block diagram in Fig. 9.12. Here we see that the device acts like a group of programmable peripheral ICs attached to the system address and data buses. They include an 82C042 keyboard/mouse controller, 146818A real-time clock, 16C452 dual UART and parallel printer port, and IDE hard disk interface. The operation of the various peripheral functions implemented within the 82341 are set up under software control. Addresses applied to the SA_0 through SA_{15} inputs are decoded by the port address decoder to select a control register within the device. Then programming information or data are written into the selected register over system data bus lines SD_0 through SD_7. Status, configuration information, and data can also be read from a register within the device.

Inputs and Outputs of the 82341 and Their Connection in the PC/AT Microcomputer System

Now that we have introduced the peripheral functions implemented with the 82341 peripheral combo, let us look at its signal interfaces and how they are used in a PC/AT microcomputer system. The signals of the 82341 are listed in Fig. 9.13. Earlier we pointed out that the operation of the peripheral functions of the 82341 are programmed over the system data bus. The system address bus and data bus signals are listed in the *common bus I/O* group in Fig. 9.13. Here we find that I/O bus cycles are used to read or write to the 82341. The control signals for these data transfers are AEN, ALE, $\overline{IOR}$, and $\overline{IOW}$. Looking at sheet 5 of Fig. 9.3, we find that these control inputs are supplied by ISA bus signals, AEN, BALE, $\overline{IOR}$, and $\overline{IOW}$, respectively. These four signals are produced by the 82344 ISA bus controller.

> **EXAMPLE 9.23**
>
> Is the IOCHRDY signal an input or output of the 82341? Where is this signal connected in the PC/AT microcomputer of Fig. 9.3?
>
> **SOLUTION** Figure 9.13 identifies IOCHRDY as an output of the 82341. In the circuits on sheet 4 of Fig. 9.3, we see that IOCHRDY is input to the microcomputer at pin 126 of the 82344 ISA bus controller.

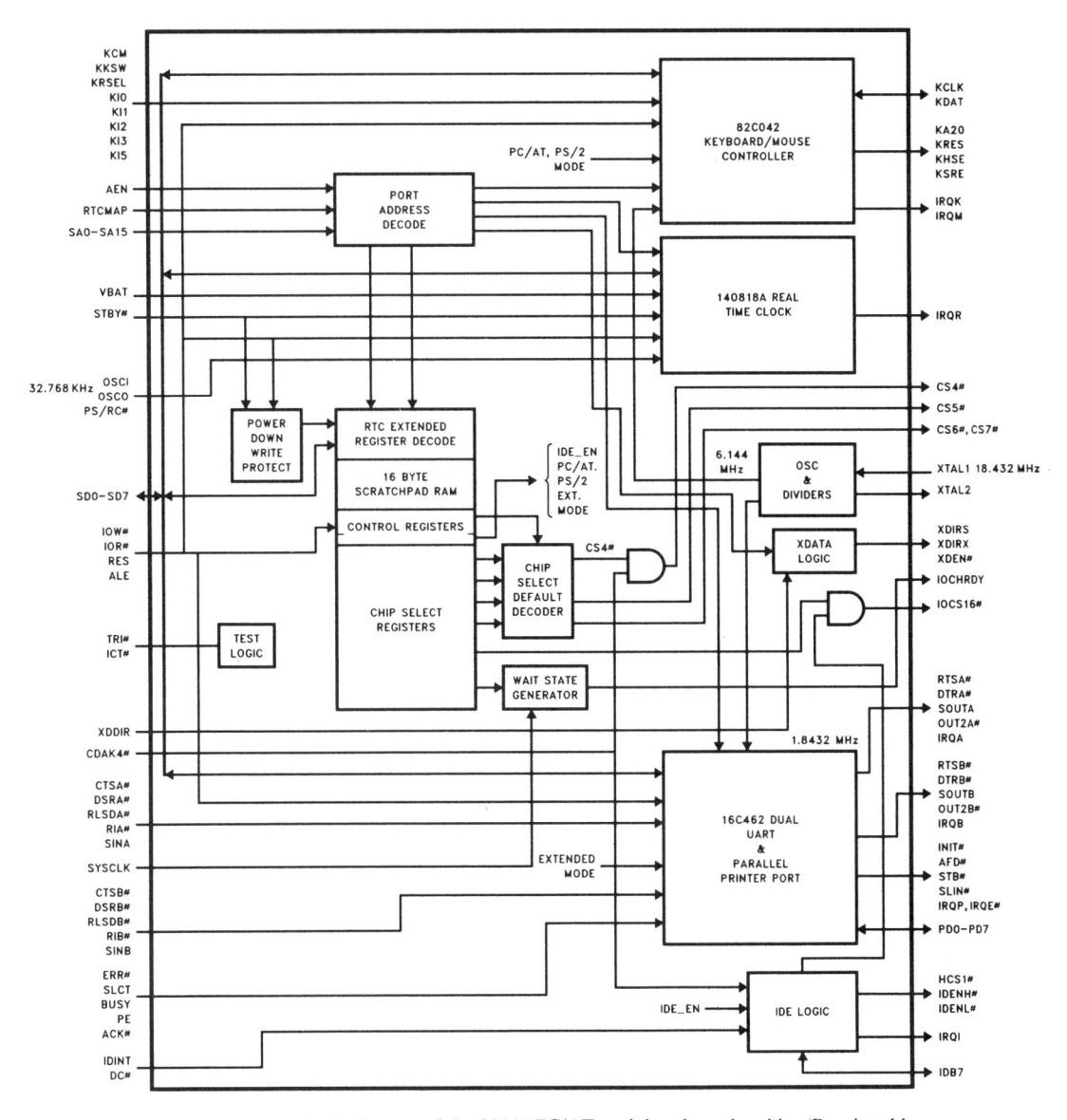

Figure 9.12 Block diagram of the 82341 PC/AT peripheral combo chip. (Reprinted by permission of Intel Corp. Copyright Intel Corp. 1989.)

We will continue with the asynchronous communication interfaces that are implemented with the UARTs of the 82341. Notice on sheet 5 of Fig. 9.3 that the signals of the 82341 that are used to implement the serial interfaces are grouped together and marked as COM A and COM B. Tracing the COM A outputs, *request to send port A* ($\overline{\text{RTSA}}$), *data terminal ready* ($\overline{\text{DTRA}}$), and *serial data output port A* (SOUTA), shows that they are buffered to RS-232C-compatible voltage levels

Sec. 9.8   82341 High-Integration Peripheral Combo

by an MC1488 line driver and then output at COM A connector J_{53}. Inputs from this RS-232C port are the signals *data set ready port A* ($\overline{DSRA}$), *clear to send port A* ($\overline{CTSA}$), *ring indicator port A* ($\overline{RIA}$), *serial input port A* (SINA), and *receive line signal detect port A* ($\overline{RLSDA}$). The mnemonic, pin number, and type for each of the signals of the 82341's communication ports are listed in Fig. 9.13.

SIGNAL DESCRIPTIONS

Signal Name	Pin Number	Signal Type	Signal Description
COMMUNICATIONS PORT A			
RTSA#	44	O1	Request to Send, Port A
DTRA#	45	O1	Data Terminal Ready, Port A
SOUTA	46	O1	Serial Data Output, Port A
CTSA#	79	I4	Clear to Send, Port A
DSRA#	78	I4	Data Set Ready, Port A
RLSDA#	77	I4	Receive Line Signal Detect, Port A
RIA#	76	I4	Ring Indicator, Port A
SINA	75	I4	Serial Input, Port A
IRQA	39	O6	Interrupt Request, Port A
OUT2A#	42	O1	Output 2, Port A
COMMUNICATIONS PORT B			
RTSB#	47	O1	Request to Send, Port B
DTRB#	48	O1	Data Termina Ready, Port B
SOUTB	49	O1	Serial Data Output, Port B
CTSB#	89	I4	Clear to Send, Port B
DSRB#	88	I4	Data Set Ready, Port B
RLSDB#	87	I4	Receive Line Signal Detect, Port B
RIB#	86	I4	Ring Indicator, Port B
SINB	85	I4	Serial Input, Port B
IRQB	37	O6	Interrupt Request, Port B
OUT2B#	43	O1	Output 2, Port B
PARALLEL PRINTER PORT			
PD0	59	IO5	Printer Data Port, Bit 0
PD1	58	IO5	Printer Data Port, Bit 1
PD2	57	IO5	Printer Data Port, Bit 2
PD3	56	IO5	Printer Data Port, Bit 3
PD4	54	IO5	Printer Data Port, Bit 4
PD5	53	IO5	Printer Data Port, Bit 5
PD6	52	IO5	Printer Data Port, Bit 6
PD7	51	IO5	Printer Data Port, Bit 7
INIT#	63	O4	Initialize Printer Signal
AFD#	62	O4	Autofeed Printer Signal
STB#	61	O4	Data Strobe to Printer
SLIN#	64	O4	Select Signal from Printer
ERR#	70	I4	Error Signal from Priner
SLCT	71	I4	Select Signal from Printer

Figure 9.13 82341 input/output signal descriptions. (Reprinted by permission of Intel Corp. Copyright Intel Corp. 1989.)

SIGNAL DESCRIPTIONS (Continued)

Signal Name	Pin Number	Signal Type	Signal Description
PARALLEL PRINTER PORT (Continued)			
BUSY	72	I4	Busy Signal from Printer
PE	73	I4	Paper Error Signal from Printer
ACK#	74	I4	Acknowledge Signal from Printer
IRQP	40	O6	Printer Interrupt Request Output
IRQE#	41	O1	Printer Interrupt Request Enable Signal
REAL TIME CLOCK PORT			
VBAT	69	NA	Standby Power—Normally 3V to 5V, battery backed
STBY#	65	I5	Power Down Control
OSCI	66	NA	Crystal Connection Input—32 kHz
OSCO	67	NA	Crystal Connection Output—32 kHz
PS/RC#	68	I5	Power Sense/RAM Clear Input
IRQR	36	O1	Real Time Clock Interrupt Request Output
RTCMAP	121	I4	High—RTC is mapped to 70H and 71H, Low—RTC is mapped to 170H and 171H
KEYBOARD CONTROLLER PORT			
KCLK	103	IO4	Keyboard Clock
KDAT	104	IO4	Keyboard Data
KCM	92	I4	General Purpose Input, Normally Color/Monochrome
KKSW	93	I4	General Purpose Input, Normally Keyboard Switch
KA20	91	O1	General Purpose Output, Normally A20 Gate
KRES	90	O1	General Purpose Output, Normally Reset
KHSE	101	O1/IO4	General Purpose Input, Normally Speed Select
KSRE	100	O1/IO4	General Purpose Output, Normally Shadow RAM Enable
IRQK	34	O1	Keyboard Interrupt Request
IRQM	35	O1	Mouse Interrupt Request
KRSEL	94	I4	General Purpose Input, Normally RAM Select
K10	99	I4	General Purpose Input, Bit 0
K11	98	I4	General Purpose Input, Bit 1
K12	97	I4	General Purpose Input, Bit 2
K13	96	I4	General Purpose Input, Bit 3
K15	95	I4	General Purpose Input, Bit 5
IDE BUS I/O			
IDENH#	2	O1	IDE Bus Transceiver High Byte Enable
IDENL#	3	O1	IDE Bus Transceiver Low Byte Enable
IDINT	122	I4	IDE Bus Interrupt Request Input
IDB7	119	IO6	IDE Bus Data Bit 7
DC#	123	I4	Floppy Disk Change Signal
HCS1#	124	O1	IDE Host Chip Select 1
IRQI#	33	O6	IDE Interrupt Request Output

Figure 9.13 (*Continued*)

SIGNAL DESCRIPTIONS (Continued)

Signal Name	Pin Number	Signal Type	Signal Description
COMMON BUS I/O			
SD0	115	IO2	System Bus Data, Bit 0
SD1	114	IO2	System Bus Data, Bit 1
SD2	111	IO2	System Bus Data, Bit 2
SD3	110	IO2	System Bus Data, Bit 3
SD4	109	IO2	System Bus Data, Bit 4
SD5	108	IO2	System Bus Data, Bit 5
SD6	106	IO2	System Bus Data, Bit 6
SD7	105	IO2	System Bus Data, Bit 7
SA0	17	I1	System Bus Address, Bit 0
SA1	18	I1	System Bus Address, Bit 1
SA2	19	I1	System Bus Address, Bit 2
SA3	20	I1	System Bus Address, Bit 3
SA4	21	I1	System Bus Address, Bit 4
SA5	22	I1	System Bus Address, Bit 5
SA6	23	I1	System Bus Address, Bit 6
SA7	24	I1	System Bus Address, Bit 7
SA8	25	I1	System Bus Address, Bit 8
SA9	26	I1	System Bus Address, Bit 9
SA10	27	I1	System Bus Address, Bit 10
SA11	28	I1	System Bus Address, Bit 11
SA12	29	I1	System Bus Address, Bit 12
SA13	30	I1	System Bus Address, Bit 13
SA14	31	I1	System Bus Address, Bit 14
SA15	32	I1	System Bus Address, Bit 15
XTAL1	82	NA	Crystal Clock Input—18.432 MHz
XTAL2	83	NA	Crystal Clock Output—18.432 MHz
IOR#	11	I1	System Bus I/O Read
IOW#	12	I1	System Bus I/O Write
RES	125	I1	System Reset
AEN	13	I1	System Bus Address Enable
ALE	14	I1	System Bus Address Latch Enable
IOCS16#	116	O8	System Bus I/O Chip Select 16
IOCHRDY	118	O8	System Bus I/O Channel Ready
SYSCLK	128	I1	System Clock—Processor Clock Divide by 2
CS4#	7	O1	Chip Select 4—Normally for External Floppy Disk Controller
CS5#	8	O1	Chip Select 5—Normally HCS0# for IDE
CS6#	9	O1	Chip Select 6—Normally for External Floppy Disk Controller
CS7#	10	O1	Chip Select 7—Normally for External Floppy Disk Controller
CDAK4#	102	I1	DMA Acknowledge Forces —CS4 Active
XDDIR	120	I1	X Data Bus Transceiver Direction
XDIRS	5	O1	Modified X Data Bus Transceiver Direction Control Signal—Excludes Real Time Clock and Keyboard Controller Decodes
XDIRX	6	O1	X Data Bus Transceiver Control Signal—Includes All CS Decodes Generated On Chip

Figure 9.13 (*Continued*)

SIGNAL DESCRIPTIONS (Continued)

Signal Name	Pin Number	Signal Type	Signal Description
COMMON BUS I/O (Continued)			
XDEN#	4	O1	X Data Bus Transceiver Enable
TRI#	126	I4	Three-State Control Input—For All Outputs to Isolate Chip for Board Tests
ICT#	127	I4	In Circuit Test Mode Control

I/O LEGEND

Pin Type	mA	Type	Comment
O1	2	TTL	
O2	24	TTL	
O4	12	TTL-OD	Open Drain, Weak Pull-Up, No VDD Diode
O6	4	TTL-TS	Three-State
O7	24	TTL-TS	Three-State
O8	24	TTL-OD	Open Drain, Fast Active Pull-Up
I1	—	TTL	
I2	—	CMOS	
I4	—	TTL	30 kΩ Pull-Up
I5	—	TTL	Schmitt-Trigger
IO2	24	TTL-TS	Three-State
IO4	12	TTL-OD	Open Drain, Slow Turn-On
IO5	12	TTL-TS	Three-State
IO6	24	TTL-TS	Three-State, 30 kΩ Pull-Up

Figure 9.13 (*Continued*)

EXAMPLE 9.24

At what pin of communication connector J_{53} is serial data input? What type of buffer is used to receive this signal?

SOLUTION Looking at J_{53} on sheet 5 of the circuit diagram in Fig. 9.3, we see that SINA is input at pin 3 of the connector. This signal is buffered by the MC1489 device U_{56}.

The serial data transfer rates of ports A and B are set by a single on-chip baud-rate generator. The frequency of the clock used for baud-rate generation is set by the 18.432-MHz crystal attached between the $XTAL_I$ ($XTAL_1$) and $XTAL_O$ ($XTAL_2$) pins of the 82341. This clock signal is scaled within the 82341 to set the receive and transmit data transmission rates.

The parallel printer port signal section is identified as LPT (line printer) on the 82341 in Fig. 9.3. These lines are used to implement a *Centronics parallel printer interface* at connector J_{56}. The meaning of the signals at this parallel printer port are identified in Fig. 9.13. Here we find that data are output in parallel over print data lines PD_0 through PD_7. The printer is signaled that data are available on the PD lines by logic 0 at the $\overline{STB}$ output. In Fig. 9.3 this signal is found to be output at pin 1 of connector J_{56}. Signals are also input to the microcomputer through the Centronics interface. For example, when the printer is busy and cannot accept additional data, it signals this fact to the microcomputer with the BUSY input. BUSY enters at pin 21 of the connector and is applied to the BUSY input of the 82341 (pin 72).

> **EXAMPLE 9.25**
>
> In the PC/AT microcomputer, what interrupt level is used to service the Centronics printer interface?
>
> **SOLUTION** From the circuits on sheet 5 of Fig. 9.3, we find that the IRQP output of the 82341 is sent to the IRQ_7 input of the 82344. Therefore, interrupt level 7 is used to service a printer attached to the parallel printer interface.

The keyboard section of the 82341 provides several functions. First, it produces three control outputs. On sheet 5 of Fig. 9.3, the signals at these outputs are identified as $\overline{RC}$, A20GATE, and TURBREQ. From the pin descriptions in Fig. 9.13, we find that they stand for reset, address bit A_{20} gate, and speed select. Signals $\overline{RC}$ and A20GATE are traced to inputs of the 82346 system controller on sheet 3 of Fig. 9.3. In Fig. 9.7 we find that a 1-to-0 transition at the $\overline{RC}$ input initiates a reset of the 80386DX MPU. Moreover, the system controller uses the A20GATE input to generate the $\overline{BLKA_{20}}$ output. Tracing the $\overline{BLKA_{20}}$ output of the 82346 in Fig. 9.3, we find that it goes two places. First, $\overline{BLKA_{20}}$ is input to the 82344 ISA bus controller. When at the active 0 logic level, it forces LA_{20} and SA_{20} to logic 0. On sheet 1 of Fig. 9.3 we find that $\overline{BLKA_{20}}$ is also input at pin 1 of AND gate U_{14}. Here it is used to gate address bit A_{20} to the 82385DX cache controller and the address latch circuit.

> **EXAMPLE 9.26**
>
> What is the signal mnemonic for the keyboard port input of the 82341 supplied with jumper J_{51}?
>
> **SOLUTION** Jumper J_{51} on sheet 5 of Fig. 9.3 controls the input at pin 92. In Fig. 9.13 the signal name for pin 92 is found to be KCM.

Keyboard entry information is input to the MPU through a synchronous serial interface made with signals keyboard clock (KBCLK) and keyboard data (KBDAT). Whenever a keycode has been read, the keyboard controller section of the 82341 issues an interrupt to the 80386DX. In Fig. 9.3 (sheet 5) we see that the interrupt request is output on the IRQ_1 line. The service routine for this interrupt initiates reading of the keycode from the keyboard controller.

The last section of the 82341 we examine is the *Integrated Drive Electronics* (IDE) hard disk bus interface. IDE is an industry standard interface for connecting hard disk drives to a microcomputer system. In the circuit diagram of Fig. 9.3 (sheet 5), the connector to the IDE interface is labeled J_{54}. Notice that EDI implements a 16-bit parallel data path between the hard disk subsystem and the microcomputer. This path consists of system data bus lines SD_0 through SD_6 and

SD$_8$ through SD$_{15}$. Data bit 7 is not returned to the MPU; instead, it is supplied to the IDB$_7$ input at pin 119 of the 82341.

The IDE interface of the 82341 controls the 74ALS245 transceivers used in the data path. For instance, in the circuit diagram, we find that the transceiver for the upper part of the data bus, U$_{52}$, is enabled by the signal $\overline{\text{IDENH}}$. The direction in which data are passed through the transceivers is determined by signal $\overline{\text{IOR}}$. In this way we see that hard disk data transfers are actually performed with input/output bus cycles.

A number of control inputs and outputs are also supplied at the IDE interface. Some examples of outputs are address strobe BALE, read/write signals $\overline{\text{IOR}}$ and $\overline{\text{IOW}}$, and address bits SA$_0$ through SA$_2$. These signals are first buffered with a 74ALS244 IC (U$_{54}$) and then supplied to the IDE interface. IDE interrupt request signal IDINT is an input from the interface.

EXAMPLE 9.27

At what pin of the IDE connector is the signal BALE output?

SOLUTION Figure 9.3 (sheet 5) shows that BALE is applied to pin 28 of the IDE connector.

9.9 82077AA FLOPPY DISK CONTROLLER

The last IC we discuss is the 82077AA *floppy disk controller*. This device is not part of the 82340 PC/AT chip set, but it is used in the microcomputer of Fig. 9.3. In IBM's original 80286-based PC/AT microcomputer, the interface to the floppy disk drive was provided by an add-in card plugged into an ISA bus slot. Due to the high integration of the 82340 ICs, space is available on the main processor board for providing the floppy disk control function.

The 82077AA is a versatile, highly integrated, 100% compatible solution for implementing the floppy disk interface in a PC/AT-compatible microcomputer. For instance, the 82077AA can control up to four drives, supports both 5.25- and 3.5-inch drives and can transfer data at four rates: 250 kilobits/sec (kbps), 300 kbps, 500 kbps, and 1 megabit/sec (Mbps). The 82077AA also contains the circuitry needed to implement a *tape drive controller*. Finally, the buffers for the microprocessor interface and drivers for the disk drive interface are built into the device. For this reason a PC/AT floppy disk interface can be completely implemented without any additional circuitry.

Block Diagram of the 82077AA

Let us begin our study of the 82077AA floppy disk controller with the block diagram of Fig. 9.14. Unlike the 82340 chips we have been examining, the 82077AA is architectured like a traditional peripheral IC. That is, it has a standard *micro-*

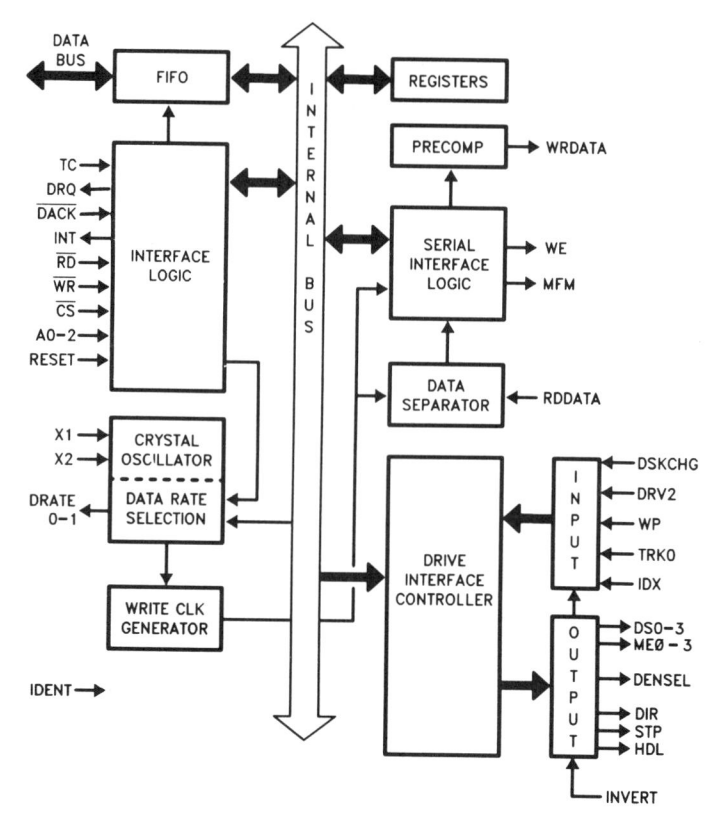

Figure 9.14 Block diagram of the 82077AA. (Reprinted by permission of Intel Corp. Copyright Intel Corp. 1989.)

processor interface. This interface consists of an eight-bit data bus, control signals such as $\overline{RD}$ and $\overline{WR}$, register select inputs A_0 through A_2, and a chip select input $\overline{CS}$. Data, configuration information, and status information are written into or read from the 82077AA's internal registers through this interface.

Notice in Fig. 9.14 that a first-in first-out (FIFO) buffer is built into the data bus interface. This FIFO is 16 bytes deep. After reset, the FIFO is disabled. Once the 82077AA is initialized, the FIFO can be either left disabled or enabled under software control. When the FIFO is enabled, all input and output transfers of data, command, or status information go through the FIFO. The FIFO has a programmable threshold level that when reached causes an interrupt request to be output. This interrupt request is output on the INT line and can be returned to the MPU to signal that data must be read from or written to the floppy disk controller. In this way we see that incoming or outgoing information is buffered within the floppy disk controller. This buffering results in better bus utilization and higher bus performance.

In Fig. 9.14 we also see that the 82077AA has an on-chip *oscillator*. This oscillator is used to synchronize the operation of the circuitry within the 82077AA and produce the data rate clock. Earlier we pointed out that four data transmission rates are supported. Notice that the data rate is set by scaling the clock in the *data rate selection* block. This scale factor is set under software control.

At the other side of the block diagram in Fig. 9.14 we find the circuitry needed to drive and control the floppy disk interface. This includes the *drive interface controller, input and output buffers*, the *data separator*, the *serial interface logic*, and *precompensation circuit*. The drive interface controller section produces all of the control signals needed to interface floppy disk drives to the PC/AT's microcomputer. Notice that all of these inputs and outputs are buffered within the 82077AA. As pointed out earlier, these on-chip buffers eliminate the need for external buffer circuitry in the floppy disk interface.

The data separator is the section of the 82077AA where read data are received. A serial bit stream of data is sent by the floppy disk drive to the RDDATA input. The phase-locked loop circuitry of the data separator locks onto this serial stream of data and takes a sample of the data during a clock period called the *data window*. The data separator reads the serial bits of data, translates them to parallel form, and loads the parallel byte of data into the FIFO buffer to await transfer to the MPU. Remember that the MPU is not signaled to read data from the FIFO until it is filled to the threshold level.

To assure that reliable serial data transfers take place between the floppy disk drive and microcomputer, the 82077AA compensates for variations in the data read frequency and has a high tolerence to bit jitter. For instance, the frequency at which read data are sent to the controller may change due to drift in the speed of the motor that drives the floppy disk. Bit jitter relates to the shifting of the data bit in the data window. The amount of bit jitter that occurs depends on both the magnetic media being read and the operating characteristics of the drive. To achieve good tolerance to read-frequency deviation and bit jitter, the 82077AA's data separator employs a dual-analog-phase locked-loop design. In fact, this design eliminates the use of the external trimming circuitry that is needed with other floppy disk controller devices.

The write precompensator block, which is identified as PRECOMP in Fig. 9.14, is in the write path from the microcomputer to the floppy disk drive. Notice that write data are output from the precompensator to the drive over the WRDATA signal line. However the function of the write precompensator actually affects the amount of bit jitter that occurs when reading data from the drive. It turns out that certain bit patterns when read from a floppy disk exhibit more shifting than others. These patterns are known to result in bit jitter. The objective of the write precompensator is to detect these patterns before they are written to the drive and to shift the bits such as to compensate for the expected bit shift. That is, certain bits are automatically made earlier or later relative to surrounding bits. In this way, the bit shift experienced when reading the data back is compensated for and a lower level of bit jitter is achieved. This technique is known as *write precompensation*.

Inputs and Outputs of the 82077AA and Their Use in the PC/AT Microcomputer System

Having introduced the 82077AA floppy disk controller and its block diagram, let us continue by examining its input and output signals and how they are connected in a PC/AT-compatible microcomputer system. The input and output signals of the 82077AA are listed in Fig. 9.15. Notice that the signals are divided into three groups: the *host interface signals, disk control signals,* and *phase-locked loop signals.*

Earlier we found that the floppy disk controller attaches to an 80386DX microcomputer through its microprocessor interface. In the microcomputer of Fig. 9.3 (sheet 5), we find that the 82077AA's microprocessor interface attaches to the ISA system bus. That is, address inputs A_0 through A_2 are driven by system address lines SA_0 through SA_2, data bus lines D_0 through D_7 are attached to system data bus lines SD_0 through SD_7, and control inputs $\overline{RD}$ and $\overline{WR}$ are supplied by $\overline{IOR}$ and $\overline{IOW}$, respectively. The function of these signals are all described in the host interface group in Fig. 9.15. For instance, the address code $A_2 A_1 A_0$ determines which one of the registers within the 82077AA is to be accessed. An example is that register address 101 must be output to the device whenever data are read from or written into the data FIFO.

EXAMPLE 9.28

What is the source of the chip select signal that enables the microprocessor interface of the 82077AA in Fig. 9.3?

SOLUTION In the circuit diagram, the $\overline{CS}$ input at pin 6 of the 82077AA is traced back to chip select output $\overline{CS}_4$ of the 82341 PC/AT peripheral combo IC (U_{50}). This output is labeled $\overline{FDCS}$ for floppy drive chip select.

The microprocessor interface supports DMA transfers of read or write data between the data bus FIFO and memory. The 82077AA's DMA handshake signals are the DMA request (DRQ) output at pin 24 and DMA acknowledge ($\overline{DACK}$) input at pin 3. If the 82077AA in the circuit of Fig. 9.3 needs to request direct access of the ISA bus, it switches the DRQ output to logic 1. This signal is sent to the DRQ_2 input at pin 124 of the 82344 ISA bus controller. When the bus is available, the DMA controller within an 82344 responds that it is ready to initiate the DMA data transfer by making its $\overline{DACK}_2$ output (pin 2) logic 0. This signal is returned through NAND gates in IC U_{501} to the $\overline{DACK}$ input (pin 3) of the floppy disk controller. This completes the DMA request/acknowledge handshake sequence. Now the DMA data transfer operation is performed.

The floppy disk drive is attached to the microcomputer at connector J_{57} on sheet 5 of the circuit diagram in Fig. 9.3. The invert ($\overline{INVERT}$) pin of the 82077AA is an option select input for this interface. If it is set to logic 1, external buffer circuitry is needed. On the other hand, setting $\overline{INVERT}$ to 0 means that the internal buffers are in use. This is the reason $\overline{INVERT}$ (pin 35) is wired to ground in the microcomputer of Fig. 9.3.

Let us next look at some of the signals provided at this interface. Earlier

Symbol	Pin #	I/O	Description
HOST INTERFACE			
RESET	32	I	**RESET:** A high level places the 82077AA in a known idle state. All registers are cleared except those set by the Specify command.
$\overline{\text{CS}}$	6	I	**CHIP SELECT:** Decodes base address range and qualifies $\overline{\text{RD}}$ and $\overline{\text{WR}}$ inputs.
A0 A1 A2	7 8 10	I	**ADDRESS:** Selects one of the host interface registers: | A2 | A1 | A0 | | Register | |---|---|---|---|---| | 0 | 0 | 0 | R | Status Register A | | 0 | 0 | 1 | R | Status Register B | | 0 | 1 | 0 | R/W | Digital Output Register | | 0 | 1 | 1 | R/W | Tape Drive Register | | 1 | 0 | 0 | R | Main Status Register | | 1 | 0 | 0 | W | Data Rate Select Register | | 1 | 0 | 1 | R/W | Data (FIFO) | | 1 | 1 | 0 | | Reserved | | 1 | 1 | 1 | R | Digital Input Register | | 1 | 1 | 1 | W | Configuration Control Register |
DB0 DB1 DB2 DB3 DB4 DB5 DB6 DB7	11 13 14 15 17 19 20 22	I/O	**DATA BUS:** Data bus with 12 mA drive
$\overline{\text{RD}}$	4	I	**READ:** Control signal
$\overline{\text{WR}}$	5	I	**WRITE:** Control signal
DRQ	24	O	**DMA REQUEST:** Requests service from a DMA controller. Normally active high, but goes to high impedance in AT and Model 30 modes when the appropriate bit is set in the DOR.
$\overline{\text{DACK}}$	3	I	**DMA ACKNOWLEDGE:** Control input that qualifies the $\overline{\text{RD}}$, $\overline{\text{WR}}$ inputs in DMA cycles. Normally active low, but is disabled in AT and PS/2 Model 30 modes when the appropriate bit is set in the DOR.
TC	25	I	**TERMINAL COUNT:** Control line from a DMA controller that terminates the current disk transfer. TC is accepted only while $\overline{\text{DACK}}$ is active. This input is active high in the AT, and PS/2 Model 30 modes and active low in the PS/2™ mode.
INT	23	O	**INTERRUPT:** Signals a data transfer in non-DMA mode and when status is valid. Normally active high, but goes to high impedance in AT, and Model 30 modes when the appropriate bit is set in the DOR.
X1 X2	33 34		**CRYSTAL 1,2:** Connection for a 24 MHz fundamental mode parallel resonant crystal. X1 may be driven with a MOS level clock and X2 would be left unconnected.

Figure 9.15 82077AA input/output signals. (Reprinted by permission of Intel Corp. Copyright Intel Corp. 1989.)

Symbol	Pin #	I/O	Description
HOST INTERFACE (Continued)			
IDENT	27	I	**IDENTITY:** Upon Hardware RESET, this input (along with MFM pin) selects between the three interface modes. After RESET, this input selects the type of drive being accessed and alters the level on DENSEL. The MFM pin is also sampled at Hardware RESET, and then becomes an output again. Internal pull-ups on MFM permit a no connect.

IDENT	MFM	INTERFACE
1	1 or NC	AT Mode
1	0	ILLEGAL
0	1 or NC	PS/2 Mode
0	0	Model 30 Mode

AT MODE: Major options are: enables DMA Gate logic, TC is active high, Status Registers A & B not available.
PS/2 MODE: Major options are: No DMA Gate logic, TC is active low, Status Registers A & B are available.
MODEL 30 MODE: Major options are: enable DMA Gate logic, TC is active high, Status Registers A & B available.
After Hardware reset this pin determines the polarity of the DENSEL pin. IDENT at a logic level of "1", DENSEL will be active high for high (500 Kbps/1 Mbps) data rates (typically used for 5.25" drives). IDENT at a logic level of "0", DENSEL will be active low for high data rates (typically used for 3.5" drives).

Symbol	Pin #	I/O	Description
DISK CONTROL (All outputs have 40 mA drive capability)			
INVERT	35	I	**INVERT:** Strapping option. Determines the polarity of **all** signals in this section. Should be strapped to ground when using the internal buffers and these signals become active LOW. When strapped to VCC, these signals become active high and external inverting drivers and receivers are required.
ME0 ME1 ME2 ME3	57 61 63 66	O	**ME0-3:** Decoded Motor enables for drives 0-3. The motor enable pins are directly controlled via the Digital Output Register.
DS0 DS1 DS2 DS3	58 62 64 67	O	**DRIVE SELECT 0-3:** Decoded drive selects for drives 0-3. These outputs are decoded from the select bits in the Digital Output Register and gated by ME0-3.
HDSEL	51	O	**HEAD SELECT:** Selects which side of a disk is to be used. An active level selects side 1.
STEP	55	O	**STEP:** Supplies step pulses to the drive.
DIR	56	O	**DIRECTION:** Controls the direction the head moves when a step signal is present. The head moves toward the center if active.
WRDATA	53	O	**WRITE DATA:** FM or MFM serial data to the drive. Precompensation value is selectable through software.
WE	52	O	**WRITE ENABLE:** Drive control signal that enables the head to write onto the disk.

Figure 9.15 (*Continued*)

Symbol	Pin #	I/O	Description
DISK CONTROL (All outputs have 40 mA drive capability) (Continued)			
DENSEL	49	O	**DENSITY SELECT:** Indicates whether a low (250/300 Kbps) or high (500 Kbps/1 Mbps) data rate has been selected.
DSKCHG	31	I	**DISK CHANGE:** This input is reflected in the Digital Input Register.
DRV2	30	I	**DRIVE2:** This indicates whether a second drive is installed and is reflected in Status Register A.
TRK0	2	I	**TRACK0:** Control line that indicates that the head is on track 0.
WP	1	I	**WRITE PROTECT:** Indicates whether the disk drive is write protected.
INDX	26	I	**INDEX:** Indicates the beginning of the track.
PLL SECTION			
RDDATA	41	I	**READ DATA:** Serial data from the disk. INVERT also affects the polarity of this signal.
HIFIL	38	I/O	**HIGH FILTER:** Analog reference signal for internal data separator compensation. This should be filtered by an external capacitor to LOFIL.
LOFIL	37	I/O	**LOW FILTER:** Low noise ground return for the reference filter capacitor.
MFM	48	I/O	**MFM:** At Hardware RESET, aids in configuring the 82077AA. Internal pull-up allows a no connect if a "1" is required. After reset this pin becomes an output and indicates the current data encoding/decoding mode (Note: If the pin is held at logic level "0" during hardware RESET it must be pulled to "1" after reset to enable the output. The pin can be released on the falling edge of hardware RESET to enable the output). MFM is active high (MFM). MFM may be left tied low after hardware reset, in this case the MFM function will be disabled.
DRATE0 DRATE1	28 29	O	**DATARATE0–1:** Reflects the contents of bits 0,1 of the Data Rate Register. (Drive capability of +6.0 mA @ 0.4V and −4.0 mA @ 2.4V)
PLL0	39	I	**PLL0:** This input optimizes the data separator, for either floppy disks or tape drives. A "1" (or V_{CC}) selects the floppy mode, a "0" (or GND) selects tape mode.

Figure 9.15 (*Continued*)

we found that read and write data signals are input and output over the RDDATA and WRDATA lines of the 82077AA, respectively. Looking at sheet 5 of Fig. 9.3, we find that RDDATA is received from the floppy disk drive at pin 30 of connector J_{57} and is input to the data separator of the 82077AA from pin 41. The interface also includes many control inputs and outputs. For example, line ME_0 is the enable signal for the motor connected as drive 0. Logic 1 is output on this line under software control to turn on the motor. An example of a control input is track zero (TRK_0). This input is at its active 1 logic level whenever the drive's read/write head is located in the track 0 position. The microcomputer can identify when the head is positioned at track 0 by polling this input.

EXAMPLE 9.29

What does logic 0 at pin 28 of J_{57} in Fig. 9.3 mean?

SOLUTION The logic 0 at pin 28 of J_{57} is applied to the WP input of the 82077AA. This means that the diskette in the floppy disk drive is not write protected.

ASSIGNMENT

Section 9.2

1. Name the three system buses of the original PC/AT.
2. Into what two other buses is the system data bus divided?
3. How much RAM is implemented on the system processor board? What size DRAMs are in use?
4. How much ROM is provided on the system processor board? What size EPROM is used?
5. What functions are performed by the clock generator block diagram?
6. What range of I/O addresses are dedicated to the master interrupt controller?
7. Which I/O addresses are reserved for DMA controller 2?
8. What I/O address range is assigned to the DMA page register?
9. Which DMA channel is used for cascading DMA controller 1 to DMA controller 2?
10. What functions are performed by timer 0? Timer 1? Timer 2?
11. Which output line is used to enable the speaker? Which input line is used to input the state of the enable signal for the speaker?
12. Over which output line is the parity check circuitry enabled? At which input is the state of the parity check signal read?
13. What are the two sources of the NMI signal?
14. What is assigned to the lowest-priority interrupt request at the master interrupt controller?
15. What is assigned to the highest-priority interrupt request at the slave interrupt controller?

Section 9.3

16. Which devices make up the core of the 80386DX microcomputer in Fig. 9.2?
17. List the numbers and names of the ICs in the 82340 PC/AT chip set.
18. What does *ISA* stand for?
19. Does the 82341 attach to the local bus of the 80386DX or the ISA bus?

Section 9.4

20. Locate the source of the CLK_2 input of the 80386DX, 80387DX, and 82385DX in the circuit of Fig. 9.3?
21. What effect will logic 0 have at the $\overline{BLKA_{20}}$ input at pin 1 of AND gate U_{14} (see sheet 1 of Fig. 9.3)?
22. To which pins of the 80387DX do the $\overline{W/R}$, $\overline{M/IO}$, and $\overline{ADS}$ outputs of the 80386DX attach?
23. What signals are applied to the $\overline{BE_0}$ through $\overline{BE_3}$ inputs of the 82385DX? What are the sources of these signals?
24. Is the 82385DX in Fig. 9.3 set up for master or slave operation?
25. What signal drives the FLUSH input of the 82385DX?

26. What SRAM device is employed in the cache memory array of the core microcomputer in Fig. 9.3?
27. What is the size of the cache memory array in the core microcomputer of Fig. 9.3?
28. What are the sources of the $\overline{\text{BROYO}}$ and $\overline{\text{BROYE}}$ inputs of the ready/wait state PAL™?
29. What determines the input conditions for which the $\overline{\text{NCA}}$ input of the 82385DX is logic 0?
30. Which IC is used to latch CPU address signals CA_{26} through CA_{31}?
31. Which latch IC produces host address signals BA_{10} through BA_{17}?
32. What output of the 82385DX produces $\overline{\text{AOE}}$ for the address latches? At what pin of the 82385DX IC is it output?
33. To which input of the 74F374s is the BACP output of the 82385DX applied? What is the pin number of this input?
34. What device provides the signal PUE?
35. Which transceiver IC interfaces host data bus lines D_{16} through D_{23} to the CPU data bus?
36. What signal drives the $\overline{G}$ inputs of the 74F646 transceivers? What is the source of this signal?

Section 9.5

37. What does PQFP stand for?
38. What information can the SD bus multiplexer switch to the SD bus?
39. What are the sources of input data for the D bus multiplexer?
40. At what pin is the signal $\overline{\text{ROMCS}}$ input to the 82345?
41. Is the signal $\overline{\text{XDREAD}}$ an input or output? What function does it provide?
42. List the names, size, and signals of each of the four data bus interfaces of the 82345.
43. Which data buses provided by the 82345 are part of the ISA bus?
44. Which pins of the 82345 provide the signals SD_0 through SD_{15}?
45. If jumper J_{40} is installed on sheet 4 of Fig. 9.3, what mode of operation is selected?
46. What is the source of the CPST_WRC input of the 82345 on sheet 2 in Fig. 9.3?
47. What does SD_1 stand for?
48. To which pin of the 82346 is the $\overline{\text{LATHI}}$ input of the 82345 attached?
49. What direction of XD bus operation is selected by $\overline{\text{XDREAD}}$ equal 1?

Section 9.6

50. How many pins are on the package of the 82346?
51. If the 80386DX in a PC/AT microcomputer is to operate at 25 MHz, what frequency clock must be applied to the $TCLK_2$ input?
52. Are the internal configuration control registers of the 82346 located in the memory or I/O address space of the PC/AT microcomputer?
53. List the output signals of the numeric coprocessor interface.
54. Which pin of the 82346 is the $D/\overline{C}$ input?

55. In the circuit of Fig. 9.3, which IC is the source of the $\overline{W/R}$, $\overline{D/C}$, and $\overline{M/IO}$ inputs of the 82346?
56. To which pin of the 82346 must output $\overline{BBE}_2$ of the 82385DX cache controller be applied?
57. How is the $\overline{CLK}_{2IN}$ input of the 82346 normally connected?
58. What does $\overline{READYI}$ stand for? What is the source of this signal in the circuit diagram on sheet 3 of Fig. 9.3?
59. What is the destination of the RESCPU output of the 82346 on sheet 3 of Fig. 9.3?
60. What is the source of the address inputs of the 82346 when the HLDA input is active?
61. Which of the on-board memory system interface signals are applied to inputs of the 82345 data buffer?
62. Which ICs are used to buffer memory address lines MA_0 through MA_{10} to produce buffered memory address lines MA_{0H} through MA_{10H}?
63. Which banks of DRAM SIMs are addressed with the buffered memory address lines?
64. Which bank of DRAM SIMs are supplied by $\overline{RAS}_3$?
65. What size DRAM SIMs can be used in the DRAM array on sheet 6 of Fig. 9.3?
66. Which SIMs make up bank 2 of the DRAM array?
67. If $CASBK_2$, LBE_0, and LBE_1 are at the 1 logic level and the other CAS and LBE signals are at logic 0, which CAS inputs to the DRAM SIM array are active? Over which data bus lines are data carried? Is a byte, word, or double-word data transfer taking place?
68. At what pin of the 82346 is the RES_{387} signal output? Where is this signal sent?
69. Over which data bus lines does the 80386DX MPU access the internal configuration registers of the 82346?

Section 9.7

70. Which circuit block of the 82344 receives the POWERGOOD signal that identifies that the power supply voltage is stable?
71. Make a list of the LSI peripheral devices that are implemented within the 82344 peripheral control block.
72. What are the outputs of the address decoder block of the 82344?
73. At what pins of the 82344 are the $\overline{BE}_0$ through $\overline{BE}_3$ signals located?
74. Is HLDA an input or output of the 82344? In the PC/AT circuit diagrams of Fig. 9.3, where is the HLDA pin of the 82344 connected?
75. If the microcomputer in Fig. 9.3 is strapped for 16-bit BIOS, which EPROM sockets are populated? What types of EPROM devices are normally used?
76. Over which data lines are information output to the 80386DX MPU when the 82344 is strapped for 16-bit BIOS?
77. If the microcomputer in Fig. 9.3 is configured to provide BIOS in two 27C256 EPROMs, what should be the setting of jumpers J_{30}, J_{31}, and J_{32}?
78. What is the purpose of the ISA bus connectors on the main processor board? How many 16-bit data bus slots are provided?
79. How large an address is implemented at the ISA bus?
80. At what pin of the ISA bus connector is the signal DRQ_0 input?

81. Which signal is output at pin B26 of the ISA bus?
82. In the microcomputer of Fig. 9.3, what signal is applied to the VBAT input of the 82344? What is the source of this signal?
83. At what pin of the 82344 is the signal SDLH/$\overline{\text{HL}}$ output?
84. What does logic 0 at $\overline{\text{XREAD}}$ tell the data buffer?
85. What type of error is signaled to the 82344 when the $\overline{\text{PCK}}$ input switches to logic 0? Which interrupt is initiated due to this input?
86. Is $\overline{\text{HIDRIVE}}$ an input or output of the 82344? What happens when jumper J_{40} on sheet 4 of the circuits in Fig. 9.3 is installed?
87. Which of the 344/346 interface signals on sheet 4 of Fig. 9.3 is used to request a refresh operation of the 82346?

Section 9.8

88. At what pin of the 82341 is the signal SINB output?
89. Which connector implements asynchronous serial port B in the PC/AT microcomputer of Fig. 9.3?
90. Which IC is used to buffer serial output SOUTB of the 82341 in the microcomputer of Fig. 9.3?
91. In the microcomputer of Fig. 9.3 at which pin of the COM B connector is the signal data set ready port B input?
92. At what pin of the 82341 is the select signal ($\overline{\text{SLIN}}$) output to the printer?
93. Is the signal $\overline{\text{ERR}}$ an input or output of the Centronics interface?
94. At what pin of the Centronics interface does the printer report the fact that a paper error has occurred?
95. Trace the destination of the TURBREQ output of the 82341.
96. Is the KKSW pin of the 82341 an input or output of the keyboard section? What does this signal stand for?
97. What level of interrupt is used to service the keyboard in the PC/AT microcomputer of Fig. 9.3?
98. What is the data path of the IDE interface in the microcomputer of Fig. 9.3?
99. Which ICs are the IDE data transceivers?
100. What level of interrupt request is used to request service for the hard disk in the microcomputer of Fig. 9.3?
101. At what pin of the IDE connector is the signal $\overline{\text{HCS}}_1$ output?

Section 9.9

102. How many floppy disk drives can be controlled by the 82077AA?
103. What physical size drives can be operated with the 82077AA?
104. List four data transfer rates supported by the 82077AA.
105. How large is the data bus FIFO of the 82077AA?
106. What is the period during which the 82077AA's data separator reads serial data called?

Chap. 9   Assignment   491

107. What technique does the 82077AA use to correct for bit jitter?

108. Is the 82077AA device in Fig. 9.3 treated as a memory-mapped or I/O-mapped peripheral?

109. What level of interrupt is used to service the floppy disk in the microcomputer of Fig. 9.3?

110. Which DMA channel is used to service the floppy disk controller in the circuit of Fig. 9.3?

111. What frequency crystal is attached to the 82077AA floppy disk controller in Fig. 9.3?

112. How many floppy disk drives are supported with the interface at connector J_{57} in Fig. 9.3?

113. At what pin of connector J_{57} on sheet 5 of Fig. 9.3 are write data output to the floppy disk drive?

114. What does logic 1 at the INDX input of the 82077AA mean? At what pin of connector J_{57} in the microcomputer of Fig. 9.3 is this signal input?

Bibliography

Bradley, David J., *Assembly Language Programming for the IBM Personal Computer.* Englewood Cliffs, NJ: Prentice Hall, 1984.

Ciarcia, Steven, The Intel 8086, *Byte*, Nov. 1979.

Coffron, James W., *Programming the 8086/8088.* Berkeley, CA: Sybex Inc., 1983.

Intel Corporation, *Components Data Catalog.* Santa Clara, CA: Intel Corporation, 1980.

Intel Corporation, *80386 Microprocessor Hardware Reference Manual.* Santa Clara, CA: Intel Corporation, 1987.

Intel Corporation, *80386 Programmer's Reference Manual.* Santa Clara, CA: Intel Corporation, 1987.

Intel Corporation, *80386 System Software Writer's Guide.* Santa Clara, CA: Intel Corporation, 1987.

Intel Corporation, *iAPX86,88 User's Manual.* Santa Clara, CA: Intel Corporation, July 1981.

Intel Corporation, *Introduction to the 80386.* Santa Clara, CA: Intel Corporation, Sept. 1985.

Intel Corporation, *MCS-86™ User's Manual.* Santa Clara, CA: Intel Corporation, Feb. 1979.

Intel Corporation, *Memory.* Santa Clara, CA: Intel Corporation, 1989.

Intel Corporation, *Microprocessor.* Santa Clara, CA: Intel Corporation, 1989.

Intel Corporation, *Peripheral.* Santa Clara, CA: Intel Corporation, 1989.

Intel Corporation, *Peripheral Design Handbook.* Santa Clara, CA: Intel Corporation, Apr. 1978.

Morse, Stephen P., *The 8086 Primer.* Rochelle Park, NJ: Hayden Book Company, Inc., 1978.

Rector, Russell, and George Alexy, *The 8086 Book*. Berkeley CA: Osborne/McGraw-Hill, 1980.

Scanlon, Leo J., *IBM PC Assembly Language*. Bowie, MD: Robert J. Brady Co., 1983.

Schneider, Al, *Fundamentals of IBM PC Assembly Language*. Blue Ridge Summit, PA: Tab Books, Inc., 1984.

Texas Instruments Incorporated, *Programmable Logic Data Book*. Dallas, TX: Texas Instruments Incorporated, 1990.

Triebel, Walter A., *Integrated Digital Electronics*. Englewood Cliffs, NJ: Prentice Hall, 1979.

Triebel, Walter A., and Alfred E. Chu, *Handbook of Semiconductor and Bubble Memories*. Englewood Cliffs, NJ: Prentice Hall, 1982.

Triebel, Walter A., and Avtar Singh, *The 8088 Microprocessor: Programming, Interfacing, Software, Hardware, and Applications*. Englewood Cliffs, NJ: Prentice Hall, 1989.

Triebel, Walter A., and Avtar Singh, *The 8086 and 80286 Microprocessors: Architecture, Software, and Interface Techniques*. Englewood Cliffs, NJ: Prentice Hall, 1990.

Triebel, Walter A., and Avtar Singh, *IBM PC/8088 Assembly Language Programming*. Englewood Cliffs, NJ: Prentice Hall, 1985.

Willen, David C., and Jeffrey I. Krantz, *8088 Assembler Language Programming: The IBM PC*. Indianapolis, IN: Howard W. Sams & Co., Inc., 1983.

Answers to Selected Assignments

CHAPTER 1

Section 1.2

1. Personal computer advanced technology.
3. A reprogrammable microcomputer is a general-purpose computer designed to run programs for a wide variety of applications, such as accounting, word processing, and languages such as BASIC or FORTRAN.

Section 1.3

9. 32-bit microprocessor.
11. Video display and printer.
15. 64K bytes; 512K bytes.

Section 1.4

17. 4-bit, 8-bit, 16-bit, and 32-bit.
19. 8086, 8088, 80186, 80188, 80286, 80386DX.
25. Real mode and protected mode.
27. Memory management, protection, and multitasking.

CHAPTER 2

Section 2.2

1. Software.
3. 80386DX machine code.
5. Mnemonic that identifies the operation to be performed by the instruction; ADD and MOV.
7. START: ADD EBX TO EAX

Section 2.3

13. Aid to the assembly language programmer for understanding the 80386DX's software operation.
15. 17.

Section 2.4

19. $FFFFF_{16}$ and 00000_{16}.
21. 1024 bytes.
23. 44332211_{16}; misaligned double word.

Section 2.5

27. (a) 7FH.
    (b) F6H.
    (c) 80H.
    (d) 01F4H.
29. -1000 = 2's complement of 1000
             = FC18H
31. (0B000H) = 27H
    (0B001H) = 01H

Section 2.6

37. 384K bytes.
39. Instructions of the program can be stored anywhere in the memory address space.

Section 2.7

43. 16 bytes.

Section 2.8

45. BX; EBX.
49. SS.
51. 16- or 32-bit data; 16-bit addresses.

Section 2.9

55. Trap flag.
57. Instructions are provided that can load the complete register or modify specific flag bits.

Section 2.10

59. Offset and segment base.
63. $021AC_{16}$.
65. 1234_{16}.

Section 2.11

67. $CFF00_{16}$.
69. Contents of address $CFEFE_{16}$ equals $11EE_{16}$.

Section 2.12

70. Register operand addressing mode.
    Immediate operand addressing mode.
    Memory operand addressing modes.
74. (a) PA = $0B200_{16}$.
    (b) PA = $0B100_{16}$.
    (c) PA = $0B700_{16}$.
    (d) PA = $0B600_{16}$.
    (e) PA = $0B900_{16}$.

CHAPTER 3

Section 3.2

1. Basic instruction set; extended instruction set.
3. System control instruction set.

Section 3.3

5. (a) Value 0110H is moved into AX.
   (b) 0110H is copied into DI.
   (c) 10H is copied into BL.
   (d) 0110H is copied into memory address DS:0100H.
   (e) 0110H is copied into memory address DS:0120H.
   (f) 0110H is copied into memory address DS:0114H.
   (g) 0110H is copied into memory address DS:0124H.
7. MOV [1010H], ES.
9. The byte of data in BL is sign-extended to 32 bits and copied into register EAX.

15. The first 32 bits of the 48-bit pointer starting at memory address DATA_F_ADDRESS is loaded into EDI and the next 16 bits are loaded into the FS register.

Section 3.4

19. ADC   EDX, 111FH.
20. SBB   AX, [BX]
23. DAA.
25. AX = FFA0H.
26. AX = 7FFFH, DX = 0000H.

Section 3.5

29. (a) DS:300H = 0AH.
    (b) DX = A00AH
    (c) DS:210H = FFFFH.
    (d) DS:220H = F5H.
    (e) AX = AA55H.
    (f) DS:300H = 55H.
    (g) DS:300H = 55H, DS:301H = 55H.
31. AND   WORD PTR [100H], 80H.

Section 3.6

37. MOV   CL, 08H
    SHL   WORD PTR [DI], CL
40. Double precision shift right.
41. The contents of the double-word memory location starting at the address DS:DI is shifted right by the number of bit positions specified by the count in CL. Bits are shifted from the LSB of EAX into the MSB of the memory location identified by DS:DI. The last bit shifted out of the LSB of the memory location is saved in the CF.

Section 3.7

43. (a) DX = 2222H, CF = 0.
    (b) DS:400H = 5AH, CF = 1.
    (c) DS:200H = 11H, CF = 0.
    (d) DS:210H = AAH, CF = 1.
    (e) DS:210H, 211H = D52AH, CF = 1.
    (f) DS:220H, 221H = AAADH, CF = 0.
44. RCL   WORD PTR [BX], 1.

Section 3.8

49. [DS:100H] = 00F7H and CF = 1.
51. BSR   EAX, DOUBLE WORD PTR [SI].

CHAPTER 4

Section 4.2

1. Executing the first instruction causes the contents of the status register to be copied into AH. The second instruction causes the value of the flags to be saved in memory at the location pointed to by DS:BX + DI.
3. STC; CLC.

Section 4.3

10. Set byte if not carry.
11. ZF = 0.
13. If the execution of the preceding instruction has set OF, FF_{16} is written to the memory location pointed to by the value of OVERFLOW.

Section 4.4

17. 8-bit; 16-bit; 16-bit; 32-bit.
21. ZF, CF, SF, PF, and OF.
23. CF = 0 and Z = 0.
25. 0100H.

29.
```
            MOV    CX, 64H
            MOV    AX, 0H
            MOV    DS, AX
            MOV    BX, A000H
            MOV    SI, B000H
            MOV    DI, C000H
    AGAIN:  MOV    AX, [BX]
            CMP    AX, 0H
            JGE    POSTV
    NEGTV:  MOV    [DI], AX
            INC    DI
            INC    DI
            JMP    NXT
    POSTV:  MOV    [SI], AX
            INC    SI
            INC    SI
    NXT:    DEC    CX
            JNZ    AGAIN
            HLT
```

Section 4.5

33. The intersegment call provides the ability to call a subroutine in either the current code segment or a different code segment. On the other hand, the intrasegment call only allows calling of a subroutine in the current code segment.
37. At the end of the subroutine a RET instruction is used to return control to the main

program. It does this by popping IP from the stack in the case of an intrasegment call and both CS and IP for an intersegment call.

39. SS:SP+1 = 10H
    SS:SP   = 00H
41. When the contents of the flags must be preserved for the instruction that follows the subroutine.

Section 4.6

45. ZF = 1 or CX = 0.
47. 63,535.

Section 4.7

51. ES.

53.
```
          MOV  AX, DATA_SEG   ;Establish Data segment
          MOV  DS, AX
          MOV  ES, AX         ;and Extra-segment to be the same
          CLD                 ;Select autoincrement mode
          MOV  CX, 64H        ;Set up array element counter
          MOV  SI, A000H      ;Set up source array pointer
          MOV  DI, B000H      ;Set up destination array pointer
          REPECMPSW           ;Compare while not end of string
                              ;and strings are equal
          JZ   EQUAL          ;Arrays are identical
          MOV  FOUND, SI      ;Save mismatch location in FOUND
          JMP  DONE
   EQUAL: MOV  FOUND, 0H      ;Arrays identical
   DONE:  ---
```

CHAPTER 5

Section 5.2

3. Defines the location and size of the global descriptor table.
5. System segment descriptor.
7. 0FFFH.
11. CR_0.
13. MP = 1, EM = 0, and R = 1.
17. 4K bytes.
19. Selector; selects a task state segment descriptor.
21. BASE and LIMIT of the TSS descriptor.
25. 00130020H.
27. Level 2.

Section 5.3

29. Selector and offset.
31. 64 TB, 16,384 segments.
37. 1,048,496 pages; 4096 bytes long.
39. Cache page directory and page table pointers on-chip.

Section 5.4

43. LIMIT = 00110H, BASE = 00200000H.
45. 00200226H.
49. Dirty bit.

Section 5.5

51. ```
    SMSW    AX
    AND     AX,F7H
    LMSW    AX
    ```

Section 5.6

53. The running of multiple processes in a time-shared manner.
57. Level 0, level 3.
59. LDT and GDT.
61. Level 0.
63. A task can access data in a data segment at the CPL and at all lower privilege levels. But it cannot access data in segments that are at a higher privilege level.
67. The call gate is used to transfer control within a task from code at the CPL to a routine at a higher privilege level.
69. Identifies a task state segment.

Section 5.7

75. Active, level 3.
77. Yes.

CHAPTER 6

Section 6.2

1. CHMOSIII.
3. INTR.
    ```

Section 6.3

5. Byte; D_0–D_7; no.
7. I/O data read.

Section 6.4

11. F12.

Section 6.5

13. 40 ns.
15. 4; 2; 80ns.
19. An extension of the current bus cycle by a period equal to one or more T states because the $\overline{READY}$ input was tested and found to be logic 1.

Section 6.6

21. 160 ns.

Section 6.7

25. Higher addressed byte.

Section 6.8

27. $M/\overline{IO}/\overline{C}W/\overline{R}$ = 111; all four; $\overline{MWTC}$.
33. $\overline{DEN}$ = 0, $DT/\overline{R}$ = 0.
35. 74F139.

Section 6.9

39. Number of inputs, number of outputs, and number of product terms.
43. 20 inputs; 8 outputs.

Section 6.10

45. When the power supply for the memory device is turned off, its data contents are not lost.
47. Ultraviolet light.
51. The access time of the 27C64 is 250 ns and that of the 27C64-1 is 150 ns. That is, the 27C64-1 is a faster device.
53. 1 ms.

Section 6.11

55. Volatile.
57. 32K × 32 bits (1M bit).

Section 6.12

63. Pin 3 of the 74F08 AND gate.
65. Pin 6 of the 74AS27 NOR gate; if any of its inputs are 1.

Section 6.13

73. 1.14 wait states.
75. Page.
77. Least recently used.

Section 6.14

79. N4.
81. The 2W/$\overline{\text{D}}$ input must be set to the 1 logic level.
85. FLUSH.
87. 8K double words.
91. 32 bits.
93. Bus watching.

CHAPTER 7

Section 7.2

1. Isolated I/O and memory-mapped I/O.
3. Memory-mapped I/O.

Section 7.3

9. Word; D_8 through D_{23}.
11. $\overline{\text{IOWR}_3}\ \overline{\text{IOWR}_2}\ \overline{\text{IOWR}_1}\ \overline{\text{IOWR}_0}$ = 0011.
13. 32.

Section 7.4

17. 80 ns.
19. Five wait states.

Section 7.5

23. Execution of this output instruction causes the value in the lower byte of the accumulator (AL) to be loaded into the byte-wide output port at address $1A_{16}$.
27. 15 bytes of data are input from the input port at address $A000_{16}$ and saved in memory locations ES:1001H through ES:100FH.
29. I/O map base; word offset 66H from the beginning of the TSS.
31. 1.

Section 7.6

33. 24.

37. $D_0 = 1$ Lower four lines of PORT C are inputs.
     $D_1 = 1$ PORT B lines are inputs.
     $D_2 = 0$ MODE 0 operation for both PORT B and the lower four lines of PORT C.
     $D_3 = 1$ Upper four lines of PORT C are inputs.
     $D_4 = 1$ PORT A lines are inputs.
     $D_6D_5 = 00$ MODE 0 operation for both PORT A and the upper four lines of PORT C.
     $D_7 = 1$ Mode being set.

41. ```
    MOV   DX, 1000H
    MOV   AL, 92H
    OUT   DX, AL
    ```

43. To enable $INTR_B$, the INTE B bit must be set to 1. This is done with a bit set/reset operation that sets bit PC_4 to 1. This command is

$$D_7 - D_0 = 0XXX1001$$

Section 7.7

47. The value at the inputs of PORT C of I/O device 0 are read into AL.
49. ```
    IN    AL, 02H      ; READ PORT A
    MOV   BL, AL       ; SAVE IN BL
    IN    AL, 06H      ; READ PORT B
    ADD   AL, BL       ; ADD THE TWO NUMBERS
    OUT   0AH, AL      ; OUTPUT TO PORT C
    ```

Section 7.8

51. $\overline{BE_0} = 0$, $A_3A_2 = 01$, $A_6A_5A_4 = 001$, and $A_{14} = 1$; 04014_{16}.

Section 7.9

55. $\overline{STB}$ Input   Signals that a byte of data is available on D_0 through D_7.
    BUSY   Output Signals the MPU that the printer is busy printing a character and is not yet ready to receive another.

Section 7.10

59. $D_7 \cdots D_0 = 01011010_2$.
61.
```
MOV  AL, 5AH
MOV  DX, 100CH
MOV  [DX], AL
```
65. 838 ns; 500 ns.
67. $N = 48_{10} = 30_{16}$.

Section 7.11

69. No.
71. 27.
76. $0F_{16}$.

Section 7.12

79. $\overline{M/IO}A_{31}A_5A_4 = 0000$.
81. D_7 through D_0.

CHAPTER 8

Section 8.2

1. Hardware interrupts, nonmaskable interrupt, internal interrupts and exceptions, software interrupts, and reset.
5. Higher priority.

Section 8.3

7. Two words; four words.
9. 16-bit base address and 16-bit offset.
11. $IP_{40} \rightarrow A0H$ and $CS_{40} \rightarrow A2H$.

Section 8.4

15. 010000_{16}; 512 bytes; 64.
17. 190H through 197H.

Chapter 8

Section 8.5

```
21. CLI        ;DISABLE INTERRUPTS AT ENTRY POINT OF
               ;NONINTERRUPTABLE SUBROUTINE
     .
     .
     .         ;BODY OF SUBROUTINE
     .
     .
    STI        ;REENABLE INTERRUPTS AT END OF
               ;SUBROUTINE
    RET        ;RETURN TO MAIN PART OF PROGRAM
```

Section 8.6

23. Interrupt acknowledge.
27. The current interrupt request has been acknowledged; Put the type number of the highest-priority interrupt on the data bus.
29. D_0 through D_7.

Section 8.7

31. 1 μs.
35. Privilege level is not changed. It remains equal to that of the interrupted code; the current stack remains active ($SS_{OLD}:ESP_{OLD}$); the old flags, old CS, and old EIP are pushed to the stack.
37. EXT = 1 = external interrupt.
    IDT = 1 = error is due to interrupt.
    TI = 1 = local descriptor table.

Section 8.8

39. $D_0 = 0$ ICW_4 not needed.
    $D_1 = 1$ Single device.
    $D_3 = 0$ Edge triggered.
    and assuming that all other bits are logic 0 gives
    $$ICW_1 = 00000010_2 = 02_{16}$$
43. MOV AL, [A004H]

Section 8.9

47. 22.
49. 64.

Section 8.10

52. Vectors 0 through 31.
54. Address of an operand access for the stack segment of memory crosses the boundary of the stack.

CHAPTER 9

Section 9.2

3. 256K bytes, 128K × 1 bit DRAMs.
5. System clock, power-on reset, and wait-state generation.
7. 0C0H through 0DFH.
11. OUT$_0$, IN$_0$.
15. Real-time clock.

Section 9.3

17. 82345 data buffer, 82346 system controller, 82344 ISA controller, and 82341 peripheral combo.
19. ISA bus.

Section 9.4

23. $\overline{CBE}_0$ through $\overline{CBE}_3$; $\overline{BE}_0$ through $\overline{BE}_3$ outputs of the 80386DX.
25. GND.
27. 8K × 32 bits.
31. U$_{21}$.
33. CK; 11.
35. U$_{28}$.

Section 9.5

37. Plastic quad flat package.
41. Input; the 82346 uses this input to signal the 82345 whether the data transfer over the XD bus is a read (input) or write (output).
43. Peripheral X-data bus and system (slot) data bus.
47. System (slot) data bus bit 1.
49. The XD bus is set for output of data.

Section 9.6

51. 50 MHz.
55. U$_{23}$.
57. To the CLK$_2$ output of the 82346.
59. RESET input at pin C9 of the 80386DX.
63. Banks 2 and 3.
67. $\overline{CAS}_8$ and $\overline{CAS}_9$; MD$_0$ through M$_{15}$; word.
69. Peripheral data bus lines XD$_0$ through XD$_7$.

Section 9.7

71. (2) 82C59A, (2) 82C37A, and (1) 82C54.
73. Pins 69, 70, 71, and 72.
75. U_{37} and U_{38}; 27C256.
79. 24 bits.
81. $\overline{DACK_2}$.
83. Pin 102.
87. OUT_1.

Section 9.8

89. J_{55}.
91. Pin 2.
93. Input.
97. Level 1.
99. U_{52} and U_{53}.

Section 9.9

103. 5.25 inch and 3.5 inch.
107. Write precompensation.
111. 24 MHz.
113. Pin 22.

Index

Index

A

Access rights (byte), 175, 179, 180, 182, 187, 189, 192, 193, 198, 380
Access time, 219, 252
Address(es), 17, 18, 22–26, 33, 34, 35, 36, 38–42, 43, 46, 124–29, 209, 214, 215, 218, 219, 227, 251, 252, 259, 264, 266, 275, 285, 286, 288, 300, 302, 307, 308, 352, 354, 358, 363, 368, 386, 390, 400, 408
  base, 31, 33, 34, 35, 42, 51, 52, 165, 175, 177, 182, 377, 379
  buffer, 233, 236
  bus, 171, 208–11, 213, 216, 218, 219, 251, 252, 276, 278, 279, 285, 308, 385, 386
  cacheable, 290, 438
  decoder, 231, 240–43, 267, 300, 301, 366
  effective, 49–52, 53, 55, 56, 57, 63, 83, 382, 383
  generating an, 38–42, 165, 172, 173–78
  index(ed), 34, 35, 36, 51, 59, 60, 62, 63
  latch, 231, 233–37, 278, 283, 285, 300, 408, 439
  logical, 38, 42, 168, 172
  noncacheable, 290, 438
  offset, 24, 36, 38, 39, 76, 126
  physical, 36, 38, 39, 40, 41, 42, 43, 47, 48, 51, 53, 56, 58, 61, 76, 133, 134, 168, 172, 175, 176, 177
  segment (base), 38, 39, 40, 41, 42, 43, 47, 49, 51
  space, 11, 22, 30, 32, 158, 159, 162, 168, 170, 171, 172, 175, 185, 188, 210, 226, 227, 283, 287, 296, 297, 298, 308, 330, 338, 377, 379
  translation, 172, 173–75, 176–78
Addressing modes, 46–63, 72, 126, 127, 133, 297
  based, 36, 52, 56–59, 175
  based–indexed, 52, 63
  direct, 50, 52–53, 56, 76, 307
  immediate operand, 46, 47–48, 72
  indexed, 36, 52, 59–63
  memory operand, 46, 48–63
  register indirect, 52, 53–56
  register operand, 46–47
Aligned double-word boundary(ies), 24–25, 227, 229
Application (software) program, 191
Arithmetic instructions, 73, 85–98
  add (ADD), 16, 17, 34, 37, 87–88, 91, 92
  add with carry (ADC), 87, 89, 91
  adjust AX after multiply (AAM), 98

511

Arithmetic instructions (cont.)
    adjust AX before divide (AAD), 98
    ASCII adjust for addition (AAA), 91, 92
    ASCII adjust for subtraction (AAS), 94
    convert byte to word (CBW), 98
    convert double word to quad word (CDQ), 98
    convert word to double word (CWD/CWDE), 98
    decimal adjust for addition (DAA), 92
    decimal adjust for subtraction (DAS), 94
    decrement (DEC), 94, 144
    divide (DIV), 95, 413
    increment (INC), 87, 89–90, 91
    integer divide (IDIV), 95, 413
    integer multiply (IMUL), 72, 95–97
    multiply (MUL), 95–97
    negate (NEG), 94
    subtract (SUB), 16, 92, 94
    subtract with borrow (SBB), 93
Arithmetic shift, 100, 101
Array, 59–60, 63, 149, 309, 310, 383, 414
ASCII-to-EBCDIC conversion, 82
Assembler, 17, 52, 125, 126
Assembly language, 15, 16, 20, 21
    general format, 17
    program, 17, 20
    statement, 16, 17, 20
Autoindexing, 150

B

Back link selector, 198, 200
Bank write control logic, 231, 233, 269, 300, 306
BASIC, 6, 7, 20, 426
Bidirectional data bus, 210, 259, 264
BIOS, 426
Bit jitter, 483
Bit test and bit scan instructions, 72, 107–9
    bit scan forward (BSF), 109
    bit scan reverse (BSR), 109
    bit test (BT), 108–9
    bit test and reset (BTR), 109
Block move program, 129–31, 145, 147, 150–51
Breakpoints, 379, 411, 413, 414
Bus arbiter, 387
Bus bandwidth, 275

Bus control logic, 231, 243, 269, 300, 368, 386, 395, 408, 410, 422
Bus cycle, 214–15, 216, 217–26, 229, 230, 270, 272, 273, 290, 300, 301, 303–6, 307, 308, 330, 354, 356, 363, 389, 438
    DMA, 354, 356
    halt/shut down, 214
    hold state, 216
    idle state, 214, 220, 352, 384, 390
    instruction fetch, 214
    interrupt acknowledge, 214, 215, 386, 387, 389, 390, 395, 396, 398, 408
    I/O read (input), 300, 301, 302, 303–5, 314, 354, 366, 408
    I/O write (output), 300, 301, 302, 303, 305–6, 314, 354, 360, 408
    misaligned data, 25–26
    nonpipelined, 218, 221–25, 226, 303–6
    pipelined, 218, 219–20, 225–26
    read, 218, 221–23, 224, 225, 226, 279, 285, 286, 330, 356, 362, 386
    time states (T), 218
    verify, 362
    wait states, 215, 220, 221, 223, 224, 270, 271, 272, 303, 305, 306, 356, 386, 389, 422, 438
    write, 213, 221, 223–24, 225, 226, 275, 279, 283, 289, 330, 356, 362, 366, 386, 390
Bus cycle indication (identification, definition), 213–14, 215, 218, 219, 300, 368, 385, 389
Bus master, 275, 289

C

C, 16, 20, 138
Cache, 270–90
    direct-mapped (one-way set associative), 273, 276, 277, 278, 283–86
    flush, 283
    hit, 272, 274, 275, 286
    hit rate, 273
    information replacement algorithm, 274, 287
    least recently used (LRU) algorithm, 275, 276, 287, 288
    memory array, 273–74, 275, 278, 286, 287, 438
    miss, 272–73, 275, 283, 286, 288
    posted write-through, 275, 276

two-way set associative, 273–74, 276, 277, 278, 287–88
Call, 131, 133–34, 350, 383
 intersegment, 133, 134
 intrasegment, 133, 134
Call gate mechanism, 195–97
Clock, 219–20, 338, 346, 347, 348, 350, 356, 422
 clock input (CLK2), 217, 219–20
 frequency, 217
 internal processor clock (PCLK), 217
Code, 16, 17, 32, 33, 39, 169, 170, 192, 193–96, 214, 270, 272, 273, 275
 object, 11, 17
 source, 17
Comment, 17
Compare and set instructions, 119–22
 byte set on condition (SETcc), 121–22
 compare (CMP), 119–21
 set byte if above (SETA), 122
 set byte if equal (SETE), 122
 set byte if parity even (SETPE), 122
Compiler, 20
Complementary high-performance metal-oxide-semiconductor III (CHMOSIII), 207
Computer, 1, 4–5
 mainframe, 4–5
 microcomputer, 1, 4–7, 8–12, 21, 207, 266–70, 421–87
 minicomputer, 1, 4, 5, 9, 11
Conforming code (segment), 391–92
Control register three (CR_3), 177
Control register zero (CR_0), 22, 77, 157, 164–65, 186, 187
 machine status word (MSW), 23, 73, 164–65, 186, 187, 200, 415
 page (PG), 165, 175
Coprocessor data channel, 415
Coprocessor interface signals, 207, 216
 coprocessor busy ($\overline{\text{BUSY}}$), 216, 384
 coprocessor error ($\overline{\text{ERROR}}$), 216, 416
 coprocessor request (PEREQ), 216
Count (counter), 34, 36, 101, 104, 310, 346, 347, 348, 350

D

Data, 6, 7, 9, 10, 16, 21, 24–30, 34–36, 39, 42, 43, 44, 46, 47, 52, 57, 59, 63, 72, 76, 139, 168, 170, 178, 192, 193, 194, 210, 211, 213, 214, 218, 219, 252, 258, 259, 264, 270, 271, 272, 275, 279, 286, 289, 296, 300, 301, 309
 base, 4, 5, 7
 bus, 208, 209–11, 216, 218, 227, 251, 252, 259, 276, 278, 279, 298, 300, 301–3, 352, 356, 368, 385, 386, 398, 408
 control, 10
 duplication, 213
 organization, 24–26
 separator, 48?
 window, 483
Data formats, 24–29
 ASCII, 28–29, 82, 87, 91, 92, 94
 bit, 7, 37, 101, 107–9
 byte, 8, 22, 25, 27, 28, 29, 30, 32, 34, 37, 40, 43, 46, 73, 76, 87, 211, 226, 251, 279, 283, 296, 297, 300, 307, 309
 decimal (BCD), 28, 87, 92, 94, 95, 98
 double word, 22, 24, 25, 26, 27, 32, 33, 34, 35, 36, 43, 44, 46, 72, 73, 76, 87, 211, 226, 279, 283, 284, 296, 297, 300, 307, 309, 377, 390
 EBCDIC, 82
 integer, 27
 packed (unpacked) decimal (BCD), 28, 87, 92, 98
 quad word, 98
 signed (integer), 27, 87, 129
 types, 27–29
 unsigned (integer), 27, 87, 129
 word, 22, 24, 25, 27, 32, 34, 35, 36, 43, 44, 46, 73, 76, 87, 211, 226, 279, 296, 297, 300, 307, 309, 377
Data registers, 22, 34, 46, 76
 accumulator (EAX), 34, 73, 76, 297, 307, 308
 base (EBX), 34, 38, 39
 count (ECX), 34, 143, 144, 145, 149, 150, 310
 data (EDX), 34, 308, 309
Data storage (memory), 7, 32, 33, 218, 258, 269–70
Data transfer instructions, 73–85
 exchange (XCHG), 73, 79–82
 load data segment (LDS), 73, 83, 84
 load effective address (LEA), 73, 83
 load extra segment (LES), 73, 83
 load register and FS (LFS), 73, 83
 load register and GS (LGS), 73, 83, 85
 load registers and SS (LSS), 73, 83–84
 move (MOV), 16, 17, 46–63, 72, 73–77
 sign-extend and move (MOVSX), 73, 78

Data transfer instructions (cont.)
  translate byte (XLAT/XLATB), 73, 82–83
  zero-extend and move (MOVZX), 73, 78, 79
Demand paged memory system, 185
Descriptor, 159, 165, 166, 170, 173, 174, 178–82, 187, 189, 192, 194, 195, 198, 391, 393
  call gate, 178, 195, 196
  interrupt (gate), 159, 162, 195, 380, 381, 391
  local descriptor table, 164, 178, 179
  segment, 159, 162, 166, 175, 178, 179, 181–82, 189
  system segment, 159, 178, 182
  task gate, 178, 195, 380
  task state segment (TSS), 165, 178, 198, 310
  trap gate, 195, 380, 381, 391
Destination operand, 17, 32, 36, 46, 47, 73
Diagnostic routine, 216
Displacement, 50, 56, 57, 59, 61, 63, 76, 124, 125, 133, 169
DMA, 216, 351–66
  acknowledge, 354
  block mode, 362, 363, 364
  cascade mode, 362
  channel, 10, 354, 358, 360, 362, 364, 365
  controller, 216, 275, 351–66, 426
  demand mode, 362–63
  interface circuitry, 354–56
  request, 354
  request/acknowledge handshake, 354
  single transfer mode, 362, 363
DMA interface signals, 207, 216
  bus hold acknowledge (HLDA), 216, 354
  bus hold request (HOLD), 216, 354
DRAM (dynamic RAM), 258, 262–66, 270, 425, 441, 442, 447–49, 457–58
  block diagram, 264
  column address, 264, 266
  density, 262
  organization, 264
  page mode, 266
  refresh controller, 266
  refreshing (refreshed), 258, 266
  row address, 264, 265
  2164B, 262
  21256, 264
  21464, 264
  421000, 264
  424256, 264
Dynamic bus sizing, 207, 211
Dynamic link, 142

E

80186 microprocessor, 9, 10
80188 microprocessor, 9, 10
80286 microprocessor, 1, 2, 3, 9, 11, 13, 22, 34, 41, 72, 207, 208, 421–27
80287 numeric coprocessor, 165, 216
80386DX microprocessor, 1, 2, 3, 5, 8, 9, 11, 13, 20, 34, 207–226
80386SX microprocessor, 9, 427
80387DX numeric coprocessor, 165, 216, 384, 415, 416, 428, 429, 438, 458
8042/80C042 keyboard controllers, 460, 473
8086 microprocessor, 1, 8, 9, 11, 15, 22, 34, 41, 72, 202, 207, 208
8088 microprocessor, 1, 2, 9, 11, 13, 72, 202, 207, 208
82340 PC/AT Chip Set, 428
  82341 peripheral combo, 428, 473–81
  82344 ISA controller, 428, 458–73
  82345 data buffer, 428, 440–46
  82346 system controller, 428, 429, 447–58
  82077AA floppy disk controller, 428, 481–87
82385DX cache controller, 275–90, 428, 429, 438, 461
  bus watching, 289
  cache memory architecture, 276
  coherency, 289
  directory, 283, 284, 285, 286, 287
  line, 284, 286, 288
  line miss, 286
  master mode, 276, 277
  page(s), 283–84, 286, 287
  set entry, 284–85, 286, 287–88
  signal interfaces, 276–83
  slave mode, 276, 277
  snoop bus, 289
  snoop hit, 289
  tag, 284, 285, 286, 287, 288, 289
  tag miss, 286
Embedded microcontroller, 10
EPROM (erasable programmable read only memory), 250–57, 269, 270, 377, 426
  block diagram, 251
  density, 252
  erasing, 251

expanding word capacity, 253, 256
expanding word length, 253, 255–56
Intelligent Programming Algorithm™, 253
organization (organized), 253
programmer, 250
programming mode, 253
Quick-Pulse Programming Algorithm™, 253
read operation, 252
standby mode, 252
storage capacity, 251, 252, 256
27C010, 252
2716, 252, 253
27128, 269
27C256, 252, 253, 255
2732, 252
27C512, 253
Error, 165, 172, 190
  code, 393
  coprocessor, 416
  IDT, 393
  page fault, 165, 185
Escape (ESC) instruction, 415
Event control, 10
Exceptions, 159, 374, 376, 377, 384, 391, 393
  general protection (fault), (GP), 310, 312, 376, 382
  page fault, 393
  service routine, 159

F

Flag control instructions, 117–19, 150
  clear carry (CLC), 118, 119
  clear direction (CLD), 150
  complement carry (CMC), 118, 119
  load AH from flags (LAHF), 117, 118
  set carry (STC), 118, 119
  set direction (STD), 150
  store AH into flags (SAHF), 117–18, 119
Flag(s), 76, 77, 87, 90, 117, 118, 119, 122, 138, 148, 391, 394
Flags register (FLAGS/EFLAGS), 22, 37–38, 138, 158, 167–68, 193, 198, 202, 215, 310, 382, 383, 390, 392, 413
  auxiliary carry (AF), 37, 38, 87, 91, 120, 147
  carry (CF), 37, 38, 87, 89, 91, 101, 105, 107, 108, 109, 119, 120, 122, 124, 127, 147

control, 38, 77
direction (DF), 38, 147, 150, 309
interrupt enable (IF), 38, 119, 215, 216, 380, 381, 382, 383, 384–85, 387, 389, 390, 392, 399, 410, 411
I/O privilege level (IOPL), 167–68, 193, 310, 312, 382
nested task (NT), 167, 168, 198
overflow (OF), 37, 38, 87, 120, 121, 124, 147, 383, 414
parity (PF), 37, 87, 120, 121, 122, 124, 127, 147
resume (RF), 167
sign (SF), 37, 38, 87, 120, 121, 147
status, 37–38, 120, 124
system, 77, 167
trap (TF), 383, 390, 392, 413
virtual 8086 mode (VM), 167, 202
zero (ZF), 37, 38, 87, 109, 120, 121, 122, 145, 147, 150, 188, 333
Floating-point arithmetic, 21, 415
Floppy disk drive, 2, 7, 21, 481, 484
Fragmentation, 176

G

Gate, 162, 178, 195, 196, 200, 380, 381, 391, 392, 394
  call, 178, 196
  interrupt, 162, 195, 380, 381, 386, 391, 394
  task, 178, 195, 200, 380
  trap, 380, 381, 391, 392, 394
General-purpose registers, 34–36, 50, 53, 57, 73, 76
Global descriptor table (GDT), 158–60, 164, 165, 166, 170, 173, 178, 179, 182, 186, 187, 195, 198, 200, 393
Global memory, 159, 164, 175, 179, 188, 215
  address space, 159, 170, 171, 188
  segment, 170

H

Halt (HLT) instruction, 131
Hard disk (drive), 2, 7, 171, 181, 185, 428, 474
High-level language, 1, 16, 20, 21, 138, 139
Hold/hold acknowledge handshake, 216

Index   515

I

Immediate operand, 43, 46, 47–48, 104, 108, 125, 127, 133
Index (registers), 22, 34–35, 36, 38, 50, 59, 60, 61, 62, 63
   destination index (DI/EDI), 34, 36, 39, 147, 148, 150, 309
   source index (SI/ESI), 34, 36, 39, 42, 147, 148, 150, 310, 383
Index scaling, 62, 63
Industry standard architecture (ISA) bus, 3, 426, 427, 428, 471–72, 474
Instruction, 5, 11, 16, 17, 21, 22, 24, 32, 33, 35, 36, 37, 38, 42, 46–63, 71–109, 168, 169, 196, 271, 275, 381, 414
Instruction set, 9, 10, 11, 38, 71–109, 117–51, 185–88, 297, 306–12, 381
   base (basic), 11, 72
   80386DX specific, 72
   extended, 72
   system control, 72–73, 159, 185–88, 192
Integrated drive electronics (IDE), 474, 480–81
Internal interrupts/exceptions, 376, 377, 384, 387, 411–16
   abort, 411, 412
   bounds check, 411, 414
   breakpoint, 379, 411, 413–14
   coprocessor error, 416
   coprocessor extension not available, 415
   coprocessor segment overrun, 415
   debug, 379, 413
   divide error, 376, 379, 411, 413
   fault, 411, 413, 414, 415, 416
   general protection, 310, 312, 376, 382
   interrupt table limit to small, 415
   invalid opcode, 414
   overflow error, 379, 411, 414
   segment overrun, 415
   stack fault, 415
   trap, 411, 412, 414
Internal registers (of the real-mode 80386DX), 16, 21, 22, 33, 34–38, 42, 46, 47, 50, 73, 76, 298
   control register zero, 22, 77
   data registers, 22, 34, 46, 76
   debug registers, 77, 413
   flags register, 22, 37–38, 138
   index registers, 22, 34–35, 36, 38, 61, 62, 63

   instruction pointer, 22, 33, 38, 39, 42, 123, 124, 126, 127, 131, 133, 379, 383, 390, 391, 411
   pointer registers, 22, 34, 35–36
   segment registers, 22, 30–33, 34, 36, 38, 42, 50, 52, 58, 73, 77
Internal registers (of the protected-mode 80386DX), 158–68
   control registers (CR_0-CR_3), 22, 77, 157, 158, 164–65, 177, 186, 187
   extended instruction pointer (EIP), 158, 169, 200, 392, 393
   flags register (EFLAGS), 158, 167–68, 193, 198, 202
   global descriptor table register (GDTR), 158–59, 187
   interrupt descriptor table register (IDTR), 158, 159–62, 377, 379, 382
   local descriptor table register (LDTR), 158, 162–64, 179
   task register (TR), 158, 165–66, 198, 200
Interrupt(s), 159, 178, 202, 214, 215–16, 346, 350, 374–416, 427
   acknowledge bus status code, 214, 215, 386, 389, 410
   context switch mechanism, 375
   controller, 210, 390, 394–408
   descriptor table (IDT), 159–62, 178, 186, 377, 379, 386, 393
   gate, 162, 195, 380, 381, 386, 391, 394
   hardware, 376, 377, 379, 384,
      hardware interrupt sequence, 387–94
   interface, 38, 207, 215–16, 385–87, 394, 408–11
   nonmaskable, 215, 216, 376, 377, 384, 387, 427
   priority, 376, 377, 379, 386, 387, 394, 395, 398, 400, 404, 406, 407, 410, 411, 427
   request/acknowledge handshake, 386, 387, 390, 391
   reset, 216, 376, 384, 447, 449
   service routine, 22, 159, 162, 215, 216, 312, 350, 375, 376, 377, 379, 380, 381, 383, 384, 385, 387, 390–91, 392, 393, 394, 396, 407, 410, 411, 413, 416
   software, 376, 377, 379, 380, 383, 384
   type number (priority level), 210, 376, 377, 379, 386, 390, 391, 394, 395, 398, 402, 407, 408, 410, 411, 416, 427

vector (pointer), 24, 377, 379, 380, 383, 386, 390, 415
vector (pointer) table, 23, 377, 383, 390, 411, 413, 415
Interrupt instructions, 381–84
  breakpoint, 413
  check array index against bounds (BOUND), 383–84
  clear interrupt (CLI), 118, 119, 192, 381, 382, 384, 399
  halt (HLT), 384
  interrupt on overflow (INTO), 383, 414
  interrupt return (IRET), 202, 383, 391, 394
  set interrupt (STI), 118, 119, 192, 381–82, 384, 385, 399
  software interrupt (INT), 383
  wait (WAIT), 384, 415
Interrupt interface signals, 207, 215–16, 385–87
  interrupt acknowledge ($\overline{\text{INTA}}$), 215, 386, 395, 398, 410
  interrupt request (INTR), 38, 208, 215, 381, 385, 386, 387, 394, 399, 410, 427
  nonmaskable interrupt (NMI), 215
  reset (RESET), 215, 216
I/O (input/output), 11, 167–68, 192–93, 208, 210, 214, 218, 295–368
  accumulator, 307
  address, 22, 210, 296, 300, 301, 302, 303, 312, 328, 333, 408
  address decoder(ing), 300, 301, 328, 331, 368
  address latch, 300, 301
  address space, 22, 210, 296, 302, 308, 310, 311, 312, 338, 354, 426
  bank select decoder, 302
  bus cycle, 300, 301, 302, 303–6, 308, 314, 330, 354, 356, 408
  data bus, 302, 303
  data bus transceiver/buffer, 300, 302
  handshaking, 333
  interface, 295, 298–303, 325–32, 366–68
  isolated, 296–97, 325–28
  memory-mapped, 296, 297–98, 329–32
  parallel, 10, 298, 312, 325–32
  polling, 333, 334
  port, 46, 214, 216, 296–98, 300, 308, 309, 312, 314, 319, 328
  protected-mode, 167–68, 192–93, 310–12
  string, 307–10

I/O bus command signals, 300
  read cycle output ($\overline{\text{IORC}}$), 300, 303, 368
  write cycle output ($\overline{\text{IOWC}}$), 300, 306, 360
I/O instructions, 193, 296, 297, 306–12
  direct, 307
  in (IN), 192, 307–9
  input string (INSB/INSW/INSD), 192, 309–10
  out (OUT), 192, 307–9
  output string (OUTSB/OUTSW/OUTSD), 192, 309, 310
  variable, 307, 308
I/O permission bit map, 198, 310–12
  base offset, 310, 311, 312

J

Jump, 350
  conditional, 109, 124, 127–29
  intersegment, 124
  intrasegment, 124
  unconditional, 124–27
Jump instructions, 122, 123–29, 194, 195, 198
  jump (JMP), 123, 124–27
  jump if equal (JE), 127, 129
  jump if zero (JZ), 38, 127, 129, 333
  jump on carry (JC), 127
  jump on not zero (JNZ), 144
  jump on parity (JP), 127
  jump on parity even (JPE), 127

K

Kernel, 191, 192
Keyboard controller, 474, 480

L

Label, 17, 126
Language, 4, 7
Large-scale integration (LSI), 298, 312, 338, 351, 386, 394
Lexical (nesting) levels, 141, 142
Limit check, 190, 191, 194

Linear address, 165, 172, 175, 176, 177, 178
  directory field, 177, 178
  offset field, 177
  page field, 177
  space, 162
Listing, 17
Local bus, 270, 272, 275, 276, 279, 283, 285, 422
Local descriptor table (LDT), 162–64, 166, 170, 173, 174, 178, 179, 182, 192, 195, 200, 393
  cache, 164
Local memory, 162, 175, 179, 188
  address space, 162, 170, 171, 188
  segment, 170
Logical shift, 100, 101
Logic instructions, 73, 99–100
  AND (AND), 34, 99, 100
  exclusive-OR (XOR), 99, 100
  inclusive-OR (OR), 99, 100
  NOT (NOT), 100
Lookup-table, 82
Loop, 34, 143, 144–45, 271, 275, 333
Loop instructions, 143–46
  loop (LOOP), 144–45
  loop while equal (LOOPE), 145
  loop while not equal (LOOPNE), 145
  loop while not zero (LOOPNZ), 145
  loop while zero (LOOPZ), 145

M

Machine code (instruction), 16, 20, 21, 52, 72, 76
Machine language (instruction), 16, 17
Machine language (statement), 20
Machine status word (MSW), 22, 73, 164–65, 186, 187, 200, 415
  emulate processor extension (EM), 77, 165, 415
  extension type (R), 165
  math present (MP), 164–65, 415
  protected mode enable (PE), 22, 73, 157, 164, 187
  task switch (TS), 73, 165, 200, 415
Macroassembler, 17
Masked out (masking), 377, 387, 390, 404, 407, 411
Mask programming, 250
Memory, 6–7, 11, 15, 16, 17, 21, 30–33, 34, 36, 38, 39, 41, 42–46, 48–63, 73, 76, 158–78, 208, 214, 216, 226–33, 250–290, 356, 377, 390

dedicated use, 22–24
general use, 22, 24
interface circuitry, 231–90
paged model, 168, 170, 175–78
segmented model, 168, 170–75
Memory (address) space, 22–26, 32, 172, 226–31, 296, 297, 298, 330, 338, 377
Memory/IO interface signals, 207, 208–15
  address strobe ($\overline{\text{ADS}}$), 214
  bus lock indication ($\overline{\text{LOCK}}$), 215, 283, 385, 387
  bus size 16 ($\overline{\text{BS16}}$), 211, 221, 224
  byte enables ($\overline{\text{BE}}_0$-$\overline{\text{BE}}_3$), 210, 212, 213, 214, 227, 385, 386
  data/control indication (D/$\overline{\text{C}}$), 213–14
  memory/input–output indication (M/$\overline{\text{IO}}$), 208, 213–14
  next address request ($\overline{\text{NA}}$), 214, 215, 225, 226
  read/write indication (W/$\overline{\text{R}}$), 213–14
  transfer acknowledge ($\overline{\text{READY}}$), 214–15, 220, 221, 224, 226, 303, 304, 306, 356, 385, 386
Memory management, 11, 72, 157, 158, 162, 191, 207
Memory manager, 171, 176, 178, 181, 185, 190
Memory management unit (MMU), 168, 169, 170, 172, 174, 175, 190
Memory organization, 226–31
Micro Channel Architecture, 4
Microcomputer, 1, 4–7, 8–12, 21, 207, 266–70, 421–87
  architecture, 5–7
  8-bit, 7
  4-bit, 7, 8
  general-purpose, 9, 10–11, 422
  input unit, 5, 6
  memory unit, 5, 6–7
  microprocessor unit (MPU), 5–6, 16, 208, 270, 271
  multichip, 8, 13
  output unit, 5, 6
  primary storage, 6–7, 10
  secondary storage, 6–7, 10, 171
  single-chip, 10
  16-bit, 1, 7, 8–9
  special-purpose, 10
  32-bit, 1, 7, 9
Microcontroller, 10, 11
Microprocessor, 1, 5, 7–11, 21
Microprocessor unit (MPU), 5–6, 16, 208, 270, 271

518   Index

Mnemonic, 16, 77, 122, 127, 208, 306
Mouse controller, 474
Multiprocessor (multiprocessing), 170, 215, 275, 387
Multitasking, 11, 72, 157, 188, 189, 197, 202, 207

N

Nested task, 168, 198
Nibble, 8, 37
Nonconforming code segment, 391–92
Nonvolatile, 7, 250
Numeric coprocessor, 164–65, 216, 387, 415, 416, 428, 429, 438, 458

O

Opcode (operation code), 16, 47, 48, 52, 76, 133, 382, 414
Object-code compatible, 11
Offset, 24, 33, 36, 38, 39, 40, 42, 44, 49, 52, 53, 55, 76, 126, 133, 169, 175, 308, 379
Operand(s), 16, 17, 32, 33, 34, 35, 36, 38, 43, 46, 47, 48, 52, 73, 104, 108, 124, 125, 127, 133, 173, 175, 178, 191, 198, 271
Operating system, 7, 11, 171, 181, 185, 191, 192, 193, 312, 425
  OS/2, 157
  PC DOS, 7, 202
  UNIX™, 157, 202
Overflow, 121, 129, 383, 414

P

Page, 165, 172, 176
Page directory, 165, 177, 178, 182
  base address, 165, 177, 182
  base register (PDBR), 165, 178
  entry, 178, 182, 184, 185
Page frame, 178, 182–84
  address, 178, 182
Page table, 165, 177, 178, 182–84
  entry, 178, 182–83, 184, 185
Page translation, 165, 172, 176–78
Page 0, 296, 307
Paging, 157, 164, 165, 172, 175–78, 189

Parallel printer interface (port), 333, 334–37, 428, 473, 474, 479–80
Parameters, 36, 39, 134, 135, 139, 196
Parity, 37, 122, 127
PC/AT microcomputer system hardware, 421–87
  architecture, 421–28
  core microcomputer, 428–40
  82340 high-integration peripheral ICs, 427–28, 440–87
Peripheral, 312, 333–66, 368, 394–408, 427–28, 440–87
Personal computer, 1, 2–5, 11, 21
  PC/AT, 1, 2–5, 7, 21, 421–87
  PCXT, 2
  Personal System/2, 4
Physical memory, 158, 159, 164, 165, 171, 181, 185, 190
Physical (memory) address space, 158, 171, 185, 210, 226, 283, 287, 308, 379
Pipelining (pipelined), 207, 215, 219, 231
Pointer, 22, 24, 72, 83, 126, 133, 139, 147, 377
Pointer registers, 22, 34, 35–36
  base pointer (BP/EBP), 35–36, 39, 58, 142
  stack pointer (SP/ESP), 35–36, 39, 43, 44, 58, 60, 133, 136, 137, 142, 390, 392, 394
Privilege level(s), 168, 181, 189, 191, 192, 193, 194, 195, 196, 202, 380, 384, 391, 392, 394
  current (CPL), 193, 194, 195, 197, 198, 200, 310, 312, 382, 391, 394
  descriptor (DPL), 181, 189, 192, 194, 195, 197, 198, 380, 392
  input/output (IOPL), 167–68, 193, 310, 312, 382
  requested (RPL), 166, 169, 188, 193, 194, 197, 200, 392, 394
Program, 5, 6, 7, 10, 11, 16, 17, 20, 21, 24, 33, 38, 120, 124, 127, 131, 171, 172, 193, 194, 200, 202, 250, 270, 272, 273, 375–76, 383, 390, 391, 394, 410, 411
Programmable array logic (PAL™), 247, 366, 368, 428
Programmable direct memory access controller (82C37A/8237A), 351–66, 426, 427, 460
  address register, 358, 360
  block diagram, 356
  command register, 358, 360
  count registers, 358, 360, 362

Index 519

Programmable direct memory access controller (*cont.*)
  DMA interface, 354–56
  first/last flip-flop, 358, 360
  fixed priority, 356, 358, 361
  internal architecture, 356–66
  mask register, 358, 364, 365
  microprocessor interface, 352–54
  mode register, 362
  priority, 356, 358, 361
  request register, 364
  rotating priority, 356, 358, 361
  status register, 358, 365, 366
  temporary register, 366
  terminal count, 363, 365
Programmable interrupt controller (82C59A/8259A PIC), 366, 367, 394–411, 426, 460
  automatic end of interrupt, 403
  block diagram, 394–97
  buffered mode, 403–4
  cascade address, 390
  cascade buffer/comparator, 397, 398
  cascaded mode, 396–97, 400, 403, 404, 411
  cascading (bus) interface, 396
  command register, 399
  data bus buffer, 397, 398
  end of interrupt, 400, 403, 406
  fully nested mode, 400, 404, 407
  host processor interface, 394–95
  identification (ID) code, 396, 398, 403
  initialization command words, 398, 400–4
  in-service register, 397, 398, 404, 406, 407
  internal architecture, 397–98
  interrupt mask register, 397, 398, 404
  interrupt (request) inputs, 396, 400, 404, 406, 409–10, 411
  interrupt request register, 397, 398
  master, 390, 396, 397, 398, 403, 404, 411
  master/slave configuration, 390, 396, 404, 411
  microprocessor mode, 403
  operational command words, 398, 404–8
  poll command, 406–7
  poll word, 407
  priority resolver, 397, 398
  priority schemes, 406
  programming, 398–400
  read register, 406
  read/write logic, 397, 398
  rotate on specific EOI, 406
  rotation, 406
  slave, 390, 396, 397, 398, 403, 411
  special fully nested mode, 404
  special masked mode, 400, 407
  specific level, 406
Programmable interval timer (82C54/8254 PIT), 298, 338–51, 366, 367, 426–27, 460
  block diagram, 338
  control word, 339, 340
  control word register, 338–39, 340
  count register, 340, 342–43
  hardware-triggered strobe (MODE 5), 351
  interrupt on terminal count (MODE 0), 345, 346
  microprocessor interface, 338
  operating modes, 345–51
  programmable one-shot (MODE 1), 338, 345, 347
  rate generator (MODE 2), 338, 348
  read-back command, 344
  read-back mode, 343–45
  software-triggered strobe (MODE 4), 349–50
  square-wave generator (MODE 3), 338, 349
Programmable logic array (PLA), 243–50
  architecture, 244–246
  block diagram, 244
  registered outputs with, 246
  standard devices, 246–50
Programmable peripheral interface (82C55A PPI), 300, 312–25
  bit set/reset, 321–23
  bit set/reset flag, 321
  block diagram, 312
  control registers, 314
  isolated parallel I/O ports, 325–28
  memory-mapped parallel I/O, 329–32
  microprocessor interfaces, 312, 314
  MODE 0 (simple I/O operation), 314, 315–18, 323
  MODE 1 (strobed I/O operation), 314, 315, 319–21, 323
  MODE 2 (strobed bidirectional I/O), 315, 321–23
  Mode set flag, 315
Programmable read-only memory (PROM), 250, 251, 252
Program (storage) memory, 7, 32, 76, 134, 218, 250, 255, 269, 270, 379, 383, 410
Protected (-address) mode, 11, 15, 22, 73, 157–202, 207, 209, 226, 310–12, 382, 384, 391

Protection, 11, 73, 157, 188–97, 207, 391
   checks, 189, 191, 193, 194, 196
   level, 191, 192, 194
   mechanism, 189
   model, 188, 191, 193
   violation, 190–91
Protection (privilege) level check, 193–94, 196

Q

Queue, 33, 47, 48, 220

R

Random access read/write memory (RAM), 7, 10, 258–70, 279, 377, 422, 425–26, 438
Read, 48, 211, 214, 216, 218, 219, 221, 224, 225, 226, 227, 259, 272, 279, 283, 285, 286, 287, 289, 330, 356, 362, 386, 390
Read-frequency deviation, 483
Read-only memory (ROM), 7, 10, 250–57, 422, 426, 442, 461, 470
Real (-address) mode, 11, 15, 22, 30, 33, 35, 36, 37, 38, 41, 43, 72, 157, 202, 207, 209, 210, 211, 227, 296, 310, 377, 383, 387, 411
Real time, 20–21
Real-time clock, 474
Repeat prefixes, 149–50
   REP, 149, 309, 310
   REPE, 150
   REPNE, 150
   REPNZ, 150
   REPZ, 150
Reprogrammable microcomputer, 4, 10–11
Rotate, 34
Rotate instructions, 73, 104–7
   rotate left (ROL), 104, 105
   rotate left through carry (RCL), 104, 107
   rotate right (ROR), 104, 105–6
   rotate right through carry (RCR), 104, 107

S

Scale factor, 50, 62
Segment(s), 30, 40, 49, 159, 162, 166, 169, 170, 171, 172, 173, 175, 178, 180, 189, 190, 192, 391, 415
   code, 31, 33, 39, 48, 124, 179, 180, 181, 191, 192, 193, 196, 391–92
   data (D, E, F, G), 31, 36, 52, 181, 191, 192, 193
   stack, 31, 42–46, 59, 135
Segmentation, 30–33, 36, 172, 178, 184, 189, 192
Segment base address, 24, 30, 175, 178–79, 198
Segment descriptor, 180–81, 189
   available (AVL), 179
   base address, 175
   conforming code (C), 195
   granularity (G), 179
   limit, 175
   present (P), 181, 189, 380
   programmer available (AVL), 179
   type (field), 182, 189
Segment descriptor cache registers, 173
   code, 180, 193
   data, 174
Segment override prefixes (SEG), 39, 52, 53, 58, 83
Segment (selector) registers, 22, 30–33, 34, 36, 38, 42, 50, 52, 58, 73, 76, 166, 169, 173, 175, 190, 392
   code segment (selector), (CS), 31, 32, 33, 39, 42, 123, 126, 127, 131, 169, 173, 175, 178, 180, 190, 194, 195, 200, 379, 383, 390, 391, 393, 411
   data segment (selector), (DS), 22, 31, 32, 34, 39, 52, 53, 58, 76, 147, 148, 173, 175, 178, 190, 193
   data segment F (selector), (FS), 31, 33, 39, 72, 173, 175, 190, 193
   data segment G (selector), (GS), 31, 33, 39, 72, 173, 175, 190, 193
   extra segment (selector), (ES), 31, 33, 34, 39, 147, 148, 173, 175, 190, 193, 309, 310
   stack segment (selector), (SS), 31, 33, 34, 35, 36, 39, 43, 45, 46, 59, 72, 142, 173, 175, 178, 190, 194, 392, 394, 415
Segment translation, 172, 173–75
Selector, 164, 165, 166–67, 168–69, 170, 173, 174, 188, 193, 195, 198, 200, 393
Serial communication channel (port), 428, 473, 475–77
Shift, 34

Shift instructions, 73, 100–103
    double precision shift left (SHLD), 104
    double precision shift right (SHRD), 104
    shift arithmetic left (SAL), 101
    shift arithmetic right (SAR), 101, 103–4
    shift logical left (SHL), 101
    shift logical right (SHR), 101
Sign bit, 27, 38, 78
Sign extend, 78
Single (stepping) step, 38, 390, 413
Software, 2, 4, 5, 15, 16–21, 33, 34, 38, 109, 159, 164, 165, 168, 170, 172, 181, 184, 188, 191, 192, 193, 197, 202, 302, 333, 352, 356, 364, 377, 384, 391, 394, 398, 407, 413
Software architecture, 15–65, 157–202
    protected-mode, 157–202
    real-mode, 15–65
Software model, 21–22, 158–78
    protected-mode, 158–78
    real-mode, 21–22
Source operand, 17, 32, 36, 46, 47, 73
Special registers, 72, 73, 77
SRAM (static RAM), 258, 259–62, 269–70, 279, 438
    block diagram, 259
    density, 259
    4361, 261
    4362, 261
    4364, 261, 262
    43256A, 261
    organization, 259
    read cycle, 262
    write cycle, 261
Stack, 32, 36, 39, 42–46, 43, 59, 133, 134, 135–37, 138, 139, 141–42, 194, 196, 202, 390, 391, 392, 393, 411, 413, 414, 415
    bottom of, 43, 44
    end of, 43
    frame, 139, 141, 142
    top of, 36, 42, 44, 46, 136, 138, 142, 390
State machine, 246
Straight-line program, 71
String, 34, 146, 309
String instructions, 10, 36, 38, 118, 146–50, 309
    compare strings (CMPS), 146, 147–48
    load string (LODS), 146, 148
    move byte or word string (MOVS), 146, 147
    repeat string (REP), 149–50
    scan string (SCAS), 146, 147, 148
    store string (STOS), 145, 148
Subroutine, 36, 42, 118, 131, 135, 136, 139, 383
Subroutine-handling instructions, 131–42
    call (CALL), 42, 133–34, 194, 195, 198
    enter (ENTER), 139–42
    leave (LEAVE), 139–42
    pop (POP), 42, 45–46, 135, 136, 391
    pop all instruction (POPA/POPAD), 72, 138
    pop flags (POPF), 138
    push (PUSH), 42, 44–45, 72, 135–36, 390
    push all instruction (PUSHA/PUSHAD), 72, 138
    push flags (PUSHF), 138
    return (RET), 42, 132, 133, 134, 196, 200
Supervisor-level protection, 184
Swap, 172
System bus, 270, 275, 276, 354, 387, 422
System control instruction set, 185–88
    adjust RPL field of selector (ARPL), 188
    clear task switch flag (CLTS), 173
    load access rights byte (LAR), 187
    load global descriptor register (LGDT), 187
    load interrupt descriptor table register (LIDT), 187, 382
    load local descriptor table register (LLDT), 187
    load machine state word (LMSW), 187
    load segment limit (LSL), 188
    load task register (LTR), 187
    store global descriptor table (SGDT), 187
    store interrupt descriptor table register (SIDT), 187, 382
    store local descriptor table register (SLDT), 187
    store machine state word (SMSW), 187
    store task register (STR), 187
    verify read access (VERR), 188
    verify write access (VERW), 188

T

Table indicator (select), (TI), 166, 169, 170, 173, 393
Tag, 126

Task, 159, 162, 164, 165, 166, 168, 170, 171, 175, 188, 192, 193, 195, 196, 197–202, 310, 312, 380
  busy (B), 198
  descriptor cache, 166, 198
  gate, 178, 200, 380
  present (P), 198
  state segment (TSS), 165, 166, 178, 198, 200, 202, 310, 312, 392
  state selector, 198
  switch, 165, 188, 198
  switching mechanism, 188, 197–202
Transistor density, 9–10
Translation lookaside buffer, 178, 183
Trusted (privileged) instructions, 192–93, 384
Type bit (T), 380
Type check, 189, 191, 194

U

Ultraviolet light, 251
Unaligned (misaligned) double word boundary(ies), 24, 25–26, 230
Undefined opcode detection mechanism, 414
Universal asynchronous receiver/transmitter (UART), 10, 475

Upward compatible, 11
User-level protection, 184

V

Very large-scale integration (VLSI), 1, 5, 11, 275, 440, 447
Virtual address (addressing), 157, 167, 168–78, 210
Virtual memory, 170–78
Virtual (memory) address space, 169, 170–71, 189, 192, 210
Virtual 8086 mode, 11, 167, 202, 207, 312
Volatile, 7, 258

W

Wait state logic, 422
Write, 48, 210, 212, 213, 214, 216, 218, 219, 223, 224, 225, 226, 227, 259, 273, 275, 279, 283, 289, 330, 356, 362, 366, 386, 390
Write compensation, 483

Z

Zero extend, 78